Hydrogeochemistry of Aquatic Ecosystems

Edited by

Sughosh Madhav
Jamia Millia Islamia,
New Delhi,
India

Virendra Bahadur Singh
Ram Lal Anand College, University of Delhi,
New Delhi,
India

Manoj Kumar
Central University of Haryana,
India

Sandeep Singh
Indian Institute of Technology Roorkee,
India

WILEY

This edition first published 2023

Registered Office(s)
John Wiley & Sons, Inc., 111 River Street, Hoboken, NJ 07030, USA
John Wiley & Sons Ltd, The Atrium, Southern Gate, Chichester, West Sussex, PO19 8SQ, UK

Editorial Office
Boschstr. 12, 69469 Weinheim, Germany

For details of our global editorial offices, customer services, and more information about Wiley products visit us at www.wiley.com.

Wiley also publishes its books in a variety of electronic formats and by print-on-demand. Some content that appears in standard print versions of this book may not be available in other formats.

Library of Congress Cataloging-in-Publication Data applied for
9781119870531

Cover Design: Wiley
Cover Image: Courtesy of Virendra Bahadur Singh

Set in 9.5/12.5pt STIXTwoText by Straive, Pondicherry, India
Printed and bound by CPI Group (UK) Ltd, Croydon, CR0 4YY

C9781119870531_231122

Contents

List of Contributors *xii*
Editors' Biography *xvi*
Preface *xvii*

1 Fluoride in Groundwater: Distribution, Sources, Processes, Analysis, and Treatment Techniques: A Review *1*
Bedour Al Sabti, Dhanu Radha Samayamanthula, Fatemah M. Dashti, and Chidambaram Sabarathinam
1.1 Introduction *1*
1.2 Permissible Limits of Fluoride in Drinking Water *3*
1.3 Concentration of High Fluoride Around the Globe *4*
1.4 Sources of Fluoride in Groundwater *4*
1.4.1 Natural Sources *6*
1.4.2 Anthropogenic Sources *7*
1.4.3 Health Effects *8*
1.4.4 Dental Fluorosis *8*
1.4.5 Skeletal Fluorosis *9*
1.5 Hazard Index *9*
1.5.1 Analytical Techniques *11*
1.5.2 Treatment Methods *11*
1.6 High Fluoride Water Types *13*
1.6.1 Relationship of pH to Fluoride *15*
1.6.2 Thermodynamic Stability of High Fluoride Waters *17*
1.6.3 Saturation States *18*
1.6.4 Fluoride vs SI-Fluorite *18*
1.6.5 SI Fluorite and SI Calcite *18*
1.7 Statistical Techniques to Assess High Fluoride Waters *19*
1.7.1 Correlation Analysis *19*
1.7.2 Principal Component Analysis *21*
1.8 Conclusion *22*
Acknowledgments *22*
References *23*

2 Geochemical Sources, Aqueous Geochemistry, Human Health Risk of Fluoride-Enriched Groundwater, and Its Remedial Measures *33*
Paulami Sahu and Chitrangada Debsarma
2.1 Introduction *33*
2.1.1 Occurrence of Fluoride *33*
2.1.2 Dispersion of Fluoride in Groundwater *34*
2.1.2.1 International Scenario *34*
2.1.2.2 National Scenario *40*
2.1.3 Fluoride and Human Health *40*
2.2 Environmental Origin of F in Groundwater *40*
2.2.1 Geological Origins *41*
2.2.1.1 Igneous, Sedimentary, and Metamorphic Minerals *41*
2.2.1.2 Sea Water *43*
2.2.1.3 Geothermal Fluid *43*
2.2.1.4 Volcanic Sources *44*
2.2.2 Anthropogenic Origins *44*
2.2.2.1 Atmosphere *44*
2.2.2.2 Fertilizer and Irrigation Water *44*
2.3 Geochemical Mechanisms for Fluoride Occurrence in Groundwater *44*
2.3.1 Stabilization *45*
2.3.1.1 Adsorption *45*
2.3.1.2 Precipitation *46*
2.3.2 Mobilization *46*
2.3.2.1 Dissolution *46*
2.3.2.2 Desorption *46*
2.4 Technologies for F Removal from Groundwater *47*
2.5 Recommended Remedial Measures to Reduce Human Health Problems *47*
2.6 Concluding Remarks *48*
References *52*

3 Spatial Distribution of Arsenic Contamination in India: A Systematic Review *61*
Abhra Singh, Risha Singh, Binod Kumar Nath, Mohammad Hashim, Masood Ahsan Siddiqui, Hasan Raja Naqvi, and Lubna Siddiqui
3.1 Introduction *61*
3.2 Research Methodology *62*
3.2.1 Framing the Research Questions *62*
3.2.2 Constructing Keywords or Search Terms for Literature Collection *63*
3.2.3 Databases and the Range of Years Searched *63*
3.2.4 Identification of Publications *63*
3.2.4.1 Inclusion Criteria *63*
3.2.4.2 Exclusion Criteria *63*
3.3 Arsenic in Aqueous Environment *63*
3.4 Arsenic-Contaminated States *64*
3.5 Medium of Contamination *67*
3.6 Sources and Mobilization *68*
3.6.1 Holocene Sediments *68*

3.6.2 Excessive Pumping, Irrigation Practices, and Poor Economic Condition *68*
3.6.3 Oxidizing and Reducing Conditions *69*
3.6.4 Organic Matter *70*
3.6.5 Micro-organisms and Phytoplankton *70*
3.6.6 Fertilizers and Crops *70*
3.6.7 Coal Mining and Industries *71*
3.7 Probable Future Arsenic Zones *71*
3.8 Exposure to Arsenic *73*
3.9 Conclusions *74*
References *74*

4 Arsenic Contamination of Groundwater in Indo-Gangetic Plain *85*
Samikshya Panda, Vinod Kumar Tripathi, Shrinivasa DJ., and Reema Sharma
4.1 Introduction *85*
4.2 Sources and Concentration of Arsenic in Groundwater *86*
4.3 Issues Related to Elevated Arsenic Concentration *86*
4.4 Removal Techniques of Arsenic from Groundwater *88*
4.4.1 Biological Oxidation *88*
4.5 Arsenic Treatment Plant Waste Disposal *89*
4.6 Challenges in Arsenic-Affected Areas *89*
4.7 Indian Scenario Regarding Measures Taken to Mitigate the Problem of As Availability in Groundwater *90*
4.8 Conclusion *92*
References *93*

5 Soil–Water Interactions and Arsenic Enrichment in Groundwater *97*
Sonam Taneja, Shivani Yadav, Harsh Pipil, Oznur Karaca, and A. K. Haritash
5.1 Introduction *97*
5.2 Sources and Occurrence of Arsenic in Soil, Sediments, and Groundwater *98*
5.3 Global Distribution of Arsenic in Groundwater *99*
5.4 Chemical Speciation of Arsenic *101*
5.5 Soil–Water Interactions *103*
5.5.1 Oxidative Desorption *103*
5.5.2 Sulfide Oxidation *103*
5.5.3 Reductive Dissolution of Iron Hydroxides *103*
5.6 Health Effects of Arsenic *104*
5.7 Case Studies *106*
5.7.1 Arsenic-Contaminated Drinking Water in Bangladesh *106*
5.7.2 Groundwater Contamination in West Bengal, India *108*
5.7.3 Arsenic Contamination Through Ammunition and Its Testing *108*
5.8 Treatment of Arsenic-Contaminated Water *109*
5.8.1 Treatment with Activated Carbon *109*
5.8.2 Adsorption by Activated Alumina *112*
5.8.3 Coagulation and Precipitation *112*
5.8.4 Electrodialysis Method *113*
5.8.5 Reverse Osmosis Method *113*
5.8.6 Ion-Exchange Process *113*

5.9 Conclusion *114*
References *114*

6 Arsenic Contamination in Groundwater and Its Removal Strategies with Special Emphasis on Nano Zerovalent Iron *121*
Shivani Thakur, Harminder Singh, Umakant Chaudhari, Sunil Mittal, and Prafulla Kumar Sahoo
6.1 Introduction *121*
6.1.1 General Chemistry and Occurrence of Arsenic in Nature *122*
6.1.2 Release and Mobilization Mechanism of Arsenic in Groundwater *123*
6.1.2.1 Sulfide Oxidation *123*
6.1.2.2 Reductive Dissolution *124*
6.1.2.3 Alkali Desorption *124*
6.1.2.4 Microbial-Mediated Arsenic Mobilization in Groundwater *125*
6.1.3 Arsenic Status in Global Groundwater Systems *126*
6.2 Conventional Methods for the Removal of Arsenic from Water *129*
6.3 Arsenic Removal from Water Using Iron-Based Adsorbents *131*
6.4 Zerovalent Iron Nanoparticles *134*
6.4.1 Synthesis of nZVI Particles *135*
6.4.2 Application of nZVI in Arsenic Removal from Groundwater/Synthetic Water *135*
6.5 Merits and Demerits of Using nZVI *139*
6.6 Conclusions and Future Perspectives *140*
References *141*

7 Chemical Speciation of Chromium and Arsenic and Biogeochemical Cycle in the Aquatic System *155*
Nitin Verma, Neha Kanojia, Sourav Kalra, and Kamal Dua
7.1 Introduction *155*
7.1.1 Heavy Metals *156*
7.1.2 Sources *156*
7.1.3 Toxic Effects *157*
7.2 Chemistry of Arsenic in Aquatic System *160*
7.2.1 Aqueous Arsenic Distribution and Speciation *164*
7.2.2 River Water *165*
7.2.3 Lake Water *166*
7.2.4 Seawater and Estuaries *166*
7.2.5 Groundwater *166*
7.2.6 Sediment Pore Waters *166*
7.2.7 Oilfield and Different Brines *167*
7.2.8 Biogeochemical Cycle of Arsenic *168*
7.3 Chemistry of Chromium in Aquatic System *169*
7.3.1 Biogeochemical Cycle of Chromium *171*
7.3.1.1 Oxidation of Cr (III) *171*
7.3.1.2 Reduction of Cr(VI) *171*
7.4 Conclusion *172*
References *172*

8 **Occurrences and Mobility of Uranium in Soil Profile Due to Groundwater–Soil Interaction** *181*
Somenath Ganguly and Uday Bhan
8.1 Introduction *181*
8.2 Uranium Mineralogy and Mode of Occurrences *183*
8.3 Uranium Mobilization Due to Soil–Water Interaction *185*
8.4 Hydro-Geochemistry of Uranium *187*
8.5 Applicability of Uranium Mobility in Groundwater Management, Geochemical Hydrocarbon Exploration, and Earthquake Prediction *189*
8.6 Uranium Health Hazard *192*
8.7 Conclusions *193*
8.8 Case Studies *193*
References *195*

9 **Study of the Rate of CO_2 Consumption with Silicate and Carbonate Weathering in Aquatic System** *199*
Uday Bhan, Vamsi Krishna Kudapa, and Ranjit Kumar
9.1 Basic Information *199*
9.2 Silicate Weathering *200*
9.3 Weathering of Ca and Mg Silicates *202*
9.4 Carbonate Weathering *202*
9.5 Rate of Silicate Weathering and CO_2 Consumption *204*
9.5.1 Rate of Silicate Weathering *204*
9.5.2 Rate of Carbon Consumption (CCR) During Silicate Rock Weathering *205*
9.6 Importance of Weathering as a Carbon Sequestration Mechanism *206*
9.7 Case Study – Himalayan Glaciers *207*
9.8 Conclusion *208*
References *208*

10 **Carbonate Chemistry, Carbon Cycle, and Its Sequestration in Aquatic System** *213*
Mallika Vashist, Harshit Chawla, and S. K. Singh
10.1 Introduction *213*
10.2 Carbon Cycle in Aquatic Systems *214*
10.3 Reactivity of Carbonate System *215*
10.3.1 The Contemporary Marine $CaCO_3$ Cycle *217*
10.3.2 Formation and Pathway of $CaCO_3$ *219*
10.3.3 Destruction and Dissolution of $CaCO_3$ *220*
10.3.3.1 Bio-erosion *221*
10.4 Present-Day Changes in the Seawater CO_2: Carbonic Acid System Due to Anthropogenic Activities *221*
10.5 Impact on Marine $CaCO_3$ Cycle Owing to Anthropogenic Ocean Acidification *222*
10.5.1 Formation of $CaCO_3$ *222*
10.5.2 Dissolution of $CaCO_3$ *223*
10.6 Biological, Ecological, and Biogeochemical Impacts *224*
10.7 Risk Mitigation and Adaptation Strategies *225*

10.8 Conclusion *226*
References *226*

11 Recent Trends in Fate, Transport, and Transformation of Inorganic and Organic Carbon in Freshwater Reservoirs *233*
Sanchit Kumar and Anshumali
11.1 Introduction *233*
11.2 The Influx of Carbon to the Reservoir *234*
11.2.1 Autochthonous *235*
11.2.2 Allochthonous *237*
11.3 Mineralization and Biogeochemical Transformation of C in the Reservoir *237*
11.3.1 Organic Carbon *237*
11.3.2 Inorganic Carbon *241*
11.3.3 Accumulation and Mineralization of C *242*
11.4 Driving Factors Controlling the C Cycle in Reservoirs *242*
11.4.1 Biotic Factors *242*
11.4.1.1 Vegetation *242*
11.4.1.2 Indigenous Primary Production *243*
11.4.1.3 Microbial Degradation *244*
11.4.2 Abiotic Factors *244*
11.4.2.1 Watershed Area *244*
11.4.2.2 Photodegradation *245*
11.4.2.3 Hydraulic Retention Time *245*
11.4.2.4 Seasonal Variability *246*
11.4.2.5 Sedimentation *247*
11.5 Carbon Cycling in Relation with Other Nutrients (N and P) *247*
11.6 Source vs Sink *249*
11.7 Water Drawdown Area as a New Emission Hotspot *251*
11.8 Climate Change Effects on C Dynamics in Reservoirs *253*
11.9 Conclusion *253*
References *254*

12 Role of Microbes in Controlling the Geochemical Composition of Aquatic Ecosystems *265*
Shilky (ORCID: 0000-0001-8292-0385), Subhashree Patra (ORCID: 0000-0002-4624-6968), Anamika Harshvardhan (ORCID: 0000-0002-0540-1220), Amit Kumar (ORCID: 0000-0002-4582-5677), and Purabi Saikia (ORCID: 0000-0001-5481-282X)
12.1 Introduction *265*
12.2 Microbial Communities of Aquatic Bodies and Their Role in Geochemical Processes *267*
12.3 Role of Microorganisms in Water Quality Management *271*
12.4 Problems and Prospects of Using Microbes in Water Quality Management *273*
12.5 Future Research Prospects and Recommendation *273*
12.6 Conclusions *274*
References *274*

13 Impacts of Pollution on the Hydrogeochemical and Microbial Community of Aquatic Ecosystems in Bayelsa State, Southern Nigeria *283*
Sylvester Chibueze Izah, Adams Ovie Iyiola, and Glory Richard
13.1 Introduction *283*
13.2 Characteristics of Aquatic Ecosystems in Bayelsa State *284*
13.2.1 Hydrogeochemistry of Aquatic Ecosystems in Bayelsa State *284*
13.2.2 Microbial Community of Aquatic Ecosystems in Bayelsa State *289*
13.3 Anthropogenic Activities Causing Pollution of Aquatic Ecosystem in Bayelsa State *289*
13.3.1 Houseboat on the Water *289*
13.3.1.1 Oil and Gas Pollution *292*
13.3.1.2 Emissions of Carbon Dioxide *292*
13.3.1.3 Cleaning Solutions and Other Chemicals *293*
13.3.1.4 Impacts of Boat Repairs and Maintenance *293*
13.3.1.5 Human Wastes and Garbage *293*
13.3.1.6 Noise Pollution *294*
13.3.2 Agricultural Activities *294*
13.3.3 Lumbering Activities *295*
13.3.4 Transportation *295*
13.3.5 Oil and Gas Exploration *296*
13.3.6 Dredging *296*
13.3.7 Wastes *297*
13.3.8 Flooding *297*
13.3.9 Marketing Activities *297*
13.3.10 Sources of Pollution *297*
13.3.11 Point Source Pollutants *297*
13.3.12 Non-point Pollution Source *297*
13.3.13 Effects of Point and Non-point Pollutants *298*
13.4 Impacts of Pollutants on Microbial Ecology and Hydrogeochemistry in Bayelsa State *299*
13.4.1 Hydrogeochemistry of Aquatic Ecosystems *299*
13.4.2 Microbial Ecology of Aquatic Ecosystems *300*
13.5 Mitigation of Pollution on the Aquatic Ecosystem *300*
13.6 Conclusion *301*
References *301*

14 Aquatic Eco-systems Under Influence of Climate Change and Anthropogenic Activities: Potential Threats and Its Mitigation Strategies *307*
Suryadipta Mukherjee, Syed Shams Rizvi, George Biswas, Abhilash Kumar Paswan, Suantak Paolalsiam Vaiphei, Taufique Warsi, and Tarik Mitran
14.1 Introduction *307*
14.2 Aquatic Ecosystems *308*
14.2.1 Freshwater Ecosystem *308*
14.2.1.1 Lotic Ecosystem *309*
14.2.1.2 Lentic Ecosystem *310*
14.2.1.3 Wetlands (Mixed Type) *310*
14.2.2 Marine Ecosystem *310*

14.3 Emerging Issues in the Aquatic Ecosystem *311*
14.3.1 Food Security *311*
14.3.2 Fisheries *312*
14.3.3 Marine and Ocean Life *312*
14.3.4 Water Quality *313*
14.3.5 Climate Change *313*
14.3.6 Infectious Diseases *313*
14.3.7 Destructive Algae *313*
14.3.8 Emerging pollutants *314*
14.3.9 Microplastic Contamination *314*
14.3.10 Freshwater Salinization *314*
14.3.11 Calcium Deficiency *314*
14.4 Impact of Climate Change on Aquatic Ecosystems *315*
14.4.1 Climate Change and Freshwater Ecosystems *316*
14.4.2 Climate Change and Marine Ecosystem *318*
14.4.2.1 Salinity *318*
14.4.2.2 Ocean Heat and Sea Surface Temperature *318*
14.4.2.3 pH *319*
14.4.2.4 Oceanic Nutrients *320*
14.4.2.5 Dissolved Gases *320*
14.4.3 Climate Change and Wetland Ecosystem *320*
14.5 Protection, Adaptation, and Mitigation Strategies *322*
14.6 Worst Case Scenario: Evidence of Anthropogenic Impact on Bengaluru Lakes: From "City of Lakes" to the "City of Froth Lakes" 323
14.6.1 Cause of Fire in an Aquatic Environment *324*
14.6.2 Protection and Mitigation to Conserve Aquatic Ecosystem *325*
14.7 Conclusion *325*
References *325*

15 Role of Stable Isotopes in Groundwater Resource Management *333*
Syed Shams Rizvi, M. A. Mohammed-Aslam, Zameer Ahmad Shah, Taufique Warsi, Mohammad Muqtada Ali Khan, Tarik Mitran, and Shyam Kanhaiya
15.1 Introduction *333*
15.2 Background *334*
15.2.1 Isotope Systematics *334*
15.2.2 Isotope Fractionations *335*
15.2.2.1 O and D in Precipitation *336*
15.2.2.2 Deuterium Excess *337*
15.2.2.3 Continental Effect *337*
15.2.2.4 Seasonal Effect *338*
15.2.2.5 Latitude Effect *338*
15.2.2.6 Altitude Effect *338*
15.2.3 Methodology and Precautions in the Collection of Rainwater Samples *338*
15.2.3.1 Rainwater Collection Techniques *339*
15.2.3.2 Principle of Isotope Ratio Mass Spectrometer *339*
15.2.4 Application of Isotopes in Water Resource Management *339*
15.2.4.1 Interconnections Between Surface and Groundwater *340*
15.2.4.2 Tracking Sources of Groundwater Recharge *340*

15.2.4.3 Groundwater Age or Groundwater Dating *341*
15.2.4.4 Source and Mechanism of Groundwater Pollution *341*
15.2.4.5 Sanitary Landfills *341*
15.2.4.6 Agriculture Watersheds *342*
15.2.4.7 Groundwater Salinization *342*
15.3 Utilization of Water Isotopes in Hydrology: A Case Study *342*
15.3.1 Study Site *342*
15.3.2 Findings and Discussions *342*
15.3.2.1 Estimation of d-excess *346*
15.3.2.2 Estimation of the Amount Effect *346*
15.3.2.3 Estimation of Seasonal Variations *348*
15.4 Conclusion *349*
15.5 Summary and Conclusions *350*
Acknowledgements *350*
References *351*

16 Assessment of Causes and Impacts of Sand Mining on River Ecosystem *357*
Soumya Pandey, Gaurav Kumar, Neeta Kumari, and Richa Pandey
16.1 Introduction *357*
16.2 Why River Sand? *359*
16.3 Processes of Sand Mining *360*
16.4 Monitoring of Sand Mining Activities *362*
16.5 Sand Mining: A Threat to Aquatic Ecosystem *364*
16.6 Impacts of Sand Mining on Components of Riverine Ecosystem *365*
16.6.1 Impacts on River Morphology and Sediment *366*
16.6.2 Impacts on Water Quality and Aquatic Life *367*
16.6.3 Sand Bank Filtration *367*
16.7 Remediation Alternative *368*
16.8 Conclusion *370*
References *370*

17 Nutrient Dynamics in the Polar Ice Sheets and Mountain Glaciers: A Review *381*
Monica Sharma Shamurailatpam, AL. Ramanathan, and N. Janardhana Raju
17.1 Introduction *381*
17.2 Nutrient Dynamics: Transformation, Storage, and Export in the Glacial Environment *383*
17.2.1 Dissolved Organic Carbon *383*
17.2.2 Nitrogen *388*
17.2.3 Phosphorus *391*
17.3 Factors Influencing Nutrients Concentration and Flux *393*
17.3.1 Climatically Driven Export of Nutrients *393*
17.3.2 Microbial Driven Nutrients Productivity *394*
17.4 Conclusion *395*
Acknowledgments *396*
References *396*

Index *407*

List of Contributors

Bedour Al Sabti
Water Research Center, Kuwait Institute for Scientific Research, Safat, Kuwait

Anshumali
Department of Environmental Science and Engineering, Indian Institute of Technology (Indian School of Mines), Dhanbad, Jharkhand, India

Uday Bhan
Department of Petroleum Engineering & Earth Sciences, UPES, Dehradun, Uttarakhand, India

George Biswas
Department of Geology, Presidency University, Kolkata, West Bengal, India

Umakant Chaudhari
Department of Environmental Science and Technology, Central University of Punjab, Bathinda, Punjab, India

Harshit Chawla
Department of Environmental Engineering, Delhi Technological University, Delhi, India

Chidambaram Sabarathinam
Water Research Center, Kuwait Institute for Scientific Research, Safat, Kuwait

Fatemah M. Dashti
Water Research Center, Kuwait Institute for Scientific Research, Safat, Kuwait

Chitrangada Debsarma
School of Environment and Sustainable Development, Central University of Gujarat, Gandhinagar, Gujarat, India

Dhanu Radha Samayamanthula
Water Research Center, Kuwait Institute for Scientific Research, Safat, Kuwait

Kamal Dua
Discipline of Pharmacy, Graduate School of Health, University of Technology Sydney, Broadway, NSW, Australia

Somenath Ganguly
Department of Petroleum Engineering & Earth Sciences, University of Petroleum and Energy Studies, Dehradun, Uttarakhand, India

A. K. Haritash
Department of Environmental Engineering, Delhi Technological University, Delhi, India

Anamika Harshvardhan
Silviculture and Forest Management Division, Institute of Wood Science and Technology, Bengaluru, Karnataka, India

Mohammad Hashim
Department of Geography, Faculty of Natural Sciences, Jamia Millia Islamia, New Delhi, India

Adams Ovie Iyiola
Department of Fisheries and Aquatic Resources Management; Faculty of Renewable Natural Resources Management; College of Agriculture and Renewable Natural Resources, Osun State University, Osogbo, Osun State, Nigeria

Sylvester Chibueze Izah
Department of Microbiology, Faculty of Science, Bayelsa Medical University, Yenagoa, Bayelsa State, Nigeria

Sourav Kalra
Chitkara University School of Pharmacy, Chitkara University, Baddi, Himachal Pradesh, India

Shyam Kanhaiya
Department of Earth and Planetary Sciences, VBS Purvanchal University, Jaunpur, Uttar Pradesh, India

Neha Kanojia
Chitkara University School of Pharmacy, Chitkara University, Baddi, Himachal Pradesh, India

Oznur Karaca
Department of Geological Engineering, Canakkale Onsekiz Mart University, Canakkale, Turkey

Mohammad Muqtada Ali Khan
Department of Geoscience, Faculty of Earth Science, Universitii Malaysia Kelantan, Jeli, Kelantan, Malaysia

Vamsi Krishna Kudapa
Department of Chemical Engineering, UPES, Dehradun, Uttarakhand, India

Amit Kumar
Department of Geoinformatics, Central University of Jharkhand, Ranchi, Jharkhand, India

Gaurav Kumar
Civil and Environmental Engineering, Birla Institute of Technology, Ranchi, Jharkhand, India

Ranjit Kumar
Department of Chemistry, UPES, Dehradun, Uttarakhand, India

Sanchit Kumar
Department of Environmental Science and Engineering, Indian Institute of Technology (Indian School of Mines), Dhanbad, Jharkhand, India

Neeta Kumari
Civil and Environmental Engineering, Birla Institute of Technology, Ranchi, Jharkhand, India

Tarik Mitran
Soil and Land Resources Assessment Division, National Remote Sensing Centre, Balanagar, Hyderabad, Telangana, India

Sunil Mittal
Department of Environmental Science and Technology, Central University of Punjab, Bathinda, Punjab, India

M. A. Mohammed-Aslam
Department of Geology, Central University of Karnataka, Kalaburagi, Karnataka, India

Suryadipta Mukherjee
Department of Geology, Presidency University, Kolkata, West Bengal, India

Hasan Raja Naqvi
Department of Geography, Faculty of Natural Sciences, Jamia Millia Islamia, New Delhi, India

Binod Kumar Nath
Public Works (Roads) Department Office of the Chief Engineer, PWD (Roads) Chandmari, Guwahati, Assam, India

Samikshya Panda
Department of Agricultural Engineering, Institute of Agricultural Sciences, Banaras Hindu University, Varanasi, Uttar Pradesh, India

Richa Pandey
Department of Mechanical Engineering, Birla Institute of Technology, Ranchi, Jharkhand, India

Soumya Pandey
Civil and Environmental Engineering, Birla Institute of Technology, Ranchi, Jharkhand, India

Abhilash Kumar Paswan
Department of Physical Science, Academy of Scientific and Innovative Research, Ghaziabad, Uttar Pradesh, India; Department of Hydrogeochemistry, CSIR-National Geophysical Research Institute, Hyderabad, Telangana, India

Subhashree Patra
Department of Environmental Sciences, Central University of Jharkhand, Ranchi, Jharkhand, India

Harsh Pipil
Department of Environmental Engineering, Delhi Technological University, Delhi, India

N. Janardhana Raju
School of Environmental Sciences, Jawaharlal Nehru University, New Delhi, India

AL. Ramanathan
School of Environmental Sciences, Jawaharlal Nehru University, New Delhi, India

Glory Richard
Department of Community Medicine, Faculty of Clinical Sciences, Niger Delta University, Wilberforce Island, Bayelsa State, Nigeria

Syed Shams Rizvi
Department of Geology, Central University of Karnataka, Kalaburagi, Karnataka, India

Prafulla Kumar Sahoo
Department of Environmental Science and Technology, Central University of Punjab, Bathinda, Punjab, India; InstitutoTecnologico Vale, Belem, PA, Brazil

Paulami Sahu
School of Environment and Sustainable Development, Central University of Gujarat, Gandhinagar, Gujarat, India

Purabi Saikia
Department of Environmental Sciences, Central University of Jharkhand, Ranchi, Jharkhand, India

Zameer Ahmad Shah
Geological Survey of India, U.T. Jammu and Kashmir, NR-Srinagar, India

Monica Sharma Shamurailatpam
School of Environmental Sciences, Jawaharlal Nehru University, New Delhi, India

Reema Sharma
Department of Agricultural Engineering, Institute of Agricultural Sciences, Banaras Hindu University, Varanasi, Uttar Pradesh, India

Shilky
Department of Environmental Sciences, Central University of Jharkhand, Ranchi, Jharkhand, India

Shrinivasa D. J.
Department of Agricultural Engineering, Institute of Agricultural Sciences, Banaras Hindu University, Varanasi, Uttar Pradesh, India

Lubna Siddiqui
Department of Geography, Faculty of Natural Sciences, Jamia Millia Islamia, New Delhi, India

Masood Ahsan Siddiqui
Department of Geography, Faculty of Natural Sciences, Jamia Millia Islamia, New Delhi, India

Abhu Singh
Department of Geography, Faculty of Natural Sciences, Jamia Millia Islamia, New Delhi, India

Harminder Singh
School of Chemical Engineering and Physical Science, Lovely Professional University, Phagwara, India

Risha Singh
Centre for the Study of Regional Development, Jawaharlal Nehru University, New Delhi, India

S. K. Singh
Department of Environmental Engineering, Delhi Technological University, Delhi, India

Sonam Taneja
Department of Environmental Engineering, Delhi Technological University, Delhi, India

Shivani Thakur
Department of Environmental Science and Technology, Central University of Punjab, Bathinda, Punjab, India

Vinod Kumar Tripathi
Department of Agricultural Engineering, Institute of Agricultural Sciences, Banaras Hindu University, Varanasi, Uttar Pradesh, India

Suantak Paolalsiam Vaiphei
Department of Hydrogeochemistry, CSIR-National Geophysical Research Institute, Hyderabad, Telangana, India

Mallika Vashist
Department of Environmental Engineering, Delhi Technological University, Delhi, India

Nitin Verma
Chitkara University School of Pharmacy, Chitkara University, Baddi, Himachal Pradesh, India

Taufique Warsi
WOTR Centre for Resilience Studies (W-CReS), Watershed Organisation Trust (WOTR), Pune, Maharashtra, India

Shivani Yadav
Department of Environmental Engineering, Delhi Technological University, Delhi, India

Editors' Biography

Sughosh Madhav Dr. Sughosh Madhav is presently working as Dr. D. S. Kothari Postdoctoral Fellow (DSKPDF) in the Department of Civil Engineering, Jamia Millia Islamia, New Delhi, India. He has obtained his master's degree from the Department of Environmental Science, Banaras Hindu University, Varanasi, India. He earned his doctorate from Jawaharlal Nehru University, New Delhi. The area of his doctoral research is the environmental impact of industrial effluents on groundwater and soil quality. He has published various research papers and book chapters in the field of environmental geochemistry, water pollution, wastewater remediation, and climate change. He also edited seven books for Wiley, Springer, and Elsevier publications on various environmental issues.

Virendra Bahadur Singh Dr. Virendra Bahadur Singh is presently working as an assistant professor in the Department of Environmental Studies, Ram Lal Anand College, University of Delhi, New Delhi, India. He has worked as Dr. D. S. Kothari Postdoctoral Fellow (DSKPDF) in the Department of Geology, University of Delhi, Delhi, India. He has also worked as National Postdoctoral Fellow (NPDF) at the Department of Civil Engineering, Indian Institute of Technology Delhi, New Delhi, India. He obtained his doctoral degree from School of Environmental Sciences, Jawaharlal Nehru University, New Delhi, India. His research work is mainly focused on hydrogeochemistry, biogeochemistry, hydrology, hydro-meteorological correlation, and suspended sediment transport. He has published more than 30 research papers in various reputed journals and also edited 2 books for Wiley and Elsevier publications.

Manoj Kumar Dr. Manoj Kumar is presently working as an assistant professor in the Department of Environmental Studies, Central University of Haryana, India. He has obtained his master's degree and doctorate degree from Jawaharlal Nehru University, New Delhi, India. The area of his doctoral research is Solute Geochemistry and Arsenic Fate in the Aquifers of Central Gangetic Basin, India. He has published several research papers in journals of international repute and book chapters in the field of hydrogeochemistry, groundwater pollution, and health risk assessment.

Sandeep Singh Prof. Sandeep Singh is working as a professor in the Department of Earth Sciences, Indian Institute of Technology, Roorkee, India. He obtained his master's and PhD degrees from the University of Roorkee (IIT, Roorkee). His research interest includes geochronology, isotope geology, exhumation processes, and Sr-isotopes in the water system. He has published more than 70 research papers in reputed journals. He is a regular reviewer of various international journals. He has earned various reputed awards, and he is a member of numerous respected academic societies. Prof. Singh is a well-known geologist and has a wide experience in isotopic geology and hydrogeochemistry.

Preface

Hydrogeochemistry has always been a subject to solve the relationships between groundwater, surface water, and aquatic systems' chemical characteristics and their relationship with the area's geology. The phenomenon of hydrogeochemical variations depends on various geochemical processes, like weathering, evaporation, and dissolution, and diverse anthropogenic activities that control the aquatic systems' hydrogeochemistry. Almost all aquatic systems are affected by human interventions, causing several problems. Further climate change also impacts the hydrogeochemistry of different ecosystems. It is, therefore, vital to understand the current status of the aquatic systems to manage and mitigate the impacts. This book covers current issues related to hydrogeochemistry of natural aquatic systems, including river, glaciers, lake, and sub-surface ecosystems, processes involved, and human beings' interference. This book tries to address all the major impending problems related to hydrogeochemistry of surface and sub-surface water with a certain degree of solutions to manage and conserve water resources. The various chapters will help to understand the processes and factors controlling the hydrogeochemistry of water resources. The book discusses multiple hydrogeochemical techniques like nutrient geochemistry, solute acquisition processes, water–sediment interactions, enrichment of geogenic elements like arsenic, fluoride, and uranium, anthropogenic addition, and biogeochemical cycling of metals, CO_2 consumption rate in the surface water bodies, sand mining in rivers, and hydrogeochemical modeling. The other highlights of the book are carbon dynamics and its sequestration in surface water bodies, soil–water interactions, and sediment loading in aquatic ecosystems. The book encompasses knowledge on isotopic studies, microbial control in hydrogeochemistry along nutrient dynamics of the polar ice sheet and glaciers. The work contained in the book covers holistic and in-depth knowledge of the hydrogeochemistry of different aquatic ecosystems.

The chapters in the book address the impending problems with a certain degree of solutions to manage and conserve water resources. This book contains both practical and theoretical aspects of the hydrogeochemistry of aquatic systems. It will be helpful for undergraduate and graduate university students or researchers and teachers, environmental engineers, and scientists working in water resource management, environmental studies, and social issues. It will also be helpful for researchers and stakeholders from all the sectors of industries and governmental agencies who are directly or indirectly associated with water resource research and management.

We tried a humble attempt to reflect upon the various aspects of aquatic hydrogeochemistry, hoping that it would be a significant addition to the already available literature. The contributors to the book having different backgrounds provide a holistic approach to the topic imbibing diverse practices and perspectives. We express our sincere gratitude to all the contributors and publishers for producing a remarkable and meaningful edited volume on an important issue.

1

Fluoride in Groundwater: Distribution, Sources, Processes, Analysis, and Treatment Techniques: A Review

Bedour Al Sabti, Dhanu Radha Samayamanthula, Fatemah M. Dashti, and Chidambaram Sabarathinam

Water Research Center, Kuwait Institute for Scientific Research, Safat, Kuwait

1.1 Introduction

Fluoride (F^-) belongs to the halogen family and is a constituent in minerals such as fluorite, fluorspar, apatite, biotite, cryolite, and muscovite (Bretzler and Johnson 2015; Dehbandi et al. 2017), apart from its availabilities in plants, soil, and groundwater. Groundwater is one of the most important sources of drinking water and one of the fundamental human rights around the globe is an access to safe drinking. Contamination and unsustainable drinking water sources could affect human health, resulting in the transmission of diseases (WHO 2018). Fluoride is one of the ions which may lead to groundwater contamination if present in high concentrations. Although high F^- in groundwater is a major concern that is still being debatable around the globe, fluoride is essential for the growth of the dental and skeletal frame of the body. Fluoride concentration in groundwater differs from one region to another based on aquifer material, geology, weathering rate, aquifer depth, contact time, pH, rainfall, and temperature (Brunt et al. 2004; Onipe et al. 2020). The geochemical process governs fluoride mobility through leaching from soil and rocks to the groundwater. Studies suggest that exposure to high fluoride imparts a vulnerable effect on the mental ability of children. The IQ levels of children exposed to higher F^- are lower than unaffected children (Choi et al. 2012; Das and Mondal 2016). The thyroid gland is susceptible to F^-, which causes an increase in thyroid-stimulating hormone (TSH) leading to a drop in Triiodothyronine (T3) and Thyroxine (T4) levels, thereby resulting in hypothyroidism (McLaren 1976; Shashi 1988; Kumar et al. 2019). Fluorosis results from a high concentration of fluoride in drinking water and depends on other sources such as dietary habits that enhance the incidence of fluorosis (Brindha and Elango 2011; Srivastava and Flora 2020). Several countries, such as West Indies, India, Poland, China, Spain, Africa, and Italy, have been reported with high fluoride concentrations (Huang et al. 2017). The geochemical data for Cameroon, Algeria, Ghana, United Kingdom, Siri Lanka, Argentina, Canada, Tanzania, Kuwait, South Africa (Silom), India (Telangana), and Brazil were

Hydrogeochemistry of Aquatic Ecosystems, First Edition. Edited by Sughosh Madhav, Virendra Bahadur Singh, Manoj Kumar, and Sandeep Singh.

collected from the literature to understand the geochemistry of F^- (Table 1.1). Some of the published data for selected countries does not contain the complete analysis results. Based on the available ions in the analytical data, they were used for statistical analysis using Statistical Package for Social Sciences (SPSS) software. The same analytical data were used

Table 1.1 Lithology and analytes considered from the literature studies of various countries but clay minerals like Vermiculite have also reported to be a source of F^- in groundwater due to the process of Fluoride ion.

Country	Lithology type	Analytes	Reference
Algeria	Sedimenatary (sand and gravel, limestone, clay, and shale)	pH, EC, Ca^{2+}, Mg^{2+}, Cl^-, SO_4^{2-}, HCO_3^-, F^-	Messaitfa (2007)
Argentina	Thick sedimentary rock and volcanoclastic mineral	pH, EC, temperature, Na^+, K^+, Ca^{2+}, Mg^{2+}, Cl^-, SO_4^{2-}, HCO_3^-, CO_3^{2-} F^-, NO_3^-, Si, Fe^{2+}, Al^{3+}, Be^{2+}, U, B	Ganyaglo et al. (2019); Jayawardana et al. (2012); Edmunds and Smedley (2013)
Brazil	Sedimentary	pH, EC, Ca^{2+}, Mg^{2+}, Cl^-, SO_4^{2-}, HCO_3^-, F^-	Rockett et al. (2013)
Canada	—	pH, EC, temperature, Na^+, K^+, Ca^{2+}, Mg^{2+}, Cl^-, SO_4^{2-}, HCO_3^-, CO_3^{2-} F^-, NO_3^-, Si, Fe^{2+}, Al^{3+}, Be^{2+}, U, B	Ganyaglo et al. (2019); Jayawardana et al. (2012); Edmunds and Smedley (2013)
Central Africa (Cameroon)	Crystalline basement (granite)/ Tertiary sedimentary rocks	pH, EC, temperature, Na^+, K^+, Ca^{2+}, Mg^{2+}, Cl^-, SO_4^{2-}, HCO_3^-, F^-, NO_3^-	Fantong et al. (2009)
Ghana	Precambrian crystalline and igneous rocks (granite)	pH, EC, TDS, temperature, Na^+, K^+, Ca^{2+}, Mg^{2+}, Cl^-, SO_4^{2-},HCO_3^-,$CO_3^{2-}$$F^-$,$NO_3^-$	Sunkari an Abu (2019)
India (Telangana)	Igneous rock (granite)		Narsimha and Sudarshan (2017)
Kuwait	Sedimentary siliciclastic and carbonates	pH, EC, TDS, temperature, Na^+, K^+, Ca^{2+}, Mg^{2+}, Cl^-, SO_4^{2-}, HCO_3^-, CO_3^{2-} F^-, NO_3^-, B, NH4$^+$, PO_4^{3-}, SiO_2, Fe^{2+}, Al^{3+}, Ba^{2+}, Li^+, Mn^{2+}, Mo, Ni^{2+}, Zn^{2+}	Al-Senafy et al. (2011)
South Africa (Silom)	—	pH, EC, TDS, temperature, Na^+, K^+, Ca^{2+}, Mg^{2+}, Cl^-, SO_4^{2-}, HCO_3^-, CO_3^{2-} F^-, NO_3^-, PO_4^{3-}	Onipe et al. (2021)
Sri Lanka	High-grade metamorphic rock	pH, EC, temperature, Na^+, K^+, Ca^{2+}, Mg^{2+}, Cl^-, SO_4^{2-}, HCO_3^-, CO_3^{2-} F^-, NO_3^-, Si, Fe^{2+}, Al^{3+}, Be^{2+}, U, B	Ganyaglo et al. (2019); Jayawardana et al. (2012); Edmunds and Smedley (2013)
Tanzania	Volcanic rock and metamorphic		
United Kingdom	—		

for different plots developed from the output results of WATEQ4F and AQUACHEM. The objective of this review is to emphasize the global distribution, sources, analysis, and treatment strategies for excessive fluoride levels in groundwater. Also, the review presents geochemical plots, statistical techniques, thermodynamic and modeling approaches to determine processes governing the fluoride release and distribution in groundwater.

1.2 Permissible Limits of Fluoride in Drinking Water

According to the WHO (2006), the maximum permitted level of F^-in drinking water is 1.5 mg/L. While the USPHS (1987) established a range of allowable F^- concentration in drinking water for regions based on their climatic conditions, because the amount of water consumed and, the amount of F^- ingested is primarily influenced by the air temperature. The rise in air temperature decreases the concentration of F^-. The maximum permissible level in tropical climates with temperatures above 26 °C is 1.4 mg/L. In light of the Indian subcontinent's environmental and socioeconomic situation, the F^- desirable limit is established at 0.6–1.2 mg/L, and the highest allowed level in the absence of any other source is set at 1.5 mg/L for drinking water (ISI 1995). The limit was set based on the daily consumption rate of water, about 2 L/day for an adult body mass, and contains about 0.2–0.5 mg fluorine as a standard diet (WHO 1994). A range of environmental, social, cultural, economic, and other circumstances affecting possible exposure, as well as the default assumptions used to create the guideline values, will need to be taken into account when creating national drinking-water standards based on these guideline values. In addition, the environmental-based variation depends on the region, as regional diets and ambient temperature control the permissible limit (Apambire et al. 1997). Furthermore, in a country with a constant warm environment and piped water as the main drinking-water source, authorities may choose a lower health-based fluoride target than this guideline value as water consumption is predicted to be higher (Guidelines for drinking-water quality 2021). Drinking water from groundwater may be beneficial or harmful depending on the concentration level of fluoride. In recent years, countries have been developing drinking standards to decrease waterborne diseases and improve safe water resources management (Ali et al. 2019). As the concentration of F^- in drinking water is different for each country, and the amount of water consumed by a person also varies concerning the climate and availability, so each region has its own standard (Figure 1.1). Drinking high fluoride groundwater is the primary reason for endemic fluorosis in the countries such as China (Guo et al. 2007). Higher F^- concentration in groundwater, i.e. exceeding the permissible limit of WHO, is observed in countries like Japan, Cameroon, China, Turkey, India, Sri Lanka, Iran, Pakistan, Brazil, and Canada. Fluorosis, on the other hand, is caused by a high quantity of fluoride in drinking water and is exacerbated by other factors such as dietary choices (Brindha and Elango 2011).

Fluoride concentrations above the permissible limit are harmful to human health, such as dental fluorosis, crippling skeletal fluorosis, and skin lesions (Rasool et al. 2017), and it also depends on the amount of intake, duration of exposure, and gastrointestinal traces. The absorbed fluoride from the food sources in the digestive system is about 30–40% (Kumar et al. 2019).

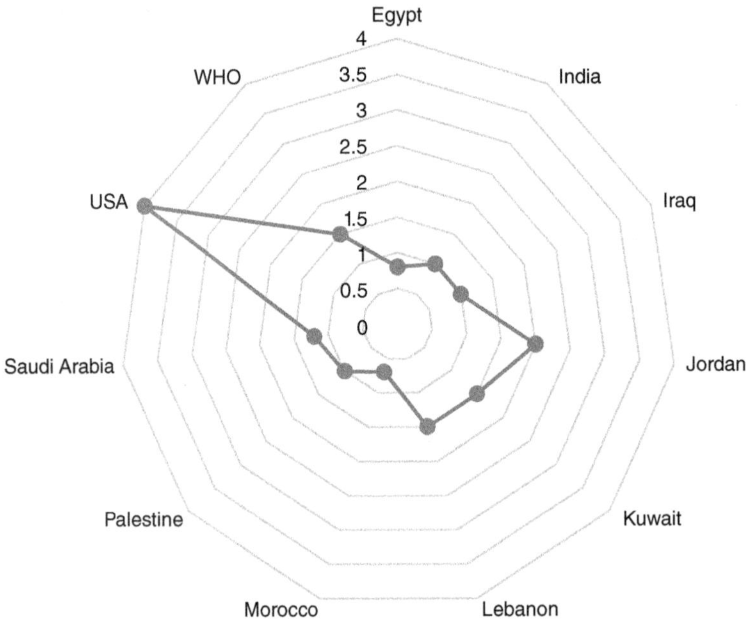

Figure 1.1 A radar chart representing the maximum fluoride concentration standard for each country as specified. Data *Source:* Guo et al. (2007), Jagtap et al. (2012), Abu-Zeid and El-Hatow (2007), and KEPA (2017).

1.3 Concentration of High Fluoride Around the Globe

Data for groundwater samples signifying maximum value fluoride concentration were collected from literature studies to represent the highest value of F^- distribution in a world map (Figure 1.2). The map represents a total of 46 regions with F^- concentration above the allowable limits, reaching up to 160 mg/L. One of the highest F^- concentration values that was found with a total of 160mg/L, found to be reported in California, North America (McMahon et al. 2020). The high fluoride concentration as it is shown on the map mostly represents the igneous and metamorphic rock formations predominantly around the tropical region. However, elevated fluoride concentration may result from various sources affecting the groundwater to be discussed in the following section.

1.4 Sources of Fluoride in Groundwater

Fluoride in groundwater occurs in variable concentrations and <1.0 mg/L is generally reported in natural waters (Hem 1985). Anion exchanges, solubility of minerals, temperature, pH, nature of the aquifers, matrix, and availability of complexing ion, residence time, and chemical composition of the groundwater are the main governing factors affecting the concentration of F^- (Apambire et al. 1997). The presence of F^- in groundwater is mostly caused by natural or geogenic contamination (Manikandan et al. 2014). Chemical

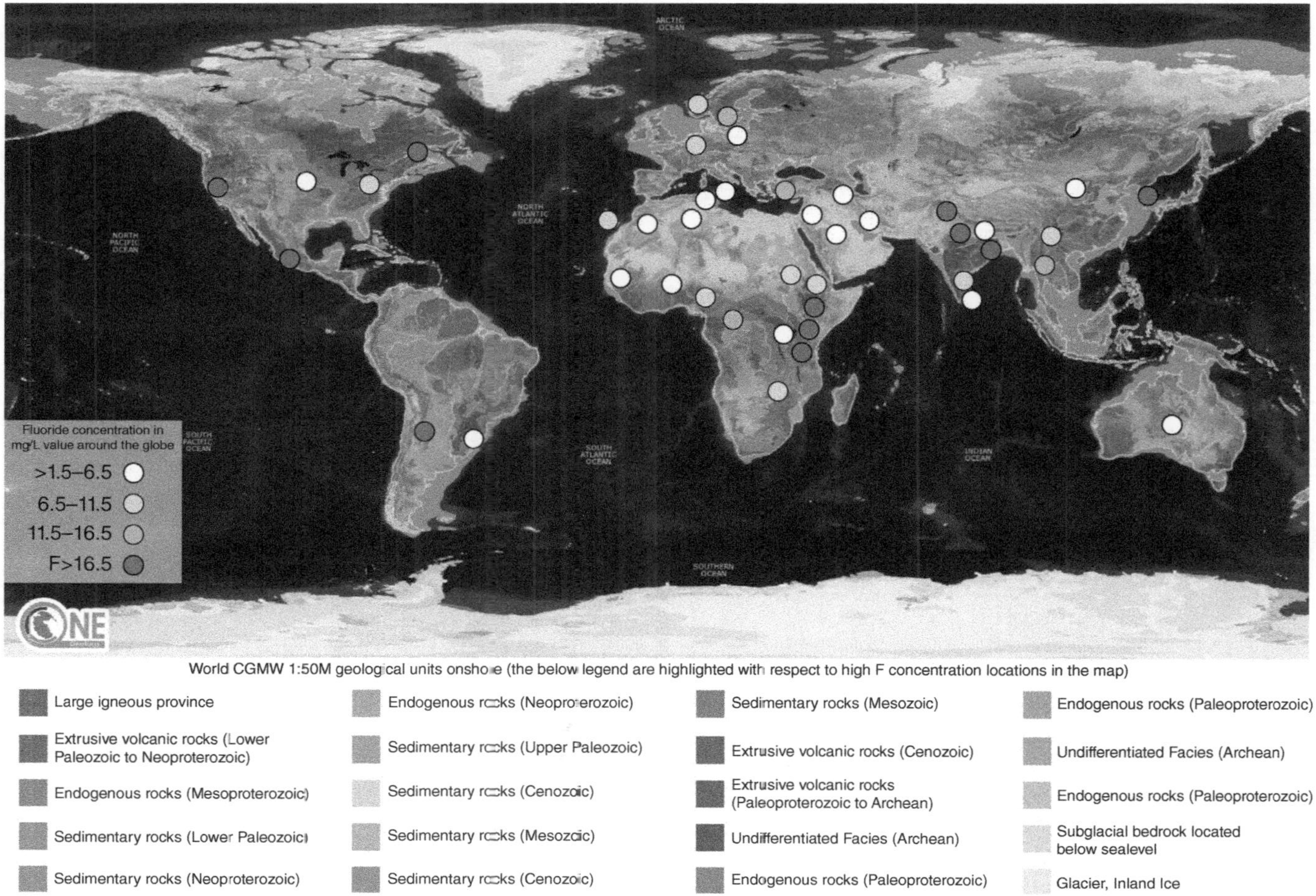

Figure 1.2 Global map represents the maximum fluoride values concentration which is above the standard limits in different regions. The lithology legend highlights only the areas of high fluoride concentration in groundwater. Source: Reproduced with the permission of the OneGeology.

weathering, atmospheric dust from continental sources, industrial pollutants, and magmatic emissions are also potential sources of F^- (Gizaw 1996). In other words, F^- reaches groundwater due to input from natural or anthropogenic sources or both. Fluorides can be transported to groundwater via water–rock interaction by different fluoride-bearing minerals (Narsimha and Sudarshan 2017). This interaction is used to examine the geogenic input of fluoride in groundwater aquifers from various forms of fluoride minerals.

1.4.1 Natural Sources

The parent rock serves as the primary source of fluoride in groundwater. Thus, various geochemical studies have been performed on various aspects of fluoride in groundwater. Some rock types showed a higher fluoride concentration than others with groundwater, such as volcanic zones with geothermal sources, sedimentary basins, and crystalline basement aquifers (Edmunds and Smedley 2005). An average F^- concentration of 715 mg/kg is observed in igneous rocks globally (Gizaw 1996). These rocks are likely to be formed in volcanic areas with magmas enriched with fluoride concentration, and the sodium plagioclase is most likely to form soft groundwater that allows higher fluoride production when reaching equilibrium (Ozsvath 2006). Fluorine can be present in different igneous rock-forming minerals such as mica, apatite, amphiboles, pegmatite, and certain types of clay (Ayoob and Gupta 2006). The predominant minerals are depicted in Table 1.2, but clay minerals like Villiaumite have also reported to be a source of F^- in groundwater due to the process of ion exchange (Handa 1975; Pickering 1985; Wenzel and Blum 1992; Bardsen et al. 1996; Subba Rao and Devadas 2003). Fluoride concentrations, on the other hand, are frequently related to water–rock interaction to a certain degree because fluoride is mostly from lithology (Manikandan et al. 2014). The hydrogeochemical studies in groundwater had reported several mechanisms of F^- release based upon the composition of the aquifer matrix, the physio chemical conditions, and the rock–water interaction. They have also inferred that the residence time in the aquifer and temperature also play a major role in release of F^- from the rock matrix to groundwater. Some researchers have studied the

Table 1.2 Classification of fluoride minerals in different rock types with their chemical composition.

S.No.	Mineral	Rocks	Chemical composition
1	Fluorite	CaF_2 Pegmatite, Metamorphosed, Limestone	CaF_2
2	Fluorspar	Pegmatite Pneumatolitic, Deposits	$CaF_2.3Ca_3(PO_4)_2$
3	Tremolite Actinolite	Clay	$Ca_2(MgFe^{2+})_5(Si_8O_{22})(OHF)_2$
4	Rock Phosphate	Limestone, Fossils	$NaCa_2(MgFe^{2+})_4(AlFe^{3+})(SiAl)_8O_{22}(OHF)_2$
5	Lepidolite	Gabbros, Dolerites	$K_2(Li, Al)_5(Si_6Al_2)O_{20}(OHF)_4$

Source: Adapted from Yadav and Khan (2010).

relationship between water–rock interaction and F^- concentration in various aquifers having different geologic settings (Saxena and Ahmed 2003; Gaciri and Ad Davis 1993; Edmunds and Smedley 1996; Nordstrom et al. 1989; Nordstrom and Jenne 1977). Groundwater development and management have encountered a serious issue due to the prevalence of F^- when harvesting or recharging structure areas located in granitic terrain. Fluorite is one of the minerals responsible for the presence of fluorine in nature, and it acts as an accessory mineral in granitic rocks (Ayoob and Gupta 2006). Deeper wells in granitic and gneissic aquifers often have higher fluoride contents. Fluorite (CaF_2) present in the granitic formation is a dominant source of F^- (Deshmukh et al. 1995). It is to be noted that the rate of dissolution and stability of the mineral in freshwater is relatively low (Nordstrom and Jenne 1977). Hence, few studies have indicated that biotite mica is present as an accessory mineral in the granitic formation, to be the prime source of F^- in hosted groundwater for their favorable F^-, $(OH)^-$ exchange (Saxena and Ahmed 2001). A pattern that has been documented in several inferences on groundwater collected from granitic regions (Yun et al. 1998a; White et al. 1963). The studies on samples from deep groundwater have indicated that the dissolution of F^- minerals in the rock matrix was governed by the depth of well, residence time, and temperature (Saxena and Ahmed 2003; Nordstrom et al. 1989). Higher residence time is generally found in massive hard rock aquifers and lesser in porous sedimentary formations. This is also an important fact that higher F^- in groundwater is more prevalent in hard rock aquifers. Fluoride can mix with clay by displacing the hydroxide from its surface in acidic water. F^- is adsorbed on the clay surface but desorbed from solid phases in alkaline water. The evaporation process can also cause fluoride enrichment in groundwater; however, the evaporation process is usually characterized by the preponderance of Cl^- ion in groundwater (Manikandan et al. 2014). Because of its granitic-type composition (alkaline), augen gneiss has the largest potential for higher groundwater fluoride. The augen gneiss is metamorphosed granite that contains fluoride-bearing minerals like biotite, hornblende, and distinctive Na-feldspar mega crystals that serve as a secondary source of sodium for fluoride enrichment (Addison et al. 2020). One of the important fluoride-bearing minerals is muscovite, with fluorine in its hydroxyl groups and crystal lattices that can act as fluoride substitutes due to their similar ionic charges. Among the micas apart from muscovite, biotite is also reported to have exchangeable F^- of octahedral sheet at the hydroxyl sites (Chidambaram et al. 2007). The lithological influences of F^- in groundwater were investigated by Singaraja et al. (2013), where the study stated that elevated F^- concentrations in groundwater were due to charnockite, hornblende biotite gneiss marine alluvium, and by flood plain alluvium. Higher concentration of F^- in hydrological systems also depends on the rate of recharge and the transport of the ion along with the groundwater flow. There is also a probability of reduction of F^- along the flow direction either due to mineral precipitation or removal by adsorption or by ion exchange process. The studies on the thermodynamic properties and the relative F^- enrichment in groundwater in the different geochemical environments are still scanty.

1.4.2 Anthropogenic Sources

Pollution that results from industries such as fabrication of aluminum, fertilizers, and steel has been reported to generate higher F^- through anthropogenic means to the natural environment (Datta et al. 2014). The use of phosphate fertilizer in agricultural operations also results in the probable presence of fluoride in groundwater (Thivya et al. 2015) where phosphate fertilizers generally contain fluorine ranging from 1.3 to 3.0% (Mclaughlin et al. 1996). Apart from the above anthropogenic sources, pesticides, sewage and sludge, groundwater table depletion, and other agricultural practices have also been linked to a rise in F^{-} concentrations in groundwater (Srinivasamoorthy et al. 2007). F^{-} may infiltrate streams through adjacent uneven joints, cracks, and through damaged pipelines transporting household, agricultural, and industrial effluents (Datta et al. 2014). Farooqi et al. (2009) study illustrated that familiar source of F^{-} in the environment including air pollution from coal combustion and phosphate fertilizers, whereas soluble F^{-} in fertilizers ranged from 60 to 255 mg/kg, with a mean of 175 mg/kg. In contrast, total F^{-} in coal samples ranged from 5.12 to 20.1 mg/kg, with an average value of 10.2 mg/kg (Farooqi et al. 2009). However, only a small portion of overall fluoride exposure is typically attributed to air (Fawell et al. 2006). Another study by Loganathan et al. (2006) reported that fluoride concentrations in New Zealand agricultural surface soils varied from 212 to 617 µg F^{-}, which was significantly lower than the fluoride values reported in the literature for sites polluted with fluoride from industrial sources.

1.4.3 Health Effects

The association between human health and fluoride was first explored in the late nineteenth century when chemists discovered varying levels of fluorine in human bones, tissues, and teeth (Kanduti et al., 2016). Fluoride is vital to human health for developing strong bones and teeth, but only in trace amounts ranging between 0.5 and 1 mg/L worldwide (Addison et al. 2020; Ozsvath 2006). However, fluoride is also known to be a very toxic element to human health if it exceeds the average standard limits published by WHO (Narsimha and Sudarshan 2017). Fluoride operates as hydrofluoric acid and is absorbed by humans and animals through their skin, where it travels quickly through the circulatory system and is stored in teeth and bones, according to Hodge and Smith (1977). According to another study, fluorosis is widespread in more than 200 million people globally (Ayoob and Gupta 2006). High fluoride levels can result in severe health issues in both adults and children, including dental fluorosis, bone deformation, and skeletal fluorosis (Narsimha and Sudarshan 2017).

1.4.4 Dental Fluorosis

Dental fluorosis considers as an irreparable toxic condition that affects the human teeth cells as an early-stage sign of having a fluoride increase in the body (Ayoob and Gupta 2006). Any fluoride ingestion in water with a concentration above the standard limits may cause dental fluorosis (Apambire et al. 1997). A case study in India illustrated that out of 258 children in the study area were examined, and 84 of them were diagnosed with dental fluorosis with ages ranging between 3 and 17 years old. The dental signs of the patients were identified from their teeth, such as chalky white color stain, browning of the tooth surface, yellowish stain, deep brown or black discoloration, pitting, and teeth loss (Jha et al. 2009). Opaque white patches on teeth characterize the diagnosis of early-stage dental fluorosis, unlike the advanced stages where the teeth are displayed as pitting on the surface of the teeth and brown to black staining (Figure 1.3) (Apambire et al. 1997). In addition, it was also suggested that calcium fluoroapatite crystals replaced the calcium in the teeth during

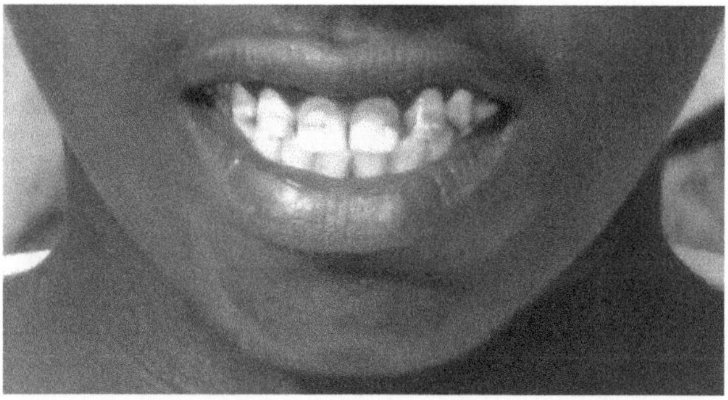

Figure 1.3 Effect of dental fluorosis due to high concentration of F^-.

a process known as mineralization. This process happens when the fluoride accumulates and combines with the calcium, and further on, the calcium will be reduced and lost from the teeth (Ayoob and Gupta 2006).

1.4.5 Skeletal Fluorosis

The consumption of water with F^- concentration greater than the permissible limit that ranges between 4.0 and 8.0 mg/L and higher (Apambire et al. 1997), will lead to severe and long-term damage to the human skeleton, a disease known as skeletal fluorosis (Ayoob and Gupta 2006). The lowest limit that has been noticed, which causes skeletal fluorosis, is 3.0 mg/L (Manikandan et al. 2014). Moreover, higher the duration and amount of fluoride entering the human body, the higher the severity of the disease (Ayoob and Gupta 2006). The symptoms shown on people with a possibility of having skeletal fluorosis vary in severity, with excess fluoride concentration consumption by Sutter from muscle and joint pain or mild rheumatic pain. In addition, to severe pain in the cervical spine region, stiffness and rigidity in joints, and crippling; skeletal fluorosis can occur once the fluoride concentration in drinking water exceeds 10 mg/L (Apambire et al. 1997; WHO 1970). Likewise, skeletal fluorosis can also be witnessed in the X-ray by increasing the bone density in adults (Mohammadi et al. 2017). Fluoride can disrupt the equilibrium of bone mineral metabolism by affecting the accretion and desorption of bone tissue (Ayoob and Gupta 2006). A study by Mohammadi (2017) analyzed two categories one with high fluoride concentration above the standard limit was diagnosed in a total of 445 people, and the other one containing low fluoride concentration below the standard limit was observed in a total of 470 people. The first area with high F^- showed that 21.1% of the people have skeletal fluorosis, only 3% from the second area suffered from skeletal fluorosis. The rates of the fluoride concentration in this study ranged from 0.22 to 10.33 mg/L (Mohammadi et al. 2017).

1.5 Hazard Index

The Hazard Index or Hazard Quotient (HQ) is the ratio of a single drug exposure level (dose or concentration) during a defined time period to the Rf_D computed for the same substance over the same time period. HQ, ratio greater than one indicates that the chemical

Table 1.3 Calculated and standard values of the parameters needed for the HQ calculation. The mean value of F^- from Kuwait's groundwater samples.

Variable	Description	Value (Women)	Value (Men)
C (mg/L)	Mean F concentration in groundwater	1.98	1.98
IR (L/day)	Daily water intake rate	2	2
EF (days/year)	Exposure frequency	365	365
ED (Year)	Exposure duration	76.4	74.6
Lt	Lifetime expectancy	75.4	75.4
BW (kg)	Average body weight	78.3	88.1
AT (Lt × 365)	Averaging time of exposure	27 521	27 521
EDI	Estimated daily intake of F^- from groundwater	0.05	0.04
RfD (mg/kg)	Oral toxicity reference	0.06	0.06
HQ	Hazard index	0.85	0.74

Source: Data from Worldbank – Kuwait Data (2021).

concentration is enough to induce persistent noncarcinogenic effects. In this review, samples of Kuwait (Table 1.3) are considered to check the effect of high concentration of fluoride using the following equation:

$$HQ = \frac{EDI}{RfD} \tag{1.1}$$

This value is according to the USEPA's Integrated Risk Information System (USEPA 2014). The estimated daily intake (EDI) of F from groundwater is used to calculate the F exposure dosage. The EDI values were calculated using Eq. 1.9 from the US Environmental Protection Agency. EDI can be determined using the following equation:

$$EDI = C \times IR \times EF \times EDBW \times AT \tag{1.2}$$

The optimum level of oral intake was assessed to depend on the baby weight and annual mean maximum temperature (Galagan et al. 1957)

$$F\left(mg / L\right) = 0.34 / E \tag{1.3}$$

E is the estimated average daily water intake for children <10 years old.

$$\text{Optimal fluoride intake}\left(mg / L\right) = \frac{0.022}{\left(0.0104 + 0.000724\right) * AMMT^{\circ}C} \tag{1.4}$$

AMMT is for a period of five years but due to malnutrition the amount of water intake is more and then the equation is modified (Ramadan and Hilmi 2014) as

$$= \frac{\left(0.022 * 0.56\right)}{\left(0.0104 + 0.000724\right) * AMMT^{\circ}C} \tag{1.5}$$

The mean fluoride concentration in Kuwait groundwater samples is 1.98 mg/L. HQ was calculated to indicate the impact of high intake fluoride concentration for the lifetime. Using the data from Table 1.2, HQ was calculated for women and men are above 1.0 indicating the noncarcinogenic affects in future if the groundwater is utilized for drinking purpose. Also, it indirectly affects through food chain as groundwater is being used for agricultural purposes.

1.5.1 Analytical Techniques

The standard analytical methods adopted for F^- determination in groundwater are Orion fluoride ion electrode model, ion-selective electrode, ion chromatography, Spectrophotometer method by zirconyl oxychloride, SPADNS (Sodium 2-(parasulfophenylazo)-1,8-dihydroxy-3, 6-naphthalenedisulfonate), and ion-sensitive electrodes (Table 1.3) (Srinivasamoorthy et al. 2008; Singaraja et al. 2012; Addison et al. 2020; Aravinthasamy et al. 2019; Shyam and Kalwania 2012; Dehghani et al. 2019; Sreedevi et al. 2006).

1.5.2 Treatment Methods

High fluoride in groundwater is treated through several processes like adsorption, sorption, electro dialysis, reverse osmosis, electrocoagulation, coagulation, precipitation, ion exchange, and by using nano membranes. These techniques have been tested on different scales at lab and field conditions (Table 1.4).

Table 1.4 Common treatment methods for the removal of fluoride preferred in different countries.

Removal method type	Materials used	Type of water	Reference
Adsorbent Column	Charcoal Fly-ash Brick, Red soil, Serpentine	Rural water	Chidambaram et al., 2013
Adsorbents	Apophy, China clay, llitestilbite, clinoptilolite, Fuller's earth, gibbsite, goethite, halloysite, kaolinite, Magnesite, bentonite, vermiculite, zeolite(s), serpentine, alkaline soil, acidic clay, kaolinitic clay, natrolite, aiken soil, diatomaceous earth and Ando soil	Water	Bower and Hatcher,1967; Bjorvartn and Bårdsen, 1997; Singano et al 1997
Adsorbents	Low grade Assam coal	Groundwater	Borah and Dey, 2009
Adsorbent	Alumina Bauxite, Alumn impregnate Alumina, H_3PO_4 and lime Flyash Synthetic Zeolites	Drinking water	Chidambaram et al., 2013

(*Continued*)

Table 1.4 (Continued)

Removal method type	Materials used	Type of water	Reference
Adsorption	Bone Char	Groundwater	Mukarjee and Singh, 2018
Adsorption	Polymer Aluminium Chloride	Contaminated water	He et al. 2016
Nano type	Hydroxy Apatite nanowires		
Adsorption and de fluoridation	Bentonite clay using lanthanum, magnesium and manganese	Drinking water	Kamble et al., 2009
Adsorption and de fluoridation	Bentonite clay was modified chemically by $MgCl_2$	Drinking water	Thakre et al., 2010
Adsorption & Co precipitation	Acid-modified raw laterite	Groundwater	Bibi et al., 2015
Bio sorption	Chitin and Chitosan	Groundwater	Kamble et al., 2009
Bio sorption	Tamrind fruit shell carbon, Orange juice residue, Moringa activated carbon and neem leaf powder	Aqueous solution	Bharali and Bhattacharyya 2015a, b
Sorption	Acid treated moringa oleifera leaves powder	Synthetic water	Dan et al., 2018
Sorption	Aluminium fumerate metal organic frame work	Groundwater	Bibi et al., 2015
Adsorption	Magnesite–Chitosan composite	Drinking water	
Bio sorption	Tea waste loaded with Al/Fe oxides	Drinking water	Cai et al., 2015
Bio sorption	Zirconium impregnate fibrous protein	Drinking water	Cai et al., 2015
Adsorption	Lanthanum carbon	Drinking water	Dong and Wang., 2016
Adsorption	Lanthanum-loaded magnetic cationic hydrogel	Drinking water	Dong and Wang., 2016
Chemisorption	Chemically activated cotton nut shell carbon	Drinking water	Mariappan et al., 2015
Chemisorption	Industrial based adsorbent (cement)	Ground water	Bibi et al., 2015
Adsorption	Activated charcoal and activated alumina	Groundwater	Chidambaram et al., 2003; Chauhan et al., 2007
Adsorption	Brick, bentonite, fly ash, kaolinite, serpentine, red mud, waste mud, rice husk, charfines, bone char, ceramic	Groundwater	Brinda and Ellango, 2011
Adsorbent	Bauxite	Groundwater	Cherukumilli et al. (2017)

Table 1.4 (Continued)

Removal method type	Materials used	Type of water	Reference
Adsorbent	Aluminium Metal Embedded ThujaOccidentalis Leaves Carbon	Water	Vaddi et al., 2021
Adsorbtion	Hybrid process combining adsorption and Donnan dialysis	Groundwater	Garrmes et al. (2002)
Electro dialysis	—	Groundwater	Chidambaram et al., 2013
Adsorbent	Brick powder	Groundwater	Yadav et al., 2006
	Calcined MgAl-CO_3 layered double hydroxides	Drinking water	Lv et al., 2007
Reverse Osmosis	—		Dubey et al., 2018; Moussa et al., 2017; Owusu-Agyeman et al., 2019; Malakootian et al., 2011 Mukherjee and Halder, 2018; Kimambo et al., 2019
Electro Coagulation	—	Drinking water	
Coagulation	Fe and Al, $Al(OH)_3$ and $Al_2(SO_4)_3$ H_2O		
Precipitation	CaF_2		
Ion exchange	Hydroxy apatite, Al_2O_3		
Nano Membrane	—		
Nalgonda technique	—	Drinking water	

1.6 High Fluoride Water Types

Hydrogeochemical facies interpretation helps determine the flow pattern and origin of groundwater masses' chemical histories. A study on groundwater aquifer indicates that fluoride concentrations less than 0.5 mg/L are limited to Ca-Mg-Na-HCO_3-Cl facies type, and above 1.0 mg/L is Na-Mg-Ca-Cl-HCO_3 facies type which means young groundwater entered the aquifers due to precipitation and diluted the F^-.

The dominant facies of Na-Mg-Ca-Cl-HCO_3 show that Ca^{2+} and Mg^{2+} are replaced by Na^+ and Cl^- and HCO_3^- enrichment. So hard facies change from hard calcium bicarbonate to sodium bicarbonate, and chloride type favors higher mobility of F^- greater than 1.5 mg/L into groundwater systems (Srinivasamoorthy et al. 2008). It is worth noting that with Mg-Na-HCO3, greater F^- was seen in the Mg^{2+} dominated water type. Hydrolysis of Mg^{2+} minerals in charnockites such as hornblende, biotite, or weathering of hydroxyl apatite or apatite could be the source of F^- in these waters. The fluoride and sodium ions are often abundant in Na-HCO_3 type groundwater, while Ca^{2+} is removed by cation exchange and calcite precipitation (Kim and Yun 2005; Chae et al. 2007). When comparing Na-CO_3 and Ca-HCO_3 groundwater, the latter is known to contain less fluoride (Lee et al. 1997), although HCO_3-type has high fluoride in rainy seasons facilitated by weathering and dissolution (Lee et al. 1997). Because of the slow dissolution of carbonate minerals or Ca^{2+} containing plagioclase, resulted in the increase of Ca^{2+} concentration with increasing TDS (Yun et al. 1998a; Yun et al. 1998b; Kim 2001; Sujatha 2003). Low Ca^{2+} concentrations in rocks and soils, as well as high Na^+ levels in soil water, have been discovered to be crucial variables supporting fluoride presence in this environment. Paliwal et al. (1969), Rao and Bhaskaran (1964), and Thergaonkar and Nawlakhe (1971) conducted similar studies on fluoride distribution and hydrogeochemistry (Thergaonkar and Nawlakhe 1971). The distribution of the water types with high fluoride concentration (Figure 1.4) shows that they are reported in all water types. The samples of Cameroon are predominantly represented by Ca-HCO_3 type: United Kingdom and Brazil are noted in the Na-Cl-HCO_3 and Na-HCO_3-SO_4 type, respectively; Tanzania, Canada, and Argentina represent Na-HCO_3. The samples of India, Sri Lanka,

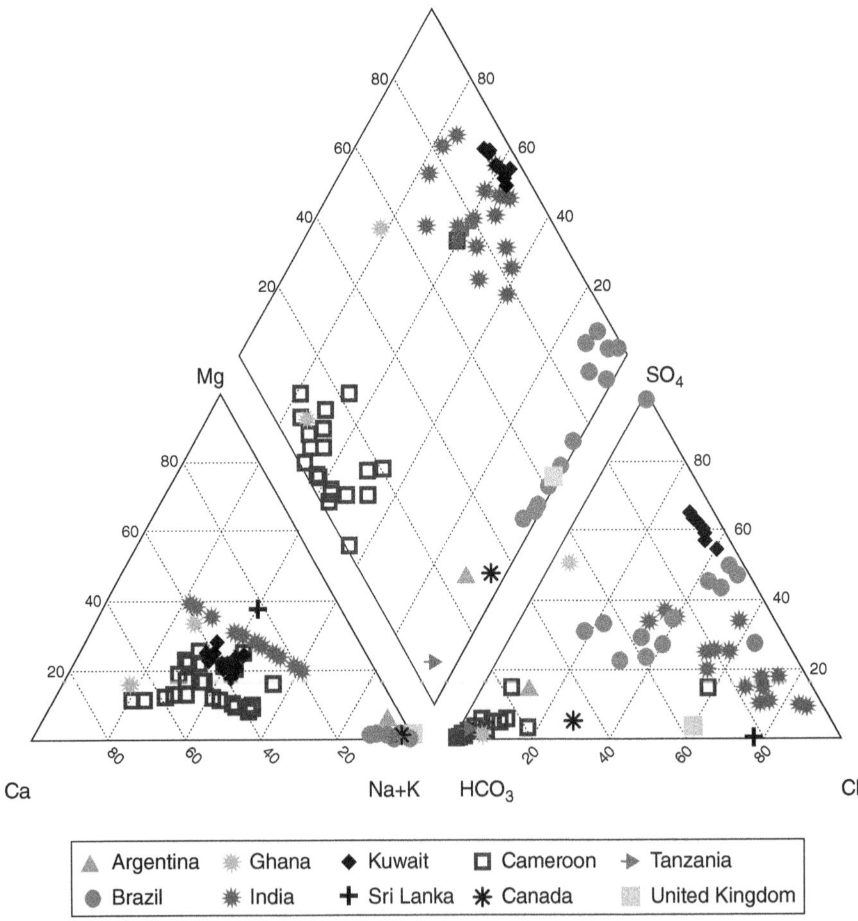

Figure 1.4 Distribution of groundwater samples in a piper diagram plot.

and Kuwait are observed to be Na-Ca-Mg-Cl, Na-Mg-Ca–Cl-HCO_3, and Na-Ca-Mg-SO_4-Cl. Ghana samples are observed in two different water types Ca-SO4-HCO_3 and Ca-Mg-Na-HCO_3. This helps to infer that though the samples are distributed throughout the plot, more samples are distributed to the Na-dominated regions in cations and Cl+SO_4-dominated regions in anions. Thus, F^- gets enriched in the recently recharged waters representing Ca-HCO_3 water types by dissolution/weathering or by the ion exchange process represented by the Na-HCO_3 water type. Nevertheless, the studies reveal that the saline/water with longer residence time in the aquifer matrix or the geothermal waters is noted to have a very high F^- concentration in water with Na-Cl or Na-Cl-SO_4 water types.

1.6.1 Relationship of pH to Fluoride

Speciation, pH, desorption and adsorption processes, precipitation, and dissolution reactions all affect F^- solubility in soil and solution (Luther et al. 1996). The solubility, plant absorption, and movement of F^- in vadose zone are all controlled by the above stated. F^- solubility is enhanced by acidic nature and low calcium carbonate concentrations, which promotes root uptake and migration to the surface and groundwater (Weinstein and Alscher-Herman 1982; Smith and Chapman 1983). F^- may be fixed as insoluble calcium fluoride (CaF_2) in soil containing considerable levels of calcium carbonate under alkaline conditions, reducing its bioavailability and mobility (Reddy and Gloss 1993). F^- is present in four primary components in soils:

1) It has been dissolved in a soil solution.
2) Sorbed to carbonates, aluminum, iron, and manganese oxides, as well as hydroxides.
3) Fluorite and fluorophlogopite solid phases.
4) Correlated to organic compounds (Selim 2003).
5) In other F^- containing rock-forming minerals.

F^- in soil water has been observed to form complexes with Al^+ and Fe^{2+}, which can ulter F^- absorption. According to certain investigations, the minerals AlF_3 and CaF_2 will precipitate at greater F^- concentrations and under very acidic conditions, whereas the mineral $Ca_5(PO_4)_3F$ will precipitate under strongly alkaline conditions. The solubility of F^- is controlled by surface adsorption to Al^+ and Fe^{2+} oxides and hydroxides in slightly acidic and neutral soils with low Ca^{2+} concentrations (Selim 2003). Considering the current dataset, the high fluoride samples show a bimodal distribution in terms of pH. Most of the samples of Cameroon, India, Kuwait, and Algeria were represented in near-neutral conditions, and those of Canada, Tanzania, Brazil, United Kingdom, South Africa, and Argentina are noted in alkaline conditions. However, Ghana samples are observed in both alkaline and acidic conditions (Figure 1.5). Few phosphatic minerals with F^- show higher dissolution in acidic conditions.

$$Ca_{10}\left(PO_4\right)_{10}F_2 + 12H^+ \rightarrow\leftarrow 10Ca^{2+} + 6H_2PO_4^- + 2F^- \qquad (1.6)$$

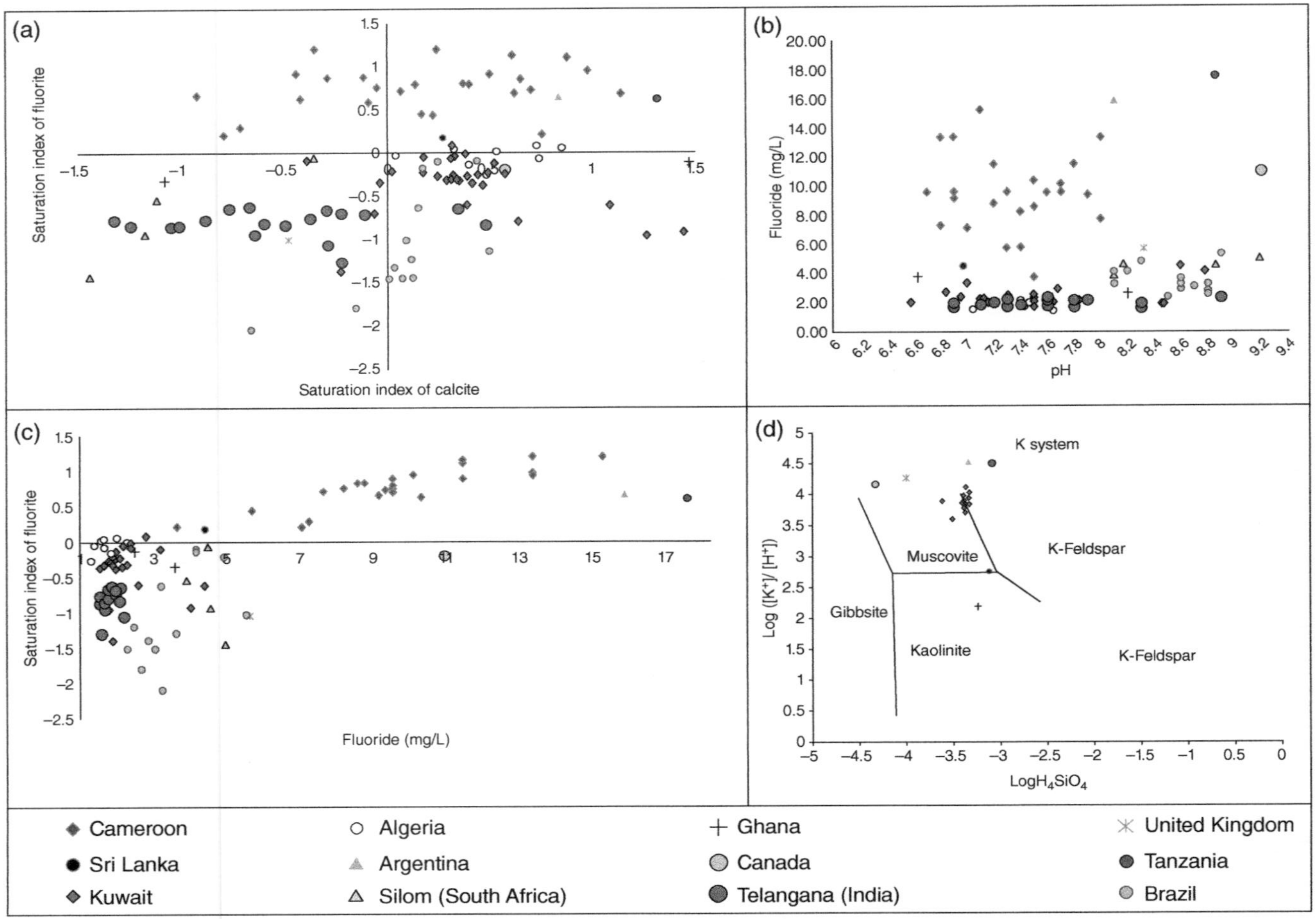

Figure 1.5 Thermodynamic plots representing the fluoride in groundwater between 12 countries. (a) The relationship between SI of Calcite and SI of fluorite (b) Relationship between pH and fluoride (mg/L). (c) The relationship between fluoride and SI of fluorite. (d) Thermodynamic plot of K-system.

$Ca(OH)^{2+}$ complexes are subjected to hydration after $(OH)^-$ – F exchange at basic to near-neutral pH. The dissolution of fluorine minerals is predominantly reported in an alkaline pH condition (Chaïrat et al. 2007). The increase in the residence period of groundwater in the aquifer matrix increases the pH and thereby the silica concentration along with the F^- from the pyriboles in the aquifer matrix (Chidambaram et al. 2013). Hence, the favorable condition for fluoride dissolution is alkaline pH (Narsimha and Sudarshan 2017), leading to higher bioavailability solubility of F^- changes with soil type and pH (Singaraja et al. 2012).

1.6.2 Thermodynamic Stability of High Fluoride Waters

Samples with higher F^- concentration were plotted in a diagram with log $[K]^+/[H]^+$ against log H_4SiO_4. The plot demonstrates that the samples fall into particular fields or along field boundaries based on their position in the plot's interpretations of the thermodynamic states of water (Figure 1.5). As the values rise from −5 to −1, the silica concentration rises to the right. Weathering causes the water to become saturated with Gibbsite at first, and then becomes stable with other fields as silica and ion concentrations rise (Chidambaram et al. 2020). The availability of plots indicates the position of high fluoride samples for the data collected. The analytical data of the countries with F^- and associated K^+, pH, and silica are considered. The samples from Argentina, Tanzania, and Kuwait are represented in the "Muscovite field," reflected a long residence time indicating higher [aK/pH] ratio and silica. Further, the samples of Ghana and Sri Lanka show similar silica values at lesser pH, indicating the dissolution in acidic to a near-neutral condition as the composition is relatively stable with "Kaolinite field." In general, samples of the "Gibbsite field" represent lesser silicate weathering. However, it could be inferred that the Canadian sample with higher pH and lesser silica could be due to the ion exchange process. There are also high fluoride samples with low and high pH; however, in general, the acidic pH group has an abnormally high concentration of F^-. Weathering to Vermiculite occurs in the vicinity during the processes of dissolution with waters having higher H_2CO_3 reacting with the minerals in the rock matrix to release F^-, H^+, HCO_3^-, and cations. Hornblende Biotite Gneiss at low pH acts on the mineral and releases H^+, F^-, cation, and HCO_3^-.

$$\begin{aligned}&(Na)_2(Ca,Mg,Fe)_3\,Al_2Si_6O_{20}(OH)_4 + H_2O + H_2CO_3 \rightarrow\\&(Ca,Mg_3Fe_3)Al_3Si_5O_{20}(OH) + F^- + Na^+ + H^+ + HCO_3^-\end{aligned} \tag{1.7}$$

The stated reaction (Eq. 1.7) is more prominent at acidic or near neutral conditions especially during the wet seasons. Conversely at alkaline environment, the increase of F^- along with more cations is facilitated by anthropogenic factors (Chidambaram et al 2017). At higher pH, hydrolysis occurs where $(OH)^-$ is released into the system with more F^- and cations.

$$\begin{aligned}&(Na)_2(Ca,Mg,Fe)_3\,Al_2Si_6O_{20}(OH)_4 + H_2O \rightarrow\\&(Mg_3Fe_3)Al_3Si_5O_{20}(OH) + OH^- + F^- + Na^+ + Ca^{2+} + Mg^{2+} + HCO_3^-\end{aligned} \tag{1.8}$$

In the above-mentioned process, hydrolysis is also a key contributor to the release of F^- in the system.

1.6.3 Saturation States

The saturation indices measure the deviation from equilibrium in terms of dissolved minerals. In terms of mineral precipitation, the saturation index (SI) of groundwater samples is stated as:

$$SI = \log \frac{IAP}{Ksp} \tag{1.9}$$

I_{AP} is known as ion activity of the solution and K_{SP} is the equilibrium constant of the reaction at temperature (T). The value of SI is zero if the water is saturated with the dissolved mineral, similarly positive and negative values of SI imply that water is super and unsaturated. A study by Srinivasamoorthy et al. (2012) regarding the influence of minerals on F^- specifies that saturation indices were oversaturated with apatite, fluoroapatite in equilibrium, and under-saturated with fluorite. During wet seasons, all F^- minerals appear under saturation, indicating the process of dilution. Therefore, apart from anthropogenic activities, dissolution and precipitation, as well as mixing, have a significant impact on the fluoride chemistry of the studied region. Fluoride minerals in the study area interacted with groundwater on a long-term basis and increased F^- concentrations, according to the association of F^- ion to the SI of different F^--bearing minerals like fluorite [CaF_2], hydroxylapatite [$Ca_5(PO_4)_3 3F$], and fluoroapatite [$Ca_5(PO_4)_3 (OH)$] (Singaraja et al. 2018). Fluoride-bearing minerals, such as hornblende, biotite, apatite, and charnockites, help to raise fluoride levels in the body (Srinivasamoorthy et al. 2008).

1.6.4 Fluoride vs SI-Fluorite

The SI of fluorite is determined by the available concentrations of both ions in water as well as other dependent elements such as pH, temperature, and the presence of other ions in the water. The calcic minerals plagioclase feldspar, calcite, dolomite, and pyriboles are the main sources of Ca^+. Essentially, F^- can come from a different or the same source. The saturation states are then governed by availability, as well as the criteria mentioned above.

$$Ca^{2+} + 2F^- \rightarrow CaF_2 \tag{1.10}$$

The increase in F^- concentration shows a relative increase of SI_f, where other stated parameters favor both elements. Such condition of super saturation is noted in samples of Cameroon. There is a linear trend of SI_f and F^- concentrations shown in samples of Cameroon. The under-saturated groups in India and Brazil samples form a cluster representation, but a linear trend is found in a sample from India. Further near-saturated clusters are also noted for both Kuwait and Algeria samples.

1.6.5 SI Fluorite and SI Calcite

To better understand F^- dissolution in groundwater, relationship between SI fluorite and SI calcite was considered. The aqueous geochemistry of enhanced F^- ion pollution in groundwater caused by increased $CaCO_3$ precipitation is controlled by solubility.

Calcite and fluorite, on the other hand, serves as a key mineral in the rock matrix influencing the elevated concentration of F^- and related hydrochemistry (Hamzaoui-Azasa et al. 2009). It also reflects that the summer has a lesser concentration of F^- and higher SI calcite, indicating the influence of F^- dissolution from the rock types. The variation of SI_f with respect to SI_c depends on the preference of Ca^{2+} ions to F^- or CO_3^{2-}. The decrease of SI_f with respect to SI_c was noted in Kuwait indicating the preferential association of Ca^{2+} with CO_3^{2-} rather than SI_f. However, on the contrary, the samples of Algeria and Brazil show an increasing trend indicating the availability of the sample of Ca^{2+} to saturate both fluorite and calcite compositions. Such dependent relationship is generally known as a common ion effect (CIE). The Ca^{2+} and carbonate are the main associates of fluoride, and interdependence could be noted (Figure 1.4). The samples of India or Cameroon do not show any significant trend or variation in fluoride saturation concerning calcite.

$$CaF_2 + H_2O + CO_2 \rightarrow CaF_2 + HCO_3 \rightarrow CaCO_3 + F^- + H_2O + CO_2 \quad (1.11)$$

Direct dissolution of CaF_2 in an alkaline environment may also increase the SI_c and decrease the SI_f. Though calcium acts as a common ion for calcite dolomite, aragonite, gypsum, anhydrite, apatite, hydroxyl apatite, and fluorite, the current review focuses only on calcite and fluorite due to the lack of complete analytical data from other countries

1.7 Statistical Techniques to Assess High Fluoride Waters

1.7.1 Correlation Analysis

Considering the collected data based on the literature, F^- shows a poor positive correlation with temperature and pH; moderate positive correlation with HCO_3^- (Table 1.5). A strong correlation was observed between F^--Na^+, F^--SiO_2, and F^--Mg^{2+}, reflecting that floriferous groundwater originated from granitic rocks (Manikandan et al. 2014). Apart from the data discussed in the current chapter, reports have also indicated that F^- shows a correlation with silica and K^+ due to release of F^- from Muscovite or with Ca^{2+} and PO_4^{3-} if derived from Calcite or Apatite. The weathering of Mica/Amphibole (i.e.) with (OH^-, F^-) group relates F^- with H_4SiO_4, HCO_3^-, and K^+ indicating the weathering of Muscovite Mica (Srinivasamoorthy et al. 2007). The reaction may be represented as

$$\text{Muscovite Mica} \rightarrow K^+ + F^- + HCO_3^- + H_4SiO_4 + \text{Kaolinite} \quad (1.12)$$

Literature studies show that fluoride is generally associated with ions Na^+, Mg^{2+}, Ca^{2+}, K^+, HCO_3^-, Cl^-, PO_4^{3-}, SiO_2 trace metals like As^{3+}, B^{3+}, and Li^+. Studies reveal a relation between pH and fluoride, implying that the high alkaline character of the water promotes fluoride leaching (Salve et al. 2008; Srinivasamoorthy et al. 2008). Hydrogeochemical interactions of F^- in groundwater with the mineral rocks exchanges Ca^{2+} by Na^+ thereby, pH increases and enhances the dissolution of F^-. F^- occurrence is also associated strongly with As^{3+}, Na^+, Cl^-, and HCO_3^-, SO_4^{2-} (Kanchan et al. 2015; Ali et al. 2016). Interaction of

Table 1.5 Statistical analysis representation (A) Correlation calculation (B) Rotated Component Matrix analysis applied to groundwater samples using four iterations converged rotation and a VARIMAX rotation with Kaiser Normalization method.

A													B			
	EC	Temp	pH	Ca^{2+}	Mg^{2+}	Na^+	K^+	Cl^-	SO_4^{2-}	HCO_3^-	F^-	NO_3^-		PCA1	PCA2	PCA3
EC	1.000	−.148	−.031	.637	**.732**	**.910**	.928	**.898**	**.797**	.411	−.013	.446	EC	**.913**	.290	−.046
Temp		1.000	−.155	−.167	−.066	−.116	−.163	−.091	−.259	.092	.324	−.119	Temp	−.086	−.177	**.846**
pH			1.000	−.379	−.360	−.151	−.107	−.139	−.162	.375	.344	.079	pH	−.277	**.648**	.004
Ca^{2+}				1.000	**.923**	**.771**	**.594**	**.760**	**.811**	−.120	−.182	.015	Ca^{2+}	**.866**	−.339	−.101
Mg^{2+}					1.000	**.820**	**.683**	**.836**	**.772**	−.039	−.156	.042	Mg^{2+}	**.914**	−.290	−.007
Na^+						1.000	**.869**	**.950**	**.788**	.274	.067	.390	Na^+	**.959**	.135	.036
K^+							1.000	**.827**	**.779**	.347	−.036	.470	K^+	**.880**	.267	−.085
Cl^-								1.000	**.705**	.193	−.074	.324	Cl^-	**.935**	.058	−.011
SO_4^{2-}									1.000	.272	−.133	.335	SO_4^{2-}	**.872**	.098	−.192
HCO_3^-										1.000	.492	**.686**	HCO_3^-	.212	**.864**	.285
F^-											1.000	−.015	F^-	−.062	**.393**	**.753**
NO_3^-												1.000	NO_3^-	.318	**.720**	−.163
													% of Variance	49.85	18.81	12.08
													Cumulative %	49.85	68.66	80.73

calcite and fluorite when they are in contact with water reflected the correlation between bicarbonate and fluoride (Kundu et al. 2001). In northern China, high Ca^{2+} and Mg^{2+}, weak alkaline conditions are associated with high F^-. Calcium bicarbonate-type facies contain low F^-, while sodium bicarbonate-type facies have higher F^- concentration in groundwater (Fantong et al. 2010). Groundwater with F^- is related to alkaline environments, enriched in Na^+ and HCO_3^- (Fantong et al. 2010). Major amounts of evaporated (O^{18} enriched) surface run-off water and rainwater infiltrate into the groundwater system, coupled with nitrate and fluoride salts from the soil, indicating that enriched $\delta^{18}O$ is directly associated with high F^- levels (Datta et al. 1999). B^+ and Li^+ have a direct relationship with F^- (Desbarats 2009; Kim et al. 2011). The presence of PO_4^{3-} also could lead to an increase in F^-. Low magnesium and calcium concentrations tend to discharge fluorine into groundwater, whereas high magnesium and calcium concentrations tend to trap fluoride (Brindha et al. 2011).

1.7.2 Principal Component Analysis

The factors that affect the fluoride concentrations are pH, EC, total hardness, total dissolved solids, ions, namely Na^+, Mg^{2+}, Cl^-, Ca^{2+}, K^+, HCO_3^-, PO_4^{3-}, and SO_4^{2-} (Yadav et al. 2020). However, few cations and anion play a dominant role and the geochemistry of high F^- groundwater, HCO_3^- is most commonly related to the F^- concentration. The F^--rich groundwater is contaminated by nitrate, indicating the influence of urban regions, apart from the usage of F^- in phosphatic fertilizers (Table 1.4). This may indicate the human impact on F^- in groundwater sources. Conversely, a connection or association of F^- with NO_3^- often suggests the influence of contaminated water (Manivannan et al. 2010). Few studies have reported that contaminated groundwater observed in urban or agricultural regions also expedites the weathering and releasing of F^- from the aquifer matrix. In such cases, the association of F^- is observed with NO_3^-, HCO_3^-, and H_4SiO_4 apart from K^+ (Manivannan et al. 2010). The F^--affected groundwater undergoes ion exchange in alkaline environments. The third factor loadings of principal component analysis indicate the relationship of temperature and fluoride concentration. Fluoride concentration in water has been related to temperatures in volcanic rocks and regional flows. High groundwater temperatures increase the fluoride in wells up to 38 °C (Gallegos et al. 2004). The results from a study show that hot springs with a maximum temperature of 60 °C reflected with fluoride 14.2mg/L (Kundu et al. 2001). Higher temperatures may dissolve the fluoride from its mineral present in the aquifer or any surface waters. Below 20 °C temperature, the fluoride concentration decreases with temperature (Shamsollahi et al. 2015). A study by Frazão et al. (2011) focused on the optimum level of fluoride concentration using mean, maximum, and daily temperatures in different regions of United States. Temperatures in different parts of the United States ranged from 10.9 to 29.6 °C, resulting in an ideal fluoride content range of 0.7–1.2 mg/L. Hence, the solubility of fluorite present in groundwater aquifer is temperature-dependent. It is also reported that temperature causes a drop in fluorite solubility, which is represented in a shift in the equilibrium constant, $K_{fluorite}$ changed from 10–10.57 at 25 °C to 10–10.8 at 10 °C (Edmunds and Smedley 2013). The factors representation is identified as from weathering silicate minerals and mobilizing F^- (H_4SiO_4, HCO_3^- and F^-) to groundwater. If Ca^{2+} and pH show negative representation, it gives the idea of ion exchange where $(OH)^-$ is exchanged for F^- ions in the mineral as they are of the same ionic radii 1.4 °A (Romo and Rustum 1957). This causes a decrease in pH. The Ca^{2+} starts precipitating with F^- released by this process.

1.8 Conclusion

The data review provides an overview of fluoride in terms of different sources, health effects, distribution in groundwater, its association with different ions, and different treatment methods adopted from various studies. The spatial orientation of fluoride concentration globally reflects the zone of high fluoride concentration in groundwater. The distribution of fluoride in groundwater from different countries is mapped with respect to lithology to reveal the significance of mineral sources and the water–rock interaction process. Fluorite-bearing minerals, igneous, metamorphic, sedimentary rock interactions, and anthropogenic activities in groundwater samples lead to fluoride toxicity. In most of the countries, groundwater is a key source of drinking water. When fluoride is used for drinking at a higher concentration greater than 1.5 mg/L, it can lead to adverse health implication. The temporal and spatial concentrations of the ion vary depending on the source, process, and nature of recharge. The common ions that influence fluoride are Na^+, K^+, Mg^{2+}, Ca^{2+}, HCO_3^-, Cl^-, PO_4^{3-}, and SO_4^{2-}. The review identified that groundwater with higher fluoride concentrations was observed in several countries that spiked to 79.20 mg/L. Excess fluoride in drinking water has serious health consequences, including dental and skeletal fluorosis. Natural and anthropogenic interactions govern groundwater containing fluoride. The hazard health risk index for fluoride was calculated for the Kuwait groundwater samples, which reflects values greater than 1.0 and preventive measures to be applied by different treatment techniques. The increased levels of fluoride in groundwater are due to geogenic reasons combined with anthropogenic inputs. The weathering of Hornblende Biotite Gneiss and the weathering of apatites in charnockite are other key geogenic factors. The pH of the groundwater plays an important influence in the dissolution and leaching of F^-. Surprisingly, the hydrogeochemical active zones match the geographical distribution of Fluoride content in the studied areas. The overview of this study provides and helps the authorities to undertake the decisions for sustainable management of high F^- groundwater.

Acknowledgments

The authors would like to extend their appreciation to Water Resource Center (WRC), Kuwait Institute for scientific research (KISR), for their help and support. The authors also like to acknowledge the OneGeology providing necessary permission to utilize their resources.

References

Abu-Zeid, K, & El-Hatow, L, 2007, Impact of Fluoride Content in Drinking Water, In: Arab Water Healthy Conference Egypt, Cairo.

Addison, M J, Rivett, M O, Phiri, P, Mleta, P, Mblame, E, Banda, M, Phiri, O, Lakudzala, W, & Kalin, R M 2020, 'Identifying groundwater fluoride source in a weathered basement aquifer in Central Malawi', *Human Health and Policy Implications Applied Sciences*, 10(14), 5006. https://doi.org/10.3390/app10145006.

Ali, S, Thakur, SK, Sarkar, A, & Shekhar, S 2016, 'Worldwide contamination of water by fluoride', *Environmental Chemistry Letters*, 14(3), 291–315. https://doi.org/10.1007/s10311-016-0563-5.

Ali, S, Fakhri, Y, Golbini, M, Thakur, SK, Alinejad, A, Parseh, I, Shekhar, S, & Bhattacharya, P. 2019, 'Concentration of fluoride in groundwater of India: A systematic review, meta-analysis and risk assessment', In: *Groundwater for Sustainable Development* (Vol. 9). Elsevier B.V. https://doi.org/10.1016/j.gsd.2019.100224.

Al-Senafy, M, Fadlelmawla, A, Al-Khalid, A, Al-Fahad, K, Bhandary, H, & Al-Haddad, A 2011, 'An application of ecological risk assessment principles to identify and prioritize groundwater pollution sources at a petrochemical factory in Kuwait', *Kuwait Journal of Science and Engineering*, 38(1A), 173–199.

Aravinthasamy, P, Karunanidhi, D, Subramani, T, Anand, B, Roy, PD, & Srinivasamoorthy K 2019, 'Fluoride contamination in groundwater of the Shanmuganadhi River Basin (south India) and its association with other chemical constituents using geographical information system and multivariate statistics', *Geochemistry*. https://doi.org/10.1016/j.chemer.2019.125555.

Apambire, WB, Boyle, DR, & Michel, FA, 1997, 'Geochemistry, genesis, and health implications of fluoriferous groundwaters in the upper regions of Ghana', *Environmental Geology*, 33(1), 13–24.

Ayoob, S, & Gupta, AK, 2006, 'Fluoride in drinking water: a review on the status and stress effects', *Critical Reviews in Environmental Science and Technology*, 36, 433–487.

Bardsen, A, Bgorraton, K, & Selving, AK 1996, 'Variability in fluoride content of subsurface water reservoirs', *Acta Odontologica Scandinavica*, 54, 343–347. https://doi.org/10.3109/00016359609003549.

Bharali, RK, & Bhattacharyya, KG 2015, 'Biosorption of fluoride on Neem (*Azadirachta indica*) leaf powder', *Journal of Environmental Chemical Engineering*, 3(2), 662–669.

Bibi, S, Kamran, MA, Sultana, J, & Farooqi, A 2017, 'Occurrence and methods to remove arsenic and fluoride contamination in water', *Environmental Chemistry Letters*, 15(1), 125–149.

Björvatn, K, & Bårdsen, A 1997, 'Use of activated clay for defluoridation of water', In: Proceedings of the First International Workshop on Fluorosis and Defluoridation of Water, 18–22 October 1995, Tanzania. *The International Society for Fluoride Research*, Auckland, 40–45.

Borah, L, & Dey, NC 2009, 'Removal of fluoride from low TDS water using low grade coal', *Indian Journal of Chemical Technology*, 16, 361–363.

Bower, CA, & Hatcher, JT 1967, 'Adsorption of fluoride by soils and minerals', *Soil Science*, 103(3), 151–154.

Bretzler, A, & Johnson, CA 2015, 'The geogenic contamination handbook: addressing arsenic and fluoride in drinking water', *Applied Geochemistry*, 63, 642–646. https://doi.org/10.1016/j.

Brindha, K, & Elango, L 2011, 'Fluoride in groundwater: Causes, implications and mitigation measures', In: Monroy, SD (Ed.), *Fluoride Properties, Applications and Environmental Management* (Vol. 1). Nova Publisher, pp. 111–136. https://www.novapublishers.com/catalog/product_info.php?products_id=15895

Brindha, K, Rajesh, R, Murugan, R, & Elango, L 2011, 'Fluoride contamination in groundwater in parts of Nalgonda district, Andhra Pradesh, India', *Environmental Monitoring and Assessment*, 172(1–4), 481–492. https://doi.org/10.1007/s10661-010-1348-0.

Brunt, R, Vasak, L, & Griffioen, J 2004, 'Fluoride in groundwater: Probability of occurrence of excessive'. IGRAC International Groundwater Resources Assessment Centre Report nr. SP 2004-2

Cai, HM, Chen, GJ, Peng, CY, Zhang, ZZ, Dong, YY, Shang, GZ, & Wan, XC 2015, 'Removal of fluoride from drinking water using tea waste loaded with Al/Fe oxides: A novel, safe and efficient biosorbent', *Applied Surface Science*, 328, 34–44.

Chae, G, Yun, S, Mayer, B, Kim, K, Kim, S, & Kwon, J 2007, 'Fluorine geochemistry in bedrock groundwater of South Korea', *Science of The Total Environment*, 385(1–3), 272–283. https://doi.org/10.1016/j.scitotenv.2007.06.038.

Chaïrat, C, Schott, J, Oelkers, EH, Lartigue, JE, & Harouiya, N 2007, 'Kinetics and mechanism of natural fluorapatite dissolution at 25 °C and pH from 3 to 12', *Geochimica et Cosmochimica Acta*, 71, 5901–5912.

Chauhan, VS, Dwivedi, PK, & Iyengar, L 2007, 'Investigations on activated alumina based domestic defluoridation units', *Journal of Hazardous Materials*, B139, 103–107.

Cherukumilli, K, Delaire, C, Amrose, S, & Gadgil, AJ 2017, 'Factors governing the performance of bauxite for fluoride remediation of groundwater', *Environmental Science & Technology*, 51(4), 2321–2328.

Chidambaram, S, Ramanathan, AL, & Vasudevan, S 2003, 'Fluoride removal studies in water using natural materials', *Water SA*, 29(3), 339–344.

Chidambaram, S, Prasanna, MV, Vasu, K 2007, 'Study on stable isotopic signatures in ground water of Gadilam river basin, Tamilnadu, India', *Indian Journal of Geochemistry*, 22(2), 209–221.

Chidambaram, S, Manikandan, S, Ramanathan, AL, Prasanna, MV, Thivya, C, Karmegam, U, & Rajkumar, K 2013, 'A study on the defluoridation in water by using natural soil', *Applied Water Science*, 3(4), 741–751.

Chidambaram, S, Thilagavathi, R, Thivya, C, Karmegam, U, Prasanna, MV, Ramanathan, AL, Tirumalesh, K, & Sasidhar, P 2017. "A study on the arsenic concentration in groundwater of a coastal aquifer in south-east India: an integrated approach," Environment, Development and Sustainability: A Multidisciplinary Approach to the Theory and Practice of Sustainable Development, Springer, vol. 19(3), pages 1015–1040, June.

Chidambaram, S, Harish, B, & Khaled, H 2020, 'CHIDAM – A software for chemical interpretation of the dissolved Ions in aqueous media', *Groundwater for Sustainable Development*, 100496. https://doi.org/10.1016/j.gsd.2020.100496.

Choi, AL, Sun, G, Zhang, Y, & Grandjean, P 2012, 'Developmental fluoride neurotoxicity: a systematic review and meta-analysis', *Environmental Health Perspectives*, 120, 1362–1368.

Dan, S, & Chattree, A 2018, 'Sorption of fluoride using chemically modified *Moringa oleifera* leaves', *Applied Water Science*, 8(2), 1–8.

Das, K, & Mondal, NK 2016, 'Dental fluorosis and urinary fluoride concentration as a reflection of fluoride exposure and its impact on IQ level and BMI of children of Laxmisagar, Simlapal Block of Bankura District, WB, India,' *Environmental Monitoring and Assessment*, 188(4), 1–14.

Datta, AS, Chakrabortty, A, De Dalal, SS, Lahiri, SC, 2014, 'Fluoride contamination of underground water in west Bengal, India. Research report', *Fluoride*, 47(3), 241–248.

Datta, PS, Tyagi, SK, Mookerjee, P, Bhattacharya, SK, Navindu, G, Bhatnagar, PD 1999, 'Groundwater NO_3 and F contamination processes in Pushkar Valley, Rajasthan as reflected from ^{18}O isotopic signature and ^{3}H recharge studies', *Environmental Monitoring and Assessment*, 56(2), 209–219. https://doi.org/10.1023/a:1005903619718.

Dehbandi, R, Moore, F, & Keshavarzi, B 2017, 'Provenance and geochemical behavior of fluorine in the soils of an endemic fluorosis belt, central Iran,' *Journal of African Earth Sciences*, 129, 56–71.

Dehghani, MH, Zarei, A, Yousefi, M, Asghari, FB, & Haghighat, GA 2019, 'Fluoride contamination in groundwater resources in the southern Iran and its related human health risks', *Desalination and Water Treatment*, 153, 95–104. doi:https://doi.org/10.5004/dwt.2019.23993;Desbarats, AJ 2009, 'On elevated fluoride and boron concentrations in groundwaters associated with the Lake Saint-Martin impact structure, Manitoba', *Applied Geochemistry* 24(5):915–927. https://doi.org/10.1016/j.apgeochem.2009.02.016.

Deshmukh, AN, Shah, KC, Sriram, A 1995, 'Coal Ash: a source of fluoride pollution, a case study of Koradi thermal power station, District Nagpur, Maharashtra', *Gondwana Geological Magzine*, 9, 21–29.

Dong, S, & Wang, Y 2016 'Characterization and adsorption properties of a lanthanum-loaded magnetic cationic hydrogel composite for fluoride removal', *Water Research*, 88, 852–860.

Dubey, S, Agarwal, M, Gupta, A 2018, 'Recent developments in defluoridation of drinking water in India', *Environmental Pollution*. Springer, pp. 345–356.

Edmunds, WM, & Smedley, PL 1996, 'Groundwater geochemistry and health: An overview', *Geological Society, London, Special Publications*, 113, 91–105. https://doi.org/10.1144/gsl.sp.1996.113.01.08.

Edmunds, WM, & Smedley, PL 2005, Fluoride in natural waters. In: Selinus, O (Ed.), *Essentials of Medical Geology*, Elsevier, Burlington, pp 301–329

Edmunds, WM, & Smedley, PL 2013, 'Fluoride in natural waters', In: Selinus, O., Alloway, B., Centeno, J.A., Finkelman, R.B., Fuge, R., Lindh, U. and Smedley, P.L. (Eds.), *Essentials of Medical Geology, Second Edition*. Springer, pp. 311–336.

Fantong, WY, Satake, H, Ayonghe, SN, Suh, EC, Adelana, SM, Fantong, EB, & Zhang, J 2010, 'Geochemical provenance and spatial distribution of fluoride in groundwater of Mayo Tsanaga River Basin, Far North Region, Cameroon: implications for incidence of fluorosis and optimal consumption dose', *Environmental Geochemistry and Health*, 32(2), 147–163.

Farooqi, A, Masuda, H, Siddiqui, R, & Naseem, M 2009, 'Sources of arsenic and fluoride in highly contaminated soils causing groundwater contamination in Punjab, Pakistan', *Archives of Environmental Contamination and Toxicology*, 56(4), 693–706. https://doi.org/10.1007/s00244-008-9239-x.

Fawell, J, Bailey, K, Chilton, J, Dahi, E, Fewtrell L, & Magara Y, 2006, Fluoride in Drinking-Water. World Health Organization, IWA Publishing, London.

Frazão, P, Peres, MA, & Cury, JA 2011, 'Drinking water quality and fluoride concentration', *Revista de Saude Publica*, 45(5) 1–10.

Gaciri SJ, & Ad Davis TC 1993, 'The occurrence and geochemistry of fluoride in some natural waters of Kenya', *Journal of Hydrology* 143, 395–412.

Galagan DJ, Vermillion JR, Nevitt GA, Stadt ZM, & Dart RE, 1957, 'Climate and fluid intake', *Public Health Report*,72(4), 4–90.

Gallegos, G, Medellin, MP, & Passo EMS 2004, 'Extraccio´n de Agua Subterra´nea a Partir de Sistemas de Flujo Regional en San Luis Potosı', *Fluoruro y Geologı´a del Subsuelo. XXXIII Congreso IAH, 7 Congreso ALHSUD. Zacatecas Me´xico*.

Ganyaglo, SY, Gibrilla, A, Teye, EM, Owusu-Ansah, EDGJ, Tettey, S, Diabene, PY, & Asimah, S 2019, 'Groundwater fluoride contamination and probabilistic health risk assessment in fluoride endemic areas of the Upper East Region, Ghana', *Chemosphere*. https://doi.org/10.1016/j.chemosphere.2019.05.276.

Garmes, H, Persin, F, Sandeaux, J, Pourcelly, G, & Mountadar, M 2002, 'Defluoridation of groundwater by a hybrid process combining adsorption and Donnan dialysis', *Desalination*, 145(1–3), 287–291.

Gizaw, B, 1996, 'The origin of high bicarbonate and fluoride concentrations in waters of the Main Ethiopian Rift Valley, East African Rift system', *Journal of African Earth Sciences*, *22*(4), 391–402.

Guo, Q, Wang, Y, Ma, T, & Ma, R 2007. 'Geochemical processes controlling the elevated fluoride concentrations in groundwaters of the Taiyuan Basin, Northern China', *Journal of Geochemical Exploration*, 93(1), 1–12. https://doi.org/10.1016/j.gexplo.2006.07.001.

Hamzaoui-Azasa, F, Bouhlila, R, & Moncef, G 2009, 'Geochemistry of fluoride and major ion in the groundwater samples of triassic aquifer (South Eastern Tunisia), through multivariate and hydrochemical', *Journal of Applied Sciences Research*, 5(11), 1941–1951.

Handa, BK 1975, 'Geochemistry and genesis of fluoride containing groundwaters in India' *Groundwater*, 13, 275–281.

He, J, Zhang, K, Wu, S, Cai, X, Chen, K, Li, Y, & Liu, J 2016, 'Performance of novel hydroxyapatite nanowires in treatment of fluoride contaminated water', *Journal of Hazardous Materials*, 303, 119–130.

Hem, JD, 1985, Study and interpretation of the chemical characteristics of natural water. USGS Water Supply Paper 2254, USGS, Reston.

Hodge, H, & Smith, FA, 1977, 'Occupational fluoride exposure', *Journal of Occupational Medicine*, *19*(1), 12–39.

Huang, D, Yang, J, Wei, X, Qin, J, Ou, S, Zhang, Z, & Zou, Y 2017, 'Probabilistic risk assessment of Chinese residents' exposure to fluoride in improved drinking water in endemic fluorosis areas', *Environmental Pollution*, 222, 118–125.

ISI 1995, Indian standard specification for drinking water, IS, 10500-1995, ISI, New Delhi.

Jagtap, S, Yenkie, MK, Labhsetwar, N, & Rayalu, S 2012, 'Fluoride in drinking water and defluoridation of water', *In Chemical Reviews*, 112(4), 2454–2466. https://doi.org/10.1021/cr2002855.

Jha, SK, Nayak, AK, & Sharma, YK, 2009, 'Fluoride occurrence and assessment of exposure dose of fluoride in shallow aquifers of Makur, Unnao district Uttar Pradesh, India', *Environmental Monitoring and Assessment*, 156(1), 561–566.

Jayawardana, DT, Pitawala, HMTGA, & Ishiga, H 2012, Geochemical assessment of soils in districts of fluoride-rich and fluoride-poor groundwater, north-central Sri Lanka', *Journal of Geochemical Exploration*, 114, 118–125.https://doi.org/10.1016/j.gexplo.2012.01.004.

Kamble, SP, Dixit, P, Rayalu, SS, & Labhsetwar, NK 2009, 'Defluoridation of drinking water using chemically modified bentonite clay', *Desalination*, 249(2), 687–693.

Kanchan, S, Kumar, V, Yadav, KK, Gupta, N, Arya, S, Sharma, S 2015, 'Effect of fly ash disposal on ground water quality near Parichha thermal power plant, Jhansi: A case study', *Current World Environment*, 10 (2), 572–580.

Kanduti, D, Sterbenk, P, & Artnik, B, 2016, 'Fluoride: a review of use and effects on health', *Mater Sociomed*, 28, 133–137. https://doi.org/10.5455/msm.2016.28.133-137.

KEPA Guidelines for Potable Water, 2017, In: Kuwaitalyawm. Issue No. 1355. https://kuwaitalyawm.media.gov.kw/. Accessed 22 Mar 2019.

Kim, JH 2001, 'Hydrogeochemical Study on Shallow Groundwater at the Coastal Area of Kimje', Seoul National University: Seoul.

Kim, K, Yun, S-T. 2005, 'Buffering of sodium concentration by cation exchange in the groundwater system of a sandy aquifer', *Geochemical Journal*, 39, 273–284.

Kim, Y, Kim, JY, Kim, K 2011, 'Geochemical characteristics of fluoride in groundwater of Gimcheon, Korea: Lithogenic and agricultural origins', *Environmental Earth Sciences* 63(5),1139–1148. https://doi.org/10.1007/s12665-010-0789-7.

Kimambo, V, Bhattacharya, P, Mtalo, F, Mtamba, J, & Ahmad, A 2019, 'Fluoride occurrence in groundwater systems at global scale and status of defluoridation – state of the art', *Groundwater for Sustainable Development*, 9, 100223. https://doi.org/10.1016/j.gsd.

Kumar, S, Singh, R, Venkatesh, AS, Udayabhanu, G, & Sahoo, PR 2019, 'Medical geological assessment of fluoride contaminated groundwater in parts of Indo-Gangetic Alluvial plains', *Scientific Reports*, 9(1), 16243. https://doi.org/10.1038/s41598-019-52812-3.

Lee JU, Chon HT, & John YW, 1997, 'Geochemical characteristics of deep granitic groundwater in Korea', *Journal of Korean Society of Groundwater Environment*, 4, 199–211 (in Korean).

Loganathan, P, Gray, CW, Hedley, MJ, & Roberts, AHC, 2006, 'Total and soluble fluorine concentrations in relation to properties of soils in New Zealand', *European Journal of Soil Science*, *57*(3), 411–421.

Luther, SM, Poulsen, L, Dudas, MJ, & Rutherford, PM, 1996, 'Fluoride sorption and mineral stability in an Alberta soil interacting with phosphogypsum leachate', *Canadian Journal of Soil Science*, *76*(1), 83–91.

Lv, L, He, J, Wei, M, Evans, DG, & Zhou, Z 2007, 'Treatment of high fluoride concentration water by MgAl-CO_3 layered double hydroxides: kinetic and equilibrium studies', *Water Research*, 41(7), 1534–1542.

Malakootian, M, Yousefi, N, & Fatehizadeh, A 2011, 'Survey efficiency of electrocoagulation on nitrate removal from aqueous solution', *International Journal of Environmental Science & Technology*,. 8, 107–114.

Manikandan, S, Chidambaram, S, & Ramanathan, AL 2014, 'A study on the high fluoride concentration in the magnesium-rich waters of hard rock aquifer in Krishnagiri district, Tamilnadu, India', *Arabian Journal of Geosciences*, 7(1), 273–285. https://doi.org/10.1007/s12517-012-0752-x.

Manivannan, R, Chidambaram, S, & Srinivasamoorthy, K 2010, 'A statistical approach to study the spatial and temporal variation of Electrical conductivity of groundwater in

Dindigul district, Tamilnadu-using GIS', In: Chidambaram, S (Ed), *Recent Trends in Water Research, Remote Sensing and General Perspectives*. IK International, New Delhi, pp. 186–196.

Mariappan, R, Vairamuthu, R, & Ganapathy, A 2015, 'Use of chemically activated cotton nut shell carbon for the removal of fluoride contaminated drinking water: Kinetics evaluation', *Chinese Journal of Chemical Engineering*, 23(4), 710–721.

McLaren, JR 1976, 'Possible effects of fluoride on the thyroid', *Fluoride* 9, 105–116.

McLaughlin, MJ, Tiller, KG, Naidu, R, & Stevens, DP, 1996, 'The behaviour and environmental impact of contaminants in fertilizers', *Soil Research*, *34*(1), 1–54.

McMahon, PB, Brown, CJ, Johnson, TD, Belitz, K, & Lindsey, BD, 2020, 'Fluoride occurrence in United States groundwater', Science of the Total Environment, 732, 139217

Messaïtfa, A 2007, 'Fluoride contents in groundwaters and the main consumed foods (dates and tea) in Southern Algeria region', *Environmental Geology*, 55(2), 377–383. https://doi.org/10.1007/s00254-007-0983-4.

Mohammadi AA, Yousefi M, Yaseri M, Jalilzadeh M, & Mahvi AH, 2017, 'Skeletal fluorosis in relation to drinking water in rural areas of West Azerbaijan, Iran', *Scientific Reports*, 7(1), 17300.

Moussa, DT, El-Naas, MH, Nasser, M, & Al-Marri, MJ 2017, 'A comprehensive review of electrocoagulation for water treatment: Potentials and challenges', *Journal of Environmental Management* 186, 24–41.

Mukherjee, S, & Halder, G 2018, 'A review on the sorptive elimination of fluoride from contaminated wastewater', *Journal of Environmental Chemical Engineering*, *6*, 1257–1270.

Narsimha, A, & Sudarshan, V 2017, 'Contamination of fluoride in groundwater and its effect on human health: A case study in hard rock aquifers of Siddipet, Telangana State, India', *Applied Water Science*, 7(5), 2501–2512.

Nordstrom, DK, & Jenne, EA 1977, 'Fluorite solubility equilibria in selected geothermal waters', *Geochimica et Cosmochimica Acta*, 41, 175–188. https://doi.org/10.1016/0016-7037(77)90224-1.

Nordstrom, DK, Ball, JW, & Donahoe, RJ, & Whittemore, D 1989, 'Groundwater chemistry and water-rock interactions at Stripa', *Geochimica et Cosmochimica Acta*, 53, 1727–1740. https://doi.org/10.1016/0016-7037(89)90294-9.

Onipe, T, Edokpayi, JN, & Odiyo, JO 2020, 'A review on the potential sources and health implications of fluoride in groundwater of Sub-Saharan Africa', *In Journal of Environmental Science and Health – Part A Toxic/Hazardous Substances and Environmental Engineering*. https://doi.org/10.1080/10934529.2020.1770516.

Onipe, T, Edokpayi, JN, & Odiyo, JO 2021, 'Geochemical characterization and assessment of fluoride sources in groundwater of Siloam area, Limpopo Province, South Africa', *Scientific Reports*, 11(1), 1–19.

Owusu-Agyeman, I, Reinwald, M, Jeihanipour, A, Schäfer, AI, 2019, 'Removal of fluoride and natural organic matter from natural tropical brackish waters by nanofiltration/ reverse osmosis with varying water chemistry', *Chemosphere*, 217, 47–58.

Ozsvath, DL, 2006, 'Fluoride concentrations in a crystalline bedrock aquifer Marathon County, Wisconsin', *Environmental Geology*, *50*(1), 132–138.

Paliwal, KV, Mehta, KK, & Gandhi RP 1969, 'Fluorine in irrigation waters of Bhilwara district of Rajasthan,' *Indian Journal of Agricultural Science*, 39, 1083–1087.

Pickering, WF 1985, 'The mobility of soluble fluoride in soils', *Environmental Pollution Series B, Chemical and Physical* 9, 281–308. https://doi.org/10.1016/0143-148x(85)90004-7.

Ramadan A, & Hilmi Y, 2014, 'The influence of climate on the determination of the upper permissible fluoride level in potable water in Sudan', *Fluoride*, 47 (2), 170–180.

Rao, NVR, & Bhaskaran CS 1964, 'Endemic fluorosis study of distribution of fluoride in water resources of Kurnool district of Andhra Pradesh', *Indian Journal of Medical Research*, 52, 180–186.

Rasool, A, Farooqi, A, Xiao, T, Ali, W, Noor, S, Abiola, O, Ali, S, & Nasim, W 2017, 'A review of global outlook on fluoride contamination in groundwater with prominence on the Pakistan current situation', *In Environmental Geochemistry and Health* (Vol. 40, Issue 4, pp. 1265–1281). Springer Netherlands. https://doi.org/10.1007/s10653-017-0054-z.

Reddy, KJ, Gloss, SP 1993, 'Geochemical speciation as related to the mobility of F, Mo and SE in soil leachates', *Applied Geochemistry*, 8, 159–163. https://doi.org/10.1016/s0883-2927(09)80029-2.

Rockett, GC, Ketzer, JMM, Ramírez, A, & Van den Broek, M 2013, 'CO_2 storage capacity of Campos Basin's oil fields, Brazil', *Energy Procedia*, *37*, 5124–5133.

Romo, LA, & Rustum, R, 1957, 'Studies of the substitution of OH– by F– in various hydroxylic minerals',. *American Mineralogist*, *42*(3–4), 165–177.

Salve, RP, Maurya, A, Kumbhare, PS, Ramteke, DS, & Wate, SR 2008, 'Assessment of groundwater quality with respect to fluoride', *Bulletin of Environmental Contamination and Toxicology*, 81(3), 289–293. https://doi.org/10.1007/s00128-008-9466-x.

Saxena, VK, & Ahmed, S 2001, 'Dissolution of fluoride in groundwater: a water–rock interaction study', *Environmental Geology*, 40, 1084–1087.

Saxena, V, & Ahmed, S 2003, 'Inferring chemical parameters for the dissolution of fluoride in groundwater', *Environmental Geology*, 43(6), 731–736.

Selim, HM 2003, 'Retention and runoff losses of atrazine and metribuzin in Soil', *Journal of Environmental Quality*, 32, 1058–1071.

Shamsollahi, HR, Zolghadr, Z, Mahvi, AH, Hosseini, SS, & Mossavi, SN 2015, 'The effect of temperature, water hardness, and exposure time on fluoride toxicity in the aquatic environment', *Fluoride*, 48(4), 338.

Shashi, A 1988, 'Biochemical effects of fluoride on thyroid gland during experimental fluorosis', *Fluoride* 21, 127–130.

Shyam, R, Kalwania, GS 2012, 'Health risk assessment of fluoride with other parameters in ground water of Sikar city (India)', *Environmental Earth Sciences*, 65(4), 1275–1282. https://doi.org/10.1007/s12665-011-1375-3.

Singano, JJ, Mashauri, DA, Dahi, E, & Mtalo, FW 1997, 'Effect of pH on defluoridation of water by magnesite', In: Proceedings of the First International Workshop on Fluorosis and Defluoridation of Water Tanzania. *The International Society for Fluoride Research, Auckland*, 30–34.

Singaraja, C, Chidambaram, S, Anandhan, P, Prasanna, MV 2013, 'A study on the status of fluoride ion in groundwater of coastal hard rock aquifers of south India', *Arabian Journal of Geosciences* 6(11), 4167–4177. https://doi.org/10.1007/s12517-012-0675-6.

Singaraja, C, Chidambaram, S, Anandhan, P, Prasann, MV, Thivya, C, & Thilagavathi, R, 2012, 'A study on the status of fluoride ion in groundwater of coastal hard rock aquifers of south India', *Arabian Journal of Geosciences*, 6, 4167–4177. doi:https://doi.org/10.1007/s12517-012-0675-6.

Singaraja, C; Chidambaram, S; Jacob, Noble, Johnson Babu, G, Selvam, S, Anandhan, P, Rajeevkumar, E, Balamurugan, K, Tamizharasan, K 2018, 'Origin of high fluoride in groundwater of the Tuticorin district, Tamil Nadu, India', *Applied Water Science*, 8(2), 54 https://doi.org/10.1007/s13201-018-0694-x.

Smith, L, & Chapman, DS 1983, 'On the thermal effects of groundwater flow: 1. Regional scale systems', *Journal of Geophysical Research*, 88, 593. https://doi.org/10.1029/jb088ib01p00593.

Sreedevi, PD, Ahmed, S, Made, B, Ledoux, E, & Gandolfi, JM 2006, 'Association of hydrogeological factors in temporal variations of fluoride concentration in a crystalline aquifer in India', *Environmental Geology*, 50(1), 1–11. https://doi.org/10.1007/s00254-005-0167-z.

Srinivasamoorthy, K, Chidambaram, S, Vasanthavigar, M, Prasanna, MV, & Peter, J 2007, 'Geochemistry of fluorides in groundwater of Salem district, Tamilnadu, India', *Indian Journal of Geochemistry*, 22(2), 237–246.

K. Srinivasamoorthy, Chidambaram, S, & Vasanthavigar, M 2008, 'Fluorides in groundwater, Tamilnadu, India', *Journal of Environmental Hydrology*, 16(25), 1–10.

Srinivasamoorthy, K, Vijayaraghavan, K, Vasanthavigar, M, Sarma, S, Chidambaram, S, Anandhan, P, Manivannan, R 2012, 'Assessment of groundwater quality with special emphasis on fluoride contamination in crystalline bed rock aquifers of Mettur region, Tamilnadu, India', *Arabian Journal of Geosciences* 5(1), 83–94. https://doi.org/10.1007/s12517-010-0162-x.

Srivastava, S., & Flora, S. J. S 2020, 'Fluoride in drinking water and skeletal fluorosis: A review of the global impact,' *Current Environmental Health Reports*, 7(2), 140–146.

Subba Rao, N, Devadas, DJ 2003, 'Fluoride incidence in groundwater in an area of peninsular India', *Environmental Geology*, 45, 243–251. https://doi.org/10.1007/s00254-003-0873-3.

Sujatha, D 2003, 'Fluoride levels in the groundwater of the south-eastern part of Ranga Reddy district, Andhra Pradesh, India', *Environmental Geology*, 44, 587–591. https://doi.org/10.1007/s00254-003-0795-0.

Sunkari, ED, & Abu, M 2019, 'Hydrochemistry with special reference to fluoride contamination in groundwater of the Bongo district, Upper East Region, Ghana', *Sustainable Water Resources Management*. https://doi.org/10.1007/s40899-019-00335-0.

Thakre, D, Rayalu, S, Kawade, R, Meshram, S, Subrt, J & Labhsetwar, N 2010, 'Magnesium incorporated bentonite clay for defluoridation of drinking water' *Journal of Hazardous Materials*, 180(1-3), 122–130.

Thergaonkar, VP & Nawlakhe, WG 1971, 'Activated magnesia for fluoride removal', *Indian Journal of Environment and Health*, 16, 241.

Thivya, C, Chidambaram, S, Thilagavathi, R, Prasanna, MV, Adithya, VS, & Nepolian, M, 2015, 'A multivariate statistical approach to identify the spatio-temporal variation of geochemical process in a hard rock aquifer', *Environmental Monitoring and Assessment*, *187*(9), 1–19.

USEPA, I 2014, '*Integrated Risk Information System of the US Environmental Protection Agency*'.

USPHS 1987, Drinking water standards. United States Pubic Health Services Publication, Washington, DC

Vaddi, DR, Mushini, VSR, & Mudumba, PSM 2021, 'Use of aluminium metal embedded Thuja occidentalis leaves carbon (AMETLC) for fluoride removal from water: equilibrium and kinetic studies', *Journal of Environmental Treatment Techniques*, 9(2), 480–490.

Weinstein, LH, Alscher-Herman, R 1982, 'Physiological responses of plants to fluorine', In: Unsworth, MH, Ormrod, DP (Eds.), *Effects of Gaseous Air Pollution in Agriculture and Horticulture*. Butterworths, London.

Wenzel, WW, & Blum, WEH 1992, 'Fluorine speciation and mobility in F-contaminated soils', *Soil Science* 153, 357–364. https://doi.org/10.1097/00010694-199205000-00003.

White, DE., Hem, JD, & Waring, GS 1963, '*Chemical composition of subsurface waters*,' (No. 440-F).

WHO, 1970, 'Fluorides and Human Health', WHO Monograph Series no. 59, 364 pp.

WHO 1994, Fluorides and oral health. Report of a WHO Expert Committee on Oral Health Status and Fluoride Use, World Health Organization (WHO Technical Report Series 846), Geneva.

WHO 2018, Drinking-water. World Health Organization fact sheets. https://www.who.int/en/news-room/fact-sheets/detail/drinking-water.

World Bank Open Data I Data, 2021. https://data.worldbank.org

World Health Organization 2006, 'Printed by TJ International (Ltd), Padstow, Cornwall, UK'.

Yadav, AK, & Khan, P 2010, 'Fluoride and flurosis status in groundwater of Todaraisingh Area of District Tonk (Rajasthan, India): a case study', *International Journal of Chemical, Environmental and Pharmaceutical Research*, 1(1), 6–11.

Yadav, AK, Kaushik, CP, Haritash, AK, Kansal, A & Rani, N 2006, 'Defluoridation of groundwater using brick powder as an adsorbent', *Journal of Hazardous Materias.*, 128, 289–293.

Yadav, K, Raphi, M, & Jagadevan, S 2020, 'Geochemical appraisal of fluoride contaminated groundwater in the vicinity of a coal mining region: spatial variability and health risk assessment', *Geochemistry*. https://doi.org/10.1016/j.chemer.2020.125684.

Yun ST, Chae GT, Koh YK, Kim SR, Choi BY, & Lee BHKSY 1998a, 'Hydrogeochemical and environmental isotope study of ground-waters in the Pungki area', *Journal of Korean Society Groundwater Environment*, 5,177–191.

Yun, ST, Koh, YK, Choi, HS, Youm, SJ, & So, CS (1998b). 'Geochemistry of geothermal waters in Korea: environmental isotope and hydrochemical characteristics II. Jungwon and Munkyeong areas'. *Economic and Environmental Geology*, 31(3), 201–213.

2

Geochemical Sources, Aqueous Geochemistry, Human Health Risk of Fluoride-Enriched Groundwater, and Its Remedial Measures

Paulami Sahu and Chitrangada Debsarma

School of Environment and Sustainable Development, Central University of Gujarat, Gandhinagar, Gujarat, India

2.1 Introduction

The surface and sub-surface or groundwater demand is snowballing around the globe by multifold with the increasing population growth. Though almost 70% of the global surface is covered by water, hardly a little portion of it is fit for drinking, irrigation, and other domestic purposes. The key sources of potable water are surface water of rivers, swamps, and lakes (0.014%) and sub-surface or groundwater (1.7%). But the natural and anthropogenic activities deteriorate the surface as well as groundwater water quality. Serious health issues are caused by the presence of manganese, iron, nitrate, arsenic, fluoride, and other heavy metals, above permissible limit. In India, arsenic and fluoride are the two chief contaminants present in sub-surface aquifers with a wide-spreading range in terms of human consumption and area of distribution (Sahu 2019; Liu et al. 2021; Onipe et al. 2021). In the present chapter, effort has been made to cover various aspects of fluoride contaminants in groundwater in a nutshell.

2.1.1 Occurrence of Fluoride

Fluoride, denoted as "F," is the lightest member of the halogen group in the periodic table. It has the maximum electromagnetic and most reactive chemical properties among all the elements in the periodic table (Gillespie et al. 1989). F has an atomic weight of 18.998, atomic number 9, oxidation state −1, and exists in both inorganic as well as organic forms. Among all the elements present in the crust of mother Earth, F places 13th most abundant element (0.06–0.09%) (Weinstein and Davison 2004; Fawell et al. 2006). The ions of F have a nearly similar radius and similar charge as the ions of hydroxide and may exchange each other in the mineral structures (Hem 1985).

Hydrogeochemistry of Aquatic Ecosystems, First Edition. Edited by Sughosh Madhav, Virendra Bahadur Singh, Manoj Kumar, and Sandeep Singh.

2.1.2 Dispersion of Fluoride in Groundwater

2.1.2.1 International Scenario

Many researchers working on groundwater around the globe have reported the occurrence of F contamination in groundwater. As per World Health Organization (WHO 2019), the permissible limit for the concentration of F in sub-surface water is 1.5 mg/L and its concentration above the permissible limit is found in numerous regions of the globe. Individuals in more than 25 countries, numbering almost 10 million round the world, are suffering from health issues related to ingestion of F-enriched water (Brindha and Elango 2011). Researchers from various nations like India, Sri Lanka, China, Japan, Pakistan, Afghanistan, Turkey, Ohio, South Korea, Kenya, South Carolina, Wisconsin, Iran, Ghana, Brazil, Norway, North Jordan, Malawi, Italy, Ethiopia, Mexico, Canada, and Southern Algeria specified the existence of F contamination in groundwater of their regions (Oruc 2008; Indermitte 2010; Alabdulaaly et al. 2013; Tahir and Rasheed 2013; Aminu and Amadi 2014; Jacks 2016; Olaka et al. 2016; Rao et al. 2020). Endemic fluorosis is reported extensively in various zones of the globe, for instance, in India (Rao et al. 2020), China (Wang et al. 2011), Africa (Firempong et al. 2013), and Mexico (Carrillo-Rivera et al. 2002). Table 2.1 shows the groundwater F concentration reported in the literature from various locations around the globe.

Table 2.1 List of fluoride concentrations in groundwater around the globe.

Continent	Country	Location	Fluoride concentration range (mg/L)	References
Asia	India	Sattenapalli (Andhra Pradesh)	1.3–12.9	Rao et al. (2020)
		Nalgonda District (Andhra Pradesh)	0.40–20.00	Jacks (2016)
		Nuzendla, Guntur District (Andhra Pradesh)	0.5–16.0	Rao and Rao (2017)
		Bihar	0.60–8.00	Jha et al. (2013)
		Birbhum, Bardaman, Bankura (West Bengal)	1.50–13.00	Mukherjee and Singh (2020)
		Cambay (North Gujarat)	0.00–10.00	Gupta et al. (2005)
		Churu (Rajasthan)	Traces-30.00	Jacks (2016)
		Jaipur (Rajasthan)	4.50–28.10	Jacks (2016)
		Nagaur (Rajasthan)	Traces-44.0	Jacks (2016)
		Delhi	0.20–32.50	Jacks (2016)
		Sonbhadra (Uttar Pradesh)	0.23–1.91	Pandey et al. (2016)
		Varanasi (Uttar Pradesh)	0.20–2.10	Raju et al. (2009)
		Gujarat	0.10–40.00	Jacks (2016)

Table 2.1 (Continued)

Continent	Country	Location	Fluoride concentration range (mg/L)	References
		Mehsana (Gujarat)	1.5–5.6	Dhiman and Keshari (2006)
		Haryana	0.17–24.70	Jha et al. (2013)
		Siwani Block (Western Haryana)	16.60–18.50	Ali et al. (2018)
		Punjab	0.44–6.00	Jha et al. (2013)
		Hyderabad (Telangana)	0.38–4.00	Sreedevi et al. (2006)
		Bagalkot (Karnataka)	0.30–6.50	Tirumalesh et al. (2007)
		Kolar District (Karnataka)	0.36–3.34	Mamatha and Rao (2010)
		Tumkur District (Karnataka)	0.78–5.35	Jha et al. (2013)
		Palghat (Kerala)	0.20–5.75	Shaji et al. (2007)
		Kerala	0.20–2.50	Jha et al. (2013)
		Alleppey (Kerala)	0.7–2.9	Raj and Shaji (2017)
		Jammu and Kashmir	0.05–4.21	Jha et al. (2013)
		Kancheepurum (Tamil Nadu)	1.00–3.24	Dar et al. (2011)
		Kanjhwala, Najafgarh (Assam)	0.40–100	Jha et al. (2013)
		Karbi Anglong (Assam)	0.40–20.60	Jacks (2016)
		Phulbani, Koraput, Dhenkanal (Odisha)	0.60–5.70	Jha et al. (2013)
		Shivpuri (Madhya Pradesh)	0.20–6.40	Ayoob and Gupta (2006)
		Maharashtra	0.11–10.20	Jha et al. (2013)
		Yavatmal (Maharashtra)	0.30–13.41	Madhnure et al. (2007)
	China	Shanxi Province	0.50-9.20	Wang et al. (2007)
		Taiyuan Basin	0.40–2.40	Li et al. (2011)
		Zhuiger Basin, Kuitun area	0.00–21.50	Wang et al. (2011)
	Indonesia	Asembagus	<0.10–3.20	Heikens et al. (2005)
		Bunyuputih	<0.10–4.20	Heikens et al. (2005)

(*Continued*)

Table 2.1 (Continued)

Continent	Country	Location	Fluoride concentration range (mg/L)	References
		Jangkar	<0.10–2.50	Heikens et al. (2005)
Asia	Iran	Ardebil	0.11–0.95	Heikens et al. (2005)
		East Azarbaijan	0.10–0.95	Mesdaghinia et al. (2010)
		Gilan	<0.02–1.56	Mesdaghinia et al. (2010)
		Hormozgan	0.17–3.15	Mesdaghinia et al. (2010)
		Maku area	0.46–5.96	Asghari Moghaddam and Fijani (2008)
		Mazandaran	<0.02–2.40	Mesdaghinia et al. (2010)
		Nationwide	<0.02–5.00	Mesdaghinia et al. (2010)
		Qazvin	0.03–2.41	Mesdaghinia et al. (2010)
		Qom	<0.02–1.51	Mesdaghinia et al. (2010)
		Sistan & Baluchestan	<0.02–1.52	Mesdaghinia et al. (2010)
		Tehran	<0.02–1.88	Mesdaghinia et al. (2010)
		West Azarbaijan	0.17–5.00	Mesdaghinia et al. (2010)
		Yazd	0.15–2.70	Mesdaghinia et al. (2010)
	Pakistan	Balochistan Provinces	0.01–24.48	Tahir and Rasheed (2013)
		Kalalanwala	2.47–21.10	Farooqi et al. (2007)
		Nagar Parkar	1.13–7.85	Naseem et al. (2010)
		North West Frontier Provinces	0.05–2.09	Tahir and Rasheed (2013)
		Punjab Provinces	0.00–19.70	Tahir and Rasheed (2013)
		Sialkot	0.41–0.99	Ullah et al. (2009)
	Saudi Arabia	Al Baha	0.31–1.70	Alabdulaaly et al. (2013)

Table 2.1 (Continued)

Continent	Country	Location	Fluoride concentration range (mg/L)	References
		Al Jouf	0.44–3.00	Alabdulaaly et al. (2013)
		Eastern province	0.50–5.00	Alabdulaaly et al. (2013)
		Hadwed Shamalyah	0.80–2.40	Alabdulaaly et al. (2013)
		Hail	0.30–4.00	Alabdulaaly et al. (2013)
		Madina	0.65–2.00	Alabdulaaly et al. (2013)
		Qassim	0.10–5.40	Alabdulaaly et al. (2013)
		Riyadh	0.12–4.90	Alabdulaaly et al. (2013)
	South Korea	Anuradhapura District	<0.02–13.70	Chandrajith et al. (2012)
		Dry Zone	0.02–5.30	Chandrajith et al. (2012)
		Hambantota District	<0.02–8.90	Chandrajith et al. (2012)
		Kurunegala District	<0.02–5.30	Chandrajith et al. (2012)
		Matale District	<0.02–9.00	Chandrajith et al. (2012)
		Monaragala District	<0.02–9.0	Chandrajith et al. (2012)
		Polonnaruwa District	<0.02–7.50	Chandrajith et al. (2012)
		Puttalam District	<0.02–13.00	Chandrajith et al. (2012)
		Ratnapura District	<0.02–7.50	Chandrajith et al. (2012)
		Vavuniya District	0.20–3.00	Chandrajith et al. (2012)
	Turkey	Isparta Province	1.50–6.00	Oruc (2008)
		Muradiye Town-Van Province	5.70–15.20	Oruc (2008)
	Yemen	Alburayhi Basin and Hidhran and Taiz City	1.08–10	Al-Amry (2009)

(*Continued*)

Table 2.1 (Continued)

Continent	Country	Location	Fluoride concentration range (mg/L)	References
Africa	Algeria	Biskra District	1.20–1.68	Messaïtfa (2008)
		El-Oued	1.18–1.32	Messaïtfa (2008)
		Ghardaia	0.38–0.40	Messaïtfa (2008)
		Onargla Districts	1.07–2.30	Messaïtfa (2008)
	Benin	Central Benin	0.54–0.72	Silliman et al. (2007)
	Cameroon	Mayo Tsanaga River Basin	0.19–15.2	Fantong et al. (2010)
	Egypt	Nile Valley, Luxor	0.11–0.45	Ahmed (2014)
		Saint Katherine Protectorate	0.79–2.00	Abdel-Azeem et al. (2009)
	Ethiopia	Central Ethiopian Rift and adjacent Highlands	0.00–75.00	Ayenew et al. (2008)
		Ethiopian Aquifers	0.00–64.00	Ayenew et al. (2008)
		Southeastern Ogaden region outside Rift Valley	<1.00–7.00	Tekle-Haimanot et al. (2006)
		Oromia	8.67–40.00	Alemu et al. (2015)
		Rift Valley	1.00–13.00	Tekle-Haimanot et al. (2006)
		Ziway-Shala Basin	1.10–68.00	Rango et al. (2012)
	Ghana	Bongo township and nearby villages	0.95–2.36	Firempong et al. (2013)
		Keta Basin	0.00–282.29	Yidana et al. (2010)
		Northern region	0.00–11.60	Salifu et al. (2012)
	Ivory Cost	Adjouffou, Port-Bouet	0.07–0.66	Osemwegie et al. (2013)
		Anoumabo, Marcory	0.01–16.78	Osemwegie et al. (2013)
	Kenya	Bondo-Rarieda	0.17–2.09	Wambu et al. (2014)
		Central Kenya Rift	0.09–74.98	Olaka et al. (2016)
		Gilgil- Elementaita	0.026–21.5	Wambu and Muthakia (2011)
		Nairobi	0.29–10.50	Coetsiers et al. (2008)
	Libya	Al Zawia and Al Zahara	10.20–11.00	Nair et al. (2006)
		Alagilat City	0.80–3.20	Elmabrok (2015)
	Malawi	Chikhwawa	0.10–4.80	Grimason et al. (2013)

Table 2.1 (Continued)

Continent	Country	Location	Fluoride concentration range (mg/L)	References
		Machinga District	0.24–7.51	Sajidu et al. (2008)
		Zomba District	0.26–6.51	Sajidu et al. (2008)
	Nigeria	Abeokuta	1.00–1.30	Gbadebo (2012)
		Bomo States	0.02–2.42	Waziri et al. (2012)
		Lagos	0.10–3.20	Gbadebo (2012)
		Zango, Katsina State	0.10–7.16	Aminu and Amadi (2014)
	Rwanda	Huye Districts	0.002–0.48	Nsengimana et al. (2012)
		Kicukiro District	0.076–0.29	Nsengimana et al. (2012)
		Musanze District	0.37–0.47	Nsengimana et al. (2012)
	South Africa	Eastern Cape, KwaZulu-Natal Limpopo, and Western Capeprovinces	0–20	Ncube and Schutte (2004)
	Sudan	Karrary Locality	0.1–0.3	Salim et al. (2014)
	Togo	Hahotee-Kpogame	0.15–1.39	Tanouayi et al. (2016)
	Tunisia	Gabes (South-eastern Tunisia)	0.38–3.78	Ketata et al. (2011)
	Uganda	Kisoro Districts	2.5	Rwenyonyi et al. (2001)
	Zimbabwe	Gokwe District	4.4-11	Mamuse and Watkins (2016)
North American	Mexico	Comarea Lagunera	<0.50–3.70	Armienta and Segovia (2008)
		Arid Zone Chihuahua	1.05–2.38	Rentería-Villalobos et al. (2017)
	USA	Texas	0.3–4.3	Segreto et al. (1984)
Latin America	Argentina	Chasicó locality	1.98–2.02	Puntoriero et al. (2014)
	Brazil	Paraiba State	0.1–2.3	Whitford et al. (1999)
Europe	Sweden	Southeast Sweden	0.30–4.20	Berger (2016)
	Estonia	Harju County	0.01–2.06	Indermitte (2010)
		Valga County	0.06–1.58	Indermitte (2010)

Source: Adapted from Kimambo et al. (2019).

2.1.2.2 National Scenario

As per Teotia and Teotia (1994), the Earth's crust contains F reserves of almost 85 million tons and out of which, 12 million tons are found in India, indicating the possibility of geogenic contamination of F in groundwater is prevalent in this country. Shortt et al. (1937) reported predominancy of endemic fluorosis in India. According to Central Groundwater Board Report (Marwaha et al. 2017), 235 districts distributed in 19 states of India are reported with F-contaminated groundwater far off permissible limit (>1.5 mg/L), causing major health issues. Locations of fluoride-contaminated groundwater in India are illustrated in Figure 2.1 and the state-wise distribution of F in groundwater is summarized in Table 2.2.

2.1.3 Fluoride and Human Health

Being an essential micronutrient for humans, F strengthens the teeth and apatite matrix of skeletal muscles. In human beings, bones and teeth contain around 95% of the total body F (Maithani et al. 1998; Liu et al. 2021). From the suitability for human consumption perspective, it is suggested by WHO (1984) that groundwater F content needs to be lesser than 1.5 mg/L. WHO re-evaluated the F guideline in 2006 and concluded that there is no requirement for revision of the value of F in the guideline. Bailey et al. (2006) suggested that the F value of 1.5mg/L in the guideline of WHO is not a "fixed" value and may vary according to the local circumstances such as diet and water consumption. The advisable groundwater F concentration as per Indian standard (IS 10500:2012) is 1.0 mg/L and in case of lack of substitute water resource, the limit is 1.5 mg/L. Ayoob and Gupta et al. (2006) reported that elevated F (>1.5 mg/L) in potable water led to skeletal and dental fluorosis, neuronal and renal disorders along with myopathy. The dose and period of exposure are the best ways to predict the effects of F. For instance, in hot and humid provinces where additional drinking water is ingested rather than the temperate climatic regions, the permissible limit is also changing accordingly (Fawell et al. 2006). Intake of water having a higher or lower concentration of F as prescribed is deleterious to human health. Hence, constant monitoring and evaluation of groundwater quality regarding F content are very essential.

2.2 Environmental Origin of F in Groundwater

During the course of magmatic crystallization, fluorine, being an incompatible lithophile element, separated as silicates and remained in the melts. Accumulation of F in uppermost continental crust where the mean abundance of it is 611 mg/kg is determined by the geochemical behavior of F (Wedepohl 1995). The Earth's crust contains about 0.06–0.09% of fluorine and because of its reactivity, F naturally exists in combinations with other minerals (WHO 2019). All the natural waters contain some concentrations of F. Surface waters like rivers and lakes normally contain F of lesser than 0.05 mg/L, while in seawater, the concentration is typically about 1 mg/L. However, in case of groundwater, F can exist in both high and low concentrations depending on the existence of F-bearing minerals and the nature of the rocks. In groundwater, F can be either anthropogenic or geogenic origin. Igneous, sedimentary, and metamorphic rocks contain fluorite and weathering of such rocks is the predominant and geogenic source of F in groundwater.

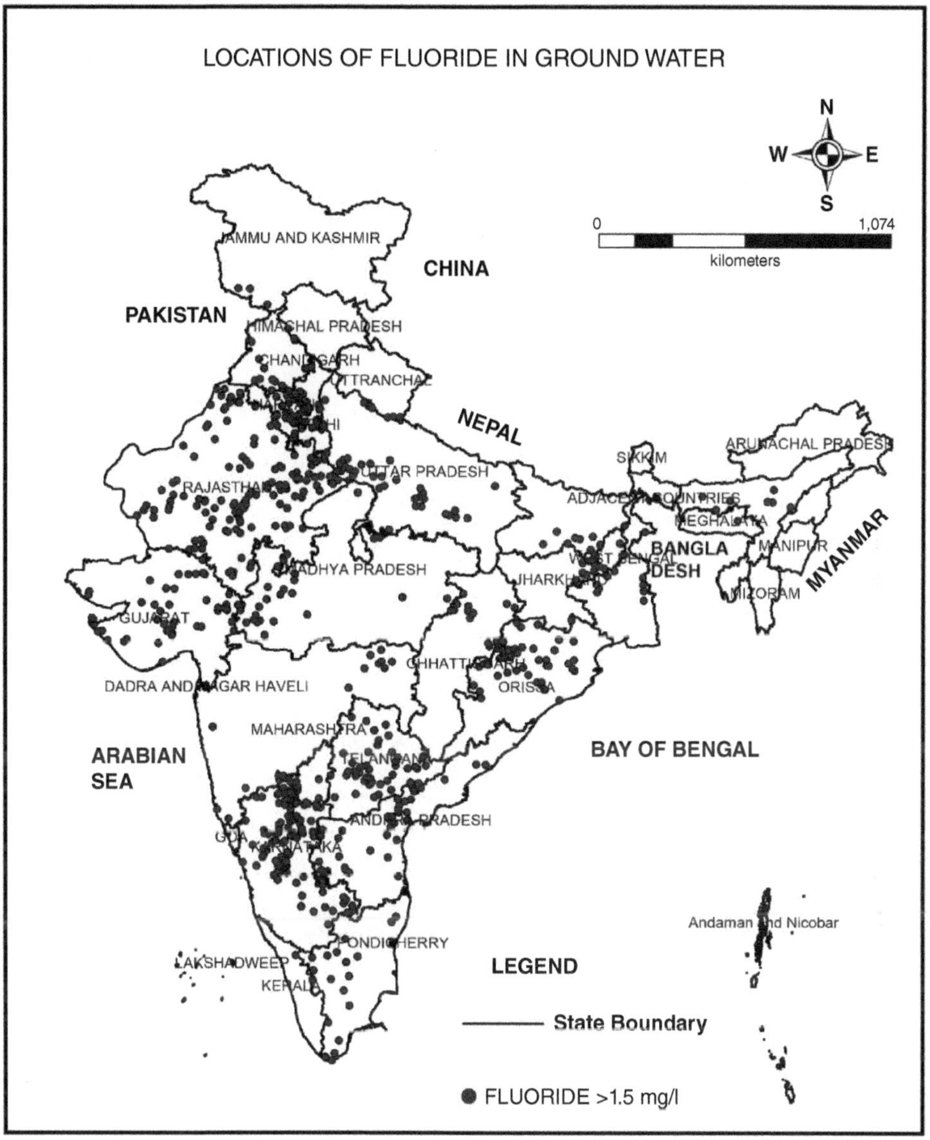

Figure 2.1 Locations of fluoride-contaminated groundwater in India. *Source:* Central Ground Water Board (CGWB). http://cgwb.gov.in/WQ/WQMAPS/FLUORIDE.jpg.

2.2.1 Geological Origins

2.2.1.1 Igneous, Sedimentary, and Metamorphic Minerals

The major geological formations are gneiss, schist, and pegmatite layer in granite rocks that contribute to F contamination in groundwater. The gneiss and schist contain fluorite, quartz, hornblende, potash-feldspar, and biotitic, whereas granite contains fluorite, feldspar, and quartz. In groundwater, the interaction of F-bearing minerals with water provides

Table 2.2 List of Indian states containing F in groundwater reported by CGWB (5 March 2017).

Sl. No.	State	No. of districts affected	Name of districts (in parts) with fluoride (>1.5 mg/L) in groundwater
1	Andhra Pradesh	19	Adilabad, Anantpur, Chitoor, Guntur, Hyderabad, Karimnagar, Khammam, Krishna, Kurnool, Mehboobnagar, Medak, Nalgonda, Nellore, Prakasham, Rangareddy, Vishakhapatnam, Viziangaram, Warangal, West Godawari
2	Assam	04	Goalpara, Kamrup, Karbi Anglong, Naugaon
3	Bihar	09	Aurangabad, Banka, Buxar, Jamui, Kalmur, Munger, Nawada, Rohtas, Supaul
4	Chhattisgarh	12	Bastar, Bilaspur, Dantewara, Janjgir-Champa, Jashpur, Kanker, Korba, Koriya, Mahasamund, Raipur, Rajnandgaon, Suguja
5	Delhi	07	Central, South, West, East, South -west, North West, North East zones
6	Gujarat	18	Ahmedabad, Amreli, Anand, Banaskantha, Bharuch, Bhavnagar, Dahod, Junagarh, Kachchh, Mahesana, Narmada, Panchmahals, Patan, Rajkot, Sabarkantha, Surat, Surrendranagar, Vadodara
7	Haryana	14	Bhiwani, Faridabad, Gurgaon, Hisar, Jhajjar, Jind, Kaithal, Kurkshetra, Mahendragarh, Panipat, Rewari, Rohtak, Sirsa, Sonepat
8	Jammu & Kashmir	02	Rajauri, Udhampur
9	Jharkhand	06	Bokaro, Giridih, Godda, Gumla, Palamau, Ranchi
10	Karnataka	20	Bagalkot, Bangalore, Bellary, Belgaum, Bidar, Bijapur, Chamarajnagar, Chikmagalur, Chitradurga, Devangere, Dharwar, Gadag, Gulbarga, Haveri, Kolar, Koppala, Mandya, Mysore, Raichur, Tumkur
11	Kerala	02	Palghat, Allepy
12	Maharashtra	08	Amrawati, Chandrapur, Dhule, Gadchiroli, Gondia, Jalna, Nagpur, Nanded,
13	Madhya Pradesh	19	Bhind, Chhatarpur, Chhindwara, Datia, Dewas, Dhar, Guna, Gwalior, Harda, Jabalpur, Jhabua, Khargone, Mandsaur, Rajgarh, Satna, Seoni, Shajapur, Sheopur, Sidhi
14	Orissa	18	Phulbani, Koraput, Dhenkenal, Angur, Boudh, Nayagarh, Puri, Balasore, Bhadrak, Bolangir, Ganjam, Jagatpur, Jajpur, Kalahandi, Koenjhar, Kurda, Mayurbhanj, Rayagada
15	Punjab	11	Amritsar, Bhatinda, Faridkot, Fatehgarh-Saheb, Firozpur, Gurdaspur, Mansa, Moga, Muktsar, Patiala, Sangrur
16	Rajasthan	32	All districts

Table 2.2 (Continued)

Sl. No.	State	No. of districts affected	Name of districts (in parts) with fluoride (>1.5 mg/L) in groundwater
17	Tamil Nadu	16	Coimbatore, Dharmapuri, Dindigul, Erode, Karur, Krishnagiri, Namakkal, Perambalur, Puddukotai, Ramnathpuram, Salem, Shivaganga, Theni, Thiruvannamalai, Vellore, Virudunagar
18	Uttar Pradesh	10	Agra, Aligarh, Etah, Firozabad, Jaunpur, Kannauj, Mahamayanagar, Mainpuri, Mathura, Maunathbhanjan
19	West Bengal	08	Bankura, Bardhhman, Birbhum, Dakhin Dinajpur, Malda, Nadia, Purulia, Uttar Dinajpur

F. F may exist in mineral being as a principal element or grime. The key sources of F include fluorine-containing minerals like apatite [$Ca_5(Cl, F, OH) (PO_4)_3$], fluorite (CaF_2), amphiboles [$A^{0-1}B_2C_5T^{8-}O_{22}(OH, F, Cl)$], micas [$AB^{2-3}(X, Si)_4O_{10}(O, F, OH)_2$], sellaite ($MgF_2$), cryolite ($Na_3AlF_6$), and topaz [$Al_2SiO_4(F, OH_2)$] (Pickering 1985; Datta et al. 1996; Ali et al. 2016). The common F bearing mineral is fluorite (CaF_2). Fluorite exists with both igneous and sedimentary rocks and has low solubility. Nearly all forms of F exist in natural aquifers in small amounts and poorly soluble conditions. The F ions may replace hydroxyl ions (OH^-) in the octahedral sheet of mineral structures as they have nearly a similar radius and same charge (Brigatti and Guggenheim 2002). The continental crust has a concentration of F of 611 mg/kg. Other categories of rock consist of F at diverse quantities, such as granites contain 810 μg/g; basalt, 360 μg/g; sandstone carries 180 μg/g; limestone, 220 μg/g; oceanic sediments bear 730 μg/g; shale, 800 μg/g; and soils, 285 μg/g (cgwb.gov.in). Pan and Fleet (2002) reported that apatite group of minerals, for instance, fluorapatite ($Ca_{10}(PO_4)6F_2$), an example of F-bearing minerals related mostly to rocks of igneous category and exists as accessory phase and the main source of F contamination.

2.2.1.2 Sea Water

The commonly found F-ions found in seawater are F^- and MgF^+, contributing around 51 and 47%, respectively, of the whole F content of ocean water. F-ions are incorporated into marine sediments when they are removed from seawater. They exist in marine sediments either by authigenic mineral precipitation, particularly carbonate fluorapatite (Van Cappellen and Berner 1988), or partially in the body parts of oceanic organisms, in fish bones or shells (Carpenter 1969).

2.2.1.3 Geothermal Fluid

This type of fluid includes water vapor, hot water comprising gases as well as dissolved solutes. Alkali-chloride solutions with pH values of nearly neutral are the main types of geothermal fluid (Edmunds and Smedley 2013). In geothermal fluid, the concentration of F is nearly linked with the F solubility equilibria and relies on the fluid temperature (Nordstrom and Jenne 1977).

2.2.1.4 Volcanic Sources

Volcanic activity along with fumarolic gases usually contains fluoride. Symonds et al. (1988) reported that active volcanoes emitted F in gaseous forms by 60,000–6 million tons globally per year. During a volcanic eruption, hydrogen fluoride gas releases as being soluble in the magmas (D'Alessandro 2006). Then the emitted gas produces very narrow salt coatings after immediate interaction with the ash particles of the plume of the volcano. Delmelle et al. (2007) reported that the produced salt coating is made up of halide salts, soluble salts along with hardly soluble compounds of F (AlF_3, CaF_2, and $Ca_5(PO_4)_3F$). Besides, it is investigated by Rosi et al. (2003) that the F-rich lavas have been released from (calc-) alkaline volcanoes, which are situated at island arcs or continental margins.

2.2.2 Anthropogenic Origins

Aluminum smelters, activities in industries for instance, ceramic firing, fertilizer factories, vehicle emissions, cement production, iron and steel industry, and glass-making industries are the anthropogenic origins of F (Haidouti 1991; WHO 2002).

2.2.2.1 Atmosphere

Although natural sources of F in atmosphere are vast like volcanic eruption, rock dust, aerosol spray from oceans, a major portion of F is anthropogenic in origin. The key origins of F in atmosphere are industrial activities such as fossil fuel burning, production of brick and ceramic, aluminum smelting, and manufacturing of iron and steel. These sources release F as various forms of particulate matter (e.g. CaF_2, NaF, and Na_2SiF_6) or gaseous state (e.g. HF, SiF_4, F_2, and H_2SiF_4) (Ozsvath 2009).

2.2.2.2 Fertilizer and Irrigation Water

Farooqi et al. (2007) reported phosphate fertilizers to contribute to groundwater F. Consecutive irrigation in agricultural lands enhances F in groundwater of that location (Young et al. 2011).

2.3 Geochemical Mechanisms for Fluoride Occurrence in Groundwater

The key origin of F in groundwater is natural or geogenic. The incidence, design of spreading, and mobilization of F in some liquid solution for instance surface, sub-surface or groundwater, completely rely on the amalgamations of various geochemical affairs carried on concurrently in natural aqueous setup. If the host rocks contain F-bearing minerals and when they weather or leach, F enrichment takes place in any solution. The geochemical mechanism or processes of F mobilization from its origin to the sub-surface water is influenced by several environmental factors or conditions for instance, water temperature, water pH, chemical constituent of water, water genesis, host-rock lithology, oxidation–reduction activities, proportion of soluble and insoluble F in host rocks, and water interaction with F-containing rocks. (Genxu and Guodong 2001). The enrichment of F in groundwater is also influenced by geological features such as dyke (Ramesam and

Rajagopalan 1985). The existence of several doleritic dykes, performing as natural barricades for flowing groundwater and making it more sluggish in cracks of the rocks for distant future, offer better ionization conveniences to F-minerals in basic dyke rocks and consequently, groundwater becomes contaminated with F.

Geochemical actions that stabilize F in the rock are adsorption and precipitation. Furthermore, the processes that mobilize the F to groundwater are dissolution and desorption.

2.3.1 Stabilization

2.3.1.1 Adsorption

Adsorption is a process in which ions (adsorbate) are transported from liquid mode to solid mode (adsorbent). This takes place by various mechanisms including surface precipitation and physical or chemical adsorption. Physio-sorption is a comprehensive adsorption affair whither outer-sphere composites are assembled in the middle of aqueous ions and surface due to electrostatic exchanges. Conversely, chemisorption outcomes in inner-sphere composites through a limited radius of bonding between the solute complex and surface oxygens. However, accurate way for F mobilization in groundwater is inexact. The general operation of the mobilization of F is the dislocation of F^- by OH when F-containing minerals interact with highly alkaline water (Khan et al. 2013). Figure 2.2a shows that in the occurrence of water, initially, the surface metal ions tend to harmonize with the molecules of water and Figure 2.2b depicts that the isolative chemisorption between the molecules of water elicits the outside of hydroxyl groups which are very sensitive to reaction. The attraction between ion and surface can exceed electrostatic repulsion if the ion and surface have similar charges. Especially, an anion may adsorb in contact with a negatively charged surface, if there is a very strong chemical contact (Stumm 1992; García and Borgnino 2015).

F adsorption onto metal hydroxides mostly rely on pH and greater adsorption takes place in robust acidic situation. The report by Farrah et al. (1987) discloses that under robust acidic situations (pH conditions 3-8) aluminum oxides disintegrate (with sufficient temperature), create F–Al composites. García and Borgnino (2015) had also reported that F adsorption onto iron hydroxides, in which situation, F is exchanged by surface hydroxyl groups. The freed F then adsorbed and created inner-sphere composites. Calcite sites have a vigorous attraction for F, consequently, F is deposited and associated with calcium carbonate or calcite in sediments. Clay minerals (for instance kaolinite and dickite), having

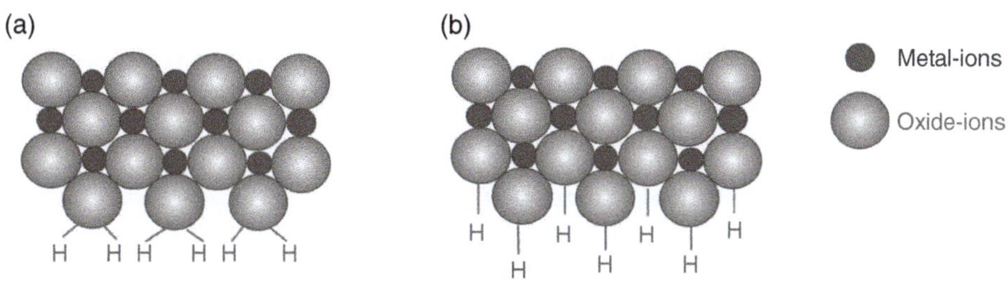

Figure 2.2 (a) In the presence of water, initially the metal ions on surface have tendency to harmonize with the molecules of water. (b) The isolative chemisorption between the molecules of water elicits the outside of hydroxyl groups. *Source:* Adapted from Sahu (2019).

phyllosilicate crystal structures, are the chief origin of F in the natural soil (Robison and Edigton 1946). In soils, clay is typically allied with iron hydroxides coverings and may adsorb F. Dissolution of F is governed by the amount of clay as it reduces hydraulic conductivity, consequently upsurges residence time of groundwater and contact time of F-containing minerals and groundwater in the aquifer (Rao et al. 2017).

2.3.1.2 Precipitation

In aqueous environments, key authigenic F-containing minerals which may precipitate from saturated solutions are fluorite, fluorapatite, and carbonate fluorapatite (García and Borgnino 2015).

2.3.2 Mobilization

2.3.2.1 Dissolution

Fluorite (fluorspar), topaz, fluorapatite, muscovite, cryolite, biotite, lepidolite, etc., are significant F-bearing minerals. Edmunds and Smedley (2013) surveyed various regions of the world that the chief cause of enrichment of elevated levels of fluorite in groundwater is dissolution of fluorite (CaF_2). Sodium bicarbonate-type lithofacies are much vulnerable to loosening F from CaF_2 minerals (Handa 1975). CaF_2 dissolution occurs as follows:

$$CaF_2 + 2NaHCO_3 \rightarrow CaCO_3 + 2Na + 2F + H_2O + CO_2$$

The vital factors that manage F dissolution in groundwater are moderate EC (electrical conductivity) and high alkalinity (Saxena and Ahmed 2001). Evaporation is the other influencing factor (Tirumalesh et al. 2007). Hem (1985) reported that in some minerals, for example, in mica and amphiboles containing F^-, F can be replaced by OH^-, if the groundwater is extremely alkaline. Because F-ions have approximately the same range of radius as OH ions and have similar charges, they could substitute for each other (Brigatti and Guggenheim 2002). For example, as per Singh and Mukherjee (2015) in muscovite, the substitution process is go-along with:

$$KAl_3Si_3O_{10}\left(OH;F\right)_2 + CO_2 + 2.5H_2O \rightarrow 1.5Al_2Si_2O_5 + K + HCO_3^- + 2F^-$$

Frengstad et al. (2001) reported the existence of calcium generates saturation ceiling conditions of F and its elimination from that system aids in the dissolution of F in groundwater. Calcium-enriched groundwater is less prone to the dissolution of F from mineral into liquid medium in comparison to sodium-enriched groundwater (Guo et al. 2007). Guo et al. (2007) showed dissolution of fluorite in groundwater is facilitated by high HCO_3 content, by the following reaction:

$$CaF_2 + 2HCO_3 \rightarrow CaCO_3 + 2F^- + H_2O + CO_2$$

2.3.2.2 Desorption

Secondary origin of F in groundwater, under alkaline conditions, is the desorption of aluminum, iron, and manganese oxyhydroxides (Borgnino et al. 2013). Under arid to semi-arid climatic conditions, high temperatures facilitate the chemical weathering of

CaF_2-bearing rocks like gneiss, granite, shale, and sandstone (Khan et al. 2013) and it frees the F which is adsorbed earlier.

2.4 Technologies for F Removal from Groundwater

The global inspection babbled about the hastily snowballing F contamination in groundwater. Globally, over 200 million individuals depend on F-contaminated drinking water beyond the permissible limit of WHO (UNESCO 2007). Nevertheless, it is recommended to bypass the high F-contaminated groundwater sources, but always it is not feasible for everyone to have. Hence, numerous scientists have established several approaches for treating groundwater having high F content and decreasing its F concentration. Water defluoridation techniques are considered as simple, efficient, and cost-effective methods to remove F from water and each technique has its specific advantages and disadvantages (Kimambo et al. 2019). These techniques include

- Precipitation/coagulation technique including Nalgonda technique, aluminum hydroxide, and calcium hydroxide.
- Electrocoagulation (EC) techniques.
- Membrane techniques comprising reverse osmosis (RO), nanofiltration (NF), electrodialysis (ED), etc.
- Adsorption techniques such as activated alumina; activated carbons; other natural, and synthetic adsorbents, etc.
- Ion-exchange techniques include strong basic anion-exchange resin with quaternary ammonium functional groups.

Figure 2.3 illustrates the list of methods or technologies for F removal from groundwater. However, the effective subtraction of F by these methods relies on the type of absorbents, their size, initial concentration of fluoride, pH of target groundwater, and the contact time.

Detail mechanisms of different defluorination techniques for groundwater are summarized in Table 2.3

Among all the techniques, adsorption is considered the most suitable method because it is eco-friendly, has low maintenance cost, and has a simple design with a variety of adsorbents and operational ease. The coagulation/precipitation, EC, ion exchange, and membrane techniques are highly effective techniques with high selectivity for F and can remove other contaminants also. But all these techniques have high installation and maintenance costs (Kimambo et al. 2019).

2.5 Recommended Remedial Measures to Reduce Human Health Problems

The ingestion of F-enriched drinking water causes carcinogenic and non carcinogenic health issues in humans such as dental and skeletal fluorosis, mottling of teeth, spinal column bending, ligament deformation, kidney problem, and neuronal disorder along with myopathy causing bilateral lameness. Hence, it is indispensable to provide safe consumable water to the

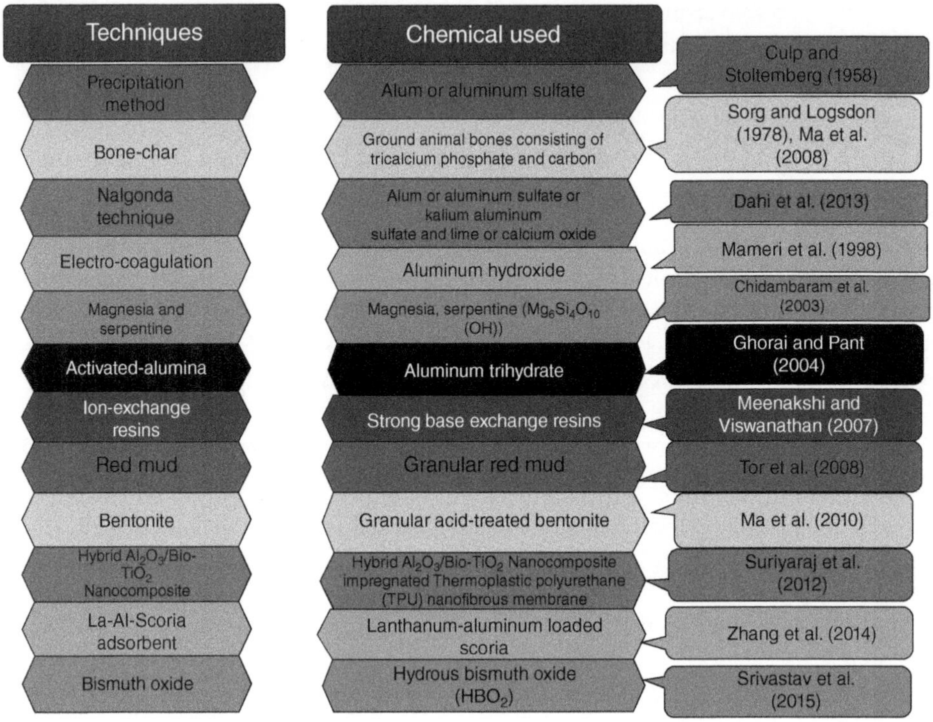

Figure 2.3 List of technologies of F removal. *Source:* Adapted from Sahu (2019).

residents. Substitution of elevated F-contaminated water source or supply of piped water from other sources is recommended at the first stage. Besides blending of high F water with low F water, use of low F water for drinking and cooking purposes and high F water for other domestic purposes is also suggested. To minimize the health vulnerability among all the inhabitants triggered by an elevated amount of F in drinking water, *de-fluoridation* filters should be provided, where no other alternate water sources are available. The residents should necessarily intake *calcium- and phosphorous-rich diet* to beat health issues caused by high F concentration in potable water, which reduces the degree of deposition of F in the human body. Additionally, *rainwater harvesting techniques* should be installed at several places which would dilute the F^- content in groundwater by elevating the local groundwater levels and acts as a quality source of groundwater recharge. Furthermore, the spread of *public awareness* on health vulnerability of elevated F content in consumable water should be highlighted over different media is the need of the moment. Otherwise, in near future, the increment frequencies of serious health issues in the local population are inevitable (Kimambo et al. 2019).

2.6 Concluding Remarks

The contamination of F in groundwater is unavoidable owing to its geogenic origin. Anthropogenic origins of F mostly from chemical fertilizers of agricultural runoff and industries are also infiltrating in groundwater which is tough to cease. It is not

Table 2.3 Defluorination techniques for groundwater.

Technologies	Description of techniques	Example	References
1) Precipitation/ coagulation technique	□ Addition of alum salts, lime, and bleaching powder to conventional water treatment. □ The foundation of the technique is adsorption of F on the flocs and subsequent elimination of F from water.	Nalgonda technique It includes the elimination of F that depends on the flocculation, sedimentation, and filtration of F with the addition of aluminum sulphate $Al_2(SO_4)_3$ or aluminum chloride and lime (CaO).	Karabulut and Atasoy (2019)
2) Electrocoagulation (EC) technique	□ A conventional EC method comprises electrode, reactor, and power supply. In this technique, direct current (DC) is generally applied to generate electric current through the electrodes. Aluminum (Al) and iron (Fe) electrode types are more preferred in EC. □ EC involves in situ generation of coagulants by EC of aluminum electrodes. Aluminum cations are generated in anode and hydrogen gas is evolved at cathode. □ During the dissolution of Al at the anode, various aqueous aluminum species are produced. The aluminum cations are transformed into polymeric species and form $Al(OH)_3$ precipitate. □ The fluoride removal by EC is carried out by means of a chemical substitution in which F^- replaced OH^- group from $Al(OH)_3$ flocs.		Karabulut and Atasoy (2019)
3) Membrane technique	□ Membrane filtration process is advanced water treatment technology that has been mainly employed in the treatment of pure and ultra-pure water.	Reverse osmosis (RO) In RO, pressure greater than the natural osmotic pressure is applied to the concentrated side of the membrane.	Karabulut and Atasoy (2019)
		Nanofiltration (NF) Defluoridation can be achieved by altering the operating variables of the NF process while concurrently keeping the required F content in the water.	Karabulut and Atasoy (2019)

(*Continued*)

Table 2.3 (Continued)

Technologies	Description of techniques	Example	References
		Electrodialysis (ED) In ED, a DC is applied as a driving force to passage ions through ion exchange membranes. ED process is used for the concentration of electrolyte solutions and for the dilution or desalination of water.	Karabulut and Atasoy (2019)
4) Adsorption technique	□ Adsorption is a surface process and the surface properties of adsorbent are modified physically or chemically to change the surface properties of adsorbents. □ Fluoride adsorption from water is controlled by pH, temperature, sorption kinetics, sorbent particle size, sorbent/sorbate concentrations, and adsorbent surface area. □ Various types of adsorbents are used for F removal, such as activated alumina, carbon, bone charcoal, activated alumina coated silica gel, calcite, activated saw dust, magnesia, serpentine, tricalcium phosphate, activated soil sorbents, carbion, defluoron, and other synthetic ion exchange resins.	Activated Alumina Activated alumina can remove anions, such as fluoride and arsenate. Activated alumina is aluminum oxide (Al_2O_3) grains that generate sorptive surface. When the water passes through a packed column of activated alumina, pollutants and other components in the water are adsorbed onto the surface of the grains.	Feenstra et al. (2007)
		Clay Clay is an earthy sedimentary material composed mostly of fine particles of hydrous aluminum silicates and other minerals and impurities. Clay is fine-textured, plastic when moist, retains hard when dried, and sinters hard when fired. Both clay powder and fired clay are capable of sorption of F as well as other pollutants from water. Mn^{2+} added Bentonite clay has potential application for F removal from groundwater by adsorption technique. Activated Red mud is also used as permeable reactive barrier (PRB) material for F removal from groundwater.	Feenstra et al. (2007)

		Bone Charcoal Bone charcoal is a blackish, porous, granular material. The major components of bone charcoal are calcium phosphate 57–80%, calcium carbonate 6–10%, and activated carbon 7–10%. In contact with water, bone charcoal is able, to some extent, to absorb a wide range of pollutants such as color, taste, and odor components. Bone charcoal has the specific ability to take up F from water.	Feenstra et al. (2007)
		Hybrid Al_2O_3/Bio-TiO_2 Nanocomposite This hybrid nanocomposite is impregnated onto electrospun thermoplastic polyurethane (TPU) nanofibers. The adhesion of the nanoparticles on the nanofibers is achieved through a silane functionalization of the materials.	Suriyaraj et al. (2015)
		La-Al-Scoria adsorbent The La^{3+}-Al^{3+}-loaded scoria (La-Al-Scoria) is used as adsorbent for F removal from groundwater. The great quantity of La-Al-O-composite-oxide present onto the surface of La-Al-Scoria is the immanent source for the outstanding adsorption capacity of F ion.	Zhang et al. (2014)
		Bismuth oxide Hydrous bismuth oxide-based adsorbents have high adsorption capacities towards nitrate, fluoride, and arsenic in water.	Ranjan and Singh (2020)
5) Ion-exchange technique	□ Ion exchange technique removes F by adsorption rather than exchanging ions. Cation exchange resins are more selective for F removal than anion exchange resins. □ Defluoridation capacity and selectivity for F are dependent on the type of resin. The loading of metal ions influences the F removal considerably, due to differences in their properties.	Most common types of ion exchange removal methods using activated alumina, clay, and bone charcoal are described earlier.	Feenstra et al. (2007)

recommended to remove F completely from drinking water as F is an essential micronutrient for humans and a tiny concentration of F is indispensable for the growth, formation, and functions of bones and teeth. Rather than total eradication, exploration of several defluorination techniques (both in situ and ex situ) is useful for the mitigation of F in groundwater. Amidst various techniques, however, RO technique has been designed as an user-friendly way and can be utilized for domestic purposes, but it also eliminates the essential ions from the human body. Several on-site techniques, for instance, rainwater harvesting methods by utilizing check dam or percolation tank, can decrease the F concentration in groundwater by diluting water within the aquifer system, are also very efficient. But all these techniques are over costly. Each of these techniques or the fusion of these needs to be designed for any targeted F-concerning location in such a manner that eventually, the technique can provide lesser F-bearing groundwater in a sustainable as well as economic means.

References

Abdel-Azeem, A. M., Abdel-Moneim, T. S., Ibrahim, M. E., Saleh, M. Y., Saleh, S. Y., & Abdel-Moneim, A. M. O. (2009). Microbiological and physicochemical analysis of groundwater and its biological effect on population in Saint Katherine protectorate, Egypt. In *Water Technology Conference* (pp. 1491–1513).

Ahmed, A. A. (2014). Fluoride in quaternary groundwater aquifer, Nile Valley, Luxor, Egypt. *Arabian Journal of Geosciences*, 7(8), 3069–3083.

Alabdulaaly, A. I., Al-Zarah, A. I., & Khan, M. A. (2013). Occurrence of fluoride in ground waters of Saudi Arabia. *Applied Water Science*, 3(3), 589–595.

Al-Amry, A. S. (2009). Hydrogeochemistry and origin of fluoride in groundwater of Hidhran & Alburayhi Basin, northwest Taiz City, Yemen. *Delta Journal of Science*, 33(1), 10–20.

Alemu, Z. A., Mengesha, S. D., Alemayehu, T. A., Serte, M. G., Kidane, A. W., & Teklu, K. T. (2015). Retrospective study of fluoride distribution in Ethiopian drinking water sources. *Asian Journal of Applied Science and Engineering*, 4(2), 127–136.

Ali, S., Thakur, S. K., Sarkar, A., & Shekhar, S. (2016). Worldwide contamination of water by fluoride. *Environmental Chemistry Letters*, 14(3), 291–315.

Ali, S., Shekhar, S., Bhattacharya, P., Verma, G., Chandrasekhar, T., & Chandrashekhar, A. K. (2018). Elevated fluoride in groundwater of Siwani Block, Western Haryana, India: a potential concern for sustainable water supplies for drinking and irrigation. *Groundwater for Sustainable Development*, 7, 410–420.

Aminu, T., & Amadi, A. N. (2014). Fluoride Contamination of Shallow Groundwater in Parts of Zango Local Government Area of Katsina State, Northwest Nigeria.

Armienta, M. A., & Segovia, N. (2008). Arsenic and fluoride in the groundwater of Mexico. *Environmental Geochemistry and Health*, 30(4), 345–353.

Asghari Moghaddam, A., & Fijani, E. (2008). Distribution of fluoride in groundwater of Maku area, northwest of Iran. *Environmental Geology*, 56(2), 281–287.

Ayenew, T., Demlie, M., & Wohnlich, S. (2008). Hydrogeological framework and occurrence of groundwater in the Ethiopian aquifers. *Journal of African Earth Sciences*, 52(3), 97–113.

Ayoob, S., & Gupta, A. K. (2006). Fluoride in drinking water: a review on the status and stress effects. *Critical Reviews in Environmental Science and Technology*, 36(6), 433–487.

Bailey, K., Chilton, J., Dahi, E., Lennon, M., Jackson, P. and Fawell. J. (Eds) (2006) Fluoride in Drinking-Water. World Health Organization, Geneva.

Berger, T. (2016). *Fluoride in surface water and groundwater in southeast Sweden: sources, controls and risk aspects* (Doctoral dissertation, Linnaeus University Press).

Borgnino, L., Garcia, M. G., Bia, G., Stupar, Y. V., Le Coustumer, P., & Depetris, P. J. (2013). Mechanisms of fluoride release in sediments of Argentina's central region. *Science of the Total Environment*, 443, 245–255.

Brigatti, M. F., & Guggenheim, S. (2002). Mica crystal chemistry and the influence of pressure, temperature, and solid solution on atomistic models. *Reviews in Mineralogy and Geochemistry*, 46(1), 1–97.

Brindha, K., & Elango, L. (2011). Fluoride in groundwater: causes, implications and mitigation measures. *Fluoride Properties, Applications and Environmental Management*, 1, 111–136.

Carpenter, R. (1969). Factors controlling the marine geochemistry of fluorine. *Geochimica et Cosmochimica Acta*, 33(10), 1153–1167.

Carrillo-Rivera, J. J., Cardona, A., & Edmunds, W. M. (2002). Use of abstraction regime and knowledge of hydrogeological conditions to control high-fluoride concentration in abstracted groundwater: San Luis Potosı basin, Mexico. *Journal of Hydrology*, 261(1–4), 24–47.

Chandrajith, R., Padmasiri, J. P., Dissanayake, C. B., & Prematilaka, K. M. (2012). Spatial distribution of fluoride in groundwater of Sri Lanka. *Journal of the National Science Foundation of Sri Lanka*, 40(4), 303–309.

Coetsiers, M., Kilonzo, F., & Walraevens, K. (2008). Hydrochemistry and source of high fluoride in groundwater of the Nairobi area, Kenya/Hydrochimie et origine des fortes concentrations en fluorure dans l'eau souterraine de la région de Nairobi, au Kenya. *Hydrological Sciences Journal*, 53(6), 1230–1240.

D'Alessandro, W. (2006). Human fluorosis related to volcanic activity: a review. *Environmental Toxicology.*, 1, 21–30.

Dar, M. A., Sankar, K., & Dar, I. A. (2011). Fluorine contamination in groundwater: a major challenge. *Environmental Monitoring and Assessment*, 173(1), 955–968.

Datta, P. S., Deb, D. L., & Tyagi, S. K. (1996). Stable isotope (^{18}O) investigations on the processes controlling fluoride contamination of groundwater. *Journal of Contaminant Hydrology*, 24(1), 85–96.

Delmelle, P., Lambert, M., Dufrêne, Y., Gerin, P., & Óskarsson, N. (2007). Gas/aerosol–ash interaction in volcanic plumes: new insights from surface analyses of fine ash particles. *Earth and Planetary Science Letters*, 259(1–2), 159–170.

Dhiman, S. D., & Keshari, A. K. (2006). Hydrogeochemical evaluation of high-fluoride groundwaters: a case study from Mehsana District, Gujarat, India. *Hydrological Sciences Journal*, 51(6), 1149–1162.

Edmunds, W. M., & Smedley, P. L. (2013). Fluoride in natural waters. In O. Selinus et al. (eds.) Essentials of Medical Geology (pp. 311–336). Springer, Dordrecht.

Elmabrok, F. M. (2015). Study of fluoride level in the ground water of Alagilat City, Libya: correlation with physicochemical parameters. *International Research Journal of Pharmacy*, 6(9), 616–622.

Fantong, W. Y., Satake, H., Ayonghe, S. N., Suh, E. C., Adelana, S., Fantong, E. B. S., & Zhang, J. (2010). Geochemical provenance and spatial distribution of fluoride in groundwater of Mayo Tsanaga River Basin, Far North Region, Cameroon: implications for incidence of fluorosis and optimal consumption dose. *Environmental Geochemistry and Health*, 32(2), 147–163.

Farooqi, A., Masuda, H., & Firdous, N. (2007). Toxic fluoride and arsenic contaminated groundwater in the Lahore and Kasur districts, Punjab, Pakistan and possible contaminant sources. *Environmental Pollution*, 145(3), 839–849.

Farrah, H., Slavek, J., & Pickering, W. F. (1987). Fluoride interactions with hydrous aluminum oxides and alumina. *Soil Research*, 25(1), 55–69.

Fawell, J., Bailey, K., Chilton, J., Dahi, E., & Magara, Y. (2006). Fluoride in Drinking-Water. IWA publishing.

Feenstra, L., Vasak, L., & Griffioen, J. (2007). Fluoride in groundwater: overview and evaluation of removal methods, International Groundwater Resources Assessment Centre. Report nr. SP. 2007-1.

Firempong, C. K., Nsiah, K., Awunyo-Vitor, D., & Dongsogo, J. (2013). Soluble fluoride levels in drinking water-a major risk factor of dental fluorosis among children in Bongo community of Ghana. *Ghana Medical Journal*, 47(1), 16–23.

Frengstad, B., Banks, D., & Siewers, U. (2001). The chemistry of Norwegian groundwaters: IV. The pH-dependence of element concentrations in crystalline bedrock groundwaters. *Science of the Total Environment*, 277(1–3), 101–117.

García, M. G., & Borgnino, L. (2015). Fluoride in the context of the environment. *Royal Society of Chemistry*, 3–21.www.rsc.org

Gbadebo, A. M. (2012). Groundwater fluoride and dental fluorosis in southwestern Nigeria. *Environmental Geochemistry and Health*, 34(5), 597–604.

Genxu, W., & Guodong, C. (2001). Fluoride distribution in water and the governing factors of environment in arid north-west China. *Journal of Arid Environments*, 49(3), 601–614.

Gillespie, R. J., Humphries, D. A., Baird, N. C., & Robinson, E. A. (1989). Chemistry. Second ed. Allyn and Bacon, Boston.

Grimason, A. M., Beattie, T. K., Masangwi, S. J., Jabu, G. C., Taulo, S. C., & Lungu, K. K. (2013). Classification and quality of groundwater supplies in the Lower Shire Valley, Malawi–Part 2: classification of borehole water supplies in Chikhwawa, Malawi. *Water SA*, 39(4), 573–582.

Guo, Q., Wang, Y., Ma, T., & Ma, R. (2007). Geochemical processes controlling the elevated fluoride concentrations in groundwaters of the Taiyuan Basin, Northern China. *Journal of Geochemical Exploration*, 93(1), 1–12.

Gupta, S. K., Deshpande, R. D., Agarwal, M., & Raval, B. R. (2005). Origin of high fluoride in groundwater in the North Gujarat-Cambay region, India. *Hydrogeology Journal*, 13(4), 596–605.

Gupta, S., Banerjee, S., Saha, R., Datta, J. K., & Mondal, N. (2006). Fluoride geochemistry of groundwater in Nalhati-1 block of the Birbhum district, West Bengal, India. *Fluoride*, 39(4), 318.

Haidouti, C. (1991). Fluoride distribution in soils in the vicinity of a point emission source in Greece. *Geoderma*, 49(1–2), 129–138.

Handa, B. K. (1975). Geochemistry and genesis of fluoride-containing ground waters in india. *Groundwater*, 13(3), 275–281.

Heikens, A., Widianarko, B., Dewi, I. C., De Boer, J. L., Seinen, W., & van Leeuwen, K. (2005). The impact of the hyperacid Ijen Crater Lake. Part I: concentrations of elements in crops and soil. *Environmental Geochemistry and Health*, 27(5), 409–418.

Hem, J. D. (1985). *Study and interpretation of the chemical characteristics of natural water* (Vol. 2254). Department of the Interior, US Geological Survey.

Indermitte, E. (2010). Exposure to fluorides in drinking water and dental fluorosis risk among the population of Estonia. Dissertationes Geographicae Universitatis Tartuensis 41.

Jacks, G. (2016). Fluoride in groundwater: mobilization, trends, and remediation. In M. Thangarajan and Vijay P. Singh Groundwater Assessment, Modeling, and Management (pp. 339–349). CRC Press.

Jha, S. K., Singh, R. K., Damodaran, T., Mishra, V. K., Sharma, D. K., & Rai, D. (2013). Fluoride in groundwater: toxicological exposure and remedies. *Journal of Toxicology and Environmental Health, Part B*, 16(1), 52–66.

Karabulut, B. Y., & Atasoy, A. D. (2019). Comparison of electrocoagulation process and other treatment technologies in fluoride removal from groundwater. *Academic Perspective Procedia*, 2(3), 1275–1282.

Ketata, M., Hamzaoui, F., Gueddari, M., Bouhlila, R., & Ribeiro, L. (2011). Hydrochemical and statistical study of groundwaters in Gabes-south deep aquifer (south-eastern Tunisia). *Physics and Chemistry of the Earth, Parts A/B/C*, 36(5–6), 187–196.

Khan, N., Hussain, S. T., Saboor, A., Jamila, N., Khan, S. N., & Kim, K. S. (2013). Chemical investigation of the drinking water Sources from Mardan, Khyber Pakhtunkhwa, Pakistan. *World Applied Sciences Journal*, 27(1), 112–122.

Kimambo, V., Bhattacharya, P., Mtalo, F., Mtamba, J., & Ahmad, A. (2019). Fluoride occurrence in groundwater systems at global scale and status of defluoridation–state of the art. *Groundwater for Sustainable Development*, 9, 100223.

Li, X., Hou, X., Zhou, Z., & Liu, L. (2011). Geochemical provenance and spatial distribution of fluoride in groundwater of Taiyuan basin, China. *Environmental Earth Sciences*, 62(8), 1635–1642.

Liu, J., Wang, X., Xu, W., Song, M., Gong, L., Zhang, M., & He, L. (2021). Hydrogeochemistry of fluorine in groundwater in humid mountainous areas: a case study at Xingguo County, Southern China. *Journal of Chemistry*, 2021, 5567353.

Madhnure, P., Sirsikar, D. Y., Tiwari, A. N., Ranjan, B., & Malpe, D. B. (2007). Occurrence of fluoride in the groundwaters of Pandharkawada area, Yavatmal district, Maharashtra, India. *Current Science*, 92, 675–679.

Maithani, P. B., Gurjar, R., Banerjee, R., Balaji, B. K., Ramachandran, S., & Singh, R. (1998). Anomalous fluoride in groundwater from western part of Sirohi district, Rajasthan and its crippling effects on human health. *Current Science*, 74, 773–777.

Mamatha, P., & Rao, S. M. (2010). Geochemistry of fluoride rich groundwater in Kolar and Tumkur Districts of Karnataka. *Environmental Earth Sciences*, 61(1), 131–142.

Mamuse, A., & Watkins, R. (2016). High fluoride drinking water in Gokwe, northwest Zimbabwe. *Journal of Water, Sanitation and Hygiene for Development*, 6(1), 55–64.

Marwaha, S., Kar, A., Sinha, S., Sekhar, S., Gupta, S.C., Gupta, S., Kaur, P. (2017). Groundwater Year Book- India 2017-2018. *Central Groundwater Board* http://cgwb.gov.in/Ground-Water/Groundwater%20Year%20Book%202017-18.pdf

Mesdaghinia, A., Vaghefi, K. A., Montazeri, A., Mohebbi, M. R., & Saeedi, R. (2010). Monitoring of fluoride in groundwater resources of Iran. *Bulletin of Environmental Contamination and Toxicology*, 84(4), 432–437.

Messaïtfa, A. (2008). Fluoride contents in groundwaters and the main consumed foods (dates and tea) in Southern Algeria region. *Environmental Geology*, 55(2), 377–383.

Mukherjee, I., & Singh, U. K. (2020). Fluoride abundance and their release mechanisms in groundwater along with associated human health risks in a geologically heterogeneous semi-arid region of east India. *Microchemical Journal*, 152, 104304.

Nair, G. A., Bohjuari, J. A., Al-Mariami, M. A., Attia, F. A., & El-Touml, F. F. (2006). Groundwater quality of north-east Libya. *Journal of Environmental Biology*, 27(4), 695–700.

Naseem, S., Rafique, T., Bashir, E., Bhanger, M. I., Laghari, A., & Usmani, T. H. (2010). Lithological influences on occurrence of high-fluoride groundwater in Nagar Parkar area, Thar Desert, Pakistan. *Chemosphere*, 78(11), 1313–1321.

Ncube, E. J., & Schutte, C. F. (2004). Relation of dental fluorosis to ground water fluoride in South Africa. In *Thailand: 4th international workshop on fluorosis prevention and defluoridation of water.*

Nordstrom, D. K., & Jenne, E. A. (1977). Fluorite solubility equilibria in selected geothermal waters. *Geochimica et Cosmochimica Acta*, 41(2), 175–188.

Nsengimana, H., Masengesho, F., & Nyirimbibi, D. K. (2012). Some physico-chemical characteristics of ground water in Rwanda. *Rwanda Journal*, 25, 101–109.

Olaka, L. A., Wilke, F. D., Olago, D. O., Odada, E. O., Mulch, A., & Musolff, A. (2016). Groundwater fluoride enrichment in an active rift setting: Central Kenya Rift case study. *Science of the Total Environment*, 545, 641–653.

Onipe, T., Edokpayi, J. N., & Odiyo, J. O. (2021). Geochemical characterization and assessment of fluoride sources in groundwater of Siloam area, Limpopo Province, South Africa. *Scientific Reports*, 11(1), 1–19.

Oruc, N. (2008). Occurrence and problems of high fluoride waters in Turkey: an overview. *Environmental Geochemistry and Health*, 30(4), 315–323.

Osemwegie, I., Oga, Y. M. S., Ahoussi, K. E., Koffi, Y. B., Kouassi, A. M., & Biémi, J. (2013). Influence of anthropogenic activities of groundwater from hand dug wells within the precarious settlements of Southern Abidjan, Côte d'Ivoire: case of the slums of Anoumabo (Marcory) and Adjouffou (Port-Bouët). *Journal of Water Resource and Protection*, 5(4), 427.

Ozsvath, D. L. (2009). Fluoride and environmental health: a review. *Reviews in Environmental Science and Bio/Technology*, 8(1), 59–79.

Pan, Y., & Fleet, M. E. (2002). Compositions of the apatite-group minerals: substitution mechanisms and controlling factors. *Reviews in Mineralogy and Geochemistry*, 48(1), 13–49.

Pandey, H. K., Duggal, S. K., & Jamatia, A. (2016). Fluoride contamination of groundwater and it's hydrogeological evolution in District Sonbhadra (UP) India. *Proceedings of the National Academy of Sciences, India Section A: Physical Sciences*, 86(1), 81–93.

Pickering, W. F. (1985). The mobility of soluble fluoride in soils. *Environmental Pollution Series B, Chemical and Physical*, 9(4), 281–308.

Puntoriero, M. L., Volpedo, A. V., & Fernández Cirelli, A. (2014). Arsenic, fluoride, and vanadium in surface water (Chasicó Lake, Argentina). *Frontiers in Environmental Science*, 2, 23.

Raj, D., & Shaji, E. (2017). Fluoride contamination in groundwater resources of Alleppey, southern India. *Geoscience Frontiers*, 8(1), 117–124.

Raju, N. J., Dey, S., & Das, K. (2009). Fluoride contamination in groundwaters of Sonbhadra district, Uttar Pradesh, India. *Current Science*, 96, 979–985.

Ramesam, V., & Rajagopalan, K. (1985). Fluoride ingestion into the natural waters of hard-rock areas, Peninsular India. *Journal of Geological Society of India (Online archive from Vol 1 to Vol 78)*, 26(2), 125–132.

Rango, T., Kravchenko, J., Atlaw, B., McCornick, P. G., Jeuland, M., Merola, B., & Vengosh, A. (2012). Groundwater quality and its health impact: an assessment of dental fluorosis in rural inhabitants of the Main Ethiopian Rift. *Environment International*, 43, 37–47.

Ranjan, M., & Singh, P. K. (2020). Concurrent removal of nitrate, fluoride and arsenic by mixed hydrous bismuth oxide from water. *Journal of Water Supply: Research and Technology-Aqua*, 69(5), 478–499.

Rao, G. B., & Rao, P. N. (2017). Geochemistry of fluoride bearing groundwater from Nuzendla Mandal, Guntur district, Andhra Pradesh, India. *International Journal of Earth Sciences and Engineering*, 10(3), 518–527.

Rao, N. S., Rao, P. S., Dinakar, A., Rao, P. N., & Marghade, D. (2017). Fluoride occurrence in the groundwater in a coastal region of Andhra Pradesh, India. *Applied Water Science*, 7(3), 1467–1478.

Rao, N., Ravindra, B., & Wu, J. (2020). Geochemical and health risk evaluation of fluoride rich groundwater in Sattenapalle Region, Guntur district, Andhra Pradesh, India. *Human and Ecological Risk Assessment: An International Journal*, 26(9), 2316–2348.

Rentería-Villalobos, M., Covarrubias-Muños, A., Pinedo-Álvarez, A., & Manjon-Collado, G. (2017). Hydrochemistry and ^{222}Rn concentrations in spring waters in the arid zone El Granero, Chihuahua, Mexico. *Geosciences*, 7(1), 12.

Robison, W. D., & Edigton, G. (1946). Fluorine in soil. *Soil Science*, 61, 341–353.

Rosi, M., Papale, P., Lupi, L., & Stoppato, M. (2003). Volcanoes: A Firefly Guide. Firefly Books Ltd. Milan, 336, 8.

Rwenyonyi, C. M., Birkeland, J. M., Haugejorden, O., & Bjorvatn, K. (2001). Dental caries among 10-to 14-year-old children in Ugandan rural areas with 0.5 and 2.5 mg fluoride per liter in drinking water. *Clinical Oral Investigations*, 5(1), 45–50.

Sahu, P. (2019). Fluoride pollution in groundwater. In P. K. Sikdar Groundwater Development and Management (pp. 329–350). Springer, Cham.

Sajidu, S. M. I., Masamba, W. R. L., Thole, B., & Mwatseteza, J. F. (2008). Groundwater fluoride levels in villages of Southern Malawi and removal studies using bauxite.

Salifu, A., Petrusevski, B., Ghebremichael, K., Buamah, R., & Amy, G. (2012). Multivariate statistical analysis for fluoride occurrence in groundwater in the Northern region of Ghana. *Journal of Contaminant Hydrology*, 140, 34–44.

Salim, L. H., Elhassan, B. M., & Elawad, M. A. (2014). Physicochemical analysis of surface and ground water in Karrary locality, Sudan. *International Journal of Engineering, Science and Technology Research*, 3, 436–438.

Saxena, V., & Ahmed, S. (2001). Dissolution of fluoride in groundwater: a water-rock interaction study. *Environmental Geology*, 40(9), 1084–1087.

Segreto, V. A., Collins, E. M., Camann, D., & Smith, C. T. (1984). A current study of mottled enamel in Texas. *Journal of the American Dental Association (1939)*, 108(1), 56–59.

Shaji, E., Viju, J., & Thambi, D. S. (2007). High fluoride in groundwater of Palghat District, Kerala. *Current Science*, 92, 240–245.

Shortt, H. E., McRobert, G. R., Barnard, T. W., & Mannadi Nayar, A. S. (1937). Endemic fluorosis in the Madras Presidency. *Indian Journal of Medical Research*, 25, 553–68.

Silliman, S. E., Boukari, M., Crane, P., Azonsi, F., & Neal, C. R. (2007). Observations on elemental concentrations of groundwater in central Benin. *Journal of Hydrology*, 335(3–4), 374–388.

Singh, C. K., & Mukherjee, S. (2015). Aqueous geochemistry of fluoride enriched groundwater in arid part of Western India. *Environmental Science and Pollution Research*, 22(4), 2668–2678.

Sreedevi, P. D., Ahmed, S., Madé, B., Ledoux, E., & Gandolfi, J. M. (2006). Association of hydrogeological factors in temporal variations of fluoride concentration in a crystalline aquifer in India. *Environmental Geology*, 50(1), 1–11.

Stumm, W. (1992). Chemistry of the Solid-Water Interface: Processes at the Mineral-Water and Particle-Water Interface in Natural Systems. John Wiley & Son Inc.

Suriyaraj, S. P., Bhattacharyya, A., & Selvakumar, R. (2015). Hybrid Al_2O_3/bio-TiO_2 nanocomposite impregnated thermoplastic polyurethane (TPU) nanofibrous membrane for fluoride removal from aqueous solutions. *RSC Advances*, 5(34), 26905–26912.

Symonds, R. B., Rose, W. I., & Reed, M. H. (1988). Contribution of C1-and F-bearing gases to the atmosphere by volcanoes. *Nature*, 334(6181), 415–418.

Tahir, M. A., & Rasheed, H. (2013). Fluoride in the drinking water of Pakistan and the possible risk of crippling fluorosis. *Drinking Water Engineering and Science*, 6(1), 17–23.

Tanouayi, G., Gnandi, K., Ouro-Sama, K., Aduayi-Akue, A. A., Ahoudi, H., Nyametso, Y., & Solitoke, H. D. (2016). Distribution of fluoride in the phosphorite mining area of Hahotoe–Kpogame (Togo). *Journal of Health and Pollution*, 6(10), 84–94.

Tekle-Haimanot, R., Melaku, Z., Kloos, H., Reimann, C., Fantaye, W., Zerihun, L., & Bjorvatn, K. (2006). The geographic distribution of fluoride in surface and groundwater in Ethiopia with an emphasis on the Rift Valley. *Science of the Total Environment*, 367(1), 182–190.

Teotia, S. P. S., & Teotia, M. (1994). Dental caries: a disorder of high fluoride and low dietary calcium interactions (30 years of personal research). *Fluoride*, 27, 59–59.

Tirumalesh, K., Shivanna, K., & Jalihal, A. A. (2007). Isotope hydrochemical approach to understand fluoride release into groundwaters of Ilkal area, Bagalkot District, Karnataka, India. *Hydrogeology Journal*, 15(3), 589–598.

Ullah, R., Malik, R. N., & Qadir, A. (2009). Assessment of groundwater contamination in an industrial city, Sialkot, Pakistan. *African Journal of Environmental Science and Technology*, 3(12) 429–446.

UNESCO (2007). (Internet source, status January 2007) Trace elements in Groundwater and public health. http://www.iah.org/briefings/Trace/trace.pdf. Accessed 9 February, 2017.

Van Cappellen, P., & Berner, R. A. (1988). A mathematical model for the early diagenesis of phosphorus and fluorine in marine sediments; apatite precipitation. *American Journal of Science*, 288(4), 289–333.

Wambu, E. W. & Muthakia, G. K. (2011). High FluorideWater in the Gilgil Area of Nakuru County, Kenya. Fluoride, 44(1), 37–41.

Wambu, E. W., Agong, S. G., Anyango, B., Akuno, W., & Akenga, T. (2014). High fluoride water in Bondo-Rarieda area of Siaya County, Kenya: a hydro-geological implication on public health in the Lake Victoria Basin. *BMC Public Health*, 14(1), 1–8.

Wang, S. X., Wang, Z. H., Cheng, X. T., Li, J., Sang, Z. P., Zhang, X. D., Wang, Z. Q. (2007). Arsenic and fluoride exposure in drinking water: children's IQ and growth in Shanyin county, Shanxi province, China. *Environmental Health Perspectives*, 115(4), 643–647.

Wang, Y., Chen, N., Wei, W., Cui, J., & Wei, Z. (2011). Enhanced adsorption of fluoride from aqueous solution onto nanosized hydroxyapatite by low-molecular-weight organic acids. *Desalination*, 276(1-3), 161–168.

Waziri, M., Musa, U., & Hati, S. S. (2012). Assessment of fluoride concentrations in surface waters and groundwater sources in Northeastern Nigeria. *Cancer*, 5, 6.

Wedepohl, K. H. (1995). The composition of the continental crust. *Geochimica et Cosmochimica Acta*, 59(7), 1217–1232.

Weinstein, L. H., & Davison, A. (2004). Fluorides in the Environment: Effects on Plants and Animals. Cabi.

Whitford, G. M., Sampaio, F. C., Arneberg, P., & Von der Fehr, F. R. (1999). Fingernail fluoride: a method for monitoring fluoride exposure. *Caries Research*, 33(6), 462–467.

WHO. (1984). International Standards for drinking water. 3rd guidelines for drinking water quality. Vol. 2: Health criteria and other supporting information. Geneva, World Health Organization.

WHO (2002). Fluorides. Environmental Health Criteria 227, 268. World Health Organization, Geneva.

World Health Organization (2019). *Preventing disease through healthy environments: inadequate or excess fluoride: a major public health concern* (No. WHO/CED/PHE/EPE/19.4. 5). World Health Organization.

Yidana, S. M., Banoeng-Yakubo, B., & Akabzaa, T. M. (2010). Analysis of groundwater quality using multivariate and spatial analyses in the Keta basin, Ghana. *Journal of African Earth Sciences*, 58(2), 220–234.

Young, S. M., Pitawala, A. H. M. T. G., & Ishiga, H. (2011). Factors controlling fluoride contents of groundwater in north-central and northwestern Sri Lanka. *Environmental Earth Sciences*, 63(6), 1333–1342.

Zhang, S., Lu, Y., Lin, X., Su, X., & Zhang, Y. (2014). Removal of fluoride from groundwater by adsorption onto La (III)-Al (III) loaded scoria adsorbent. *Applied Surface Science*, 303, 1–5.

3

Spatial Distribution of Arsenic Contamination in India: A Systematic Review

Abhra Singh[1], *Risha Singh*[2], *Binod Kumar Nath*[3], *Mohammad Hashim*[1], *Masood Ahsan Siddiqui*[1], *Hasan Raja Naqvi*[1], *and Lubna Siddiqui*[1]

[1] *Department of Geography, Faculty of Natural Sciences, Jamia Millia Islamia, New Delhi, India*
[2] *Centre for the Study of Regional Development, Jawaharlal Nehru University, New Delhi, India*
[3] *Public Works (Roads) Department Office of the Chief Engineer, PWD (Roads) Chandmari, Guwahati, Assam, India*

3.1 Introduction

Arsenic (As), a naturally occurring metalloid and ranking 14th in element abundance, exists ubiquitously in the Earth's crust in the form of sulfides associated with Copper (Cu), Nickel (Ni), Iron (Fe), Lead (Pb), Zinc (Zn), and Cobalt (Co) and is toxic (Nag et al. 1996; Patel et al. 2005; He and Charlet 2013). Arsenic compounds are of both organic and inorganic nature and occur in four (−3, 0, +3, +5) oxidation states (Patel et al. 2005). Inorganic arsenic compounds, in nature, are mainly found as As(III) and As(V) species, whereas predominant organic ones are monomethylarsonic acid (MMHA) and dimethylarsinic acid (DMAA) (Pal et al. 2002). As(III) is mostly found in anoxic conditions and As(V) in oxic conditions (Upadhyay et al. 2019). Biomethylation of inorganic compounds by bacteria and fungi produces organic arsenic compounds (Ye et al. 2012). Toxicity of these compounds (organic and inorganic) of arsenic decreases in the following order Arsine (AsH_3) > Arsenite (As(III)) > Arsenate (As(V)) > MMHA > DMAA (Nag et al. 1996).

The World Health Organization, in the year 1997, addressed water contaminated with arsenic as "major public health issue" and required attention on an "emergency basis" (WHO 1999; Pandey et al. 2006). Mexico, Argentina, Chile, Taiwan, China, Vietnam, and Bangladesh have experienced casualties due to arsenic in groundwater (Guha Mazumder et al. 1992; Chatterjee and Das 1993). Sources of contamination were either the sedimentary deposits by rivers or the volcanic sediments. Industrial arsenic contamination was reported from Australia (as lead arsenate in pesticides), Thailand, Nakajo town of Japan (as arsenic sulfide in chemicals), and Toronto city in Canada (in lead smelters). Few gold mining areas, like Obusai in Ghana, were found arsenic positive. Moreover, reports of arsenic contamination in food and beverages came into picture from England, Germany (soy sauce and milk powder), and Japan (arsenic-contaminated wine) (Mandal and Suzuki 2002).

Hydrogeochemistry of Aquatic Ecosystems, First Edition. Edited by Sughosh Madhav, Virendra Bahadur Singh, Manoj Kumar, and Sandeep Singh.

India is not an exception to these arsenic contamination episodes. The first arsenic contamination case surfaced in 1976 in Chandigarh (Datta and Kaul 1976). Then the catastrophic series of arsenicosis in humans and arsenic contamination in water, sediments, crops, and soil followed. The cases were prominent in selected pockets whereas obscure in other places. Such observations not only are intriguing but also highlight the need for proper attention.

This study aims to provide a comprehensive review of the distribution of arsenic episodes with respect to the administrative regions of India from 1988 to 2021. Further, it gives the reasons for this distribution with respect to the natural settings or the human activities in that region. In other words, sources of arsenic contamination and mobilization in water, sediments, soil, and crops at those locations have been reviewed. This study can be particularly beneficial for researchers who are further interested in finding out the potential areas for arsenic contamination.

3.2 Research Methodology

We undertook a systematic approach following the methodological guidelines of Kitchenham et al. (2009) and Dutt et al. (2017) while reviewing the literature. The methodology functional for this study is detailed below (Figure 3.1).

3.2.1 Framing the Research Questions

The questions to be addressed by the review are as follows:

i) What are the locations of arsenic contamination in India?
ii) What are the mediums (water, soil, sediments, crops, etc.) in which contamination was found at those locations?
iii) What are the sources or triggering factors responsible for arsenic mobilization at those locations?
iv) What will be the new probable locations of arsenic contamination in future?

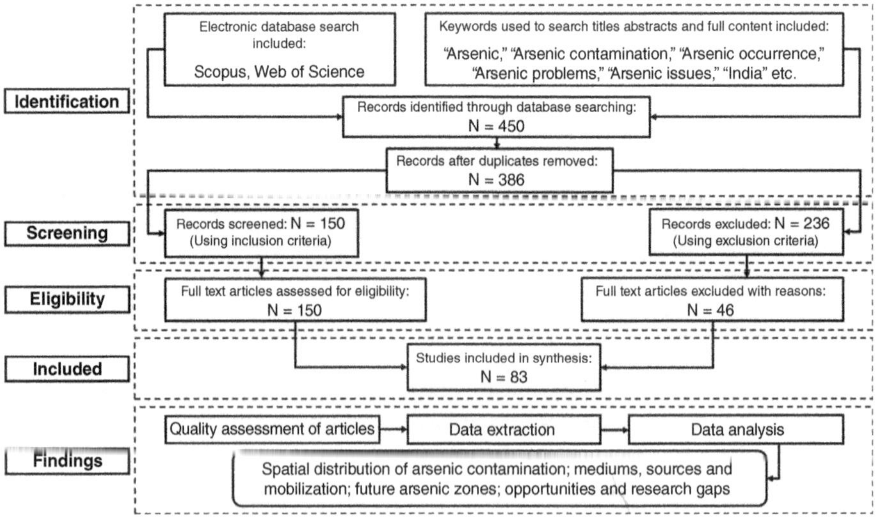

Figure 3.1 Methodology functional for the review.

3.2.2 Constructing Keywords or Search Terms for Literature Collection

The keywords or the search terms are defined and chosen in such a way that they are specific to the research questions. They are "Arsenic," "Arsenic Contamination," "Arsenic Occurrence," "Arsenic Problems." "Arsenic Issues," "India," "Trace metals," "Heavy Metals," "Arsenic Mobilization," "Groundwater Arsenic," "Arsenic Toxicity," "West Bengal," "Lower Ganga Plain," "Middle Ganga Plain," "Himalayas," "Arsenic Geochemistry." Boolean operators "AND" and "OR" are used along with the above-mentioned strings for literature search.

3.2.3 Databases and the Range of Years Searched

To obtain the relevant and peer-reviewed articles, electronic databases, namely "Scopus" and "Web of Science," were utilized. Articles published in "Nature" and "Science" were also searched. The search considered publications on arsenic contamination in India from 1992 to 2021.

3.2.4 Identification of Publications

Articles were selected keeping in vision the pre-specified research questions. So, the specific inclusion and exclusion criteria adopted for the review are listed in the following sections.

3.2.4.1 Inclusion Criteria

i) Articles that deliberated on arsenic contamination in drinking water, soil, sediments, and health problems in the wake of contamination.
ii) Articles discussing arsenic as contaminants in different regions of India only.
iii) Articles studying only arsenic as toxicant trace metals.

3.2.4.2 Exclusion Criteria

i) Review articles are not included in this review.
ii) Articles that are not specific to arsenic contamination but discussing the mitigation and removal methods of arsenic.
iii) Articles published in a language other than English.

3.3 Arsenic in Aqueous Environment

Natural events (weathering, erosion, dissolution, and geothermal) and anthropogenic activities introduce arsenic in aqueous environment (Koley 2021). It ranges from ppb to ppm in aqueous system and present in +3 and +5 oxidation states (Duker et al. 2005). The reduced state, As(III), is more toxic and mobile. Some common arsenic-bearing minerals in Earth's crust include Arsenopyrite (FeAsS), Energite (CuAsS), Orpiment (As_2S_3), Niccolite (NiAs), Realgar (As_4S_4), Cobaltite (CoAsS), and Scorodite ($FeAsO_4.2H_2O$) (Jha and Tripathi 2021). In India, Himalayan rivers carry arsenic in sediments in the adsorbed form on ferrous oxide and hydroxides (Raju 2021). The sedimentation of mid-Holocene period, rich in organic matter and found in lower stretch of river basins, is a source of arsenic in surface and groundwater (Raju 2021). Under reducing conditions, arsenic is released into the environment by reductive dissolution of ferrous oxides and hydroxides (Smedley and Kinniburgh 2002). Anthropogenic activities adding arsenic in water include excessive application of pesticides, fertilizers, mining, and coal combustion in thermal power plants (Ahamad et al. 2020; Madhav et al. 2020).

3.4 Arsenic-Contaminated States

Following are the locations identified through the selected papers (Figure 3.2):

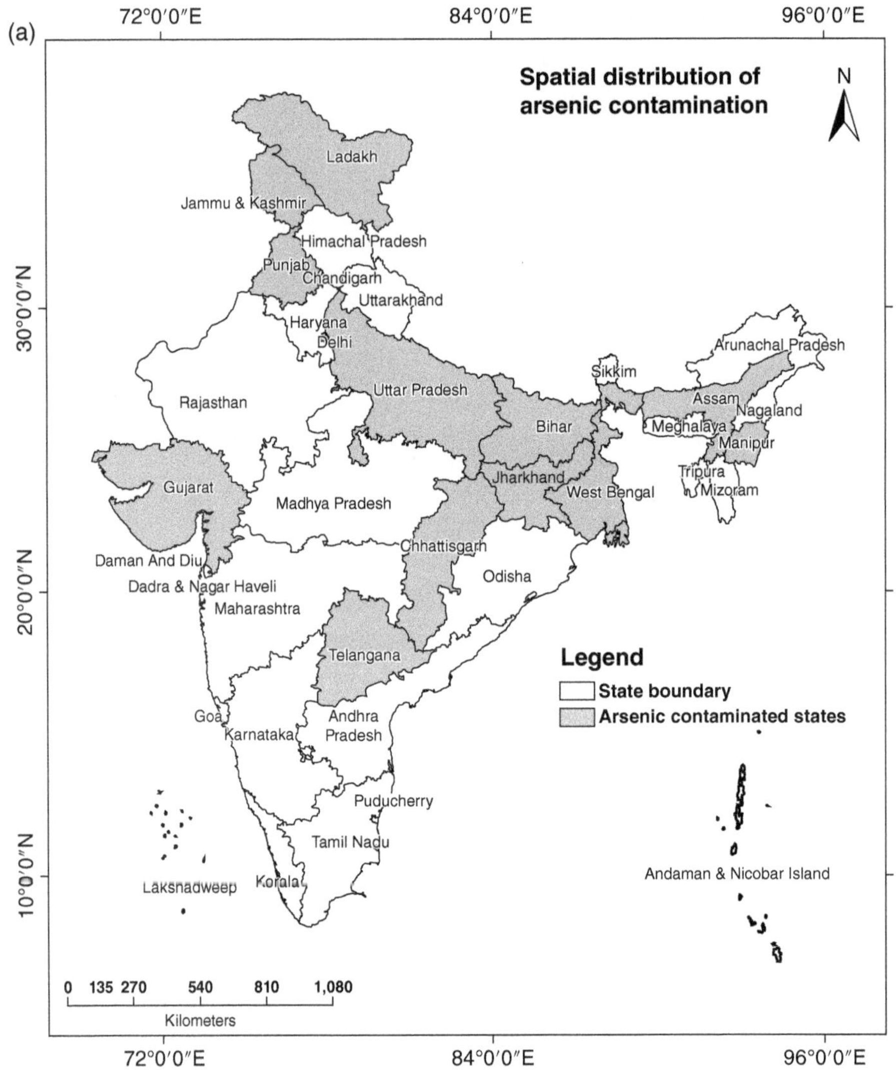

Figure 3.2 Arsenic-contaminated locations of India as mentioned in the reviewed articles. (a) Map exhibiting the arsenic-contaminated states in India as mentioned in the reviewed articles. (b) Bar diagram for the selected states showing their frequency of being chosen for study in the reviewed articles. (c) Table showing the studied districts within the chosen states in the reviewed articles.

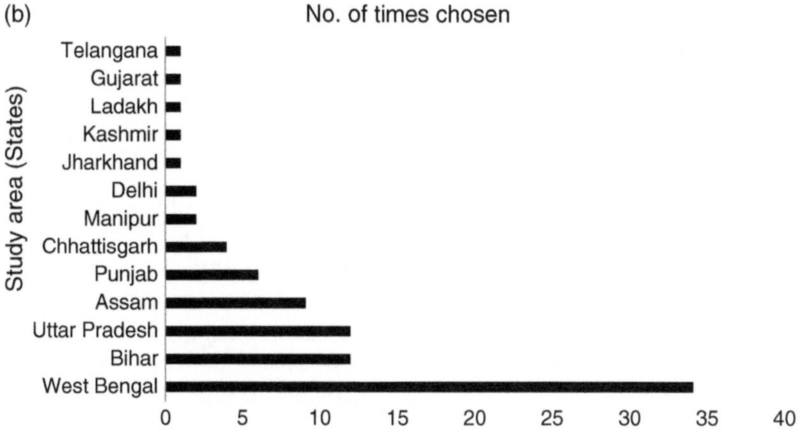

(c)

States	Chosen districts
West Bengal	Kolkata, 24 Parganas (N&S), Nadia, Barddhaman, Murshidabad, Malda, Nadia, Purbasthali, Bankura, Barasat, Murshidabad, Hoogli, Sunderbans, Darjeeling, Barasat
Bihar	Bhojpur, Bhagalpur, Vaishali, Buxar, Ara, Samastipur, Begusarai, Lakhisarai, Munger, Khagaria,
Uttar Pradesh	Ballia, Ghazipur, Varanasi, Allahabad, Patna, Faizabad, Gonda, Basti
Assam	Golaghat, Cachar, Karimganj, Hailakandi, Silchar, Darrang, Jorhat, Diphu, Udalguri, Majuli
Punjab	Rupnagar, Kapurthala, Hoshiarpur, Jalandhar, Nawanshahr, Patiala, Amritsar, Gurdaspur, Bathinda, Taran
Chhattisgarh	Rajnandgaon, Kanker
Manipur	Thoubal, Imphal
Delhi	New Delhi
Jharkhand	Sahibganj
Kashmir	Vale of Kashmir
Ladakh	Ladakh
Gujarat	Ankaleshwar
Telangana	Hyderabad

Figure 3.2 (Continued)

West Bengal: South Kolkata (Behala) was the first location identified with arsenic in groundwater whereby a factory producing copper acetoarsenite (Paris-green) had intoxicated (5–58 mg/L As) its vicinity by discharging its wastes (Guha Mazumder et al. 1992; Chatterjee and Das 1993). Residents in the nearby areas were found with the symptoms of keratosis, anemia, peripheral neuropathy, and liver and gastrointestinal disease. In 1994, six districts namely, Murshidabad, Bardhman, N-24 Parganas, S-24 Parganas, Malda, and Nadia have impermissible limits of arsenic in their groundwater (Das et al. 1996; Mukherjee et al. 2005; Mazumder et al. 2010). Villages and cities, namely Chakdah, Baruipur, Purbasthali, Beldanga, Haringhata, Kaliachak, English Bazar, Moyna, Adrebok, Barasat, Debagram, and Nabipur, of these districts were found contaminated with high levels of arsenic in the later years (Nag et al. 1996; Guha Mazumder et al. 1998; Pal et al. 2002; Ravenscroft et al. 2005; Purkait et al. 2008; Sengupta et al. 2008; Maity et al. 2010; Bhowmick et al. 2013; Chowdhury et al. 2015; Mondal and Pal 2015). Groundwater with high levels of arsenic was also confirmed in Sunderbans (Mandal et al. 2009).

Uttar Pradesh-Bihar: Researchers are of the view that the groundwater of eastern part of Uttar Pradesh and western part of Bihar is rich in arsenic with levels up to 576 μg/L (Acharyya et al. 2005; Gaur et al., 2005; Ahamed et al. 2006; Chakraborti et al. 2009; Chandra et al. 2011; Chauhan et al. 2009; Chowdhury et al. 2015; Janardhana Raju 2012; Kumar et al. 2016; Saha et al. 2011; Shah 2008; Srivastava and Sharma 2013; Shah 2015; Suman et al. 2019; Kumar et al. 2019, 2020; Rahman et al. 2020; Chattopadhyay et al. 2020; Kumar et al. 2021). This regions include Ballia, Gazipur, Mirzapur, Varanasi, Faizabad, Gonda, Basti, Vaishali, Kanpur of UP and Ballia, Bhojpur, Patna, Buxar, Begusarai, Munger, Bhagalpur, Khagaria, and Lakhisarai of Bihar. These locations are in proximity of Ganga River where arsenic enrichment is prevailing (>50 μg/L).

Assam: Episodes in Assam were reported in Golaghat district of Assam in 2011. Six blocks of Golaghat were affected of which Gamariguri was worst affected. Each sample taken from Gamariguri displayed high arsenic concentrations (above 0.01 ppm) (Chetia et al. 2011). Other districts of Arsenic incidence that came into picture were Cachar, Karimganj, Jorhat, and Diphu (Mahanta et al. 2015). These districts reported arsenic levels up to 100 μg/L (Shah 2012; Sailo and Mahanta 2013; Das and Kumar 2015; Choudhury et al. 2017). Majuli Island of Assam has significant concentration of arsenic in the groundwater (137 μg/L) and sediments (0.61 mg/kg) (Goswami et al. 2020)

Chhattisgarh: Chhattisgarh came in highlights through research in 2005 when its districts, namely Ambargarh-Chowki, Rajnandgaon, and Kanker, that were confined within rift zones reported cases of arsenicosis. The groundwater of these districts showed of having arsenic concentrations more than 250 μg/L (Acharyya et al. 2005; Patel et al. 2005; Pandey et al. 2006; Shukla et al. 2010).

Punjab: South-western, northern, central, and sub-mountainous districts (Amritsar, Kapurthala, Jalandhar, Ludhiana, Gurdaspur, Nawanshehar, Rupnagar, and Hoshiarpur) of Punjab have an elevated arsenic content in their water with levels up to 29.5 ppb (permissible limit: 10 ppb) (Hundal et al. 2009; Sidhu et al. 2012; Hundal et al. 2013). Villages in southwestern districts are worst affected where cases of cancer have been reported. In the newer alluvium of Ravi River, higher concentration of arsenic (255.9 μg/L) was detected (Kumar and Singh 2019).

Manipur: More than 50 μg/L of arsenic was detected in four districts located in Manipur valley in 2007. These four districts that constituted 60% of Manipur population were Thoubal, Bishnupur, Imphal West, and Imphal East (Chakraborti et al. 2008; Singh et al. 2013). According to an investigation in 2012, Imphal West recorded 73% of its groundwater exceeding 0.01 μg/L (Singh et al. 2013).

Delhi: Yamuna flood plains in Delhi were found contaminated by arsenic (max 180 ppb) in 2012 (Dubey et al. 2012; Mishra et al. 2014). Concentrations above 10 ppb were detected in the vicinity of coal-based thermal power plants in Badarpur and Rajghat (Dubey et al. 2012).

Telangana: Patancheru, one of the industrial regions of Telangana, is notorious for numerous cases of arsenic contamination. Levels above the impermissible limit have been revealed around the industrial regions of Hyderabad (Sekhar et al. 2003).

Jharkhand: Sahibganj, Udhawa, and Rajmahal blocks of Sahibganj district in Jharkhand are severely affected by arsenic contamination with levels touching up to 300 μg/L (Nayak et al. 2008). Children and adults were found suffering from arsenical skin lesions. A case of arsenical skin lesion resulting in cancer was also reported in this district.

Gujrat: Enrichment of sediments with arsenic was discovered in Ankaleshwar Industrial Estate of Gujrat. It has been developed on the flood plains of Narmada River and comprises fertilizers, dyes, chemicals, pharmaceutical, and glass industries (Shirke and Pawar 2015). Arsenic enrichment in this region was up to 18.87 mg/kg (Shirke and Pawar 2015). Higher concentration of arsenic has also been discovered in the Kachchh and Banas Kantha districts (Wu et al. 2021).

Ladakh & Kashmir: In Ladakh, around 70% of the groundwater samples had arsenic concentration more than 10 μg/L. The concentration varied between 1.1 and 25 μg/L in the younger alluvium, 1.3–86 μg/L in the cold springs and the groundwater wells, and 181–2186 μg/L in the thermal springs of the Upper Indus Basin of the Ladakh region (Lone et al. 2019). In the Upper Jhelum Basin of the Kashmir Valley, the level of arsenic ranged from 0.14 to 192 μg/L in the groundwater (Jeelani et al. 2020).

3.5 Medium of Contamination

Arsenic as a contaminant is found in water, sediments, soil, and crops and enters the food chain through these mediums, and finally the human body. This section explores the medium chosen by the researchers in their papers to study arsenic. This will give an idea of the adversely affecting medium. It will also highlight the extent of the study done on different mediums. Figure 3.3 reveals the percentage share of different mediums of contamination studied in the selected papers. This analysis is based on the samples taken by the authors to detect the noxious level of arsenic. Figure 3.3 speaks that on an average, half of the reviewed

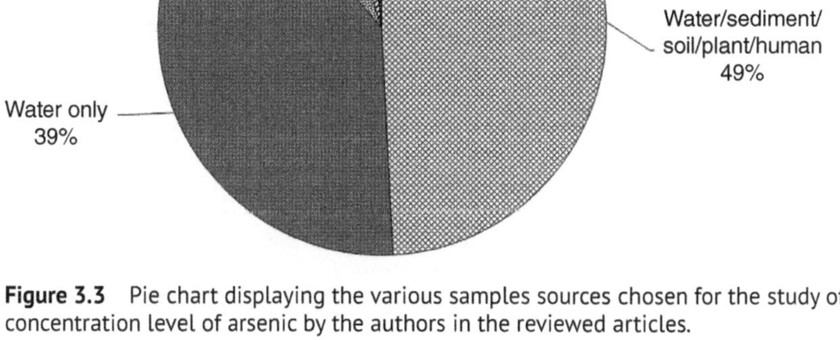

Figure 3.3 Pie chart displaying the various samples sources chosen for the study of the concentration level of arsenic by the authors in the reviewed articles.

literature sampled water, sediments, soil, crops, and the human body in combinations and not solely. Among these combinations, water with sediments (borehole and aquifers) and water with biological samples (hair, blood, urine, and nails) are picked up the most. This reveals that there exists a relation in arsenic in water bodies and sediments in the vicinity. Second, it portrays that in these locations arsenic has been accumulated in the human body at toxic levels. Remaining reviewed literature was focused on sole mediums among which majority studies are done taking water as a sample. This analysis brings to the fact that water, either surface water or groundwater, is sampled in almost 90% of the reviewed papers. That means water is the common medium of arsenic intake in the human body at those locations. Although sediments in the neighborhood of those locations contain arsenic, lesser studies have focused on sediments (near-surface and deep) when compared with those on water.

3.6 Sources and Mobilization

3.6.1 Holocene Sediments

Researchers have accredited the Himalayas as a source of parent rocks rich in arsenic, which on erosion supplies arsenic adsorbed on iron-rich minerals in the rivers (Podgorski and Berg 2020; UNICEF 2018; Burgess et al. 2010; Nath et al. 2008; Purkait and Mukherjee 2008; Chauhan et al. 2009; Shah 2014; Shah 2015). Rivers having their source in Himalayas transport these eroded minerals from hills and deposit them in plains (Smedley and Kinniburgh 2002; McArthur et al. 2004; Ravenscroft et al. 2005; Chandra et al. 2011; Das et al. 2015). In India, the highest incidences of arsenic contamination have been seen in Ganga plains (mainly in LGP). Ganga, originating from Gangotri glacier in the Himalayas and merging in the Bay of Bengal, deposits its sediments along its course in paleo-channels, meanders, back swamps, cut-offs, and abandoned channels. According to the researchers, the sediments that were deposited during late Pleistocene–Holocene (i.e. newer alluvium) in these depositional features are rich in arsenic (Ravenscroft et al. 2005; Nath et al. 2008; Bhattacharyya and Mukherjee 2009; Fendorf et al. 2010; Raju 2012; Jeelani et al. 2020; Sahu and Saha 2020). This newer alluvium is fine grained (sand, silt, and clay), organic rich, grey to black colored, and argillaceous sediments (Burgess et al. 2010; Chandra et al. 2011; Shah 2012; Shah 2015).

3.6.2 Excessive Pumping, Irrigation Practices, and Poor Economic Condition

Mid-1960s marked an important milestone in India, i.e. the Green Revolution. There was a transition from traditional agriculture to the introduction of high-yielding variety seeds and new agricultural techniques. They were employed first in Punjab, Haryana, and parts of western Uttar Pradesh. The second wave of Green Revolution was in the 1980s (Fujita 2012). This revolution aimed self-sufficiency in food grains by the continuous expansion of farming areas, double cropping in existing farmland, and using seeds with improved genetics. Double cropping meant having two crop seasons per year instead of one crop season. Earlier one crop season was in trend based on the monsoon season in India. However, this double cropping season necessitated massive irrigation projects (Ameen and Raza 2017). In West Bengal, Bihar, Assam, and Eastern Uttar Pradesh, Boro rice was cultivated in traditional manner. With the intensification in the demand and consequent production of Boro rice, surface water was unable to meet the accelerating demands of irrigation. Therefore, there was a shift in use from surface water to groundwater for irrigation (Mukherjee et al. 2007).

Prior to the Green Revolution, water from ponds, lakes, and dug wells were consumed for the needs. However, in the coming years after the Green Revolution, the surface water was found unsuitable for consumption due to excessive use of fertilizers and pesticides in the fields. Eventually, farmers started drawing underground water that was easily available at 8–10 m depth, cold in summer and lukewarm in winter (Chakraborti et al. 2009). Regions like Bihar with a cropping intensity of 176%, groundwater was catering the 80% of the irrigation demand (Saha et al. 2010). Punjab, a region of agrarian importance to India, witnessed an increase in the number of tube wells from 55,000 in 1967 to one million in 2008 (Hundal et al. 2009). It also creates perfect ground for microbial degradation and addition of dissolved organic carbon (DOC) to the surface and below by percolating downward (Farooq and Chandrasekharam 2015). This accelerated pace of the increasing the number of tube wells or hand pumps was also attributed to poor economic conditions of farmers. As the cost of installation of tube wells of shallow depth is meagre and affordable, it led to the installation of a multitude of tube wells at shallow depths without following any regulations and checks (Chakraborti et al. 2009). This excessive exploitation of groundwater from shallow depths was the keystone in mobilizing arsenic from its natural sources which otherwise was in resting mode and not as an active intoxicant (Singh et al. 2014).

Green Revolution also encouraged the growth of crops that require more irrigation water in those regions that receive less rainfall. In Punjab, there was a shift toward the cultivation of paddy crops employing flood irrigation practices (Hundal et al. 2013). These practices led toward stagnant water for longer duration in the agricultural fields creating anaerobic conditions favorable for the mobilization of arsenic in water. Excessive withdrawal and insufficient recharging of groundwater caused a lowering of the water table. This, in turn, led to oxidizing conditions in vodase zone and reducing conditions in aquifers (Pal et al. 2002) (separately discussed in Section 5.3) and exposed the arsenic-containing minerals for its mobilization.

3.6.3 Oxidizing and Reducing Conditions

Immoderate withdrawal of groundwater has caused aeration or oxidation of arsenopyrites in vadose zone (Nickson et al. 1998). Arsenopyrite, when comes into contact with atmospheric oxygen, releases water-soluble compounds of arsenic (As(III), As(V)) (Adel 2005). These compounds percolate downward to the groundwater making it toxic. Excessive withdrawal has also created anaerobic conditions in the aquifers (Pal et al. 2002; Adel 2005). Ferric oxyhydroxide coatings reduce from Fe^{3+} to Fe^{2+}, in reducing conditions, and release arsenic into water (Nickson et al. 2000; McArthur et al. 2001; Smedley and Kinniburgh 2002; Brammer and Ravenscroft 2009). Such cases of oxidation and reduction were observed in West Bengal, Bihar, Assam, and Punjab (Ravenscroft et al. 2001; Adel 2005; Brammer and Ravenscroft 2009; Hundal et al. 2009; Burgess et al. 2010; Kumar et al. 2010; Chetia et al. 2011; Sidhu et al. 2012; Choudhury et al. 2017; Kumar and Singh 2019; Goswami et al. 2020).

3.6.4 Organic Matter

Holocene sediments predominantly are under stagnant water and heavy vegetative cover (Mukhopadhyay et al. 2006; Saha et al. 2010; Bhowmick et al. 2013). Such conditions are favorable for microbial degradation of the organic matter (peat deposits), that consequently increases the DOC in groundwater (Nickson et al. 2000; McArthur et al. 2001; Ravenscroft et al. 2001; McArthur et al. 2004; Nath et al. 2008; Sailo and Mahanta 2013; Choudhury et al. 2017; Jeelani et al. 2020). The DOC mobilizes arsenic from sediments into the groundwater, which subsequently percolates deeper (Shah 2008; Fendorf et al. 2010; Kar et al. 2010). The aquifers that are recharged through these zones become vulnerable to arsenic contamination due to the percolating DOC and consequent reducing conditions (Ravenscroft et al. 2001; Wu et al. 2021; Acharyya and Shah 2007b). Researchers have detected contaminated water and sediments due to such conditions in West Bengal, Uttar Pradesh, Bihar, and Assam (Nickson et al. 2000; Ravenscroft et al. 2001).

3.6.5 Micro-organisms and Phytoplankton

Seasonal variations in the arsenic content in water have been attributed to micro-organisms by some authors (Oremland and Stolz 2003; Islam et al. 2004; Malasarn et al. 2004; Chakraborti et al. 2009; Kar et al. 2010; Mandal et al. 2009; Mukhopadhyay et al. 2006; Purkait and Mukherjee 2008; Farooq and Chandrasekharam 2015). Some micro-organisms growing in the swampy conditions of Middle and Lower Ganga Plains intake arsenic for their growth during monsoon season reducing the concentration in the season (Mukhopadhyay et al. 2006). Similarly, phytoplankton and foraminifera in Sunderbans absorb arsenic dissolved in water during their blossoming season and reduce its concentration (Mandal et al. 2009). However, the arsenic that was consumed by them is again released in the environment with their decomposition during post-monsoon (Purkait and Mukherjee 2008).

3.6.6 Fertilizers and Crops

With the advent of High Yielding Variety (HYV) seeds, there was growth in the use of fertilizers and pesticides (Shirke and Pawar 2015). Pesticides have lead arsenate and copper arsenate in them that on application introduce arsenic into the soil (Shirke and Pawar 2015). In addition, fertilizers have a high content of phosphate that too helps in the mobilization of arsenic (Kar et al. 2010; Hundal et al. 2013; Sailo and Mahanta 2013; Kumar and Singh 2019; Kumar et al. 2020; Rahman et al. 2020). Since arsenic has a tendency of being adsorbed on iron coatings, if phosphate is present in surroundings, it competes for these adsorption sites on iron-enriched minerals (Hundal et al. 2013), helping arsenic to remain in solution. In addition, the runoff from these agricultural fields pollutes the nearby surface water (Kumar and Singh 2019). Not only the surface, but they also pollute the groundwater on percolation. Crops grown on such fields intake arsenic from the soil.

Intake of arsenic by crops is not only through the contaminated soil but also through the contaminated water (Brammer and Ravenscroft 2009; Khan et al. 2009; Sharma et al. 2014). During pre-monsoons when most of the region of India experiences dry and hot weather, most of the sources of surface water dry up or reduce their level. To maintain the productivity of their land, farmers refill them by pumping groundwater at shallow depths that is already rich in arsenic. When crops are irrigated through this contaminated water, arsenic content increases in the crops (Brammer and Ravenscroft 2009; Chowdhury et al. 2015). This is why rice is highly vulnerable to arsenic contamination owing to significant irrigation requirement (Sidhu et al. 2012). However, each crop and even different parts of the same crop do not equally retain arsenic (Rahaman et al. 2013; Biswas 2019). Researchers have mentioned that the tendency of accumulation is more in roots and less in fruits. Among fruits, those fruits with more water content have a higher accumulation of arsenic than those with less juicy ones (Mishra et al. 2014). Their research found tomatoes with higher arsenic concentration than brinjal. Radish (root) had the highest concentration among all (Joseph et al. 2015). In paddy crops, the highest concentration was observed in roots followed by straw, husk, and grain. Punjab, Assam, Manipur, and Chhattisgarh have seen such mobilization of arsenic.

3.6.7 Coal Mining and Industries

The coal occurring naturally is not toxic itself. When it is mined, exposed to weathering and leaching, the concentrations of heavy metals in it are mobilized including arsenic (Acharyya and Shah 2007a; Shukla et al. 2010; Dubey et al. 2012). For example, Amjhore mine (Shah 2015) and Gondwana coal fields in Peninsular India (Acharyya and Shah 2007a). Second, when coal undergoes combustion in thermal power plants, fly ash, one of the by-products, is generated. The fly ash has a high content of arsenic in it which not only contaminates air but also contaminates water with arsenic upon settling (Dubey et al. 2012). Moreover, in India, bituminous coal has been used for electricity generation. This coal on burning releases several by-products of which 40% is fly ash (Journal and Chemistry 2017). In Delhi, arsenic-contaminated water, soil, and sediment were found in nearby thermal coal plants (Dubey et al. 2012). In addition, industries like pharmaceutical, adhesives, and semiconductors employ arsenic in manufacturing their products (Patel et al. 2005; Chakraborti et al. 2008; Shirke and Pawar 2015). Wastes generated from these industries are dumped in water or nearby areas and that with time become arsenic-contaminated zones (Chatterjee and Mukherjee 1999; Gaur et al. 2005; Farooq et al. 2011; Dubey et al. 2012). Water percolating from these zones contaminates the groundwater and aquifers. River Yamuna is an example of it wherein wastes from industries are dumped. Farmers cultivating vegetables along its plain use this polluted water for irrigation that consequently leads to arsenic accumulation in vegetables (Mishra et al. 2014).

3.7 Probable Future Arsenic Zones

On studying different locations and triggers of arsenic mobilization simultaneously, we observe some trends and patterns in the behavior of contamination (Figure 3.4). States namely, Uttar Pradesh, Bihar, Jharkhand, West Bengal, Assam, and Manipur have the same nature of arsenic contamination. These states lie in the plains formed by the rivers, Ganga and Brahmaputra, descending from the Himalayas. Hence, Holocene sediments deposited by these rivers in entrenched channels, meanders, back swamps, paleo-channels, and oxbow lakes are rich in arsenic. There is an increasing trend in the incidences from Uttar Pradesh to West Bengal owing to more depositional activity in the lower plains of Ganga. Similarly, Assam that lies in the lower plains of Brahmaputra has the same fate. In Manipur, which embraces both hills and valley regions, incidences are seen in the valley due to the depositional activity of the rivers. Further, excessive withdrawal through tube wells and the presence of abundant organic matter in those depositional features activate arsenic present

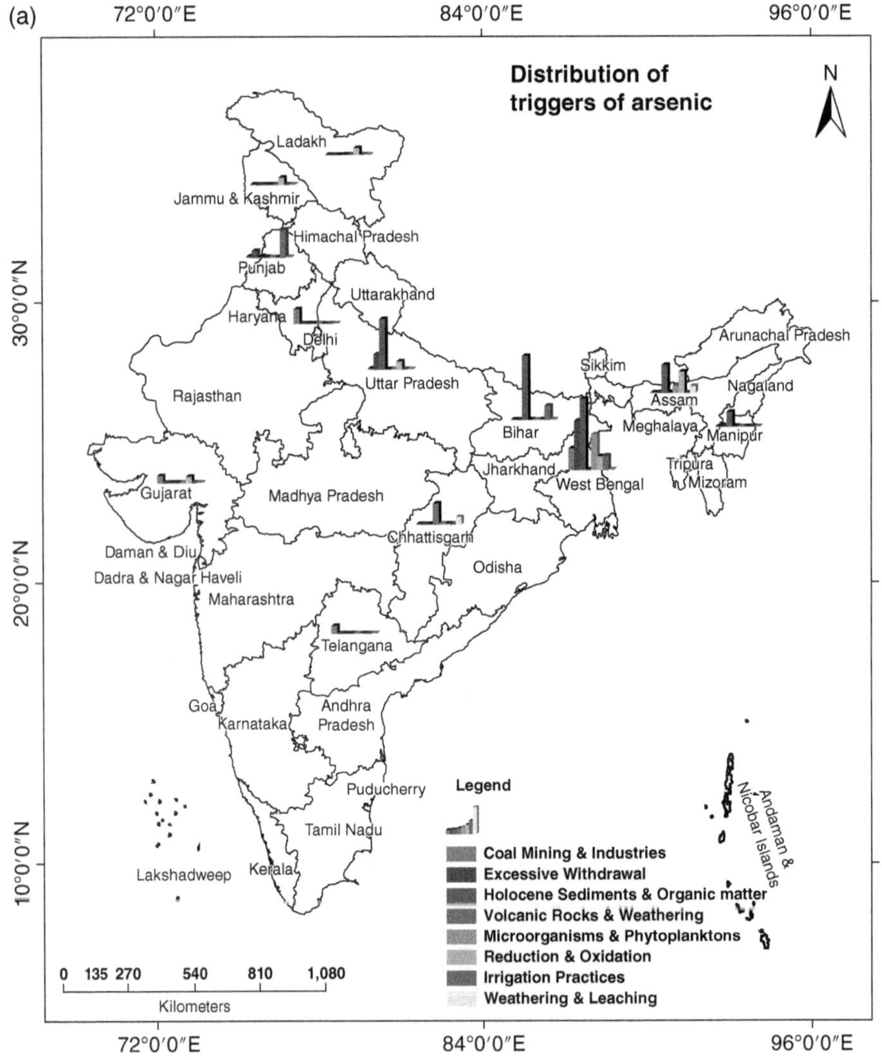

Figure 3.4 Sources and triggers of arsenic mobilization discussed in the reviewed articles. (a) Map displaying the distribution of various triggers of arsenic mobilization in the arsenic-contaminated states in India as mentioned in the reviewed articles. (b) Bar diagram showing the frequency of the discussed reasons behind arsenic mobilization and contamination by the authors in the reviewed articles.

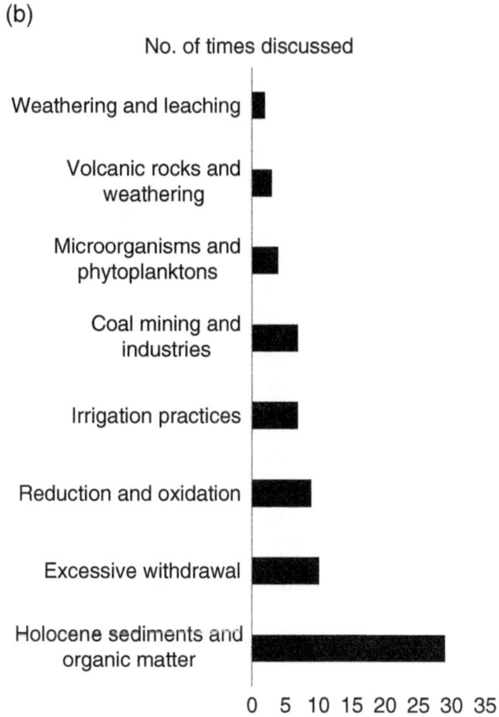

Figure 3.4 (Continued)

in the Holocene sediments. Therefore, we can say that the whole Ganga and Brahmaputra basins, where depositional and human activities are intense, are probable zones of arsenic. In addition to this, regions that are employing excessive fertilizers, pesticides, and wrong irrigation practices hold potentials of arsenic contamination like Punjab, Haryana, and western Uttar Pradesh.

Delhi, Gujrat, and Telangana also reflect the common nature of arsenic mobilization. These are among the industrial hubs of India that dump a huge amount of untreated wastes every year. In majority, arsenic is released from tanning, metal and alloys, textile and dyeing, fertilizer and pesticides, pharmaceutical, and coal-based thermal industries. Thus, the vicinity of such industrial locations has probabilities of arsenic exposure. Lastly, coal mining and volcanic regions, like Chhattisgarh, of Peninsular India are future zones of arsenic episodes in India.

3.8 Exposure to Arsenic

An overlong consumption of arsenic through food and water has detrimental health effects on circulatory system, liver, kidney, and skin (Singh et al. 2021). Classified as Class I Human Carcinogen, it causes cancer of kidney, lung, liver, bladder, gastrointestinal tract, and skin (International Agency for Research on Cancer 2004; Zhou and Xi 2018).

Arun et al. have related the high occurrence of cancer with the arsenic-contaminated Gangetic plains. Arsenic uptake in the human body causes cardiovascular problems, hypertension, genetic imbalance, neurological, and pregnancy issues (Brinkel et al. 2009; Rahman et al. 2009; Koley 2021). Consumption of elevated concentrations of arsenic in drinking water for a longer duration leads to acute toxicity causing dermal lesions (hyperpigmentation, hyperkeratosis), peripheral vascular disease, and peripheral neuropathy (Saha et al. 1999; Duker et al. 2005; Ahamad et al. 2020; Madhav et al. 2020). The case is worst when the socioeconomic conditions, awareness, and dietary habits are poor (Koley 2021).

3.9 Conclusions

In India, there is heterogeneity in the distribution and sources of arsenic contamination. The north-eastern region comprising Uttar Pradesh, Bihar, West Bengal, Assam, and Manipur has more arsenic-affected districts. These are followed by the central part (Chhattisgarh and Jharkhand) and the northern part (Punjab and Delhi). Least arsenic episodes are from southern (Telangana) and western India (Gujrat). Southern and western regions experience arsenic episodes owing to the industrial activities. The central part is arsenic positive due to coal mining and volcanic activities. The northern part is contaminated due to excessive use of fertilizers and coal-based thermal plants. Holocene sediments in the depositional landforms of rivers, excessive withdrawal of groundwater, and microbial degradation are the sources of arsenic enrichment in north-eastern states.

Human interferences in the natural state of arsenic are the root cause of the surfacing of its toxicity. Over-exploitation of groundwater with the advancement of the Green Revolution in the 1960s was the starting point of arsenic episodes in India. These were further increased by industrial advancement in cities coupled with coal mining activities. The situation worsened with the cultivation shift from less irrigation requirement crops to flooding irrigation crops. Above all, poor socioeconomic conditions of developing India aided in aggravating the situation.

References

Acharyya, S. K., & Shah, B. A. (2007a). Arsenic-contaminated groundwater from parts of Damodar fan-delta and west of Bhagirathi River, West Bengal, India: influence of fluvial geomorphology and Quaternary morphostratigraphy. *Environmental Geology*, 52(3), 489–501. https://doi.org/10.1007/s00254-006-0482-z

Acharyya, S. K., & Shah, B. A. (2007b). Groundwater arsenic contamination affecting different geologic domains in India – A review: Influence of geological setting, fluvial geomorphology and Quaternary stratigraphy. *Journal of Environmental Science and Health – Part A Toxic/Hazardous Substances and Environmental Engineering*, 42(12), 1795–1805. https://doi.org/10.1080/10934520701566744

Acharyya, S. K., Shah, B. A., Ashyiya, I. D., & Pandey, Y. (2005). Arsenic contamination in groundwater from parts of Ambagarh-Chowki block, Chhattisgarh, India: source and release mechanism. *Environmental Geology*, 49(1), 148–158. https://doi.org/10.1007/s00254-005-0074-3

Adel, M. M. (2005). The background state leading to arsenic contamination of Bengal Basin groundwater. *Journal of Water and Health*, 3(4), 435–452. https://doi.org/10.2166/wh.2005.053

Ahamad, A., Madhav, S., Singh, A. K., Kumar, A., & Singh, P. (2020). Types of water pollutants: Conventional and emerging. In: *Sensors in Water Pollutants Monitoring: Role of Material*, 21–41. Springer, Singapore.

Ahamed, S., Kumar Sengupta, M., Mukherjee, A., Amir Hossain, M., Das, B., Nayak, B., Pal, A., Chandra Mukherjee, S., Pati, S., Nath Dutta, R., Chatterjee, G., Mukherjee, A., Srivastava, R., & Chakraborti, D. (2006). Arsenic groundwater contamination and its health effects in the state of Uttar Pradesh (UP) in upper and middle Ganga plain, India: A severe danger. *Science of the Total Environment*, 370(2–3), 310–322. https://doi.org/10.1016/j.scitotenv.2006.06.015

Ameen, A., & Raza, S. (2017). International journal of advances in scientific research green revolution: A review QR code *Correspondence Info. *International Journal of Advances in Scientific Research*, 3(12), 129–137. https://doi.org/10.7439/ijasr

Bhattacharyya, D., & Mukherjee, P. K. (2009). Contamination of shallow aquifers by arsenic in upper reaches of Tista river at Siliguri-Jalpaiguri area of West Bengal, India. *Environmental Geology*, 57(7), 1687–1692. https://doi.org/10.1007/s00254-008-1450-6

Bhowmick, S., Nath, B., Halder, D., Biswas, A., Majumder, S., Mondal, P., Chakraborty, S., Nriagu, J., Bhattacharya, P., Iglesias, M., Roman-Ross, G., Guha Mazumder, D., Bundschuh, J., & Chatterjee, D. (2013). Arsenic mobilization in the aquifers of three physiographic settings of West Bengal, India: Understanding geogenic and anthropogenic influences. *Journal of Hazardous Materials*, 262, 915–923. https://doi.org/10.1016/j.jhazmat.2012.07.014

Biswas, A., (2019). A systematic review on arsenic bio-availability in human and animals: special focus on the rice-human system. *Reviews of Environmental Contamination and Toxicology*. https://doi.org/10.1007/398_2019_28

Brammer, H., & Ravenscroft, P. (2009). Arsenic in groundwater: A threat to sustainable agriculture in South and South-east Asia. *Environment International*, 35, 647–654. https://doi.org/10.1016/j.envint.2008.10.004.

Brinkel, J., Khan, M. H., & Kraemer, A. (2009). A systematic review of arsenic exposure and its social and mental health effects with special reference to Bangladesh. *International Journal of Environmental Research and Public Health*, 6(5), 1609–1619. https://doi.org/10.3390/ijerph6051609

Burgess, W. G., Hoque, M. A., Michael, H. A., Voss, C. I., Breit, G. N., & Ahmed, K. M. (2010). Vulnerability of deep groundwater in the Bengal Aquifer System to contamination by arsenic. *Nature Geoscience*, 3, 83–87.

Chakraborti, D., Singh, E. J., Das, B., Shah, B. A., Hossain, M. A., Nayak, B., Ahamed, S., & Singh, N. R. (2008). Groundwater arsenic contamination in Manipur, one of the seven North-Eastern Hill states of India: A future danger. *Environmental Geology*, 56(2), 381–390. https://doi.org/10.1007/s00254-007-1176-x

Chakraborti, D., Ghorai, S. K., Das, B., Pal, A., Nayak, B., & Shah, B. A. (2009). Arsenic exposure through groundwater to the rural and urban population in the Allahabad-Kanpur track in the upper Ganga plain. *Journal of Environmental Monitoring*, 11(8), 1455–1459. https://doi.org/10.1039/b906584a

Chandra, S., Ahmed, S., Nagaiah, E., Singh, S. K., & Chandra, P. C. (2011). Geophysical exploration for lithological control of arsenic contamination in groundwater in Middle Ganga Plains, India. *Physics and Chemistry of the Earth*, 36(16), 1353–1362. https://doi.org/10.1016/j.pce.2011.05.009

Chatterjee, A., & Das, D. (1993). A study of ground water contamination by arsenic in the residential area of Behala, Calcutta due to industrial pollution. *Environmental Pollution*, 80(1), 57–65.

Chatterjee, A., & Mukherjee, A. (1999). Hydrogeological investigation of ground water arsenic contamination in South Calcutta. *Science of the Total Environment*, 225(3), 249–262. https://doi.org/10.1016/S0048-9697(98)00367-2

Chattopadhyay, A., Singh, A. P., Singh, S. K., Barman, A., Patra, A., Mondal, B. P., & Banerjee, K. (2020). Spatial variability of arsenic in Indo-Gangetic basin of Varanasi and its cancer risk assessment. *Chemosphere*, 238, 124623. https://doi.org/10.1016/j.chemosphere.2019.124623

Chauhan, V. S., Nickson, R. T., Chauhan, D., Iyengar, L., & Sankararamakrishnan, N. (2009). Ground water geochemistry of Ballia district, Uttar Pradesh, India and mechanism of arsenic release. *Chemosphere*, 75(1), 83–91. https://doi.org/10.1016/j.chemosphere.2008.11.065

Chetia, M., Chatterjee, S., Banerjee, S., Nath, M. J., Singh, L., Srivastava, R. B., & Sarma, H. P. (2011). Groundwater arsenic contamination in Brahmaputra river basin: A water quality assessment in Golaghat (Assam), India. *Environmental Monitoring and Assessment*, 173(1–4), 371–385. https://doi.org/10.1007/s10661-010-1393-8

Choudhury, R., Mahanta, C., Verma, S., & Mukherjee, A. (2017). Arsenic distribution along different hydrogeomorphic zones in parts of the Brahmaputra River Valley, Assam (India). *Hydrogeology Journal*. https://doi.org/10.1007/s10040-017-1584-2

Chowdhury, A. N., Samanta, S., Manna, S. K., Sharma, A. P., Bandopadhyay, C., Pramanik, K., Sarkar, S., & Mohanty, B. P. (2015). Arsenic in freshwater ecosystems of the Bengal delta: status, sources and seasonal variability. *Toxicological and Environmental Chemistry*, 97(5), 538–551. https://doi.org/10.1080/02772248.2015.1053482

Das, A., & Kumar, M. (2015). Arsenic enrichment in the groundwater of Diphu, Northeast India: Coupled application of major ion chemistry, speciation modeling, and multivariate statistical techniques. *Clean – Soil, Air, Water*, 43(11), 1501–1513. https://doi.org/10.1002/clen.201400632

Das, N., Patel, A.K., Deka, G., Das, A., Sarma, K.P. & Kumar, M. (2015). Geochemical controls and future perspective of arsenic mobilization for sustainable groundwater management: A study from Northeast India. *Groundwater for Sustainable Development*, 1(1-2), 92–104. https://doi.org/10.1016/j.gsd.2015.12.002

Das, D., Samanta, G., Mandal, B. K., Chowdhury, T. R., Chanda, C. R., Chowdhury, P. P., Basu, G. K., & Chakraborti, D. (1996). Arsenic in groundwater in six districts of West Bengal, India. *Environmental Geochemistry and Health*, 18(1), 5–15. https://doi.org/10.1007/BF01757214

Datta, D.V. & Kaul, M.K. (1976). Arsenic content of drinking water in villages in Northern India: A concept of arsenicosis. *Journal of the Association of Physicians of India*, 24(9), 599–604.

Dubey, C. S., Mishra, B. K., Shukla, D. P., Singh, R. P., Tajbakhsh, M., & Sakhare, P. (2012). Anthropogenic arsenic menace in Delhi Yamuna Flood Plains. *Environmental Earth Sciences*, 65(1), 131–139. https://doi.org/10.1007/s12665-011-1072-2

Duker, A. A., Carranza, E. J. M., Hale, M. (2005). Arsenic geochemistry and health. *Environment International*, 31, 631–641. https://doi.org/10.1016/j.envint.2004.10.020

Dutt, A., Ismail, M. A., & Herawan, T. (2017). A systematic review on educational data mining. *IEEE Access*, 5, 15991–15993. https://doi.org/10.1109/ACCESS.2017.2654247

Farooq, S. H., & Chandrasekharam, D. (2015). Surface Generated organic matter: an important driver for arsenic mobilization in Bengal Delta plain. In: Ramanathan, A.L., Johnston, S., Mukherjee, A., Nath, B. (eds.), *Safe and Sustainable Use of Arsenic-Contaminated Aquifers in the Gangetic Plain*, 179–196. Springer. https://doi.org/10.1007/978-3-319-16124-2_12

Farooq, S. H., Chandrasekharam, D., Norra, S., Berner, Z., Eiche, E., Thambidurai, P., & Stüben, D. (2011). Temporal variations in arsenic concentration in the groundwater of Murshidabad District, West Bengal, India. *Environmental Earth Sciences*, 62(2), 223–232. https://doi.org/10.1007/s12665-010-0516-4

Fendorf, S., Michael, H. A., & Geen, A. V. (2010). Spatial and temporal variations of groundwater arsenic in South and Southeast Asia. *Science*, 308 (5982), 1123–1127. https://doi.org/10.1126/science.1172974

Fujita, K. (2012). Green revolution in India and its significance in economic development: implications for Sub-Saharan Africa. *JICA Research Institute*, 15, 5–6. http://courses.essex.ac.uk/ec/ec330/lecture_presentations/Fujita_green_rev_in_india.pdf

Gaur, V. K., Gupta, S. K., Pandey, S. D., Gopal, K., & Misra, V. (2005). Distribution of heavy metals in sediment and water of River Gomti. *Environmental Monitoring and Assessment*, 102, 419–433. https://doi.org/10.1007/s10661-005-6395-6

Goswami, R., Kumar, M., Biyani, N., & Shea, P. J. (2020). Arsenic exposure and perception of health risk due to groundwater contamination in Majuli (river island), Assam, India. *Environmental Geochemistry and Health*, 42, 443–460. https://doi.org/10.1007/s10653-019-00373-9

Guha Mazumder, D. N., Das Gupta, J., Chakraborty, A. K., Chatterjee, A., Das, D., & Chakraborti, D. (1992). Environmental pollution and chronic arsenicosis in South Calcutta. *Bulletin of the World Health Organization*, 70(4), 481–485.

Guha Mazumder, D. N., Haque, R., Ghosh, N., De Binay, K., Santra, A., Chakraborty, D., & Smith, A. H. (1998). Arsenic levels in drinking water and the prevalence of skin lesions in West Bengal, India. *International Journal of Epidemiology*, 27(5), 871–877. https://doi.org/10.1093/ije/27.5.871

He, J., & Charlet, L. (2013). A review of arsenic presence in China drinking water. *Journal of Hydrology*, 492, 79–88. https://doi.org/10.1016/j.jhydrol.2013.04.007

Hundal, H. S., Singh, K., & Singh, D. (2009). Arsenic content in ground and canal waters of Punjab, North-West India. *Environmental Monitoring and Assessment*, 154(1–4), 393–400. https://doi.org/10.1007/s10661-008-0406-3

Hundal, H. S., Singh, K., Singh, D., & Kumar, R. (2013). Arsenic mobilization in alluvial soils of Punjab, North-West India under flood irrigation practices. *Environmental Earth Sciences*, 69(5), 1637–1648. https://doi.org/10.1007/s12665-012-1999-y

International Agency for Research on Cancer. (2004). IARC monographs on the evaluation of carcinogenic risks to humans: Some drinking water disinfectants and contaminants, including arsenic. World Health Organization International Agency for Research on Cancer, Lyon, France.

Islam, F.S., Gault, A.G., Boothman, C., Polya, D.A., Charnock, J.M., Chatterjee, D., Lloyd, J.R. (2004). Role of metal-reducing bacteria in arsenic release from Bengal delta sediments. *Nature*, 430, 68–71. https://doi.org/10.1038/nature02638

Jeelani, G., Lone, S. A., Nisa, A. U., Mukherjee, A., & Deshpande, R. D. (2020). Sources and processes of groundwater arsenic mobilization in upper Jhelum basin, western Himalayas. *Journal of Hydrology*, 591, 125292. https://doi.org/10.1016/j.jhydrol.2020.125292

Jha, P. K., & Tripathi, P. (2021). Arsenic and fluoride contamination in groundwater: a review of global scenarios with special reference to India. *Groundwater for Sustainable Development*, 13, 100576. https://doi.org/10.1016/j.gsd.2021.100576

Joseph, T., Dubey, B., & McBean, E. A. (2015). A critical review of arsenic exposures for Bangladeshi adults. *Science of the Total Environment*, 527–528, 540–551. https://doi.org/10.1016/j.scitotenv.2015.05.035

Journal, I., & Chemistry, A. (2017). Fly ash in India: generation vis-à-vis utilization and global perspective. *International Journal of Applied Chemistry*, 13(1), 29–52.

Kar, S., Maity, J. P., Jean, J. S., Liu, C. C., Nath, B., Yang, H. J., & Bundschuh, J. (2010). Arsenic-enriched aquifers: occurrences and mobilization of arsenic in groundwater of Ganges Delta Plain, Barasat, West Bengal, India. *Applied Geochemistry*, 25(12), 1805–1814. https://doi.org/10.1016/j.apgeochem.2010.09.007

Khan, N. I., Owens, G., Bruce, D., & Naidu, R. (2009). Human arsenic exposure and risk assessment at the landscape level: a review. *Environmental Geochemistry and Health*, 31, 143–166. https://doi.org/10.1007/s10653-008-9240-3

Kitchenham, B., Brereton, O. P., Budgen, D., Turner, M., Bailey, J., & Linkman, S. (2009). Systematic literature reviews in software engineering-A systematic literature review. *Information and Software Technology*, 51, 7–15. https://doi.org/10.1016/j.infsof.2008.09.009

Koley, S. (2021). Future perspectives and mitigation strategies towards groundwater arsenic contamination in West Bengal, India. *Environmental Quality Management*. https://doi.org/10.1002/tqem.21784

Kumar, A., & Singh, C. K. (2019). Arsenic enrichment in groundwater and associated health risk in Bari doab region of Indus basin, Punjab, India. *Environmental Pollution*. https://doi.org/10.1016/j.envpol.2019.113324.

Kumar, P., Kumar, M., Ramanathan, A. L., & Tsujimura, M. (2010). Tracing the factors responsible for arsenic enrichment in groundwater of the middle Gangetic Plain, India: a source identification perspective. *Environmental Geochemistry and Health*, 32(2), 129–146. https://doi.org/10.1007/s10653-009-9270-5

Kumar, A., Kumar, R., Ali, M., Gahlot, V., & Ghosh, A. K. (2016). Groundwater arsenic poisoning in Buxar district, Bihar, India: Health Hazards. Arsenic Research and Global Sustainability – Proceedings of the 6th International Congress on Arsenic in the Environment, AS 2016, 378–379. https://doi.org/10.1201/b20466-178

Kumar, A., Ali, M., Kumar, R., Rahman, M. S., Srivastava, A., Chayal, N. K., Sagar, V., Kumari, R., Parween, S., Kumar, R., Niraj, P. K., Anand, G., Singh, S. K., & Ghosh, A. K. (2019). High arsenic concentration in blood samples of People of Village Gyaspur Mahaji, Patna, Bihar Drinking Arsenic-Contaminated Water *Exposure and Health*, 12, 131–140.

Kumar, A., Singh, C. K., Bostick, B., Nghiem, A., Mailloux, B., & Geen, A. V. (2020). Regulation of groundwater arsenic concentrations in the Ravi, Beas, and Sutlej floodplains of Punjab, India. *Geochimica et Cosmochimica Acta*, 276, 384–403. https://doi.org/10.1016/j.gca.2020.03.003

Kumar, A., Ali, M., Kumar, R., Kumar, M., Sagar, P., Pandey, R.K., Akhouri, V., Kumar, V., Anand, G., Niraj, P. K., Rani, R., Kumar, S., Kumar, D., Bishwapriya, A., Ghosh, A.K. (2021).

Arsenic exposure in Indo Gangetic plains of Bihar causing increased cancer risk. *Scientific Reports*, 11, 2376. https://doi.org/10.1038/s41598-021-81579-9

Lone, S. A., Jeelani, G., Mukherjee, A., & Coomar, P. (2019). Geogenic groundwater arsenic in high altitude bedrock aquifers of upper Indus river basin (UIRB), Ladakh. *Applied Geochemistry*. https://doi.org/10.1016/j.apgeochem.2019.104497

Madhav, S., Ahmad, A., Singh, A. K., Kushawaha, J., Chahuhan, J. S., Sharma, S., & Singh, P. (2020). Water Pollutants: Sources and Impact on the Environment and Human Health. In: Singh, P., Patil, S., Pooja, D., Kumar, P. (eds.), *Sensors in Water Pollutants Monitoring: Role of Material, Advanced Functional Materials and Sensors*. Springer-Verlag Singapore PTE LTD. https://doi.org/10.1007/978-981-15-0671-0_4

Mahanta, C., Choudhury, R., Basu, S., Hemani, R., Dutta, A., Barua, P. P., Borah, P. J., Bhattacharya, M., Bhattacharya, K., Alam, W., Saikia, L., Mukherjee, A., & Bhattacharya, P. (2015). Preliminary assessment of arsenic distribution in Brahmaputra River Basin of India based on examination of 56,180 public groundwater wells. In: Ramanathan, A., Johnston, S., Mukherjee, A., Nath, B. (eds.), *Safe and Sustainable Use of Arsenic-Contaminated Aquifers in the Gangetic Plain*, 57–64. Springer, Cham. https://doi.org/10.1007/978-3-319-16124-2_4

Maity, J. P., Kar, S., Jean, J. S., Liu, C. C., Banerjee, S., & Santra, S. C. (2010). Groundwater contamination by arsenic in of 24-parganas district of West Bengal, India: A health risk. ICNIT 2010–2010 International Conference on Networking and Information Technology, 497–499. https://doi.org/10.1109/ICNIT.2010.5508463

Malasarn, D., Saltikov, C. W., Campbell, K. M., Santini, J. M., Hering, J. G., & Newman, D. K. (2004). *arrA* is a reliable marker for As(V) respiration. *Science*, 306(5695), 455. https://doi.org/10.1126/science.1102374

Mandal, B.K. & Suzuki, K.T. (2002). Arsenic round the world: a review. *Talanta*, 58(1), 201–235. https://doi.org/10.1016/S0039-9140(02)00268-0

Mandal, S. K., Dey, M., Ganguly, D., Sen, S., & Jana, T. K. (2009). Biogeochemical controls of arsenic occurrence and mobility in the Indian Sundarban mangrove ecosystem. *Marine Pollution Bulletin*, 58(5), 652–657. https://doi.org/10.1016/j.marpolbul.2009.01.010

Mazumder, D. N., Ghosh, A., Majumdar, K., Ghosh, N., Saha, C., & Mazumder, R. N. (2010). Arsenic contamination of ground water and its health impact on population of district of Nadia, West Bengal, India. *Indian Journal of Community Medicine*, 35(2), 331–338. https://doi.org/10.4103/0970-0218.66897

McArthur, J. M., Ravenscroft, P., Safiulla, S., & Thirlwall, M. F. (2001). Arsenic in groundwater: testing pollution mechanisms for sedimentary aquifers in Bangladesh. *Water Resources Research*, 37(1), 109–117

McArthur, J. M., Banerjee, D. M., Hudson-Edwards, K. A., Mishra, R., Purohit, R., Ravenscroft, P., Cronin, A., Howarth, R. J., Chatterjee, A., Talukder, T., Lowry, D., Houghton, S., & Chadha, D. K. (2004). Natural organic matter in sedimentary basins and its relation to arsenic in anoxic ground water: the example of West Bengal and its worldwide implications. *Applied Geochemistry*, 19, 1255–1293. https://doi.org/10.1016/j.apgeochem.2004.02.001

Mishra, B. K., Dubey, C. S., Shukla, D. P., Bhattacharya, P., & Usham, A. L. (2014). Concentration of arsenic by selected vegetables cultivated in the Yamuna flood plains (YFP) of Delhi, India. *Environmental Earth Sciences*, 72(9), 3281–3291. https://doi.org/10.1007/s12665-014-3232-7

Mondal, D., & Pal, S. (2015). A multi-parametric spatial modeling of vulnerability due to arsenic pollution in Murshidabad district of West Bengal, India. *Arabian Journal of Geosciences*, 8(10), 8047–8054. https://doi.org/10.1007/s12517-015-1809-4

Mukherjee, S. C., Saha, K. C., Pati, S., Dutta, R. N., Rahman, M. M., Sengupta, M. K., Ahamed, S., Lodh, D., Das, B., Hossain, M., Nayak, B., Mukherjee, A., Chakraborti, D., Dulta, S. K., Palit, S. K., Kaies, I., Barua, A. K., & Asad, K. A. (2005). Murshidabad – one of the nine groundwater arsenic-affected districts of West Bengal, India. Part II: dermatological, neurological, and obstetric findings. *Clinical Toxicology*, 43(7), 835–848. https://doi.org/10.1080/15563650500357495

Mukherjee, A., Fryar, A. E., & Howell, P. D. (2007). Regional hydrostratigraphy and groundwater flow modeling in the arsenic-affected areas of the western Bengal basin, West Bengal, India. *Hydrogeology Journal*, 15(7), 1397–1418. https://doi.org/10.1007/s10040-007-0208-7

Mukhopadhyay, B., Mukherjee, P. K., Bhattacharya, D., & Sengupta, S. (2006). Delineation of arsenic-contaminated zones in Bengal Delta, India: a geographic information system and fractal approach. *Environmental Geology*, 49(7), 1009–1020. https://doi.org/10.1007/s00254-005-0139-3

Nag, J. K., Balaram, V., Rubio, R., Albertí, J., & Das, A. K. (1996). Inorganic arsenic species in groundwater: a case study from Purbasthali (Burdwan), India. *Journal of Trace Elements in Medicine and Biology*, 10(1), 20–24. https://doi.org/10.1016/S0946-672X(96)80004-6

Nath, B., Berner, Z., Chatterjee, D., Mallik, S. B., & Stüben, D. (2008). Mobility of arsenic in West Bengal aquifers conducting low and high groundwater arsenic. Part II: comparative geochemical profile and leaching study. *Applied Geochemistry*, 23(5), 996–1011. https://doi.org/10.1016/j.apgeochem.2007.11.017

Nayak, B., Das, B., Chandra Mukherjee, S., Pal, A., Ahamed, S., Amir Hossain, M., Maity, P., Dutta, R. N., Dutta, S., & Chakraborti, D. (2008). Groundwater arsenic contamination in the Sahibganj district of Jharkhand state, India in the middle Ganga plain and adverse health effects. *Toxicological and Environmental Chemistry*, 90(4), 673–694. https://doi.org/10.1080/02772240701655486

Nickson, R., McArthur, J., Burgess, W., Ahmed, K. M., Ravenscroft, P., & Rahmann, M. (1998). Arsenic poisoning of Bangladesh groundwater. *Nature*, 395, 338.

Nickson, R. T., McArthur, J. M., Ravenscroft, P., Burgess, W., & Ahmed, K. M. (2000). Mechanism of arsenic release to groundwater, Bangladesh and West Bengal. *Applied Geochemistry*, 15, 403–413. https://doi.org/10.1016/S0883-2927(99)00086-4

Oremland, R. S., & Stolz, J. F. (2003). The ecology of arsenic. *Science*, 300, 939–943.

Pal, T., Mukherjee, P. K., Sengupta, S., Bhattacharyya, A. K., & Shome, S. (2002). Arsenic pollution in groundwater of West Bengal, India – an insight into the problem by subsurface sediment analysis. *Gondwana Research*, 5(2), 501–512. https://doi.org/10.1016/S1342-937X(05)70738-3

Pandey, P. K., Sharma, R., Roy, M., Roy, S., & Pandey, M. (2006). Arsenic contamination in the Kanker district of central-east India: geology and health effects. *Environmental Geochemistry and Health*, 28(5), 409–420. https://doi.org/10.1007/s10653-005-9039-4

Patel, K. S., Shrivas, K., Brandt, R., Jakubowski, N., Corns, W., & Hoffmann, P. (2005). Arsenic contamination in water, soil, sediment and rice of central India. *Environmental Geochemistry and Health*, 27(2), 131–145. https://doi.org/10.1007/s10653-005-0120-9

Podgorski, J., & Berg, M. (2020). Global threat of arsenic in groundwater. *Science*, 368, 845–850.

Podgorski, J. E., Eqani, S. A. M. A. S., Khanam, T., Ullah, R., Shen, H., & Berg, M. (2017). Extensive arsenic contamination in high-pH unconfined aquifers in the Indus Valley. *Science Advances*, 3 (8), e1700935. 10.1126/sciadv.1700935.

Purkait, B., & Mukherjee, A. (2008). Geostatistical analysis of arsenic concentration in the groundwater of Malda district of West Bengal, India. *Frontiers of Earth Science in China*, 2(3), 292–301. https://doi.org/10.1007/s11707-008-0023-z

Purkait, B., Kadam, S. S., & Das, S. K. (2008). Application of artificial neural network model to study arsenic contamination in groundwater of Malda district, Eastern India. *Journal of Environmental Informatics*, 12(2), 140–149. https://doi.org/10.3808/jei.200800132

Rahaman, S., Sinha, A. C., Pati, R., & Mukhopadhyay, D. (2013). Arsenic contamination: a potential hazard to the affected areas of West Bengal, India. *Environmental Geochemistry and Health*, 35(1), 119–132. https://doi.org/10.1007/s10653-012-9460-4

Rahman, M. M., Ng, J. C., & Naidu, R. (2009). Chronic exposure of arsenic via drinking water and its adverse health impacts on humans. *Environmental Geochemistry and Health*, 31(SUPPL. 1), 189–200. https://doi.org/10.1007/s10653-008-9235-0

Rahman, A., Mondal, N. C., & Fauzia, F. (2020). Arsenic enrichment and its natural background in groundwater in the proximity of active floodplains of Ganga River, Northern India. *Chemosphere*, 265, 129096. https://doi.org/10.1016/j.chemosphere.2020.129096

Raju, N. J. (2012). Arsenic exposure through groundwater in the middle Ganga plain in the Varanasi environs, India: A future threat. *Journal of the Geological Society of India*, 79(3), 302–314. https://doi.org/10.1007/s12594-012-0044-9

Raju, N. J. (2021). Arsenic in geo-environment: a review of sources, geochemical processes, toxicity and removal technologies. *Environmental Research*, 203, 111782. https://doi.org/10.1016/j.envres.2021.111782

Ravenscroft, P., McArthur, J. M., & Hoque, B. M. (2001). *Geochemical and Palaeohydrological Controls on Pollution of Groundwater by Arsenic*. Arsenic Exposure and Health Effects IV. Elsevier.

Ravenscroft, P., Burgess, W. G., Ahmed, K. M., Burren, M., & Perrin, J. (2005). Arsenic in groundwater of the Bengal Basin, Bangladesh: distribution, field relations, and hydrogeological setting. *Hydrogeology Journal*, 13, 727–751. https://doi.org/10.1007/s10040-003-0314-0

Saha, J. C., Dikshit, A. K., Bandyopadhyay, M., & Saha, K. C. (1999). Critical reviews in environmental science and technology a review of arsenic poisoning and its effects on human health a review of arsenic poisoning and its effects on human health. *Critical Reviews in Environmental Science and Technology*, 29(3), 37–41. https://doi.org/10.1080/10643389991259227

Saha, D., Sarangam, S. S., Dwivedi, S. N., & Bhartariya, K. G. (2010). Evaluation of hydrogeochemical processes in arsenic-contaminated alluvial aquifers in parts of Mid-Ganga Basin, Bihar, Eastern India. *Environmental Earth Sciences*, 61(4), 799–811. https://doi.org/10.1007/s12665-009-0392-y

Saha, D., Sahu, S., & Chandra, P. C. (2011). Arsenic safe alternate aquifers and their hydraulic characteristics in contaminated areas of Middle Ganga Plain, Eastern India. *Environmental Monitoring and Assessment*, 175(1–4), 331–348. https://doi.org/10.1007/s10661-010-1535-z

Sahu, S., & Saha, D. (2020). Groundwater arsenic contamination and availability of safe water for drinking in Middle Ganga Plain in India. *Arsenic Water Resources Contamination, Advances in Water Security*. https://doi.org/10.1007/978-3-030-21258-2_1

Sailo, L., & Mahanta, C. (2013). Hydrogeochemical factors affecting the mobilization of As into the groundwater of the Brahmaputra alluvial plains of Assam, Northeast India. *Environmental Sciences: Processes and Impacts*, 15(9), 1775–1782. https://doi.org/10.1039/c3em00156c

Sekhar, K. C., Chary, N. S., Kamala, C. T., Rao, J. V., Balaram, V., & Anjaneyulu, Y. (2003). Risk assessment and pathway study of arsenic in industrially contaminated sites of Hyberabad: a case study. *Environment International*, 29(5), 601–611. https://doi.org/10.1016/S0160-4120(03)00017-5

Sengupta, S., McArthur, J. M., Sarkar, A., Leng, M. J., Ravenscroft, P., Howarth, R. J., & Banerjee, D. M. (2008). Do ponds cause arsenic-pollution of groundwaterin the Bengal Basin? An answer from West Bengal. *Environmental Science & Technology*, 42(14), 5156–5164.

Shah, B. A. (2008). Role of quaternary stratigraphy on arsenic-contaminated groundwater from parts of Middle Ganga Plain, UP-Bihar, India. *Environmental Geology*, 53(7), 1553–1561. https://doi.org/10.1007/s00254-007-0766-y

Shah, B. A. (2012). Role of quaternary stratigraphy on arsenic-contaminated groundwater from parts of Barak Valley, Assam, North-East India. *Environmental Earth Sciences*, 66(8), 2491–2501. https://doi.org/10.1007/s12665-011-1472-3

Shah, B. A. (2015). Arsenic contamination in groundwater in the Middle Gangetic Plain, India: its relations to fluvial geomorphologyand quaternary stratigraphy. In: AL. Ramanathan et al. (eds.) *Safe and Sustainable Use of Arsenic-Contaminated Aquifers in the Gangetic Plain*, 33–53. Capital Publishing Company. https://doi.org/10.1007/978-3-319-16124-2_3

Shah, B.A. (2014). Arsenic in groundwater, Quaternary sediments, and suspended river sediments from the Middle Gangetic Plain, India: distribution, field relations, and geomorphological setting. *Arabian Journal of Geosciences*, 7, 3525–3636. https://doi.org/10.1007/s12517-013-1012-4

Sharma, A. K., Tjell, J. C., Sloth, J. J., & Holm, P. E. (2014). Review of arsenic contamination, exposure through water and food and low cost mitigation options for rural areas. *Applied Geochemistry*, 41, 11–33. https://doi.org/10.1016/j.apgeochem.2013.11.012

Shirke, K. D., & Pawar, N. J. (2015). Enrichment of arsenic in the Quaternary sediments from Ankaleshwar industrial area, Gujarat, India: an anthropogenic influence. *Environmental Monitoring and Assessment*, 187(9). https://doi.org/10.1007/s10661-015-4815-9

Shukla, D. P., Dubey, C. S., Singh, N. P., Tajbakhsh, M., & Chaudhry, M. (2010). Sources and controls of Arsenic contamination in groundwater of Rajnandgaon and Kanker District, Chattisgarh Central India. *Journal of Hydrology*, 395(1–2), 49–66. https://doi.org/10.1016/j.jhydrol.2010.10.011

Sidhu, S. S., Brar, J. S., Biswas, A., Banger, K., & Saroa, G. S. (2012). Arsenic contamination in soil-water-plant (rice, *Oryza sativa* L.) continuum in central and sub-mountainous Punjab, India. *Bulletin of Environmental Contamination and Toxicology*, 89(5), 1046–1050. https://doi.org/10.1007/s00128-012-0799-0

Singh, E. J. K., Gupta, A., & Singh, N. R. (2013). Groundwater quality in Imphal West district, Manipur, India, with multivariate statistical analysis of data. *Environmental Science and Pollution Research*, 20(4), 2421–2434. https://doi.org/10.1007/s11356-012-1127-2

Singh, N., Singh, R. P., Mukherjee, S., McDonald, K., & Reddy, K. J. (2014). Hydrogeological processes controlling the release of arsenic in parts of 24 Parganas district, West Bengal. *Environmental Earth Sciences*, 72(1), 111–118. https://doi.org/10.1007/s12665-013-2940-8

Singh, R., Gayen, A., Kumar, S., & Dewangan, R. (2021). Geo-spatial distribution of arsenic contamination of groundwater resources in intricate crystalline aquifer system of Central India: Arsenic toxicity manifestation and health risk assessment. *Human and Ecological Risk Assessment: An International Journal*. https://doi.org/10.1080/10807039.2020.1865787

Smedley, P. L., & Kinniburgh, D. G. (2002). A review of the source, behaviour and distribution of arsenic in natural waters. *Applied Geochemistry*, 17(5), 517–568. https://doi.org/10.1016/S0883-2927(02)00018-5

Srivastava, S., & Sharma, Y. K. (2013). Arsenic occurrence and accumulation in soil and water of eastern districts of Uttar Pradesh, India. *Environmental Monitoring and Assessment*, 185(6), 4995–5002. https://doi.org/10.1007/s10661-012-2920-6

Suman, S., Sharma, P. K., Siddique, A. B., Rahman, M. A., Kumar, R., Rahman, M. M., Bose, N., Singh, S. K., Ghosh, A., Matthews, H., & Mondal, D. (2019). Wheat is an emerging exposure route for arsenic in Bihar, India. *Sciene of the Total Environment*, 703, 134774. https://doi.org/10.1016/j.scitotenv. 2019.134774

UNICEF (2018). Guidance On The Investigation & Mitigation Of Arsenic Contamination. Arsenic Primer.

Upadhyay, M. K., Shukla, A., Yadav, P., & Srivastava, S. (2019). A review of arsenic in crops, vegetables, animals and food products. *Food Chemistry*, 276, 608–618. https://doi.org/10.1016/j.foodchem.2018.10.069

WHO (World Health Organisation). (1999). Arsenic in drinking water. Fact Sheet No. 210, Geneva(WHO).

Wu, R., Podgorski, J., Berg, M., & Polya, D. A. (2021). Geostatistical model of the spatial distrbution of arsenic in groundwaters in Gujrat State, India. *Environmental Geochemistry and Health*, 43, 2649–2664. https://doi.org/10.1007/s10653-020-00655-7

Ye, J., Rensing, C., Rosen, B. P., & Zhu, Y. G. (2012). Arsenic biomethylation by photosynthetic organisms. *Trends in Plant Science*, 17(3), 155–162. https://doi.org/10.1016/j.tplants.2011.12.003

Zhou, Q., & Xi, S. (2018). A review on arsenic carcinogenesis: epidemiology, metabolism, genotoxicity and epigenetic changes. *Regulatory Toxicology and Pharmacology*, 99. Elsevier Inc. https://doi.org/10.1016/j.yrtph.2018.09.010

4

Arsenic Contamination of Groundwater in Indo-Gangetic Plain

Samikshya Panda, Vinod Kumar Tripathi, Shrinivasa DJ., and Reema Sharma

Department of Agricultural Engineering, Institute of Agricultural Sciences, Banaras Hindu University, Varanasi, Uttar Pradesh, India

4.1 Introduction

Currently, there is a large concern about the availability of safe drinking water in many developing countries, including India. The main reason behind this is the higher concentration of different toxic and poisonous solutes coming from various natural as well as anthropogenic sources (Madhav et al. 2021). There is a need for proper knowledge regarding aquifer physical characteristics and its hydraulic parameters for safe and sustainable groundwater management. Usually, the groundwater movement along its flow paths underneath the ground surface increases the chemical concentration, hence, the groundwater suitability for different uses is predicted by the groundwater chemistry and the geological feature of an aquifer (Kortatsi 2007). Many countries are dealing with water scarcity and water quality issues due to contaminated groundwater which is also degrading the drinking water quality. During the past few years, there has been the biggest groundwater calamity in the Indo-Gangetic plain of India and has been seen due to arsenic contamination. From the report of CGWB, 2018, it has been seen that several states of India like Jharkhand, Bihar, Uttar Pradesh, and West Bengal situated along the inundated areas of the Ganga river have mostly been affected due to the elevated concentration of arsenic in groundwater in which its concentration is above the permissible standard limit of 10 μg/L. Due to the availability of fertile and nutrient-rich land and sufficient water resources in the Indo-Gangetic plain occupied by these states, there is an opportunity for large agricultural production in these areas. Hence, there is an extensive use of groundwater for many domestic and irrigation purposes in this region (Bhattacharya et al. 2011). There is an imbalance of groundwater dynamics due to the excessive pumping of shallow and deep aquifers which originated an induced leakage of arsenic from shallow aquifer to deep aquifer in this area (Muralidharan 1998).

Hydrogeochemistry of Aquatic Ecosystems, First Edition. Edited by Sughosh Madhav, Virendra Bahadur Singh, Manoj Kumar, and Sandeep Singh.

4.2 Sources and Concentration of Arsenic in Groundwater

Groundwater with elevated arsenic concentration is a common problem in different parts of the world. An approximate population of 200 million people across the globe is exposed to arsenic concentration surpassing the WHO standard limits in drinking water (Shankar et al. 2014). Basically, there are two types of source for the occurrence of arsenic in groundwater which are natural and anthropogenic sources. Some natural sources are geological formations like sedimentary and volcanic rocks, geothermal activity sources, sources from coal, and volcanic activities. Major anthropogenic sources like mining and industrial sources as well as various use of arsenic-rich fertilizers and pesticides which ultimately reaching to the river basin in terms of erosion process (Rani et al. 2022). Due to this, a large population in India is affected by this problem. From various research studies, it has been shown that wetlands can influence the arsenic level in that area in shallow soil profiles and the occurrence of arsenic deposition increases with the longer existence of wetland in that area (Zhipeng Gao et al. 2022). Increasing pH can also stimulates arsenic content in the groundwater by increasing its toxicity in that aquifer (Dey et al. 2022). For various countries, different regulatory standards of arsenic content for drinking water quality are there. In 1958, World Health Organization (WHO 2011) standardizes the permissible arsenic concentration for drinking water as 200 g/L. (i.e. 200 ppb; ppb: parts μ per billion). Again it was reduced to 50g/L in 1963, after re-evaluation. In 1993, there was a rise in health hazards due to the adverse effect of arsenic-contaminated water. Hence, the standard concentration was again reduced to 10g/l. In India, the allowable limit for arsenic concentration in drinking water was 50 ppb until 2009 (IS: 10500 2009). After 2009, Bureau of Indian Standards (BIS) has set the permissible limit for arsenic concentration in drinking water as 10 ppb and in the absence of any alternative sources, "maximum permissible limit" of arsenic in drinking water has been set to 50 ppb (IS: 10500 2012). In 1974, the US Environmental Protection Agency (USEPA) issued a standard for drinking water regulation for arsenic of 50 μg/L. This standard does not protect public health as it creates a fatal cancer risk. Therefore, the permissible standard of arsenic in drinking water was reduced to 10 μg/L by USEPA in 2001. In 1998, the European standard for arsenic permissible limit for drinking water was also 10 μg/L. In Australia, the maximum limit of arsenic in drinking water is 7 μg/L and other countries like Japan, Jordan, Syria, and Mongolia are following a standard maximum limit of 10 μg/L for health risk minimization (Smedley and Kinniburgh 2002).

4.3 Issues Related to Elevated Arsenic Concentration

In India, the subsequent common features were found in the arsenic-affected areas:

Concentration of arsenic at very low level is used in homeopathic medicines to cure various diseases. But the availability of groundwater with such a limited concentration is found at rare occasions. Majority of population is affected due to severe health hazards like anorexia, nausea, dyspepsia, altered taste, pain in abdomen, enlarged liver and spleen, and ascites. The carcinogenic effect of arsenic leads to skin, bladder, liver, and lung cancer (Ahamad et al. 2020; Madhav et al. 2020).

There is a large reduction in seed germination and length of root and shoot in crops after irrigation with arsenic-rich groundwater. Many poor farmers suffer from bad quality of water as well as their agricultural yield is affected. Due to the repeated and long-term use of arsenic-rich water on agricultural soil, the arsenic concentration is increased in this soil and yield is affected (Rani et al. 2022).

The degradation of groundwater quality standard is a global issue. Arsenic is an important contaminant of groundwater in various regions of the world. From different studies, it has been reported that more than 70 countries are concerned about elevated arsenic concentration in groundwater and soil. The major countries dealing with the concern of arsenic contamination are East Asia and South-East Asia. In India, the worst-affected regions due to the increase in arsenic level of groundwater are the Gangetic plains of West Bengal, Uttar Pradesh, Bihar, and Jharkhand (Jha and Tripathi 2021).

In 2003, Rahman and Ravenscrof (2003) estimated that the arsenic level in paddy field is increased up to 46 mg/g dry weight after irrigating with arsenic-rich water from tube wells. From a large number of research studies, it has been concluded that there is a positive correlation between the arsenic contamination in agricultural farm soil and arsenic-rich irrigation water. Again in 2018, Bhattacharya reported that arsenic-contaminated groundwater used as irrigation can enhance the level of arsenic in the soil of agricultural farm up to fivefold in comparison to normal soil.

According to Elizabeth and Gillispie, irrigation with arsenic-rich groundwater is the major cause of soil pollution and yield reduction in the concerned area. Further adding to this study, normally arsenic appears in soils having background concentrations of 6–7 mg/kg. But recurrent application of arsenic-contaminated groundwater for irrigation purposes may increase its concentration up to >10 mg/kg which may create toxicity over that area as well. Arsenic doses are also found in the food grains which are being irrigated with arsenic-rich water and ultimately its creating health hazards in human who are consuming this food. Both the soil and food grains are becoming toxic by using arsenic-rich irrigation water. Agricultural soil and food crop toxification from arsenic-contaminated groundwater is a global issue in geographically dispersed countries like India. Numerous factors are responsible for increasing the arsenic concentration in the concerned areas which are the sources of the geological features, arsenic-rich chemical fertilizer residues, and industrial effluents discharging into the surface water sources, and this ultimately leads to groundwater contamination (Ahamad et al. 2020; Madhav et al. 2020).

In 2017, according to Bhatacharya, it was reported that arsenic could influence food chain also along with the irrigation water. Vegetables and arable crops grown by using arsenic-affected groundwater can increase the arsenic level in those plants and ultimately in the food chain. Mainly the tuberous and fruity vegetables such as potato, brinjal, arum, amaranth, radish, lady's finger, and cauliflower are reported to be arsenic-rich than other leafy vegetables. Also oil seeds, pulses, and high-yielding rice varieties can accumulate more arsenic than other types of food crops. This leads to arsenic toxicity in the plants and creates adverse effect on human health after consuming. Arsenic not only affects drinking water, but also impact on food chain in the concerned areas. The economy of the Indo-Gangetic region largely depends on agriculture because of the alluvial soil condition. But due to the arsenic toxicity in these areas, people are being affected both physically and economically. People having poor nutrition are a common issue in India. Numerous efforts are made to provide safe drinking water using treatment units and alternate measures. But these efforts are only limited to the drinking water supply. Arsenic enrichment in food chain is another big issue in such areas and it is very difficult to measure the impact of arsenic toxicity on the crops, and ultimately, the dietary impact of arsenic on human health after consuming it. For measuring the impact of arsenic-contaminated groundwater on the food chain, more elaborated research studies are required so that a proper assessment can be done in these regions.

Arsenic treatment plant waste disposal is another biggest issue. Arsenic-contaminated sludge removed from treatment plants has very high leaching potential in the soil which will ultimately pollute the groundwater if left untreated on the soil. This sludge may contain 5–7 kg of arsenic per cubic meter (Prasanta Mandal et al. 2016). It is very risky to discharge them into the soil directly because of the toxic effect. Hence for safe disposal of solid arsenic sludge, new technology and alternate uses of the sludge in making bricks and concrete from it can be followed. Otherwise there will be no value of any water treatment plant in that area.

4.4 Removal Techniques of Arsenic from Groundwater

For solving the issue of As contamination in groundwater, there is an immediate need of creating awareness among the affected population and a proper watershed planning with cost-effective measures should be implemented. For As removal from groundwater, various techniques have been originated and developed in various parts of the world. Some of these technologies are discussed below.

Electrocoagulation is basically used for the treatment of wastewater and it is known to be an effective technology to treat arsenic-contaminated water to reduce the As contamination up to a safe drinking water quality standards (Thakur and Mondal 2016). In 2019, Holmes et al. (2019) introduced acement-based filter medium (CBFM) for reducing the heavy metals along with As from contaminated groundwater. This technique was used as a readily available method for treating As-affected groundwater.

In 2000, Wegelin developed a technique called SORAS, which is Solar Oxidation and removal of arsenic. This was a very simple method in which contaminated water was exposed to sunlight by storing them in PET bottles to minimize the concentration of arsenic in the drinking water.

In 2001, Tahura et al. (2001) introduced a household device for arsenic removal from contaminated water. This unit was known as the Bucket Treatment Unit (BTU) which was based on the principle of coagulation, co-precipitation, and adsorption processes.

Solar oxidation is a very widely used simple method in which the arsenic-rich water is exposed to solar radiation for oxidizing the arsenic present in that water and this ultimately decreases the arsenic concentration of that water (Mukherjee et al. 2007).

4.4.1 Biological Oxidation

Some aerobic micro-organisms like *Gallionella ferruginea* and Leptothrixochracea catalyze the biotic oxidation of iron, and iron creates favorable environment for arsenic adsorption which ultimately reduces the arsenic level in groundwater (Pallier et al. 2010).

Nanofiltration membrane process is also an effective technique for arsenic removal from groundwater (Song et al. 2015). Controlled artificial groundwater recharge with less affected water can also decrease the arsenic level in groundwater. Other arsenic removal methods, including reverse osmosis, ion exchange technology, distillation, and membrane filtration are also effective technologies for reducing the groundwater arsenic concentration.

4.5 Arsenic Treatment Plant Waste Disposal

In 2016, Prasanta Mandal suggested that arsenic sludge disposal from the water treatment plant can be done by using the following methods: (i) on-site sanitation pits/landfill; (ii) mixing the sludge with concrete in a proper ratio; (iii) mixing the solid arsenic sludge with clay burnt bricks. Disposal through landfill is not a very efficient way of sludge treatment because of the groundwater contamination. Another two methods for alternative use of solid arsenic sludge can be better options to follow in the concerned areas. The developing world is facing the problem of arsenic-contaminated water. Filtration of the same water generates the toxic sludge. There is a need to develop the methodology for effective, reliable, and environmentally safe disposal of arsenic-contaminated sludge. Over the last few years, arsenic sludge technologies have improved significantly.

4.6 Challenges in Arsenic-Affected Areas

Despite the development of various arsenic removal techniques from groundwater using advance principles, there is still a lacuna in complete arsenic removal in practical basis at the field level. A large number of plants have adopted different technological interventions to deal with this situation. But a few plants are showing satisfactory performance in arsenic removal at the field level in a sustainable manner. Mostly the challenges in efficient removal of arsenic without affecting the ecosystem are removal efficiency of the filters media, maintenance, replacement of used filters, running cost of the plant, etc. Following challenges can be addressed regarding the concerned areas:

i) Field database management and data sampling from the field.
ii) Fund allocation for the development of arsenic removal treatment plant.
iii) Operation and maintenance of devices involved.
iv) Lack of skill and awareness among the local community.
v) Complete assessment of arsenic concentration, including food chain, soil, and groundwater.
vi) Disposal of arsenic sludge removed from the treatment plant in the concerned areas.
vii) Gap between the authorities and local people.
viii) Lack of coordination in the progress work.
ix) Coordination among state and central government for undertaking the work.
x) Lack of medical facility to overcome the adverse dietary effect of arsenic in food and drinking water.

4.7 Indian Scenario Regarding Measures Taken to Mitigate the Problem of As Availability in Groundwater

Arsenic was first detected in the groundwater of West Bengal in 1978 and after that many states of India particularly those situated near the Indo-Gangetic plain were reported as higher arsenic concentration that exceeds the drinking water criteria set by BIS. Hence, many state and central government initiatives were introduced to mitigate the arsenic problem in groundwater of these areas through proper planning and application measures schemes and awareness among the people. Various programs were initiated for the identification of contaminated source and construction of treatment plants near it. Also alternative drinking water supply schemes were established. Many state-wise action plans were taken to tackle the situation. Some of these state-wise initiatives are discussed below:

Assam: There are 2571 number of arsenic-contaminated habitations reported, and from these 2212, number of habitations were treated with remedial measures. Under state government schemes, 115 habitations were provided with Piped water supply schemes (PWSS) and in 2097 habitations, new dug wells were constructed on the shallow aquifer as these aquifers remain free from arsenic contamination. The state government is planning to provide Piped water supply and shallow dug wells in the remaining 359 arsenic-contaminated habitations.

Bihar: The Bihar state government has taken some short and long-term remedial measures in affected villages and districts due to arsenic contamination. Under short-term measures, dug wells were converted into sanitary wells, tube wells were replaced by hand pumps having more than 100 m depth, new solar pump-operated sanitary wells were constructed and hand pumps along with arsenic filtration units were established. Under long-term remedial measures, deeper hand pumps with proper sealing were constructed; mini water supply schemes and rainwater harvesting systems along with treatment units were established; multivillage piped water supply schemes based on surface water were reserved.

Chhattisgarh: The state government has taken many initiatives to irradicate the arsenic contamination in many districts and villages. A grouped water supply scheme is under implementation stage in 18 villages to overcome this issue. This project will be completed within three years under the collaboration of PHED (Public Health Engineering Department) and Government of Chhattisgarh.

Jharkhand: The state government has taken many initiative measures to tackle the arsenic problem in many affected areas. It has introduced "State Water Quality Task Force" to treat the increasing arsenic level in groundwater. All the habitations having high arsenic concentration were monitored and regularly tested under "Uniform Rural Drinking Water Quality Monitoring Protocol." The application and planning of corrective measures were set on the basis of permissible arsenic limit in drinking water as 10 ppb. Village-wise security and safety plan was executed under the public involvement along with Gram Panchayat and "Village Water and Sanitation Committee." A provision of household water quality testing facility was provided at doorstep for testing arsenic concentration using test kit.

Punjab: An action plan was developed for the construction of 1884 Arsenic and metal treatment units for wastewater, which was initially developed by IIT Madras. This project was undertaken by Department of Water Supply and Sanitation and Government of Punjab to set these treatment units in 1512 affected villages of Punjab.

West Bengal: In 1978, arsenic was first detected in West Bengal. After that the government has taken many short- and long-term remedial measures which include installation of deeper hand pumps and construction of community water purification plants for wastewater treatment based on chemical precipitation, adsorption, sedimentation, ion exchange method, and in situ treatment. Again in 2005–2006, the state government has installed 12 number of surface water-based pipe water supply schemes and 338 number of arsenic treatment plants in groundwater. All these projects were implemented under collaboration of Public Health Engineering Department (PHED) and Government of West Bengal during the last few years (Bhattacharya 2017). In West Bengal, the surface water sources are free from arsenic contamination. Hence, the surface water is provided through organized piped water supply system to many places in West Bengal from surface water sources with some minor additional surface water treatment where assured surface water sources are available.

Although a number of initiatives were undertaken by different state and central government in these states, the complete eradication of water crises due to arsenic elevation in these regions is inadequate. People residing in these areas are bound to live in such unhealthy condition bearing the drinking water crises and they have no alternative option to solve this crisis also. Many initiatives were proposed but are not yet functional in many areas in these states. Hence, many people suffer from numerous health crises by consuming the poisonous water.

The framework for solving the water crises issues in the arsenic-affected areas can be done by three major activities: (i) New research and technological interventions by ensuring the efficient and sustainable solutions for groundwater treatment and arsenic-contaminated sludge disposal, (ii) Providing arsenic-free potable water to local people, and (iii) Skill development capacity building in local community to deal with such situation. Under research and development activities, there is a necessity to formulate the required initial database to predict the causes and its solution by studying the mobilization and dissolution process of arsenic along with the groundwater aquifer characteristics to develop the in situ and ex situ methods for disinfecting the aquifers from arsenic. The major parameters to be considered are the technology effectiveness, cost of interventions, ecological sustainability, alternate water management options, and strategies to fulfil the water demand in arsenic-polluted vulnerable areas. Certain technologies should be developed to manage the arsenic influence in food chain due to the excess usage of arsenic-rich groundwater in terms of irrigation during the critical period. Government organizations and NGOs working on these areas should efficiently involve in the field level to demonstrate the procedure which needs to be followed effectively for ensuring the local people a good eco-friendly environment to manage their livelihoods. Research and skill development programs should be undertaken in these areas to educate and empower the local communities regarding the impact and challenges of arsenic in water and food chain (CGWB 2011).

The main aim of any treatment action in the arsenic-affected area is to reduce the arsenic concentration in existing water source and to prevent them from further arsenic accumulation in drinking water, soil, and food crops. Following strategies should be maintained to achieve this goal:

i) Substituting low-arsenic water such as rainwater or shallow water in place of arsenic-rich groundwater in daily life of people. Water having less arsenic concentration can be used for cooking and drinking with little treatment, whereas arsenic-rich water can be an option for bathing, washing clothes, etc.
ii) Installing low-cost potable water treatment unit, where major issues regarding water quality degradation are not reported.
iii) Combining low-arsenic water with high-arsenic water to find a tolerable arsenic concentration in the water.
iv) To check the further pollution of water source, strict regulatory acts should be followed by local people and industries.
v) People education and awareness are big factors to understand the issue and deal with it. Hence, skill development and awareness programs should be done in the affected areas.
vi) Long-term action plan should be made to deal with the contaminated areas.
vii) Local communities should be engaged for the successful intervention of the treatment plan.
viii) Local farmers should be aware of the arsenic enrichment of soil and food crops, so that use of arsenic-rich groundwater in irrigation can be limited and in place of that, supplemental irrigation from rainwater can be done.
ix) Safe disposal of arsenic-rich sludge from the treatment unit can be done by making their alternate use in the concrete making and clay bricks making industry so that it will not pollute the soil further.
x) Socioeconomic condition of the residing people of that concerned areas depends upon the productivity of the area. Hence, a sustainable planning for proper management of drinking water, soil condition, and people health is necessary to eradicate the issue.

4.8 Conclusion

Groundwater contaminated through arsenic accumulation is a global issue. Arsenic is a significant chemical contaminant of groundwater which is the major natural source of drinking water as well as for irrigation purposes. A large number of people are affected due to this issue in different countries including India. There is a need to address this issue for better management of water source in such regions. Growing population and rapid industrialization are creating a large pressure on freshwater sources. In this recent era, all the agricultural activities are becoming water intensive and these effluents are ultimately discharging into the rivers and ponds, which are the surface water sources. Again it creates metal toxicity in that water leading to freshwater crises in that area. People living in the lower reach of Ganga basin are chronically affected and have been exposed to contaminated drinking water. With every new survey, more people are suffering from arsenic-related crises in terms of numerous known and unknown diseases by risking their life. Numerous factors are responsible for increasing the arsenic concentration in the concerned areas which are the sources of the geological features, arsenic-rich chemical fertilizer residues, and industrial effluents discharging into the surface water sources and this ultimately leads to groundwater contamination. This chapter basically focuses on the elevated arsenic contamination of groundwater in the northern plain situated near the Ganga basin. A number of scientific methods for arsenic concentration reduction in groundwater have also been discussed here. To overcome this challenging situation, various initiatives were taken by different state governments. This chapter basically reviews various sources, the standard of water quality based on arsenic concentration, and removal techniques of arsenic from groundwater along with government initiatives taken for overcoming groundwater arsenic contamination in the Indo-Gangetic plain. The increase in arsenic concentration is a global issue that needs to be diagnosed in a proper way; and although complete prevention is not possible, different measures and regulatory act should be implemented and followed strictly, so that it will create a healthy environment and maximum water sources will remain usable. A number of projects and work plans have been developed to diagnose the prominent arsenic areas in the deltaic region of the Ganga basin and these initiatives will be successful only when the people will realize this issue and are aware of themselves to protect the water resources rather than exploiting it in many anthropogenic ways. Proper planning and management of water resources will be accomplished by engaging the local people along with NGOs and Government officials to eradicate the water crises originated due to arsenic contamination. Maximum efforts and immediate attention should be given to overcome this issue. Arsenic is becoming an unwanted global crisis and affecting the common people and aquatic life in a higher rate. It needs urgent supervision and precaution for rehabilitation and development of water resources. Numerous arsenic removal techniques and devices have been developed currently on the basis of different scientific working principles and have been installed in fields to predict the cause and amount of arsenic concentration in groundwater. But only a few of them are in working condition and providing satisfactory results; however, a large number of in situ technological devices are still unfunctional due to the lack of skill and knowledge regarding the technology. The operation and maintenance are important factors in the development of any in situ treatment unit. Involvement of the people from society in the operation and maintenance process and making the society responsible with the required knowledge can solve many problems accompanied by the water scarcity problems in the arsenic-affected areas. Finally, it is concluded that numerous works have been done to analyze and diagnose the arsenic-affected areas to save the people from the harmful effects of Arsenic. Much more developmental works are needed to be accomplished before it goes beyond our control. The residing society is the subsequent sufferer of any natural calamity and the ultimate beneficiary of any initiatives as well. Creating awareness and following the precautionary measures against this harmful situation are the ultimate solutions to protect the affected communities.

References

Ahamad A., Raju N.J., Madhav S., Khan, A.H. (2020). Trace elements contamination in groundwater and associated human health risk in the industrial region of southern Sonbhadra, Uttar Pradesh, India. *Environmental Geochemistry and Health*, 42(10): 3373–3391.

Bhattacharya A.K. (2017). Arsenic contamination in Indian groundwater. *Journal of the National Institute of Public Health India* 45(2):18–36.

Bhattacharya P., Mukherjee A., Mukherjee A.B. (2011). *Arsenic in groundwater of India:Encyclopedia of Environmental Health*, Elsevier Inc., pp.150–164.

BIS (2009). *Drinking Water-Specification (IS-10500:2009)*, 24 p.

BIS (2012). *Drinking Water- Specification (second revision). IS-10500:2012.* 11 p.

CGWB, (2011). Dynamic groundwater Resources of India (As on 31st March, 2009). 225p.

Dey A., Mridha D., Joardar M., Das A., Chowdhury N.R., Roychowdhury T. (2022). Distribution, prevalence and health risk assessment of fluoride and arsenic in groundwater from lower Gangetic plain in West Bengal, India. *Groundwater for Sustainable Development*, 16:100722. https://doi.org/10.1016/j.gsd.2021.100722.

Gao Z., Guo H., Li S., Wang J., Ye H., Han S., Cao W. (2022). Remote sensing of wetland evolution in predicting shallow groundwater arsenic distribution in two typical inland basins. *Science of the Total Environment*, 806:150496. https://doi.org/10.1016/j.scitotenv.2021.150496.

Holmes R.R., Hart M.L., Kevern J.T. (2019). Removal of arsenic from synthetic groundwater using sulfur-enhanced cement-based filter media. *Journal of Hazardous, Toxic, and Radioactive Waste*, 23(3):04019006.

Jha P., Tripathi P. (2021). Arsenic and fluoride contamination in groundwater: A review of global scenarios with special reference to India. *Groundwater for Sustainable Development*, 13:100576. /https://doi.org/10.1016/j.gsd.2021.100576.

Kortatsi B.K. (2007). Hydro chemical framework of groundwater in the Ankobra Basin, Ghana. *Aquatic Geochemistry*, 13(1):41–74.

Madhav S., Kumar A., Kushawaha J., Ahamad A., Singh P., & Dwivedi S.B. (2020). Geochemical assessment of groundwater quality in Keonjhar City, Odisha, India. *Sustainable Water Resources Management*, 6(3):1–11.

Madhav S., Raju N.J., Ahamad A., Singh A.K., Ram P., & Gossel W. (2021). Hydrogeochemical assessment of groundwater quality and associated potential human health risk in Bhadohi environs, India. *Environmental Earth Sciences*, 80(17):1–14.

Mandal P., Debbarma S.R., Saha A., Ruj B. (2016). Disposal problem of arsenic sludge generated during arsenic removal from drinking water. *Procedia Environmental Sciences*, 35(2016):943–949.

Mukherjee P., Chatterjee D., Jana J., Maity P.B., Goswami A., Saha H., Sen M., Nath B., Shome D., Sarkar M.J., Bagchi D.(2007). Household water treatment option: removal of arsenic in presence of natural Fe-containing groundwater by solar oxidation. In *Trace Metals and other Contaminants in the Environment*. Bhattacharya, P., Mukherjee, A.B., Bundschuh, J., Zevenhoven, R. and Loeppert, R.H. (eds.) 9, pp. 603–622. Elsevier B.V.

Muralidharan D. (1998). Protection of deep aquifers from arsenic contamination in Bengal Basin. *Current Science*, 75(4):351–353.

Pallier V., Feuillade-Cathalifaud G., Serpaud B., Bollinger J.C. (2010). Effect of organic matter on arsenic removal during coagulation/flocculation treatment. *Journal of Colloid and Interface Science*, 342(1):26–32.

Rahman A., Ravenscroft P. (2003). Groundwater resources and development in Bangladesh. In *Options for Arsenic Removal from Groundwater*. Firoz Mallick, A.S.M. and Das, H.K. (eds.). The University Press Ltd.

Rani L., Srivastav A.L., Kaushal J., Grewal A.S., Madhav S. (2022). Heavy metal contamination in the river ecosystem. In *Ecological Significance of River Ecosystems*, Madhav, S., Kanhaiya, S., Srivastav, A., Singh, V., and Singh, P. (eds.), pp. 37–50. Elsevier.

Shankar S., Shankar U., Shikha (2014) Arsenic contamination of groundwater: A review of sources, prevalence, health risks, and strategies for mitigation. *The Scientific World Journal.* https://doi.org/10.1155/2014/304524

Smedley P.L., Kinniburgh D.G. (2002). A review of the source, behaviour and distribution of arsenic in natural waters. *Applied Geochemistry*, 17(5):517–568.

Song J., Jia S.Y., Ren H.T., Wu S.H., Han X. (2015). Application of a high-surface-area schwertmannite in the removal of arsenate and arsenite. *International journal of Environmental Science and Technology*, 12(5):1559–1568.

Tahura S., Shahidullah S.M., Rahman T., Milton A.S. (2001). Evaluation of An Arsenic Removal Household Device: Bucket Treatment Unit(BTU). In *Technologies for Arsenic Removal from Drinking Water*, Ahmed, M.F., Ashraf Ali, M., and Adeel, Z. (eds.), p. 158–170. Bangladesh University of Engineering and Technology and The United Nations University.

Thakur L.S., Mondal, P. (2016). Simultaneous arsenic and fluoride removal from synthetic and real groundwater by electrocoagulation process: *Journal of Environmental Management*, 190:102–112.

WHO (2011). *Guidelines for Drinking Water Quality: Recommendation edn, vol 4.* World Health Organization, Geneva.

5

Soil–Water Interactions and Arsenic Enrichment in Groundwater

Sonam Taneja[1], *Shivani Yadav*[1], *Harsh Pipil*[1], *Oznur Karaca*[2], *and A. K. Haritash*[1]

[1] *Department of Environmental Engineering, Delhi Technological University, Delhi, India*
[2] *Department of Geological Engineering, Canakkale Onsekiz Mart University, Canakkale, Turkey*

5.1 Introduction

Groundwater constitutes 97% of the total freshwater and is a major source of drinking in many parts of the world. Around 1.5 billion people worldwide depend on groundwater for drinking, domestic activities, agriculture, and industrial usage (Akhtar and Rai 2019). As groundwater flows through the pore spaces of rock/soils, cracks, joints, and fractures, it interacts with the components of these geological formations, for example, sediments, and metal-bearing minerals and rocks which influences the physicochemical properties of water and may hamper the quality of groundwater. Surface water that seeps into the soil can also introduce contaminants that can deteriorate groundwater quality.

Arsenic (As) is the 20th most abundant, naturally occurring element on the Earth. It is a metalloid which means it exhibits properties of metals as well as non-metals. It is placed in group (V-A) in the periodic table, with phosphorus, nitrogen, bismuth, and antimony due to its nonmetallic properties. However, the physical properties of arsenic resemble that of metal as it is brittle in nature and gray or tin white in color; thus, it is referred to as a metalloid (Chen 2012). Like other metalloids, As also has semiconductor properties (Wenzel 2013). It has wide applications in pesticide manufacturing, agricultural additives, pharmaceuticals, and wood preservation. It is also used in mining, metallurgical, glass-making, and semiconductor industries (Masuda 2018).

Arsenic is a heavy metalloid with a density of 5.7 g/cm, atomic number 33, and atomic mass 74.9216 g/mol. Like many heavy metals, it is also considered toxic as it persists in the environment for long durations and cannot be biodegraded (Wenzel 2013). Arsenic mainly originates from soil and sediments from where they can enter the atmosphere and hydrosphere through various pathways such as runoff, dust, and interactions between soil and sub-surface waters, resulting in elevated concentration of As in groundwater. As a heavy metalloid, As cannot be degraded but can only transform into different phases. For

Hydrogeochemistry of Aquatic Ecosystems, First Edition. Edited by Sughosh Madhav, Virendra Bahadur Singh, Manoj Kumar, and Sandeep Singh.

example, As present in the atmosphere gets settled onto the sediments through wet or dry deposition, whereas As present in the liquid and solid phase interacts through adsorption and desorption. Arsenic in an aqueous medium is toxic as it can enter in drinking water and cause serious adverse health effects. For example, arsenic contaminations in drinking groundwater led to severe mass poisoning in West Bengal and Bangladesh (Chowdhury et al. 1997). This incident attracted global attention to the threats of arsenic to public health (Herath et al. 2016).

To control the presence of excessive arsenic concentration in drinking water, WHO reduced the permissible levels of arsenic from 50 to –10 μg/L. However, the maximum concentration of As in drinking water in some countries is still 50 μg/L (WHO 2001). Thus, the contamination of groundwater by arsenic is a serious concern which makes it necessary to identify the source and investigate geochemical processes that govern the dissolution and mobilization of As from sediments to groundwater (Herath et al. 2016).

5.2 Sources and Occurrence of Arsenic in Soil, Sediments, and Groundwater

Arsenic occurs as a trace component since more than 99% of As comes from rocks and minerals. Arsenic is not present as a free element and exists as compounds with other elements to form arsenides, arsenites, oxides, silicates, etc. These compounds are derived from the interactions of groundwater with sediment, rocks, and minerals. There are only three primary ores of arsenic, namely realgar (As_4S_4), orpiment (As_2S_3), and arsenopyrite (FeAsS), even though it is the main component of more than two hundred minerals (Polya and Middleton 2017). Other sources are pyrite (FeS_2) and other unspecified sulfide minerals like arsenolite (As_2O_3), cobaltite (CoAsS), olivenite (Cu_2OHAsO_4), enargite (Cu_3AsS_4), proustite (Ag_3AsS_3), and tennantite ($Cu_6[Cu_4(Fe, Zn)_2]As_4S_3$), as well as secondary As-bearing minerals (Smedley and Kinniburgh 2002). Sulfide deposits have higher concentrations of arsenic, while terrestrial arsenic concentration is around 5 mg/kg. Some arsenic-rich minerals are sedimentary iron deposits and iron and manganese ores. Phosphate-rich deposits also contain arsenic up to 2900 mg/kg (Masuda 2018).

Although arsenic is geogenic in nature, it enters the environment from natural and anthropogenic sources (Garelick et al. 2009). Weathering of rocks and minerals is the major source of arsenic in sediments and groundwater. As-sulfide ores release a mobile form of As, introduced in either air as dust or water through deposition (Polya and Middleton 2017). Other important natural sources are volcanic eruptions, forest fires and rock erosion, weathering, volcanic eruption, rock erosion, and forests (Chen 2012).

The use of arsenic for various anthropogenic activities has also increased the level of arsenic contamination in groundwater. Use of arsenic as wood preservatives for protecting timber or as pesticides and insecticides to kill termites releases toxic forms of arsenic into groundwater (Garelick et al. 2009; Chen 2012). Other major sources are mining and smelting of non-ferrous metals (Stazi et al. 2015; Polya and Middleton 2017). Processes like mining and smelting introduce arsenic into groundwater through metal leaching and atmospheric deposition of As dust (Wenzel 2013). Arsenicals like trivalent and pentavalent oxides and arsenic acid are used in the manufacture of electronics such as transistors and semiconductors (Nriagu et al. 2007). Approximately 24,000 tonnes of arsenic are emitted into the environment due to these anthropogenic activities (WHO 2001).

Excessive extraction of groundwater and burning of coal are the major sources of As-contaminated groundwater in North-East India (Shah 2015). Burning of coal at high temperatures emits As in atmosphere through volatilization of As_4O_6. The emitted As ultimately gets deposited in water and soil. Although there are many anthropogenic sources of arsenic contamination, the extent of contaminating is still less compared to the natural sources (Shankar et al. 2014). This makes it important to understand the global distribution of arsenic-rich rocks and minerals to identify their source and also their chemistry with the sub-surface water.

5.3 Global Distribution of Arsenic in Groundwater

The uneven distribution of arsenic worldwide is assumed to be because of two reasons. The first hypothesis is the natural occurrence of arsenic ores and As-bearing minerals in the alluvial sediments of the Ganga and Brahmaputra River. South Asian regions, including China, India, Bangladesh, Vietnam, and Nepal, are significantly affected by arsenic pollution (Table 5.1) (Ravenscroft et al. 2009). Among these, West Bengal (India), Bihar (India),

Table 5.1 Global distribution of Arsenic in groundwater.

Country	Location	As concentration in groundwater (µg/L)	Major sources	References
China	Mongolia	53	Presence of dissolved Fe and Mn creates reducing conditions for As-bearing minerals	Smedley et al. (2003)
	Taiwan	>10	Reductive dissolution from Fe-oxyhydroxides in deep aquifers	Liu and Wu (2019)
PR China	Hetao basin	130	Redox conditions regulate As in GW	Feng et al. (2022)
Pakistan	Sindh Province	100–2580	Reductive dissolution of As minerals	Brahman et al. (2013)
Bangladesh	Bangladesh	50–3200	Natural release of As from river sediments under strong reducing conditions	Chowdhury et al. (2000)
India	West Bengal	10–50	Natural release of As from river sediments under strong reducing conditions	Chowdhury et al. (2000)
	Bihar	50–1000	Oxidation of arsenic-rich iron sulfide from excessive withdrawal of water from wells	Chakraborti et al. (2003)

(*Continued*)

Table 5.1 (Continued)

Country	Location	As concentration in groundwater (μg/L)	Major sources	References
	Assam	194–657	Shallow depth wells on Holocene Newer Alluvium aquifers rich in As	Shah (2015)
	Chhattisgarh	50	Reductive dissolution of As-bearing minerals	Acharyya et al. (2005)
	Manipur	798–986	Shallow depth wells on Holocene Newer Alluvium aquifers rich in As	Shah (2015)
Cambodia	Mekong Delta	>50	Major source of As in aquifers are shallow sediments	Benner et al. (2008)
Nepal	Nepal	>50	Shallow depth wells on Holocene Newer Alluvium aquifers rich in As	Thakur et al. (2011)
Argentina	Cordoba Province	10–1200	Positive correlation between As and F concentrations was observed	Litter et al. (2019)
Mexico	Northwestern Mexico and Arizona	>50	Natural source from As-bearing minerals into the aquifers	Camacho et al. (2011)
Canada	Western Québec	>10	Oxidation of As-bearing sulfides	Bondu et al. (2017)
Hungary	Pannonian Basin	0.5–240	Microbially mediated Fe reduction was observed	Rowland et al. (2011)
South Africa	Namaqualand	500	Mineralization of rocks. Acid mine drainage from coal and gold mining also contribute to GW As enrichment	Abiye and Bhattacharya (2019)

and Bangladesh are the major affected areas along the Gangetic plain due to natural contamination (Chowdhury et al. 1997; Mazumder and Dasgupta 2011; Polya and Middleton 2017). Arsenic originating from geological sources is mobilized in groundwater through geochemical processes such as weathering, volcanic eruptions, oxidation, reduction, adsorption, precipitation, and methylation (Smedley and Kinniburgh 2002). Another hypothesis suggests the role of anthropogenic activities such as groundwater extraction causing arsenic contamination. It explains that excessive groundwater extraction causes oxidation of pyrites and releases toxic trivalent arsenic in groundwater, discussed in detail later.

The majority of arsenic-rich minerals are found along the Indo-Gangetic belt. South and South-East Asia have significant exposure to arsenic-contaminated groundwater compared to other regions. Bangladesh has shown maximum chronic toxicity as 25–40 million people

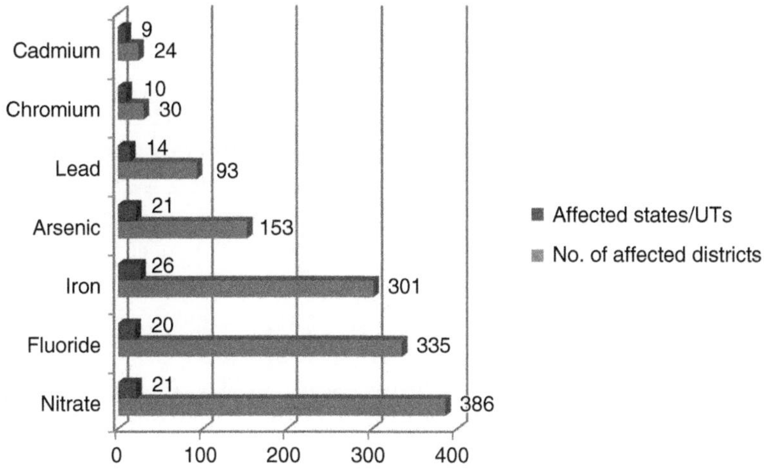

Figure 5.1 Number of affected states and districts with groundwater contaminants (metals) in India.

have been exposed to As concentrations above the permissible limit (WHO 2001). Other hotspots of arsenic have been identified in Asian countries like China, Taiwan, Cambodia, and Vietnam (Mukherjee et al. 2006; Guo et al. 2014). Around 60% of the provinces in China have shown high As toxicity in groundwater (Guo et al. 2014). North American countries, Mexico, Canada, and the United States have also demonstrated arsenic contamination in groundwater (Welch et al. 2000; Camacho et al. 2011).

In India, the alluvial plains of Indo-Gangetic basin are exposed to arsenic contamination, including states like West Bengal, Bihar, Chhattisgarh, Uttar Pradesh, and Assam (Chowdhury et al. 1997; Acharyya et al. 2005; Mukherjee et al. 2006; Kumar et al. 2021). Groundwater contamination due to various contaminants has affected many states in India (Figure 5.1). Arsenic contamination in drinking water was first reported in West Bengal when people complained about skin lesions. Several studies in the region showed that more than eight districts had arsenic beyond the permissible limit. Later, Bihar, Jharkhand, and Chhattisgarh also had high levels of arsenic as people were dependent on contaminated tube-well water for drinking. Eighteen out of thirty-eight districts of Bihar have been reported contamination of groundwater with arsenic (Kumar et al. 2021). North-East India is also exposed to arsenic contamination in states like Assam and Manipur, where most people depend on groundwater for drinking and agricultural activities (Shah 2015).

5.4 Chemical Speciation of Arsenic

Chemically, As is mainly present in four oxidation states, i.e. −3, 0, +3, and +5. Arsenic compounds can exist as organic, inorganic, or arsine gas. Arsine gas (AsH_3) is a highly toxic gas formed when arsenic reacts with acid. Organic compounds are mainly methylarsonic acid (MMAA) or dimethylarsinic acid (DMAA). Inorganic arsenic compounds like arsenites (III) and arsenates (V) are found more abundant in soil and water than organic compounds (Chowdhury et al. 1997). Arsenate (V) is a nontoxic, more stable form than arsenite (III) in aerobic conditions, whereas As(III) is a toxic, soluble, and mobile form in anaerobic conditions. As(V) has more affinity to adsorb on clay particles, organic matter, and oxides of iron and manganese. Oxidation states of As govern its mobility and toxicity in groundwater. The most critical factors controlling speciation are Redox potential (Eh) and pH. Arsenite and arsenate are usually most abundant inorganic As species in aqueous environments. At different pH, As(III) and As(V) can change their forms to H_3AsO_3, $H_2AsO_3^-$, $HAsO_3^{2-}$, and AsO_3^{3-} and H_3AsO_4, $H_2AsO_4^-$, $HAsO_4^{2-}$, and AsO_4^{3-}, respectively. Under oxidizing and slightly acidic (pH < 6.9) conditions, arsenate (V) is dominant.

Depending on the pH range, arsenate can hydrolyze to H_3AsO_4 (pH < 2.0) or to $H_2AsO_4^-$ and $HAsO_4^{2-}$ (pH 2–11). However, when pH is less than 9.2, reducing conditions persist which allow arsenite (III) species to predominate (Pagni 2006). Under these conditions, arsenious acid H_3AsO_3 reduces to $H_2AsO_3^-$, whereas, if the pH value exceeds 12, it occurs in the form of $HAsO_3^{2-}$ (Buyuktuncel et al. 1997). Apart from pH and Eh, the mobility of these species is regulated by organic carbon, clay particles, and oxides and hydroxides of iron, aluminum, and manganese. Oxides of Fe, Al, and Mn act as sinks for As in sediments because of their strong affinity to adsorb on them (Ghrefat et al. 2016). The extent of binding and adsorption depends on speciation. The Eh-pH diagram (Figure 5.2) shows the exact arsenic speciation and oxidation states at a particular pH and redox potential.

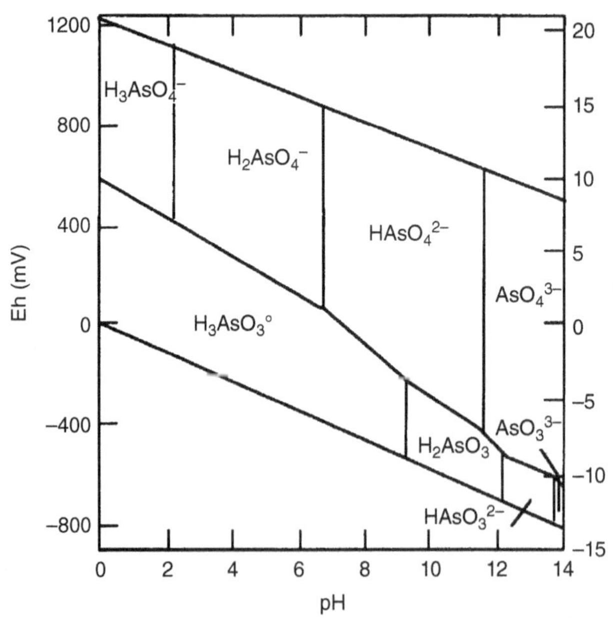

Figure 5.2 Distribution of arsenic into arsenite (As(III)) and arsenate (As(IV)) against pH in aqueous medium at 25 °C.

5.5 Soil–Water Interactions

In most As-contaminated areas, a high concentration of As in groundwater is not necessarily linked to sites with a high concentration of As in the minerals. This is due to soil–water interactions that cause geochemical triggers, resulting in groundwater's arsenic enrichment. Sorption of arsenic species on soil and sediments governs the mobilization of arsenic in the aqueous medium. In aqueous conditions, arsenic can mobilize under both oxidizing and reducing environment (Smedley and Kinniburgh 2002). Metal oxides, especially Fe, Al, and Mn, act as major binding sites for arsenic species in sediments. Some of the geochemical triggers are discussed in this section.

5.5.1 Oxidative Desorption

Under oxidizing (aerobic) and slightly acidic environment, As(V) is adsorbed by oxide minerals. However, with the increase in pH above 8.5, As desorbs from the oxides into the groundwater. Thus, desorption of As(V) in aerobic and alkaline conditions is one of the reasons for enrichment of As in groundwater. As(III), being unstable in aerobic conditions, gets oxidized to the less mobile form of As(V). Presence of Fe and Mn ions also influences As mobilization by increasing oxidation of arsenite in acidic pH. Use of deep tube-wells to aerate groundwater provides oxidizing conditions that cause widespread contamination of dissolved As in the sub-surface. Oxidized desorption or alkali desorption has caused high As levels in many parts of the world especially in the United States (Ravenscroft et al. 2009).

5.5.2 Sulfide Oxidation

Arsenic-bearing sulfides, FeS_2/FeAsS, undergo oxidation when exposed to atmosphere. Excessive extraction of groundwater, mining, or erosion can drive changes to the water table, thereby exposing sulfides to oxidizing environment. Arsenic sulfide ores are oxidized in oxidizing environment to release As(V) into the groundwater. Presence of nitrate (NO_3^-) at pH > 5.0 can act as an oxidant and mobilize As. At lower pH, arsenic tends to adsorb on the minerals. Thus, increasing pH leads to mobilization of As in groundwater. When the water table recharges, anoxic conditions occur that reduces As(V) into As(III), causing arsenic poisoning in water. Other ions such as phosphates and silicates also compete with arsenic for adsorption sites, which increase the desorption of As (Smedley and Kinniburgh 2002).

5.5.3 Reductive Dissolution of Iron Hydroxides

Reducing conditions occur in lakes, aquifers, and sediments due to oxygen consumption and carbon dioxide production on decomposition of organic matter (Smedley and Kinniburgh 2002). In these anoxic conditions, Fe(III) hydroxides are reduced to Fe(II), causing reduction of As(V) to As(III), and thus, mobilization of arsenic (Chen 2012). Since As(III) is highly mobile and has lower affinity to mineral surface, it is released into the groundwater (Guo et al. 2014). There is a strong correlation between desorption rate and pH values, as with increasing pH values, desorption rate will also increase, leading to higher levels of As in groundwater. Organic matter may also create reducing conditions in the sediment and aquifers (Guo et al. 2014)

Reductive dissolution can be seen in many regions, especially West Bengal, Bangladesh, and China. Higher levels of As in parts of Assam and North-East India are also believed to be because of reductive dissolution of ferric hydroxides. Also, high arsenic enrichment in groundwater of Indus basin, Punjab, and India was found due to mixed redox conditions (Shah 2015).

5.6 Health Effects of Arsenic

The toxicity of arsenic to humans differs depending on the different forms of arsenic. Toxicity of inorganic arsenic compounds is more compared to organoarsenicals, whereas arsenite is two to ten times more toxic than arsenate (Mandal et al. 2001). As(III) binds with thiol or sulfhydryl group suppressing the activity of more than two hundred enzymes thus affecting the functioning of various organs. The arsenic exposure of humans occurs via dermal, oral, or respiration. The exposure to As(III) and As(V) occurs through intake of water, air, or soil; where organoarsenicals enter the human body via consumption of marine animals and plants. Arsenic is almost completely absorbed in the gastrointestinal tract which is higher than other heavy metals. Absorption of arsenic is more via oral or inhalation route. The absorbed arsenic has affinity to bind to red blood cells or is deposited in other organs like liver, kidney, skin, muscles, and bones. Inorganic arsenic suppresses the activity of enzymes involved in metabolism of glutathione and synthesis of DNA, and crosses the placental barrier, affecting the nervous system of the fetus. Arsenic interferes with the oxidative phosphorylation causing inhibition of mitochondrial enzymes thus impairing the cellular respiration. Some of the effects of arsenic poisoning in humans are shown in Figure 5.3.

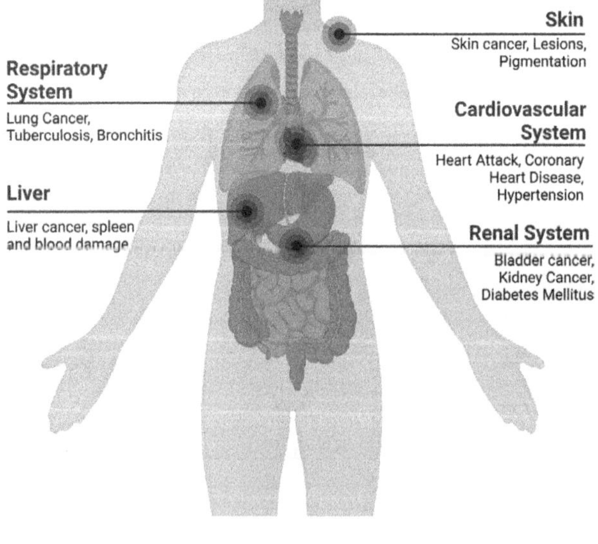

Figure 5.3 Arsenic poisoning effects on human health.

Arsenic was officially recognized as carcinogenic in 1980 (IARC 2012). It is known to cause cancer by exposure via gastrointestinal and respiratory route. Various animals and epidemiological studies reported inorganic arsenic compounds as clear carcinogens or potential carcinogens. The studies conducted across the globe, Bangladesh, India, Chile, Taiwan, the United States, and Argentina where the consumption of water having high concentration of arsenic (150 μg/L), have provided key stem evidences that supported the association of arsenic exposure and carcinogenicity (Cho et al. 2013). Though several hypotheses about carcinogenic mechanism of toxic inorganic arsenic compounds including the oxidative stress, induced chromosomal abnormalities, altered DNA repair, growth factors, and methylation of DNA have been reported, the biomolecularity mechanism is still lagging behind. The International Association for Research on Cancer (IARC) has fortified the symptoms of arsenic with cancer of lungs, bladder, and skin; however, limited reports have revealed the relationship of arsenic with kidney, liver, and prostate cancer.

Several stems of studies have reported that arsenic is associated with skin cancers because of keratinization. A cohort study of 654 residents of Taiwan reported the observed incidence rate of 14 per 1000 person-years of skin cancer. A study from European countries of Hungary, Romania, and Slovakia linked long-term exposure of arsenic in drinking water with cancer risks (Leonardi et al. 2012). Association of arsenic ingestion with lung cancer was observed by Hopenhayn-Rich et al. (1998). Kidney and bladder cancers have known to be associated with arsenic exposure in Taiwan and Bangladesh (Chiang et al. 1993). Detoxification process of arsenic is inhibited due to inhibition of DNA repair and reduction in levels of essential nutrients (Gamble et al. 2007).

Bioaccumulation of arsenic in the early growth of development among children can induce abnormal neurological behaviors during adolescence and adulthood. Arsenic toxicity is neurotoxic as it causes neuritis which affects the functioning of nerves. A study in Mexico assessed the linkages between arsenic exposure and verbal intelligence of children; it was found that the child exposed to higher arsenic concentration performed significantly lower than the child exposed to lower As concentrations (Wasserman et al. 2004). Arsenic exposure may also cause diabetes, particularly type 2 diabetes have been reported. According to studies done in Taiwan, the rate of diabetes incidence among people who consumed high levels of arsenic in their drinking water was 2–5 times greater than in non-exposed participants, indicating that arsenic has a role in diabetes incidence (Tseng et al. 2002). Inorganic arsenic is thought to be diabetogenic in humans, although further study is needed to determine the exact mechanism.

Arsenic exposure has been linked to a variety of health problems, especially skin problems all over the world (Table 5.2). Arsenic-induced keratosis is thought to be a precursor to in situ carcinoma of the skin. According to a research in Bangladesh, melanosis and keratosis were found in 36 of 167 inhabitants (13.9%) who were exposed to drinking water with low arsenic contents of 10 g/L or less as reported in a study in Bangladesh (Ahsan et al. 2000). Chronic arsenic exposure has been linked to blackfoot disease (BFD), a kind of peripheral vascular disease, in Taiwan. Arsenic has an effect on the function of thrombocytes. As(III) causes thrombocyte agglutination in the presence of thrombin, resulting in cardiovascular illness. Arsenic enhances the development of arterial thrombus in animals (Lee et al. 2002).

Arsenic poisoning has been linked to an increase in fetal death and premature delivery. During pregnancy, arsenic has been reported to influence urine excretion and metabolite

Table 5.2 Health effects of arsenic-contaminated water around globe.

N (population affected)	Country	Symptoms	References
73	Arizona	Pulmonary inflammation	Joysula et al. (2006)
10,182	Bangladesh	Melanosis, leucomelanosis, and keratosis	Argos et al. (2011)
267	India (West Bengal)	Pulmonary and systemic inflammation	Sinha et al. (2014)
834	India (West Bengal)	Decrease in lung functioning	Das et al. (2014)
4167	USA	Increased risk of coronary heart diseases (CHD)	Jones et al. (2011)
499	Texas	Hypertension, CHD, and hyperlipidemia	Gong and O'Bryant (2012)
2707	Denmark	Myocardial infarction	Monrad et al. (2017)
900	Italy	Skin infection	Profili et al. (2018)
654	India (West Bengal)	Neurobehavioral symptoms and depression	Mukherjee et al. (2014)
223	Chile	Depression symptoms	Valdés et al. (2017)
200	Bangladesh	Liver toxicity	Islam et al. (2011)
116	New Hampshire	Fetal immune dysregulation	Nadeau et al. (2014)
214	New Hampshire	Infant infection	Farzan et al. (2013)
60	Thailand	DNA impairment	Hinhumpatch et al. (2013)
124	Mexico	Epigenetic changes and DNA damage	Sampayo-Reyes et al. (2010)
572	Taiwan	Oxidative DNA damage and increased risk of urinary cancer	Chung et al. (2008)
280	Finland	Risk of bladder cancer	Michaud et al. (2004)
8086	Taiwan	Increased risk of lung cancer and urothelial cancer	Yang et al. (2013)
405	Mongolia	Hypertension	Zhang et al. (2013)
3254	Pakistan	Melanosis and keratosis	Fatmi et al. (2009)

distribution, influencing the fetus' development. Furthermore, depending on the stage of pregnancy, these impacts are said to appear in various ways; however, further research is needed in this area (Chakraborti et al. 2003).

5.7 Case Studies

5.7.1 Arsenic-Contaminated Drinking Water in Bangladesh

In Bangladesh, the contamination of groundwater by arsenic has led to the poisoning of its population. It is estimated that out of 125 million, about 35–77 million of its population is affected by arsenic by groundwater consumption (Kinniburgh and Smedley 2001). The extent of the risk involved is greatest in the history which is beyond the population affected in Bhopal gas tragedy (1984) in India and Chernobyl disaster (1986) in Ukraine. Initially, characteristic skin lesion is distinguished by pigmentation change in arms and palms, legs and sole of the feet, and upper chest. The various causes were ruled out that could have caused such pigmentation, and the sources of water were analyzed which was used by the patients. This leads to the first confirmed case in 1993 in Bangladesh in its Chapai Nawabganj District (DPHE 1993); however, arsenic-affected patients were seen earlier in India (West Bengal) in 1983. The analysis of such sources suggests that the drinking water is being extracted by the patients from the tube-wells and underground aquifers. Now, it is found that the arsenic level in groundwater in 59 of 64 districts is beyond the acceptable drinking water limit 0.01 ppm (WHO 1993).

The tube-wells are installed and are being used since 1940s in Bangladesh, but the arsenic-related issue came to light in last 25 years due to exponential increase in the number of tube-wells in Bangladesh and subsequent increase in its consumers for drinking water. More emphasis was put on tube-well and dug-well installation because of the scarcity of surface water, and microbes-contaminated surface water that has caused a significant diseases and mortality. Acute gastrointestinal disease was common in infants and children. When these wells are being installed, arsenic was not considered as a contaminant; thus, its tests were not included in the water testing. When arsenic came in light in 1993, further investigations of tube-wells were carried out by the Department of Occupational and Environmental Health of the National Institute of Preventive and Social Medicine. It was found that more than 50% of the samples had >50 μg/L of arsenic which is 5 times beyond the WHO-prescribed limit of 10 μg/L. One of the estimate suggests that about 21 million people of Bangladesh are exposed to arsenic concentration of 50 μg/L, and this number would become double if WHO's standard of 10 μg/L is considered. The extent of exposure to arsenic contamination is many times higher than the actual estimates made till date.

The sources of arsenic in drinking water in Bangladesh are believed to be naturally geological. Arsenic exists as arsenopyrite (FeAsS), realgar (As_4S_4), and arsenic trisulfide (As_2S_3). A large number of chemical reactions, oxidation, reduction, and precipitation, and biological reactions viz. methylation take place, causing the mixing of arsenic with groundwater table. One of the theories suggests the contamination of groundwater by oxidation of arsenic pyrites which is stable in reducing environment under the Earth where no oxygen is present. As water is extracted from the tube-wells, the level of water table lowers below the arsenic deposits making them exposed to oxygen, causing their oxidation. Oxidized arsenic is released from the minerals while some of it gets absorbed on iron hydroxides. When water table is recharged, the condition becomes reducing and arsenic adsorbed by iron hydroxide is released and mixes with the recharging water, making it arsenic contaminated for further use.

The exposure to arsenic contamination has caused long-term health effects which include skin lesion, skin cancer, cancer in kidney and bladder, neurological effects, hypertension, cardiovascular diseases, and pulmonary disease. Generally, the latency period in case of arsenic is about 10 years. However, children younger than 10 years also show skin lesions (Mazumder et al. 1998). The skin cancer appears to show after 20 years of exposure to arsenic-contaminated water. These health effects can be mitigated if long-term and short-term strategies are adopted. Short-term strategies include identifying the well that has lower arsenic contamination, the use of filtration system to separate out the arsenic-contaminated water, use of chemicals that remove arsenic from drinking water, and abandoning the use of highly contaminated wells. Long-term measures may include installing tube-wells below 200 m and shallow dug-well, and rainwater harvesting.

5.7.2 Groundwater Contamination in West Bengal, India

An environmental health catastrophe unfolded in lower Ganga plains of West Bengal in 1983 when K. C. Saha, Department of Dermatology, School of Tropical Medicine in Kolkata, identified the first case of arsenic-induced skin lesions. After ruling out other causes, water sources used by the patients were analyzed, and the diagnosis of arsenic-caused diseases was confirmed. In West Bengal, around 2700 villages with more than 6 million affected population from 9 districts consumed arsenic-contaminated water having arsenic concentration >0.05 mg/L. Morbidity from skin lesions due to chronic arsenic toxicity was reported in almost one-third of the study population aged more than 15 years. Table 5.1 shows the symptoms and signs of arsenicosis patients hospitalized in Bengal (Mazumder et al. 1998). The contamination of arsenic in groundwater in the region is spreading fast. With depleting groundwater table, more wells spill out water contaminated with arsenic. Shallow wells with depth less than 20 m have clean and safe water as these wells mostly contain rainwater or water flowing through sediments. However, almost all the wells of the region tapping water at a depth of 20–100 m are reported of giving excess of arsenic-contaminated water because of the presence of arsenopyrite layers in the underground aquifer strata. A report from Prof. Dipankar Chakraborti, Director, School of Environmental Sciences, Jadavpur University, Kolkata, stated that a greater number of people are at the risk of arsenic poisoning from consuming contaminated water from sunken wells. Earlier it was estimated around 150 million of the population from India and Bangladesh to be affected; however, almost 330 million people maybe at the risk. The arsenic-affected areas lie in the alluvial plains of the quaternary period. The alluvial plains of Ganga extend in northwest while alluvial plains of Brahmaputra river extend in northeast of India in the Garo-Rajmahal Gap. It consists of crystalline rocks and meta-sediments of great Himalayan thrust blocks in the north. The Shillong Plateau, Bihar, and Purulia district of West Bengal has metamorphic rocks, and granites of Pre-Cambrian era with widespread sulfide mineralization (Guha et al. 2005). According to the British Geological Survey (BGS), the position of the arsenic-affected areas is close to where Ganga River enters Bangladesh, it could be the primary source of arsenic in the Bengal alluvium. Several incidences of chronic arsenic toxicity have been reported from the adjoining states of Bihar, Uttar Pradesh, Jharkhand, Assam, Chhattisgarh, and Andhra Pradesh in the upper, middle, and lower Ganga and Brahmaputra plains. It is inferred that the entire upper, middle, and lower stretch of the Gangetic plains might be affected by arsenic (Mazumder and Dasgupta 2011).

5.7.3 Arsenic Contamination Through Ammunition and Its Testing

During the World War I (WWI), the German forces first used "mustard gas" as modern weapon warfare against the allied army which took them off guard. The yellow-green colored gas, Cl_2, is a highly corrosive poison that results in the formation of hydrochloric acid in the mucous membrane of nose and lungs. To counter attack this chemical agent, the US War

Department carried out extensive research to find a "more effective" chemical warfare. Lewisite, an arsenic-containing gas, was then discovered, named after W. Lee Lewis, who discovered it. Later on, American University of Experimental Station (AES) was established by US War Department on American University campus in 1917 in order to develop a new chemical warfare and an effective synthetic method for the production of Lewisite. Since there were no airplanes developed at that time, delivery of bombs was not a choice and mortar was chosen as the weapon of choice. Thus, a mortar-based lewisite laden delivery system was developed. The AES was then dissolved after WWI. However, later after 70 years in 1990, a buried cache of mortar shells was found in American University campus when a real estate developer was digging a hole for the basement. Investigations then reported that mortar shells and other munitions were buried by the US War Department when they left the campus in 1918. In 1998, a search and a geophysical investigation began in order to locate the buried WWI munitions using radar that could detect metallic objects below ground. Geophysical investigation results were correlated with aerial maps of 1917 era to identify those sites where munitions were used, stored, and might still remain there. Vapor containment structures were constructed over the excavated sites in case of an unintended detonation since there are chances that the buried object might be fused munitions. As the mortar was laden with lewisite, the leaking of arsenic from buried leftover munitions and shells caused contamination of surrounding soil. The American University and its surrounding areas were later declared as formerly used defense site. Around 70 mortar shells were removed by Army Corps that were buried and more than 600 dump truck loads of arsenic-contaminated soil were removed from university athletics' field. The unwilling victims, the university and its surrounding neighborhoods, have unknowingly lived with arsenic-contaminated soil for over 85 years because of the irresponsible disposal of the deadly weapons by the War Department.

5.8 Treatment of Arsenic-Contaminated Water

WHO prescribes the limit of arsenic-contaminated water for drinking purpose is 10 μg/L. However, people tend to use the easily available water sources that may contain arsenic more than the prescribed limits by WHO (WHO 1993). It is suggested to find the water source that has minimum arsenic content in that locality. But it is required to treat the arsenic-contaminated water in case of nonavailability of safer drinking water. Adsorption is the most widely used method for arsenic removal using different adsorbents (Table 5.3). Other suggested treatment methods are coprecipitation, electrodialysis, reverse osmosis, and ion-exchange, discussed in the following sections.

5.8.1 Treatment with Activated Carbon

Activated carbon is the carbon which is produced for removal of dissolved impurities, gases, odor, etc., from water by its excellent property of absorption. It is manufactured by heating carbonaceous materials like saw dust or wood in a closed chamber, and then heating it at 800 °C in an inert environment in the absence/limited supply of oxygen making it activated. This process of producing activated carbon in an inert environment is called as pyrolysis. The heating process removes the hydrocarbons leaving behind the activated

Table 5.3 Comparison of different adsorbents for arsenic removal from water and wastewater.

			Adsorption capacity (mg/kg)	
Adsorbent types	**Advantages**	**Limitations**	**As(III)**	**As(V)**
Activated carbon	High surface area; good removal efficiency; commercially viable	Expensive regeneration; remove only few milligrams of metal ions per gram of adsorbent; regeneration problem	29.9	30.48
Agricultural by-products				
Rice husks	No activation required; low cost; easily available	Required in large quantities; contact time is for longer duration	146.30	–
Coconut husk carbon	High surface area; high pore volume and distribution; high adsorption capacity	Required in large quantities; contact time is for longer duration	0.663	0.514
Industrial waste				
Red mud	Cheap; high surface area; great ability to reduce toxicity; commercially viable	Dependency on pH and temperature	–	1.081
Char carbon	Easily available; high surface area; great ability to reduce toxicity; commercially viable	Dependency on pH and temperature	89	34.46
Iron (III) oxide-loaded melted slag	High surface area; great ability to reduce toxicity; commercially viable	Dependency on pH and temperature	2.9–30.1	18.8–78.5
Soil constituents				
Clay minerals	Low cost; easily available; can be regenerated	Interference of other contaminants reduce efficiency		<0.23
Iron-coated sand	Low cost; regeneration not required; can remove both As(III) and As(V); commercially viable	Produce toxic waste	0.028	
Zeolites	Applicable for the removal of a wide range of metals; low cost	Less efficiency		65.93
Metal oxides				
Manganese oxide	Low cost; natural abundance; high surface area	Removal is pH dependent; low regeneration rate	53	22

Zirconium oxide	Low cost; natural abundance; high surface area	Removal is pH dependent; low regeneration rate	112.4	89.90
Titanium oxide	Wide practicality; easily available; commercially viable	Limited binding sites; expensive	59.93	37.46
Activated Alumina	Efficient; can be regenerated in situ	Higher efficiency is achieved at low pH; can only remove As(V)	–	11–24
Iron (III)-loaded chelating resin	Adsorption capacity independent of pH; can be regenerated in situ	Additional treatment is required to remove sulfate and nitrate interference	62.93	55.44
Iron oxides				
Ferrihydrite	High efficiency; low cost; can oxidize As(III) to As(V); commercially viable	High efficiency is achieved at low pH; cannot be regenerated	266.5	111.02
Goethite	High efficiency; low cost; can oxidize As(III) to As(V); commercially viable	High efficiency is achieved at low pH; cannot be regenerated	7.50	12.5
Zero-valent iron	High degree of reactivity with contaminants; reduces treatment time; commercially viable	Efficient over a narrow pH range; efficiency reduces with time; low selectivity of contaminants	2.47	–
Biosorbents				
Lessonia nigrescens	Naturally present; low cost; economical; can be reused for many cycles	Dependence on environmental variables; high durations of process; small-scale application	–	45.2
Eichornia cressipes	Naturally present; low cost; economical; can be reused for many cycles	Dependence on environmental variables; high durations of process; small-scale application	–	0.34
Penicillium chrysogenum	Naturally present; low cost; economical; can be reused for many cycles	Dependence on environmental variables; high durations of process; small-scale application	–	24.52
Penicillium purogenum	Naturally present; low cost; economical; can be reused for many cycles	Dependence on environmental variables; high durations of process; small-scale application	35.6	–

Source: Adapted from Mohan and Pittman (2007).

carbon which is highly porous and absorptive. It can be used in powdered, granular, or slurry form for removing the arsenic from water. The powdered form of activated carbon can be applied to arsenic-contaminated water followed by its filtration. The slurry consists of activated carbon in suspension and then it is fed into the water to be treated. The granular form of activated carbon can be used as the media of filtration through which the contaminated water is allowed to pass. The absorptive efficiency of this filter media reduces with time, and thus, its regeneration is required. Research nowadays have made the arsenic removal more efficient in which iron is impregnated to activated carbon that adsorbs arsenate and arsenite to iron hydroxide making complexes on the surface and helps in further improving its arsenic removal efficiency (Chen et al. 2007).

5.8.2 Adsorption by Activated Alumina

Arsenic can also be removed by treating it with activated alumina which is manufactured from aluminum hydroxide by dehydroxylating it. This produces a highly porous material which is capable of removing arsenic from drinking water. Arsenic exists as arsenite As^{3+} (trivalent) and arsenate As^{5+} (pentavalent). Activated alumina has the ability to remove the As(V) only, but nowadays, As(III) can also be removed from drinking water using a new grade of activated aluminum. However, the removal efficiency of the As^{5+} is affected by the presence of phosphates and silicates at $pH > 7.0$, while the presence of silicates ($pH > 5.0$) and bicarbonates ($pH > 6.0$) affects the removal efficiency of As^{3+} from drinking water. The granular bed of activated alumina is used to remove the arsenic from the contaminated water by passing water through it and safer water is collected on the other end. When activated alumina gets saturated with arsenic, it can be regenerated with 1% caustic soda (NaOH) and about 80% of the activated alumina bed can be regenerated. The wastewater produced during the regeneration of filter bed contains a very high concentration of arsenic and its disposal should be done carefully. Usually, this wastewater is treated with aluminum and iron salts for removal of arsenic by coagulation followed by sedimentation, although this is a pH-dependent process (Majumder 2018). One of the advantages of using this process is that the activated alumina is available commercially in market. Also, arsenic removal efficiency is high. On the other hand, it possesses drawbacks such as replenishment of activated alumina for its efficient use. The pH of the water needs regular monitoring since the chemical reactions involved are pH dependent. Also, the sludge produced needs trained personnel to handle it and its proper disposal is a challenge.

5.8.3 Coagulation and Precipitation

In this method, iron and aluminum salts are being used that helps in the removal of arsenic (both As^{3+} and As^{5+}) by coagulation followed by sedimentation process. This is followed by filtration that removes the suspended solids producing clear and safer water for drinking. At $pH < 7.5$, both iron and aluminum salts can be used as coagulant for removing As^{5+}, while at $pH > 7.5$, iron salts can remove both As^{3+} and As^{5+} effectively. This process requires filtration at the end when filter media is deposited with arsenic and backwashing is required using caustic soda. These chemicals are easily available in the market and can be used at domestic and community level to remove arsenic from water. Also, the capital investment for this treatment method is low. However, the sludge produced from this method contains very high concentration of arsenic and its disposal is a challenge, thus making it necessary to have a proper training and induction program for its operator (Giles et al. 2011).

5.8.4 Electrodialysis Method

Electrodialysis method uses the electricity to break the bond between H_2O molecules and ionic species present in the contaminated water. As soon as the electricity is passed through the arsenic-contaminated water, the arsenic ions get free from water molecules and they start attracting and moving toward the oppositely charged poles. Negatively charged ions will get attracted to positive pole, and vice versa. The presented impurities are segregated from the water in a different compartment when it is allowed to pass through a membrane. Such membrane is made up of ion-exchange resin that could only allow either positive or negative ion but not both. Such membranes are very thin, about 1/180 of a centimeter. Such methods are best for treating the water supplies in remote area where alternate source of safe drinking water is not available.

5.8.5 Reverse Osmosis Method

This method uses semipermeable membrane (SPM) that allows the water to pass through it but stops the arsenic contamination on the other side, thus providing safe drinking water. We know that in simple osmosis process, water flows through the membrane till the osmotic pressure on cleaner water side is equal to the osmotic pressure on contaminated side. While in reverse osmosis, the contaminated water is allowed to pass through the SPM under pressure which only allows the pure water to pass through it leaving behind the impurities. We can say that the osmotic pressure is reversed to remove the arsenic impurities from contaminated water to obtain safer drinking water. RO process has high arsenic removal efficiency and it requires a very compact room for its installation. However, it has a very high initial cost of its procurement and installation, along with high operation and maintenance cost. It also requires pretreatment of the water because SPM is susceptible to get foul due to the presence of organic matter, iron, manganese, and suspended matter. Presence of sulfates and carbonates also scales the membrane that results in its poor efficiency (Abejón et al. 2015).

5.8.6 Ion-Exchange Process

It is a process in which one or more ions causing contamination of water is exchanged with another nonobjectionable or less objectionable ion while it passes through a bed of strong cation exchange resins called zeolites. Between pH 8.0 and 9.0, as the arsenic-contaminated water is passed through it, sodium ions are released from the resins into the water and arsenic gets attached to resins, resulting in safer drinking water. The increased concentration of arsenic at the outlet of this treatment unit shows that the resins have been used and they are exhausted. The resins can be regenerated in which the ion exchange gets reversed. They can be regenerated by passing 10% brine solution through bed of resin by which the sodium gets attached back to resins and it is again ready to get reused. Arsenic released in this process can be removed by backwashing the treatment unit. This wastewater containing high concentration of arsenic is treated with iron and aluminum salts to precipitate out insoluble arsenate. pH reduction to 5.0 aids in the formation of insoluble hydroxides that can be precipitated out (Oehmen et al. 2011).

5.9 Conclusion

Presence of arsenic in groundwater is a global concern, especially in the alluvial and deltaic sediments and aquifers. Arsenic is a naturally occurring element originating from As-bearing rocks, minerals, and sediments. Soil–water interaction leads to many geochemical processes such as oxidation, reduction, adsorption, and precipitation that causes dissolution and mobilization of As into the groundwater. It is a toxic element as it can enter the food chain because of its non-biodegradability and ability to persist in environment for longer periods. Its presence in drinking water poses serious health risks as it is a carcinogen. The toxicity and mobility of arsenic in water are dependent on its speciation at different pH and redox potentials. Therefore, to mitigate the problem of arsenic contamination in groundwater, it is essential to study the chemistry of soil–water interactions and identify the sources of arsenic in water. Many techniques have been developed to combat arsenic contamination either by reducing its toxicity or by precipitating it out from the water. Treatment with alumina or iron and coagulation and precipitation remove As from groundwater by exploiting the adsorption and precipitation properties of arsenic with these metal oxides at optimum pH and redox conditions. Other alternatives to prevent arsenic poisoning can be shifting to the use of surface waters, re-claimed wastewater or rainwater harvesting, etc. Proper awareness of adverse effects of arsenic among the people can prove to be effective in monitoring and mitigating this crisis.

References

Abejón, A., Garea, A. & Irabien, A., 2015. Arsenic removal from drinking water by reverse osmosis: minimization of costs and energy consumption. *Separation and Purification Technology*, 144, pp.46–53. https://doi.org/10.1016/j.seppur.2015.02.017.

Abiye, T.A. & Bhattacharya, P. 2019. Arsenic concentration in groundwater: archetypal study from South Africa. *Groundwater for Sustainable Development*, 9, 100246.

Acharyya, S.K., Shah, B.A., Ashyiya, I.D. & Pandey, Y., 2005. Arsenic contamination in groundwater from parts of Ambagarh–Chowki block, Chhattisgarh, India: source and release mechanism. *Environmental Geology*, 49(1), pp.148–158.

Ahsan, H., Perrin, M., Rahman, A., Parvez, F., Stute, M., Zheng, Y., Milton, A.H., Brandt-Rauf, P., Van Geen, A. & Graziano, J., 2000. Associations between drinking water and urinary arsenic levels and skin lesions in Bangladesh. *Journal of Occupational and Environmental Medicine*, 42(12), pp.1195–1201.

Akhtar, N. & Rai, S.P., 2019. Heavy metals concentrations in drinking water and their effect on public health around moth block of Jhansi District, Uttar Pradesh, India. *Indian Journal of Environmental Protection*, 39(9), pp.945–953.

Argos, M., Kalra, T., Pierce, B.L., Chen, Y., Parvez, F., Islam, T., & Ahsan, H., 2011. A prospective study of arsenic exposure from drinking water and incidence of skin lesions in Bangladesh. *American Journal of Epidemiology*, 174(2), pp.185–194. https://doi.org/10.1093/aje/kwr062

Benner, S.G., Polizzotto, M.L., Kocar, B.D., Ganguly, S., Phan, K., Ouch, K. & Fendorf, S. 2008. Groundwater flow in an arsenic-contaminated aquifer, Mekong Delta, Cambodia. *Applied Geochemistry*, 23(11), pp.3072–3087.

Bondu, R., Cloutier, V., Rosa, E., & Benzaazoua, M. 2017. Mobility and speciation of geogenic arsenic in bedrock groundwater from the Canadian Shield in western Quebec, Canada. *Science of the Total Environment*, 574, pp.509–519.

Brahman, K.D., Kazi, T.G., Afridi, H.I., Naseem, S., Arain, S.S. & Ullah, N., 2013. Evaluation of high levels of fluoride, arsenic species and other physicochemical parameters in underground water of two sub districts of Tharparkar, Pakistan: a multivariate study. *Water Research*, 47(3), pp.1005–1020.

Buyuktuncel, E., Bektas, S., Salih, B., Evirgen, M.M. & Genç, Ö., 1997. Arsenic speciation in geothermal waters by HPLC/GFAAS and HPLC/HGAAS methods. *Fresenius Environmental Bulletin*, 6, pp. 494–501.

Camacho, L.M., Gutiérrez, M., Alarcón-Herrera, M.T., de Lourdes Villalba, M. & Deng, S., 2011. Occurrence and treatment of arsenic in groundwater and soil in northern Mexico and southwestern USA. *Chemosphere*, 83(3), pp.211–225.

Chakraborti, D., Mukherjee, S.C., Pati, S., Sengupta, M.K., Rahman, M.M., Chowdhury, U.K., Lodh, D., Chanda, C.R., Chakraborti, A.K. and Basu, G.K., 2003. Arsenic groundwater contamination in Middle Ganga Plain, Bihar, India: a future danger?. *Environmental Health Perspectives*, 111(9), pp.1194–1201. https://doi.org/10.1289/ehp.5966.

Chen, J.P. 2012. *Decontamination of Heavy Metals: Processes, Mechanisms, and Applications*. CRC Press. https://doi.org/10.1201/b12672.

Chen, W., Parotto, R., Zou, J., Cannon, F.S. & Dempsey, B.A., 2007. Arsenic removal by iron-modified activated carbon. *Water Research*, 41(9), pp.1851–1858. https://doi.org/10.1016/j.watres.2007.01.052.

Chiang, H.S., Guo, H.R., Hong, C.L., Lin, S.M. & Lee, E.F., 1993. The incidence of bladder cancer in the black foot disease endemic area in Taiwan. *British Journal of Urology*, 71(3), pp.274–278. https://doi.org/10.1111/j.1464-410X.1993.tb15942.x.

Cho, Y., Seo, S., Choi, S.H., Lee, S., Kim, K., Kim, H.J. & Choi, J.W., 2013. Association of arsenic levels in soil and water with urinary arsenic concentration of residents in the vicinity of closed metal mines. *International Journal of Hygiene and Environmental Health*, 216(3), pp.255–262. https://doi.org/10.1016/j.ijheh.2012.05.003.

Chowdhury, T.R., Mandal, B.K., Samanta, G., Basu, G.K., Chowdhury, P.P., Chanda, C.R., Karan, N.K., Lodh, D., Dhar, R.K., Das, D. & Saha, K.C., 1997. Arsenic in groundwater in six districts of West Bengal, India: the biggest arsenic calamity in the world: the status report up to August, 1995. In Abernathy, C.O., Calderon, R.L., Chappell, W.R. (eds) *Arsenic* (pp. 93–111). Springer, Dordrecht. https://doi.org/10.1007/978-94-011-5864-0_9.

Chowdhury, U.K., Biswas, B.K., Chowdhury, T.R., Samanta, G., Mandal, B.K., Basu, G.C. & Chakraborti, D. 2000. Groundwater arsenic contamination in Bangladesh and West Bengal, India. *Environmental Health Perspectives*, 108(5), pp.393–397.

Chung, C.J., Huang, C.J., Pu, Y.S., Su, C.T., Huang, Y.K., Chen, Y.T. & Hsueh, Y.M. 2008. Urinary 8-hydroxydeoxyguanosine and urothelial carcinoma risk in low arsenic exposure area. *Toxicology and Applied Pharmacology*, 226(1), pp.14–21. https://doi.org/10.1016/j.taap.2007. 08.021

Das, D., Bindhani, B., Mukherjee, B., Saha, H., Biswas, P., Dutta, K. & Ray, M.R. 2014. Chronic low-level arsenic exposure reduces lung function in male population without skin lesions. *International Journal of Public Health*, 59, pp.425–574. https://doi.org/10.1007/s00038-014-0567-5

DPHE (1993). Report on Arsenic Pollution in Nawabganj District. Directorate of Public Health Engineering, Dhaka.

Farzan, S.F., Korrick, S., Li, Z., Enelow, R., Gandolfi, A.J., Madan, J. & Karagas, M.R. 2013. In utero arsenic exposure and infant infection in a United States cohort: a prospective study. *Environmental Research*, 126, pp.24–30. https://doi.org/10.1016/j.envres.2013.05.001

Fatmi, Z., Azam, I., Ahmed, F., Kazi, A., Gill, A.B. & Kadir, M.M., 2009. Core Group for Arsenic Mitigation in Pakistan. Health burden of skin lesions at low arsenic exposure through groundwater in Pakistan. Is river the source? *Environmental Research*, 109, 575–581. https://doi.org/10.1016/j.envres.2009.04.002.

Feng, S., Guo, H., Sun, X., Han, S., & Li, Y. 2022. Relative importance of hydrogeochemical and hydrogeological processes on arsenic enrichment in groundwater of the Yinchuan Basin, China. *Applied Geochemistry*, 137, 105180.

Gamble, M.V., Liu, X., Slavkovich, V., Pilsner, J.R., Ilievski, V., Factor-Litvak, P., Levy, D., Alam, S., Islam, M., Parvez, F. & Ahsan, H., 2007. Folic acid supplementation lowers blood arsenic. *The American Journal of Clinical Nutrition*, 86(4), pp.1202–1209. https://doi.org/10.1093/ajcn/86.4.1202.

Garelick, H., Jones, H., Dybowska, A. & Valsami-Jones, E., 2009. Arsenic pollution sources. *Reviews of Environmental Contamination*, 197, pp.17–60

Ghrefat, H., El Waheidi, M., Batayneh, A., Nazzal, Y., Zumlot, T. & Mogren, S., 2016. Pollution assessment of arsenic and other selected elements in the groundwater and soil of the Gulf of Aqaba, Saudi Arabia. *Environmental Earth Sciences*, 75(3), p.229. https://doi.org/10.1007/s12665-015-5020-4.

Giles, D.E., Mohapatra, M., Issa, T.B., Anand, S. & Singh, P., 2011. Iron and aluminium based adsorption strategies for removing arsenic from water. *Journal of Environmental Management*, 92(12), pp.3011–3022. https://doi.org/10.1016/j.jenvman.2011.07.018.

Gong, G. & O'Bryant, S.E. 2012. Low-level arsenic exposure, AS3MT gene polymorphism and cardiovascular diseases in rural Texas counties. *Environmental Research*, 113, pp.52–57. https://doi.org/10.1016/j.envres. 2012.01.003.

Guha, M., Chattopadhyay, A. & Nagdeve, D.A., 2005. Environmental health catastrophe in eastern India: a case study of arsenic morbidity in west Bengal. *The International Journal of Third World Medicine*, 3(2), pp.1–2.

Guo, H., Wen, D., Liu, Z., Jia, Y. & Guo, Q., 2014. A review of high arsenic groundwater in Mainland and Taiwan, China: distribution, characteristics and geochemical processes. *Applied Geochemistry*, 41, pp.196–217.

Herath, I., Vithanage, M., Bundschuh, J., Maity, J.P. & Bhattacharya, P., 2016. Natural arsenic in global groundwaters: distribution and geochemical triggers for mobilization. *Current Pollution Reports*, 2(1), pp.68–89. https://doi.org/10.1007/s40726-016-0028-2.

Hinhumpatch, P., Navasumrit, P., Chaisatra, K., Promvijit, J., Mahidol, C. & Ruchirawat, M. 2013. Oxidative DNA damage and repair in children exposed to low levels of arsenic in utero and during early childhood: application of salivary and urinary biomarkers. *Toxicology and Applied Pharmacology*, 273, pp.569–579. https://doi.org/10.1016/j. taap.2013.10.002

Hopenhayn-Rich, C., Biggs, M.L. & Smith, A.H., 1998. Lung and kidney cancer mortality associated with arsenic in drinking water in Cordoba, Argentina. *International Journal of Epidemiology*, 27(4), pp.561–569. https://doi.org/10.1093/ije/27.4.561.

International Agency for Research on Cancer (IARC). (2012). Arsenic and arsenic compounds. IARC MonogrEvalCarcinog Hum 100C: 41–93.

Islam, K., Haque, A., Karim, R., Fajol, A., Hossain, E., Salam, K.A. & Hossain, K., 2011. Dose-response relationship between arsenic exposure and the serum enzymes for liver function tests in the individuals exposed to arsenic: a cross sectional study in Bangladesh. *Environmental Health*, 8, pp.10–64. https://doi.org/10.1186/1476-069X-10-64

Jones, M.R., Tellez-Plaza, M., Sharrett, A.R., Guallar, E. & Navas-Acien, A. 2011. Urine arsenic and hypertension in US adults: the 2003–2008 National Health and Nutrition Examination Survey. *Epidemiology*, 22, pp.153–161. https://doi.org/10.1097/EDE.0b013e318207fdf2

Joysula, A.B., Poplin, G.S., Kurzius-Spencer, M., McClellen, H.E., Kopplin, M.J., Stürup, S. & Burgess, J.L. 2006. Environmental arsenic 40 SINHA AND PRASAD exposure and sputum metalloproteinase concentrations. *Environmental Research*, 102, pp.283–290. https://doi. org/10.1016/j.envres.2006.01. 003

Kinniburgh, D.G. & Smedley, P. 2001. Arsenic contamination of groundwater in Bangladesh. DPHE, Govt. of Bangladesh and DFID, British Geological Survey, Final Report, vol. 2, pp. 105–107.

Kumar, A., Ali, M., Kumar, R., Kumar, M., Sagar, P., Pandey, R.K., Akhouri, V., Kumar, V., Anand, G., Niraj, P.K. & Rani, R. 2021. Arsenic exposure in Indo Gangetic plains of Bihar causing increased cancer risk. *Scientific Reports*, 11(1), pp.1–16. https://doi.org/10.1038/ s41598-021-81579-9

Lee, M.Y., Bae, O.N., Chung, S.M., Kang, K.T., Lee, J.Y. & Chung, J.H. 2002. Enhancement of platelet aggregation and thrombus formation by arsenic in drinking water: a contributing factor to cardiovascular disease. *Toxicology and Applied Pharmacology*, 179(2), pp.83–88. https://doi.org/10.1006/taap.2001.9356

Leonardi, G., Vahter, M., Clemens, F., Goessler, W., Gurzau, E., Hemminki, K., Hough, R., Koppova, K., Kumar, R., Rudnai, P. & Surdu, S., 2012. Inorganic arsenic and basal cell carcinoma in areas of Hungary, Romania, and Slovakia: a case–control study. *Environmental Health Perspectives*, 120(5), pp.721–726. https://doi.org/10.1289/ehp.1103534.

Litter, M.I., Ingallinella, A.M., Olmos, V., Savio, M., Difeo, G., Botto, L. & Ahmad, A. 2019. Arsenic in Argentina: technologies for arsenic removal from groundwater sources, investment costs and waste management practices. *Science of the Total Environment*, 690, pp.778–789.

Liu, C.W. & Wu, M.Z. 2019. Geochemical, mineralogical and statistical characteristics of arsenic in groundwater of the Lanyang Plain, Taiwan. *Journal of Hydrology*, 577, 123975.

Majumder, C. 2018. Arsenic (V) removal using activated alumina: kinetics and modeling by response surface. *Journal of Environmental Engineering*, 144(3), 04017115. https://doi. org/10.1061/(ASCE)EE.1943-7870.0001337

Mandal, B.K., Ogra, Y. & Suzuki, K.T. 2001. Identification of dimethylarsinous and monomethylarsonous acids in human urine of the arsenic-affected areas in West Bengal,

India. *Chemical Research in Toxicology*, 14(4), pp.371–378. https://doi.org/10.1021/tx000246h.

Masuda, H. 2018. Arsenic cycling in the Earth's crust and hydrosphere: interaction between naturally occurring arsenic and human activities. *Progress in Earth and Planetary Science*, 5(1), pp.1–11. https://doi.org/10.1186/s40645-018-0224-3.

Mazumder, D.G. & Dasgupta, U.B. 2011. Chronic arsenic toxicity: studies in West Bengal, India. *The Kaohsiung Journal of Medical Sciences*, 27(9), pp.360–370. https://doi.org/10.1016/j.kjms.2011.05.003

Mazumder, D.N.G., Haque, R., Ghosh, N., De, B.K., Santra, A., Chakraborty, D. & Smith, A.H., 1998. Arsenic levels in drinking water and the prevalence of skin lesions in West Bengal, India. *International Journal of Epidemiology*, 27(5), pp.871–877. https://doi.org/10.1093/ije/27.5.871

Michaud, D.S., Wright, M.E., Cantor, K.P., Taylor, P.R., Virtamo, J. & Albanes, D. 2004. Arsenic concentrations in prediagnostic toenails and the risk of bladder cancer in a cohort study of male smokers. *American Journal of Epidemiology*, 160(9), pp.853–859. https://doi.org/10.1093/aje/kwh295

Mohan, D. & Pittman Jr, C.U., 2007. Arsenic removal from water/wastewater using adsorbents – a critical review. *Journal of Hazardous Materials*, 142(1–2), pp.1–53.

Monrad, M., Ersbøll, A.K., Sørensen, M., Baastrup, R., Hansen, B., Gammelmark, A. & Raaschou-Nielsen, O., 2017. Low level arsenic in drinking water and risk of incident myocardial infarction: a cohort study. *Environmental Research*, 154, pp.318–324. https://doi.org/10.1016/j.envres.2017.01.028.

Mukherjee, A., Sengupta, M.K., Hossain, M.A., Ahamed, S., Das, B., Nayak, B., Lodh, D., Rahman, M.M. & Chakraborti, D., 2006. Arsenic contamination in groundwater: a global perspective with emphasis on the Asian scenario. *Journal of Health, Population and Nutrition*, 24, pp.142–163.

Mukherjee, B., Bindhani, B., Saha, H., Sinha, D., & Ray, M. R., 2014. Platelet hyperactivity, neurobehavioral symptoms and depression among Indian women chronically exposed to low level of arsenic. *Neurotoxicology*, 45, pp.159–167. https://doi.org/10.1016/j.neuro.2014.10.011.

Nadeau, K.C., Li, Z., Farzan, S., Koestler, D., Robbins, D., Fei, D.L. & Karagas, M.R. 2014. In utero arsenic exposure and fetal immune repertoire in a US pregnancy cohort. *Clinical Immunology*, 155, pp.188–197. https://doi.org/10.1016/j.clim.2014.09.004

Nriagu, J.O., Bhattacharya, P., Mukherjee, A.B., Bundschuh, J., Zevenhoven, R. & Loeppert, R.H., 2007. Arsenic in soil and groundwater: an overview. *Trace Metals and other Contaminants in the Environment*, 9, pp.3–60.

Oehmen, A., Valerio, R., Llanos, J., Fradinho, J., Serra, S., Reis, M.A., Crespo, J.G. & Velizarov, S., 2011. Arsenic removal from drinking water through a hybrid ion exchange membrane–coagulation process. *Separation and Purification Technology*, 83, pp.137–143. https://doi.org/10.1016/j.seppur.2011.09.027.

Pagni, R. (2006). Environmental Chemistry, (Baird, Colin; Cann, Michael). https://doi.org/10.1021/ed083p21.

Polya, D.A. & Middleton, D.R., 2017. Arsenic in drinking water: sources & human exposure. In Bhattacharya, P., Polya, D., Jovanovic, D. (Eds.), *Best Practice Guide on the Control of Arsenic in Drinking Water* (p. 15). IWA Publishing.

Profili, F., Nuvolone, D., Barbone, F., Aprea, C., Centi, L., Frazzetta, R. & Voller, F. 2018. Health effects among a cohort exposed to low-level arsenic in a geothermal area of Tuscany, Italy. *International Archives of Occupational and Environmental Health*, 91(8), pp.971–979. https://doi.org/10.1007/s00420-018-1340-5.

Ravenscroft, P., Brammer, H. & Richards, K. 2009. *Arsenic Pollution: A Global Synthesis* (Vol. 94). John Wiley & Sons.

Rowland, H.A., Omoregie, E.O., Millot, R., Jimenez, C., Mertens, J., Baciu, C. & Berg, M. 2011. Geochemistry and arsenic behaviour in groundwater resources of the Pannonian Basin (Hungary and Romania). *Applied Geochemistry*, 26(1), pp.1–17.

Sampayo-Reyes, A., Hernández, A., El-Yamani, N., López-Campos, C., Mayet-Machado, E., Rincón-Castañeda, C.B. & Ricardo Marcos, R. 2010. Arsenic induces DNA damage in environmentally exposed Mexican children and adults. Influence of GSTO1 and AS3MT polymorphisms. *Toxicological Sciences*, 117, pp.63–71. https://doi.org/10.1093/toxsci/kfq173

Shah, B.A. 2015. Status of groundwater arsenic contamination in the states of North-east India: a review. *Indian Groundwater*, 5, pp.32–37.

Shankar, S., Shanker, U. & Shikha. 2014. Arsenic contamination of groundwater: a review of sources, prevalence, health risks, and strategies for mitigation. *The Scientific World Journal, 2014*, 304524.

Sinha, D., Mukherjee, B., Bindhani, B., Dutta, K., Saha, H., Prasad, P. & Ray, M.R. 2014. Chronic low level arsenic exposure inflicts pulmonary and systemic Inflammation. *Journal of Cancer Science and Therapy*, 6(3), pp.62–69. https://doi.org/10.4172/1948-5956.1000250

Smedley, P.L. & Kinniburgh, D.G., 2002. A review of the source, behaviour and distribution of arsenic in natural waters. *Applied Geochemistry*, 17(5), pp.517–568. https://doi.org/10.1016/S0883-2927(02)00018-5.

Smedley, P., Zhang, M., Zhang, G., & Luo, R., 2003. Mobilisation of arsenic and other trace elements in fluviolacustrine aquifers of the Huhhot Basin, Inner Mongolia. *Applied Geochemistry* 18, pp.1453–1477. https://doi.org/10.1016/S0883-2927(03)00062-3

Stazi, S.R., Marabottini, R., Papp, R. & Moscatelli, M.C., 2015. Arsenic in soil: availability and interactions with soil microorganisms. *In* Sherameti, I., Varma, A. (eds) *Heavy Metal Contamination of Soils. Soil Biology* (Vol. 44, pp. 113–126). Springer, Cham. https://doi.org/10.1007/978-3-319-14526-6_6

Thakur, J.K., Thakur, R.K., Ramanathan, A.L., Kumar, M. & Singh, S.K. 2011. Arsenic contamination of groundwater in Nepal – an overview. *Water*, 3(1), pp.1–20.

Tseng, C.H., Tseng, C.P., Chiou, H.Y., Hsueh, Y.M., Chong, C.K. & Chen, C.J., 2002. Epidemiologic evidence of diabetogenic effect of arsenic. *Toxicology Letters*, 133(1), pp.69–76. https://doi.org/10.1016/S0378-4274(02)00085-1.

Valdés, M., Hanchey, A., Muñoz, M. P., Baumert, B., & Iglesias, V. 2017. Low-level arsenic exposure during pregnancy and its association with postpartum depression: a cohort study of women from Arica, Chile. *Revue D Epidemiologie Et De Sante Publique*, 65(6), pp.427–435. https://doi.org/10.1016/j.respe.2017.05.010

Wasserman, G.A., Liu, X., Parvez, F., Ahsan, H., Factor-Litvak, P., van Geen, A., Slavkovich, V., Lolacono, N.J., Cheng, Z., Hussain, I. & Momotaj, H., 2004. Water arsenic exposure and children's intellectual function in Araihazar, Bangladesh. *Environmental Health Perspectives*, 112(13), pp.1329–1333. https://doi.org/10.1289/ehp.6964

Welch, A.H., Westjohn, D.B., Helsel, D.R. & Wanty, R.B., 2000. Arsenic in ground water of the United States: occurrence and geochemistry. *Groundwater*, 38(4), pp.589–604.

Wenzel, W.W., 2013. Arsenic. In: Alloway, B. (Eds.) *Heavy Metals in Soils. Environmental Pollution*, vol 22. Springer, Dordrecht. https://doi.org/10.1007/978-94-007-4470-7_9

WHO (World Health Organization) (2001). Arsenic and arsenic compounds: Environmental Health Criteria EHC 224.

World Health Organization 1993. *Guidelines for Drinking-Water Quality*. World Health Organization.

Yang, T., Hsu, L., Chen, H., Chiou, H., Hsueh, Y., Wu, M. & Chen, C., 2013. Lifetime risk of urothelial carcinoma and lung cancer in the arseniasis- endemic area of Northeastern Taiwan. *Journal of Asian Earth Sciences*, 77, pp.332–337. https://doi.org/10.1016/j.jseaes.2013.03.023

Zhang, C., Mao, G., He, S., Yang, Z., Yang, W., Zhang, X. & Guo, X., 2013. Relationship between long-term exposure to low-level arsenic in drinking water and the prevalence of abnormal blood pressure. *Journal of Hazardous Materials*, 262, pp.1154–1158. https://doi.org/10.1016/j. jhazmat.2012.09.045

6

Arsenic Contamination in Groundwater and Its Removal Strategies with Special Emphasis on Nano Zerovalent Iron

Shivani Thakur[1], *Harminder Singh*[2], *Umakant Chaudhari*[1], *Sunil Mittal*[1], *and Prafulla Kumar Sahoo*[1,3]

[1] *Department of Environmental Science and Technology, Central University of Punjab, Bathinda, Punjab, India*
[2] *School of Chemical Engineering and Physical Science, Lovely Professional University, Phagwara, India*
[3] *InstitutoTecnologico Vale, Belem, PA, Brazil*

6.1 Introduction

All around the world, arsenic has been found in groundwater as one of the major contaminants. Many countries, such as India, Bangladesh, South America, and North America, have reported high arsenic contamination in their groundwater (Lee et al. 2015; Bentahar et al. 2016). Murcott (2012) found that more than 300 million people had suffered from this contaminant worldwide, about 13 million in the USA and 70 million people in Bangladesh. Long-term exposure of the human body to arsenic not only leads to arsenicosis but also causes various complications such as heart disease, stroke, skin lesions, hyperkeratosis, anuria, hyper pigmentation, and sometimes also leads to cancer of lungs, gall bladder, liver, kidney, and skin (Halim et al. 2009; Li et al. 2010; Lalhmunsiama et al. 2016). Currently, the WHO's permissible limit recommended for arsenic in drinking water is 10 µg/L. However, it has been recognized that at least 140 million people in 50 countries are consuming groundwater with an arsenic concentration above this permissible limit. Among these countries, Bangladesh has grabbed much attention, where in 2012, about 19 million people were found to consume arsenic-contaminated water having concentrations above their set national standards, i.e. 50 µg/L, and about 30 million people consume water having concentration above 10 µg/L (WHO 2018). Extensive research has been done for the past few decades to develop various technologies for effective As remediation from water. These include methods such as ion exchange, electrodialysis, forward osmosis, reverse osmosis, chemical precipitation, iron oxyhydroxide precipitation, electrocoagulation, iron coprecipitation, membrane filtration, and ultrafiltration (Zhu et al. 2009). The mentioned techniques required more energy, time, and space; therefore, researchers developed materials for efficient and cost-effective arsenic removal by using nanotechnology which provides a large

Hydrogeochemistry of Aquatic Ecosystems, First Edition. Edited by Sughosh Madhav, Virendra Bahadur Singh, Manoj Kumar, and Sandeep Singh.

surface area for interaction between arsenic and adsorbents. Nanoparticles (NPs) worked as a bridge between large materials and atomic size materials, having a size ranging between 1 and 100 nm (Prabhakar and Samadder 2018). Several investigations reported different NPs' capacities for removing arsenic from water. Iron-based adsorbents such as iron oxides, bimetallic iron oxides, and nano zerovalent iron (nZVI) were reported as highly proficient adsorbents for arsenic removal (Tang et al. 2011; Kumar et al. 2014).

Iron-based NPs are more favored to remove arsenic as compared to other NPs because of the abundant nature of iron. Therefore, preparing NPs from iron is pretty cost effective, and secondly, because of their high affinity for As(V) and As(III) (Daus et al. 2004). nZVI particles have become one of the most significant nanomaterials. Its use is witnessing a massive surge for environmental remediation due to its comparatively lower cost, high reactivity power, and excellent adsorption capacity (Phenrat and Lowry 2019). However, studies reported that these nZVI particles because of their high surface energy and intrinsic magnetic interaction, readily get agglomerated and oxidized when reacting with air and water, therefore, limiting the chemical reactivity of the material. Thus, these are generally capped with some inorganic or organic materials or supported on some surface stabilizers/immobilizers such as chitosan (CS), alginate, montmorillonite, activated carbon, and other porous structures to increase stability and transport characteristics (He and Zhao 2005; Li et al. 2010). This chapter summarizes the basic hydrogeochemistry of As in groundwater and emphasizes its different removal methods from water with particular reference to nZVI.

6.1.1 General Chemistry and Occurrence of Arsenic in Nature

Arsenic is an element of the periodic table occurring naturally in the environment that comprises about 0.00005% of the earth's crust. It is ranked 20th most abundant element in nature and 12th in human body. It appears as a silver-gray brittle crystalline solid having an atomic weight of 74.9 u; specific gravity 5.73 g cm^{-3}, melting point 817 °C (at 28 atm), boiling point 613 °C, and vapor pressure 1 mmHg at 372 °C. It is found higher in phosphorites and fine-grained argillaceous sediments usually in the range between 0.5 and 2.5 mg/kg (Mandal and Suzuki 2002).

In nature, arsenic exists both in organic and inorganic forms. It finds its way in water bodies both via natural sources such as volcanic emissions, weathering, and erosion of rocks and soils (Onishi 1969); and from different anthropogenic activities such as mining, petroleum, sewage sludge, ceramic industries, and coal fly ash (Viraraghavan et al. 1999). Arsenic mainly occurs in combination with sulfur, oxygen, and iron and exhibits −3, 0, +3, and +5 oxidation states (Smedley and Kinniburgh 2001). In comparison to arsenic (V), arsenic (III) behaves as hard acid and forms complexes mostly with nitrogen and oxides, while arsenic (V) forms complexes mostly with sulfides (Bodek et al. 1988). In water, arsenic can be present commonly in two forms, i.e. arsenite (AsO_3^{3-}) and arsenate (AsO_4^{3-}) (Greenwood and Earnshaw 1984). Arsenites species include $As(OH)_3$, $As(OH)_4^-$, AsO_2OH^{2-}, and AsO_3^{3-}, while Arsenate species include AsO_4^{3-}, $HAsO_4^{2-}$, and $H_2AsO_4^-$. The stability of both these trivalent and pentavalent species is dependent upon the environment's redox potential. In an oxygen-rich environment, As(V) is predominant while in moderately reducing condition, As(III) dominates (Greenwood and Earnshaw 2012). As(III) is 60 times more toxic than the oxidized As(V) (Vu et al. 2003). Inorganic forms of arsenic are reported to be found most often in water supplies, while an organic state of arsenic is found to be present in surface water predominately due to biological activity and industrial pollution. Examples of some organic compounds of arsenic are dimethyl arsenic acid ($C_2H_7AsO_2$) and monomethyl arsenic acid (CH_5AsO_3), while arsenic trioxide (As_2O_3) and sodium dihydrogen arsenate (NaH_2AsO_4) are some inorganic compounds of arsenic present in water (Smedley and Kinniburgh 2001). The pH of water also influences the availability of different arsenic species in water. The uncharged species of arsenic are predominant in water under the oxidizing condition ($pH < 9.2$). The oxyanions, including pentavalent arsenate, become more soluble in groundwater as pH increases. Even at near-neutral pH, arsenic, unlike other elements, is soluble in groundwater in higher concentrations; thus, it easily contaminates the groundwater (Smedley and Kinniburgh 2001; Vu et al. 2003).

6.1.2 Release and Mobilization Mechanism of Arsenic in Groundwater

Mobility of As in groundwater is affected by different processes such as natural weathering, volcanic eruption, geochemical reactions, microbiological phenomenon, and other anthropogenic activities (Maity et al. 2019). Four possible primary geochemical channels, including geochemical activity, reductive dissolution, sulfide oxidation, and alkali desorption (Figure 6.1) are generally considered by different studies to be responsible for natural arsenic contamination in groundwater (Chauhan et al. 2009; Shankar et al. 2014).

6.1.2.1 Sulfide Oxidation

Processes such as oxidation of As-bearing sulfide minerals mainly occur in countries like Ghana, Thailand, and the United States, where acid drainage is produced after the weathering of sulfide minerals containing different toxic inorganic contaminants in higher amounts (Blowes et al. 2013). Sulfide minerals such as realgar (As_4S_4), arsenopyrite (FeAsS), orpiment (As_2S_3), undergo oxidation and produce mobile As, thus, contaminating the groundwater. Equations 6.1 and 6.2 show oxidation of As-bearing sulfide minerals (Neil et al. 2014).

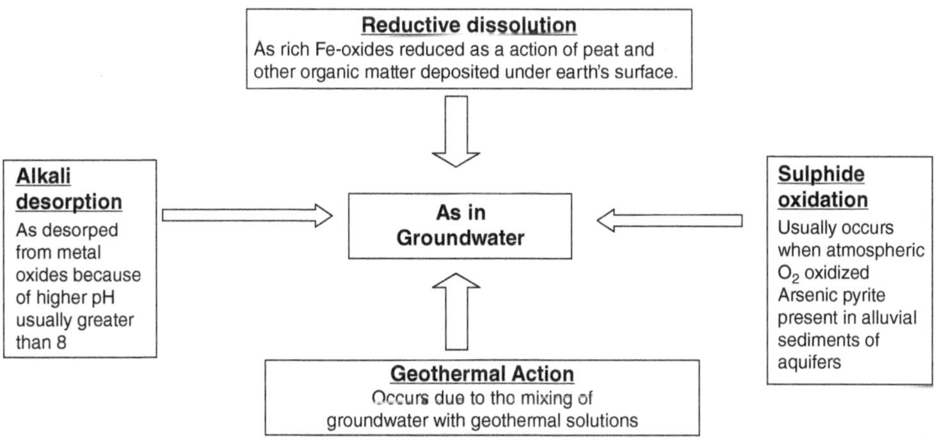

Figure 6.1 Possible mechanisms responsible for As entry and mobility in groundwater. *Source:* Adapted from Chauhan et al. (2009).

$$FeAsS(s)+1.5H_2O+2.750_2(aq)\rightarrow Fe^{2+}(aq)+H_3AsO_3(aq)+SO_4^{2-}(aq) \quad (6.1)$$

$$8Fe(As,S)_2(s)+13NO_3^-(aq)+25H_2O+10H^+(aq)\rightarrow 8Fe^{2+}(aq) \\ +8HAsO_4^{2-}(aq)+13NH^{4+}(aq) \quad (6.2)$$

Oxidation of pyrite FeS_2 takes place in a manner as shown in Eq. 6.3 (Herath et al. 2016):

$$2FeS_{2(S)}+7O_{2(g)}+2H_2O_{aq}\rightarrow 2Fe^{2+}_{(aq)}+4SO^{2-}_{4(aq)}+4H^+_{(aq)}) \quad (6.3)$$

Dissolved organic carbon and pH are responsible for the oxidation of As bearing sulfide minerals and release of As in groundwater (Islam et al. 2013). At natural pH, mostly Fe(III) exists in colloidal or solid form, which causes the release and transport of As in groundwater. *Thiobacillus ferroxidans* bacteria can catalyze the oxidation reaction by adhering to the mineral surface, thus causing the weak solubilization of As(III), As(V), and Fe(II). It also produces Fe(III) ion that can further cause oxidation of sulfide ion (Muehe et al. 2016). The release of FeAsS mainly occurs in the pH range of 1.4–6.4 (Nordstrom 2011).

6.1.2.2 Reductive Dissolution

The reductive dissolution of arsenic is the most critical geochemical process in South Asia and South-East Asia (Erbs et al. 2010). In sedimentary rocks, As(III) tends to be more commonly adsorbed on $Fe(OH)_3$. Reduction of $Fe(OH)_3$ by sedimentary organic matter causes As(III) release in groundwater. This process is called dissimilatory reduction of iron. High pH values make this desorption process easier (Guo et al. 2014). HCO_3^- and PO_4^{3-} also trigger high As mobility from As-bearing aquifer (Butaciu et al. 2017), as shown in Figure 6.2. The presence of high concentrations of aqueous Fe(II) and some other minerals such as hematite makes sulfate reduction more favorable. On reduction, these sulfides may transport from the region of their reduction to $Fe(OH)_3$ zone as depicted in (Eqs. 6.4–6.6) (Herath et al. 2016):

$$As_2S_{3(s)}+6H_2O_{(l)}\rightarrow 2H_3AsO_{3(aq)}+3HS^-_{(aq)}+3H^+_{(aq)} \quad (6.4)$$

Sulfide released from there may then release sorbed As(III) from the surface of Fe(III) $OAs(OH)_2$ via either ligand exchange reaction or reductive dissolution of Fe(III)OAs(OH) as depicted in Eqs. 6.5 and 6.6, respectively:

$$Fe(III)OAs(OH)_{2(S)}+HS^-_{(aq)}\rightarrow Fe(III)S_{(S)}+H_3AsO_{3(aq)} \quad (6.5)$$

$$Fe(III)OAs(OH)_{2(S)}+HS^-_{(aq)}\rightarrow Fe(II)_{(aq)}+H_3AsO_{3(aq)}+S^-_{(aq)} \quad (6.6)$$

6.1.2.3 Alkali Desorption

Alkali desorption causes the release of As present in alluvium and bedrock aquifer. It mainly occurs in Argentina, the USA, and China. Low iron and toxic alkali groundwater mostly favor this process. At or near-neutral pH, As is mainly adsorbed by oxide minerals which get desorbed from them when pH goes above 8.5 (Gonzalez Rodriguez et al. 2019).

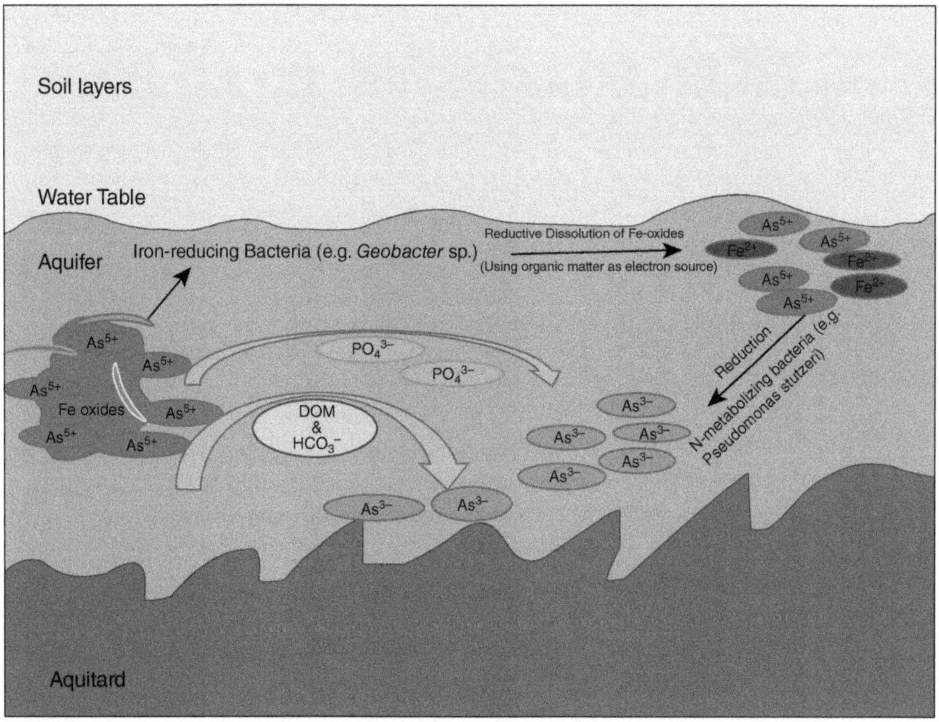

Figure 6.2 A schematic description of As mobilization in aquifers by reductive dissolution mechanism mediated by microorganisms, organic matter, phosphate, and bicarbonate ions. *Source:* Adapted from Jiang et al. (2019).

The elevated amount of As in Argentina results from the desorption of metal oxyhydroxide at a high pH (8–9.5). The mechanism of Arsenic release in countries like the USA, Chile, China, and New Zealand also happened because of geothermal activities. Metamorphic, radioactive sources, and faulting are the possible heat sources for these thermal activities. Because of the presence of a very high geothermal gradient between the upper continental crust and mantle, arsenic from the crust may be leached in the course of rising magma toward the surface. In India and Bangladesh, the reduction of Fe-oxyhydroxide from alluvial sediments and oxidation of sulfide from those sediments are responsible for As dissolution (Lord et al. 2012; Bundschuh and Maity 2015).

6.1.2.4 Microbial-Mediated Arsenic Mobilization in Groundwater

The mobility of As species gets influenced by the presence of different microbes and redox conditions present in groundwater (Huq et al. 2018a). Some iron-reducing and sulfate-reducing bacteria are well recognized to assist in arsenic's mobility. *Geobacter* sp. and *Shewanella* sp. are some iron dissimilatory bacteria that use organic carbon as sources for electrons and cause the reduction of As-containing iron oxides resulting in its release (Figure 6.2). They also cause its reimmobilization by creating secondary iron minerals such as magnetite and vivianite. On the other hand, sulfate-reducing bacteria present in aquifers generate sulfide, which causes the reduction of As-containing iron oxides to cause arsenic mobilization (Jiang et al. 2019).

Furthermore, sulfate-reducing bacteria are also responsible for reducing nitrate and Fe, which influences the As release in groundwater from the reductive dissolution of As-bearing minerals (Guo et al. 2017; Huq et al. 2018b). Some N-metabolizing bacteria species such as *Pseudomonas stutzeri* elevates the amount of As(III) in groundwater by reducing the As(V) into As(III) from sediments. Metal-reducing anaerobic bacteria are also responsible for As mobilization in groundwater. Methanogens and other bacterial remain active in the anoxic groundwater condition, thus aiding in As mobility in many ways. Another microbe (MIT-13) reduces As(V) to As(III), enhancing As release in groundwater (Huq et al. 2020).

6.1.3 Arsenic Status in Global Groundwater Systems

Over 70 nations worldwide have reported arsenic concentrations ranging between 0.5 and 5000 ppb in their groundwater (Jha and Tripathi 2021). Groundwater arsenic pollution has been reported globally, notably along the large deltas and some major rivers of South Asia and South American nations. It has also been reported from countries like Argentina, Bangladesh, Chile, China, Greece, Iran, India, Japan, Nepal, Pakistan, and Vietnam (Yadav et al. 2021). Table 6.1 describes this global scenario of arsenic pollution in groundwater. About 110 million people in Asian countries are consuming As-contaminated water. In India, arsenic contamination in groundwater was first investigated in Chandigarh and its adjacent parts in 1976. Later, it was reported in the deltaic region of West Bengal in 1984 (Bhowmick et al. 2018; Yadav et al. 2021). At present, 17 states are reported to be affected by this contamination, such as Uttar Pradesh, Jharkhand, Bihar, Assam, Manipur, Chhattisgarh, and Punjab (Yadav et al. 2021). In a study conducted in nine districts of West Bengal, where a total of 135,555 samples were analyzed, it was found that around half of them had As concentration greater than 10 ppb, with the highest concentration of 3700 ppb. About 10 million people in Bihar are at high risk because of high As concentration in their drinking water. About 22 districts of Bihar were severely affected by As contamination, where its concentrations in groundwater were greater than 10 ppb (Chakraborti et al. 2009; Kumar et al. 2020a). In 20 districts of Uttar Pradesh, As concentration was greater than 10 ppb with the highest concentration in district Ballia (Yasunori et al. 2012; Singh et al. 2018). About 12 different locations in Punjab state also had high As concentrations up to 397.5 ppb (Yadav et al. 2021).

The central river basin of China, such as Pearl, Yellow, Yangtze river, was also reported to have higher As level in groundwater (Guo et al. 2014; Zhou et al. 2017). Pakistan also had high arsenic contamination in the groundwater, especially along the Indus River and its tributaries, affecting about 50–60 million people. Other Asian countries, such as Iran, Afghanistan, Nepal, and Taiwan, have also been affected by arsenic (Rasool et al. 2016; Shahid et al. 2018). Contamination of As in groundwater is prevalent in Canada and the USA, though it is significantly lower in most places than in Asian Countries (Sorg et al. 2014; Janković 2020). A high concentration of 105 and 580 ppb was reported from Saskatchewan and British Columbia, respectively. Many parts of Europe reported a high As concentration in groundwater, including Serbia, Italy, Hungary, Romania, Greece, and Sweden. In the eastern part of Croatia, the As concentration was reported up to 610 ppb, affecting about two million people (Rowland et al. 2011; McGrory et al. 2017). Around 24.3 million people in Africa are at severe risk because of high As concentration, recorded

Table 6.1 Worldwide distribution of As in groundwater.

Country	Locations	Concentration range (ppb)	Sources	Climate condition	References
Argentina	Buenos Aires province	0–163	Mobilization from sediment		(Sierra et al. 2016)
Bangladesh	Kushtia district	6.05–590.7	Shallow Holocene aquifer enriched with As and oxidative weathering	Subtropical monsoon	(Huq et al. 2019)
China	Xinjiang Province	0.25–185	Reductive dissolution and desorption		(Zhou et al. 2017)
Greece	Lesvos Island	1.78–54.7	Silicate weathering, desorption, and reductive dissolution of Mn	Mediterranean	(Zkeri et al. 2018)
Iran	Rafsanjan plain in Kerman Province	4–278			(Rahnamarad et al. 2020)
India	Puducherry	BDL-36.88	Reductive dissolution of arsenic-bearing minerals	Humid tropical	(Sridharan and Nathan 2018)
India	Berpeta, Assam	0.1–569	Reducing conditions created by heavy sediment load and organic matter		(Jain et al. 2018)
India	Bhojpur, Bihar	>3–1805	Overexploitation of groundwater		(Chakraborti et al. 2016)
India	Patna, Bihar	>5–300	River Sone alluvium		(Singh et al. 2016)
India	Rajnandgaon, Chhattisgarh	148–985	Reductive dissolution		(Patel et al. 2017)
India	Sahibganj, Jharkhand	1–133			(Alam et al. 2016)
India	Moradabad, Uttar Pradesh	24–98	Leaching of As from waste dumping sites (hazardous materials)	Subtropical monsoon	(Saba and Umar 2016)
India	Malwa, Punjab	4.35–23.94	Overexploitation of groundwater, evaporative environmental condition, use of As-bearing agrochemicals	Semiarid	(Kaur et al. 2017)
India	Firozpur, Punjab	BDL-255.6	The presence of bicarbonate in higher amounts and nitrate in lower concentrations triggered the reductive dissolution of As-bearing minerals		(Kumar and Singh 2019)

(Continued)

Table 6.1 (Continued)

Country	Locations	Concentration range (ppb)	Sources	Climate condition	References
India	Fatehgarh, Punjab	0.02–18.4		Semiarid	(Kumar et al. 2020b)
India	Amritsar, Punjab	72–168	Excessive use of As-bearing agrochemicals	Semiarid	(Virk 2019)
India	Ropar, Punjab	2.0–11	Reductive dissolution of As-bearing minerals	Semiarid	(Sharma et al. 2018)
India	Ballia, Uttar Pradesh	0.34–75.6	Reductive dissolution of arsenic-bearing minerals	Subhumid	(Singh and Singh 2018)
India	Nadia, West Bengal	3–206	Overexploitation of groundwater		(A. Das et al. 2020)
Japan	Kumamoto region	0.1–60.6	Arsenic desorption from metal oxides		(Hossain et al. 2016)
Nepal	Bara	0.12–75.7			(Rasheed et al. 2022)
Pakistan	Jhelum	0.08–100.25		Semiarid	(Kazmierczak et al. 2022)
Vietnam	Red river delta	0–488	Presence of organic matter and reductive dissolution of iron oxyhydroxides		(Sierra et al. 2016)

up to 1000 ppb in many parts of Africa. Mostly, the dissolution of As-bearing minerals has been considered responsible for the high As concentration in their groundwater (Singh 2017; Medunić et al. 2020).

6.2 Conventional Methods for the Removal of Arsenic from Water

Although the WHO has recommended a maximum permissible limit of 10 μg/L for As in drinking water, there are many nations where people are exposed to a much higher concentration. Several methods and techniques based on a phenomenon like oxidation, precipitation, coagulation, adsorption, ion exchange, and membrane processes, have been used to tackle arsenic contamination (Figure 6.3) (Mohan and Pittman 2007). These techniques have some advantages and disadvantages, as represented in Table 6.2.

Most of these techniques are costly, therefore, unaffordable to be used in poor or rural areas. Hence, people in those areas practiced various household treatment methods, also known as point-of-use methods. These methods are not only less expensive but also quicker.

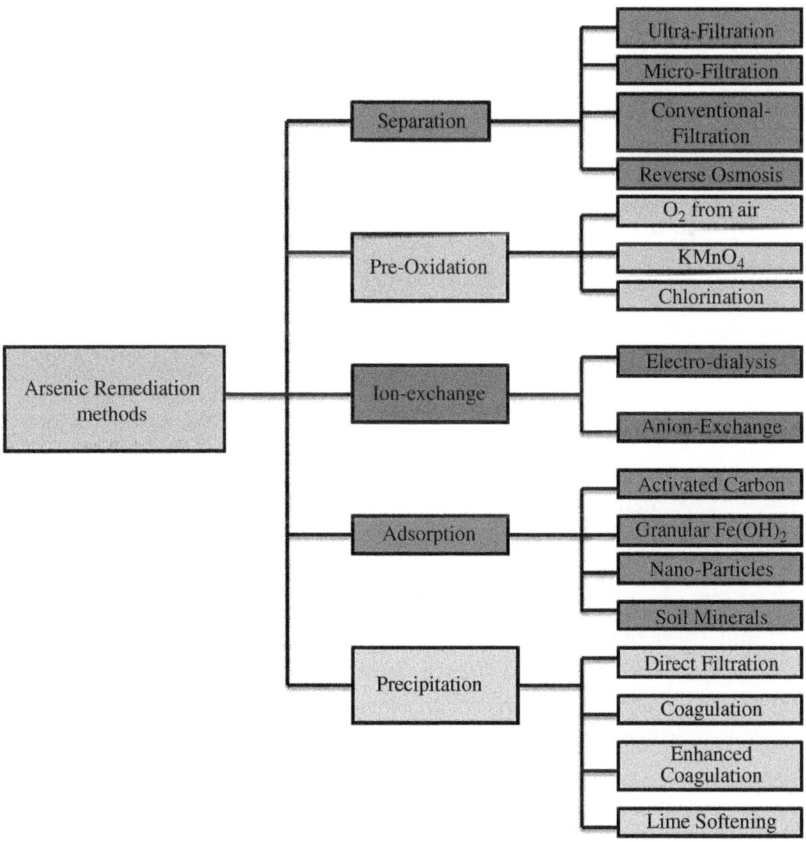

Figure 6.3 Major methods of As removal from water. *Source:* Adapted from Weerasundara et al. (2021).

Table 6.2 Comparison of different methods for arsenic removal from water.

Technologies	Advantages	Disadvantages
1) Oxidation		
a) Air oxidation	Very simple, low cost	Significantly less efficient and slow process
b) Chemical oxidation	Simple and rapid	Harmful for microbes, costly
2) Coagulation precipitation		
a) Alum	Relatively low cost	Efficiency is less for As(III), toxic sludge
b) Iron	Simple, high efficiency	Efficiency is less for As(III), toxic sludge
3) Sorption		
a) Alumina activated	Common, commercially available	Hi-tech maintenance and operation are required
b) Carbon activated	Economically efficient	A high quantity of carbon is required and produces toxic solid waste
c) By-product of industries	Cheap	May produce other pollutants
d) Zerovalent iron and iron oxide	Development scope	Cost is high
4) Stabilization	Highly efficient, the reduction rate is high	Costly, stability prediction is difficult
5) Cementation	Highly efficient	Stability of adsorbed arsenic needs to be confirmed
6) Ion exchange	Reaction is fast	produces sludge
7) Membrane	Highly efficient, well defined	Capital and running costs are high, high operation, and maintenance cost

One of the commonly used methods is Solar Oxidation and Removal of Arsenic (SORAS), a simple and economical method for treating arsenic that involves the irradiation of water with sunlight in small transparent bottles, preferably made up of polyethene terephthalate (Wegelin et al. 2000). This method involves the photochemical oxidation of As(III) followed by precipitation, and then filtration of As(V) adsorbed on Fe(III) hydroxides (Litter et al. 2012). Another most widely implemented technique, especially in Bangladesh, is the bucket treatment unit. Though cheap, both methods are not as efficient as electrocoagulation, adsorption, ion exchange, membrane filtration, etc., for removing arsenic. In the electrocoagulation method, considered as one of the simple and most efficient methods for arsenic removal in water, an electric current is allowed to dissolve a sacrificial anode generally made up of iron and aluminum in a contaminated aqueous medium with suspended, emulsified, or dissolved species leading to the production of cations such as Fe^{2+} or Al^{3+}, and some gases such as H_2 and O_2 (Chen 2004). Metals ions may react with OH^- ions generated at the cathode during the formation of H_2 gas, leading to the formation of insoluble hydroxides, which then remove pollutants from contaminated water by forming complexes and flocs due to electrostatic attraction (Kumar et al. 2004). Though this method removes As efficiently, it has some disadvantages too. It is pH-dependent and works more efficiently at pH 7 only. It also produces a large amount of As-rich coagulant requiring extra energy for its filtration and safe disposal (Ali et al. 2013; Weerasundara et al. 2021). Lewis acid–base interaction method is used to remove both As(III) and As(V), which are Lewis bases or strong ligands and can donate lone pair of electrons, by interacting them with some solid surface having Lewis acid characteristics (Park et al. 2009). On the other hand, the ion exchange method, also known as the Coulombic method, is efficient only for removing arsenic (V) as it possesses negative charges. Here, arsenic-contaminated water is allowed to pass through ion exchange resin beds, which then remove the arsenate and some other anions, particularly sulfate ions, leaving behind unnecessary pollutants on the bed. The bed after this cycle is rinsed with a brine solution (chloride exchange) for reusing ion exchange resin (Wang et al. 2000). Insoluble forms of Arsenic are removed chiefly by the precipitation method, which allows the precipitation of arsenic either in the form of calcium arsenate or ferric arsenate or As(III) sulfides (Robins 2006). In most cases, arsenic found in water is very less so reverse osmosis is a practical option for its removal (Sarkar et al. 2012), but it requires a high amount of energy and operating cost and has less efficiency (Fox et al. 2016). The forward osmosis process is another option that requires less pressure and operating cost but is significantly less efficient (Jin et al. 2012). In comparison to these conventional methods, most developing countries adopt the adsorption method to treat arsenic-contaminated water as it is considered a very efficient and economical method. This method does not require the addition of chemicals, skilled personnel, and consistent electricity though its efficiency is affected by pH, presence of other contaminants, exposure time, a dose of adsorbent, and temperature (Sarkar and Paul 2016). Mostly adsorbents such as granular adsorbents, metal oxide coated on sand, iron oxide-based adsorbents, zerovalent iron, iron/olivine composite are used to remove As from water. It has been reported that the quantitative efficiency of As removal from water by adsorption method is >95% (Kumar et al. 2019).

6.3 Arsenic Removal from Water Using Iron-Based Adsorbents

Though technologies like ion exchange, membrane filtration, and electrocoagulation can remove arsenic from wastewater to some extent, certain limitations exist. Hence, these technologies cannot meet the current standards required for the treatment of arsenic-contaminated water (Pillewan et al. 2011; Qu et al. 2013). NPs are one of the options to overcome these limitations. They have properties like high surface area, high reactivity, high specificity, catalytic potential, and the ability for self-assembly, thus making them a suitable candidate for water treatment applications (Qu et al. 2013). The most common and widely used NPs are those based on iron oxide, i.e. magnetite (Fe_3O_4), maghemite (γ-Fe_2O_3), and nZVI NPs. Both these categories of NPs have grabbed considerable interest in research focusing on engineering applications for water and wastewater treatment for the past few years (Yantasee et al. 2007; Li et al. 2012). Table 6.3 comprises the different iron-based adsorbents and their arsenic removal capacities. Tang et al. (2011) synthesized ultrafine NPs

Table 6.3 Arsenic removal capacities of different iron-based adsorbents.

Adsorbents	Specific surface area (m^2/g)	Water type	Initial concentration (mg/L)	Adsorption capacity As(III) (mg/g)	Adsorption capacity As(V) (mg/g)	Optimum condition	Adsorbent concentration	References
α-Fe_2O_3	162	Natural and synthetic water	10	95	47	pH 7, Temp 298 K	0.1	(Tang et al. 2011a, 2011b)
α-Fe_2O_3-polymer monolith	50	Synthetic water		2.7		pH 3–12	0.5	(Savina et al. 2011)
Perlite/γ-Fe_2O_3	27.7	Synthetic water			4.64	pH 7, Temp 298 K	0.05	(Thanh et al. 2011)
Fe_2O_3		Drinking water		20	4.9	pH 6, milling time 24 h	0.003–0.1	(Luther et al. 2012)
Fe_3O_4	179	Waste water	1	46.06	16.56	pH 7, Temp 300 K	0.06	(Feng et al. 2012
γ-Fe_2O_3	168.73	Synthetic water	10–200	74.83	105.25	pH 6, Temp 323 K	0.16	(Lin et al. 2012)
α-FeOOH	167.8	Synthetic water	100		76.34	pH 3, Temp 298 K	0.006	(Ghosh et al. 2012)
F400-Fe	896		1		3.25	pH 7, Temp 298 K		(Nieto-delgado and Rangel-mendez 2012)
ACZ-Fe	806				1.45	pH 7, Temp 298 K		(Nieto-delgado and Rangel-mendez 2012)
α-Fe_2O_3		Natural water	1.5	2.89	3.84	pH 4–7, Temp 298 K,	0.5	
γ-Fe_2O_3	100	Synthetic water			45	pH 7	2.15	(Kilianová et al. 2013)
Fe-Cu binary oxide	282	Synthetic water	10	122.3	82.7	pH 7, Temp 298 K	0.5	(Zhang et al. 2013)
Iron(III)-cerium(V) mixed oxide	104	Ground water	4.5 and 4.8	2.42	2.1	pH 7, Temp 303 K	2	(Basu and Ghosh 2013)
Polymer nanocomposite (Fe_3O_4)	156	Synthetic water		2.83		pH 8.6, Temp 299 K	1.7	(Vunain et al. 2013)
Fe_3O_4		Synthetic water		188.69	153.8	pH 7, Temp 300 K	1	(Lunge et al. 2014)
γ-FeOOH	135	Synthetic water	0–20		37.3	pH 7, Temp 298.5 K	0.25	(Faria et al. 2014)

Fe/Ni binary oxide	245	Synthetic water		168.6	90.1	pH 7, Temp 298 K	0.5	(Liu et al. 2015
Reduced graphene oxide-supported mesoporous Fe_2O_3/ TiO_2	275.23	Potable water	20	77.7	99.5	pH 6–7	0.2	(Babu et al. 2016)
Fe_7S_8 nanoparticles		Synthetic water	0.3	14.3	31.3	pH 4, Temp 298 K	0.01	(Cantu et al. 2016)
Fe/Cu nanoparticles	79.5	Synthetic water		19.68	21.32	pH 7, Temp 295 K	0.05	(Babaee et al. 2018)
Sulfur-doped Fe_3O_4 nanoparticles		Synthetic water		58.38	20.24	pH 2–10		(Liu et al. 2018)
Chitosan magnetic graphene oxide nanocomposites	152.38	Synthetic water	10	45		pH 7.3, Temp 298 K	5	(Sherlala et al. 2019)
Fe_3O_4@CuoandGO	159.6	Synthetic water		70.36	62.60	pH 3–10, Temp 298 K	0.12	(Wu et al. 2019)
Polymer-supported iron oxide nanoparticles		Synthetic water			28.57	pH 6.6+–0.3 Temp 297 K	0.05 to 0.5	(Saif et al. 2019)
Chitosan functionalized iron nanosheet	111.8	Synthetic water	2–60	108.6		pH 7	0.4	(Zeng et al. 2020)
Magnetic graphene oxide nanocomposite	180.84	Synthetic water			69.44	pH 3	0.4	(Thy et al. 2020)
GT-1 iron nanoparticles	37.8				19.9	pH 6, Temp 298 K	0.002	(Wu et al. 2020)
Calcined-Fe/Pd nanoparticles	335	Synthetic water	1	1.176		pH 6.2, Temp 303 K	0.4	(Lin et al. 2021)

α-Fe_2O_3 by a solvent, thermal process that removed As from natural and synthetic water. Removal capacities of As(V) and As(III) were 47 and 95 mg/g, respectively, using 0.1 g/L of adsorbent (α-Fe_2O_3). Its efficiency was good in a higher concentration of competing ions and even when arsenic is present in a meagre amount in contaminated water. Andjelkovic et al. (2014) assessed the efficiency of a novel adsorbent synthesized by mechanical–chemical treatment of Fe_3O_4 and MnO_2 in a fixed ratio of 3 : 1 for removing arsenic from water. The result reported the adsorption capacities for removing both As(V) and As(III) to be very low in both cases; thus, the synthesized adsorbent proved to be not that significant for removing arsenic from water. Attia et al. (2014) also developed a novel adsorbent by coating the magnetic gamma-Fe_2O_3 NPs with zeolite to check its efficiency in removing arsenic from water and found its adsorption capacity for As to be 19.39 mg/g from a solution with an initial concentration of 20 mg/L at an optimum pH 2.5 and temperature 297 K. Lunge et al. (2014) synthesized magnetic iron oxide NPs by using the template of tea waste (MION-Tea). Adsorption capacities for As(III) and As(V) were 188.69 and 153.8 mg/g, respectively, using 1 g/L of adsorbent at a pH 7 with an initial concentration of 2 mg/L As in solution.

Liu et al. (2015) developed Fe-Ni binary oxide NPs to remove As(III) and As(V) from water and reported these particles to be very effective in removing arsenic species from water as compared to all other arsenic adsorbents with maximum adsorption capacities of 168.6 and 90.1 mg/g for removing As(III) and As(V), respectively. These particles also have no significant effect on other competing ions present in water. Feng et al. (2012) coated Fe_3O_4 NPs with super paramagnetic ascorbic acid and found that coating of ascorbic acid increased the surface area of Fe_3O_4 from 15.63 to 179 m^2/g. Ascorbic acid was used here as a ligand which is highly soluble in water and has strong coordination with transition metals, thus increasing adsorption of As from wastewater. Maximum adsorption capacities of As(III) and As(V) were found to be 46.06 and 16.56 mg/g, respectively, at an optimum pH 7 and temperature 300 K by using 60 mg/L of adsorbent. However, adsorption of As decreased with the increasing pH of the solution. It also worked well in contaminated water having low arsenic concentration. In another experiment, Fe@Cu oxide nanocomposite was fabricated with graphene oxides which also increased the specific surface area from 83.22 to 159.6 m^2/g and also increased hydroxyl groups. Maximum adsorption capacities for As(III) and As(V) were reported to be 70.36 and 62.60 mg/g, respectively (Wu et al. 2019). Among all iron-based adsorbents listed in Table 6.3, the highest adsorption capacity of As(III) and As(V) was reported by using MION-Tea.

6.4 Zerovalent Iron Nanoparticles

Nanomaterials show distinctive properties such as smaller size, larger surface area, a higher number of reactive sites, easy separation, and catalytic potential that provide better arsenic removal efficiency than other available adsorbents. Researchers have recently shown a significant interest in removing metals and some organic contaminants from water using nZVI NPs because of its smaller size and larger surface-area-to-volume ratio than bulk zerovalent iron (ZVI), along with its high reactivity due to the high surface area. Nowadays, nZVI is used in water treatment systems and in several research and industrial sectors as catalysts, adsorbents, etc. (Huber 2005). One of the advantages of using these particles in the water system is that these particles can be transported effectively by the flow of water after modification.

6.4.1 Synthesis of nZVI Particles

The reactivity and mobility of nZVI particles depend on the size, surface, modifying/capping material, oxide layer, or support material of which nZVI is made up of. All these factors are considered during the synthesis of nZVI. For instance, to prevent the agglomeration of nZVI, NPs are supported and dispersed on solid particles. Surface modification is done using organic polymers and polyelectrolytes (Adusei-Gyamfi and Acha 2016). In general, nZVI particles can be synthesized using physical or chemical methods in two ways:

i) Top-down synthesis – In this method, the size of bulk iron is reduced to nanoscale. It includes milling processes, pulsed laser ablation, and noble gas sputtering methods.
ii) Bottom-up synthesis – In this method, NPs are produced by building up the nano-iron particles from atoms generated from ions or molecules. It includes production from dissolved iron salts, nanosized iron oxides, or iron-containing molecules or by carbonization at high temperature or by electrochemical deposition process or with the aid of reducing agents such as $NaBH_4$, N_2H_4, and $Na_2S_2O_4$.

The nZVI which is synthesized using a bottom-up approach has a typical core-shell structure in which the core consists of metallic or zerovalent iron and a mixed valent oxide shell of Fe^{2+} and Fe^{3+} gets formed as a result of oxidation of core shell (Figure 6.4).

6.4.2 Application of nZVI in Arsenic Removal from Groundwater/Synthetic Water

Groundwater remediation using ZVI(Fe^0) was first employed in early 1990 in permeable reactive barrier (PRB) systems where ZVI was tested as a PRB material to remediate groundwater for longer term (Reynolds et al. 1990; Gillham and O'Hannesin 1994). One of the drawbacks of ZVI is that it can only be used for shallow groundwater plume treatment, cannot treat source zone water, and is less effective in removing contaminants. To tackle this problem, nZVI was discovered. Due to its small size (<100 nm), high surface area, and very high reactivity, the use of nZVI has been increased for treating contaminated soil,

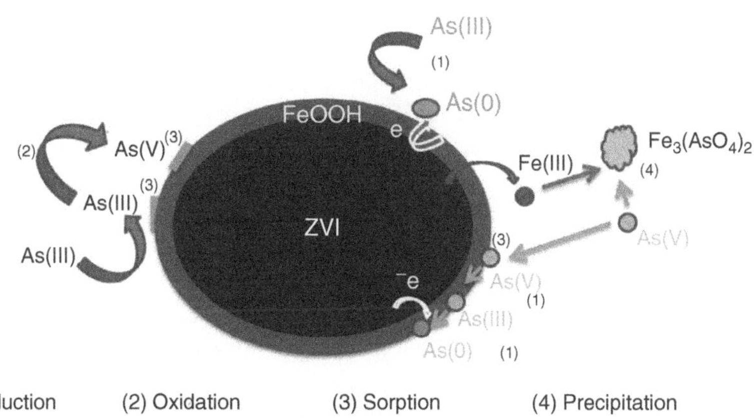

Figure 6.4 Core-shell structure of nZVI depicting mechanism for removal of As(III) and As(V) in different redox conditions. *Source:* Adapted from Yan et al. (2013).

groundwater, and wastewater. Because of its high adsorption capacity, it is considered as the most appropriate option for the rapid removal of contaminants from aqueous solution, thus, widely used for the removal of arsenic (Li et al. 2006; Morgada et al. 2009). The first reason for nZVI being used in remediation purposes lies in the electron-donating nature of ZVI (Stumm and Morgan 1996). This property is contributed by its high reduction potential. Under anaerobic conditions, oxidation of nZVI (Fe^0) mediated by water releases Fe^{2+} and H_2 (Eqs. 6.7 and 6.8). Known as the probable reducing agent, both Fe^{2+} and H_2 also make nZVI efficient for remediation purposes (Lu et al. 2016).

$$2Fe^0 + 4H^+ + O_2 \rightarrow 2Fe^{2+} + 2H_2O \tag{6.7}$$

$$2Fe^0 + 2H_2O \rightarrow Fe^{2+} + H_2 + 2OH^- \tag{6.8}$$

Secondly, these NPs may be injected directly into the region of pollutant sources (Pasinszki and Krebsz 2020). Thirdly, they are pretty cost-effective, i.e. 0.14 US dollar/g as compared to other NPs such as TiO_2 (0.18 US dollar/g), magnetite (0.44 US dollar/g), and other iron-oxide NPs (1.20 US dollar/g) (Brame et al. 2011).

Studies revealed that arsenic is removed by ZVI generally by the mechanism of adsorption, surface precipitation, coprecipitation, and reduction with various iron corrosion products such as ferrous/ferric hydroxides (Mak et al. 2009). However, the effectiveness of these mechanisms depends upon the operating conditions. Thus, to better understand the overall removal mechanism, it is mandatory to describe the effect of critical parameters, such as pH, dissolved oxygen, hardness, and humic acid (HA) on the removal of arsenic.

Kanel et al. (2005) conducted batch experiments to find out the feasibility of nZVI synthesized to remove As(III) from groundwater at different pH ranges from 3 to 12. The maximum adsorption capacity of As(III) calculated by Freundlich adsorption isotherm was found to be 3.5 mg of As(III)/g of nZVI at optimum pH 7 and temperature 25 °C. About 88.6–99.9% of As adsorption was found in the pH range from 4 to 10. Below pH 3 and above pH 10, the adsorption was decreased sharply. HCO_3^-, $H_4SiO_4^0$, and $H_2PO_4^{2-}$ were the potential interferences in the adsorption reaction of As(III). In another batch experiment, they investigated the feasibility of synthesized nZVI to remove the As(V) from groundwater at pH ranging between 3 and 11. The study reported 100% removal of As(V) (1 mg/L) with nZVI (0.1 g/L) at pH 3–7. The trend of adsorption decreasing with an increase in pH was the same as in the case of As(III) (Kanel et al. 2006).

Though the efficiency of bare nZVI to remove arsenic was good, it has been reported that direct application of nZVI for treating contaminants in water may result in rapid loss of nZVI which may further cause iron pollution in drinking water due to its tiny particle size, high mobility, high agglomeration tendency, lack of stability, and low, reducing specificity in water. It must be loaded onto some supporting material or used with surface stabilizers such as alginate, chitosan, activated carbon, and other porous structures (He and Zhao 2005). The efficiencies of these different loaded and supported nZVI in arsenic removal are compared in Table 6.4.

Zhu et al. (2009) investigated the performance of nZVI/AC synthesized by loading nZVI onto activated carbon for arsenic removal from drinking water through batch adsorption experiments. It was found that nZVI/AC was effective for arsenic removal with relatively

Table 6.4 Arsenic removal efficiency of different nZVI nanoparticles.

Adsorbents	Water type	Surface area (m^2/g)	Initial concentration (mg/L)	Adsorption capacity As(III) (mg/g)	Adsorption capacity of As(V) (mg/g)	Reaction conditions	Adsorbent dose (g/L)	References
N/S-nZVI	Synthetic water		1–100	121	125	pH 6.5, Temp 298 K	0.5	(Du et al. 2013)
nZVI/AC	Synthetic water	141	5	26.8		pH 7		(Baikousi et al. 2015)
nZVI/ montmorillonite	Synthetic water	36.97	5	59.9	45.5	pH 7, Temp 295 +–1 K	1	(Bhowmick et al. 2014)
nZVI/AC	Groundwater	69.4	2–100	18.2	12	pH 3.5, Temp 318 K	1.5	(Zhu et al. 2009)
nZVI	Synthetic water		100	1.7	0.7	pH 7	1	(Wu et al. 2017)
F-nZVI	Synthetic water	218.84		50.08	91.42		1	(Yadav et al. 2016)
Iron chitosan nanoparticles	Groundwater	69		94+-1.5	119—-2.6	pH 7, Temp 298 K	0.1	(Gupta et al. 2012)
CS-nZVI-CMβ-CD	Surface and groundwater	12.1		18.51	13.51	pH 6, 295 K	0.1	(Sikder et al. 2014)
Ca-alginate nZVI	Synthetic and groundwater					pH 7, 298 K	0.02	(Bezbaruah et al. 2013)
LnZVI	Synthetic groundwater				135	pH 5		(Dong et al. 2012)
nZVI-RGO	Synthetic water	100.65	7	35.83	29.04	pH 7, Temp 298 K	0.4	(Wang et al. 2014)
nZVI@SiO2@ cellulose	Synthetic water		300	70	70	pH 2.8–4.0, Temp 303 K		(Liu et al. 2019)
GB/nZVI	Synthetic water	60.08	40	181.5		pH 4	0.1	(Liu et al. 2020)
OL-nZVI	Synthetic water			300		pH 4–8,		(Poguberović et al. 2016)
nZVI/GS	Synthetic water	199	1–20	45.57	45.12	pH 4–8		(Liu et al. 2019)
GnZVI	Groundwater		5	36	43–49			(Das and Bezbaruah, 2021)

fast kinetics. At pH 6.5, the adsorption capacity for arsenite and arsenate calculated using Langmuir adsorption isotherm was 18.2 and 12.0 mg/g, respectively. However, it appeared to be more effective for arsenite removal than arsenate at pH >5. It was also evidenced by a study conducted (Zhu et al. 2020) on arsenic(V) removal by the same adsorbent nZVI/AC. It was observed that at adsorbent dose 1.5 mg/L, pH 3.5, and temperature 308 K, around 99% removal of As(V) occurred, but as pH increased from 3.5 to 6.5 and 9.5, the As(V) adsorption efficiency on nZVI/AC decreased by 10.1 and 73.4%, respectively. The removal efficiency of both arsenite and arsenate was found to be decreasing in the presence of phosphate and silicate, while the presence of other anions and HA did not affect their removal. On the other hand, common metal cations (Ca^{2+}, Mg^{2+}) increased arsenate adsorption but ferrous iron (Fe^{2+}) decreased the adsorption. It was also reported that regeneration of spent adsorbent nZVI/AC could be effectively achieved by elution with 0.1 M dilute NaOH. The adsorption capacity of arsenic stopped decreasing after eight cycles of adsorption–regeneration. In another study, the performance of chitosan zerovalent iron NPs (CIN) was evaluated for effective removal of both species of Arsenic, i.e. As(III) and As(V), in groundwater over different pH ranges. The result revealed that the concentration of As(III) and As(V) decreased from 2 mg/L to less than 5 μg/L in just less than 180 minutes. When injected with an initial adsorbent dose of 0.5 g/L, CIN was reported to follow pseudo-first-order kinetics with maximum adsorption capacity (as calculated by Langmuir isotherm model) of 94 ± 1.5 and 119 ± 2.6 mg/g for As(III) and As(V), respectively, at pH 7. Moreover, competing anions like phosphate and silicate were also found to have an insignificant effect on the adsorption of arsenic by CIN (Gupta et al. 2012). The bifunctional resin-ZVI composite (Du et al. 2013) was synthesized by fusing the oxidation characteristics of nZVI/O_2 with adsorption properties of iron oxide and anion exchange resin N-S. The result indicated that N-S-ZVI could effectively remove both As(III) and As(V) from water. The Langmuir isotherm model was used to calculate maximum adsorption capacity, 121 and 125 mg/g for As(III) and As(V), respectively, at pH 6.5 and temperature 25 °C. The efficiency for As(III) removal at different pH ranges was also calculated, and the order for adsorption behavior at different pH values was found to be pH 5 > pH 7 > pH 9 > pH 3. Bhowmick et al. (2014) experimented with montmorillonite-supported nanoscale zerovalent iron NPs to analyze their efficiency in removing both As(III) and As(V) from the aqueous solution. Adsorption was analyzed for both As(III) and As(V) over a wide range of pH 4–10. The result reported that adsorption kinetic followed pseudo-second-order reaction with maximum adsorption taking place between pH 4 and 8, and as pH increased above 9, the adsorption was found to be decreased. The maximum adsorption capacity calculated from the Langmuir adsorption isotherm was found to be 59.9 and 45.5 mg/g for As(III) and As(V), respectively, at pH 7.0 and temperature 22 ± 1 °C. NO_3^-, SO_4^{2-}, and HCO_3^- were found to have a trivial effect on the adsorption of both As(III) and As(V). However, PO_4^{2-} was found to reduce the adsorption of both species. Bezbaruah et al. (2014) evaluated the efficiency of calcium-alginate entrapped nZVI to remove As(V) from synthetic water by comparing it with the performance of bare nZVI. Batch mode studies were performed using the solution with initial As(V) concentrations ranging from 1 to 10 mg/L. It was observed that 85 and 100% As removal was achieved within 2 hours. The reaction rate constant obtained for both bare nZVI and entrapped nZVI was compared using a two-way variance analysis. The result indicated no significant difference between them at 95% confidence (pvalue > 0.05). However, at the end of 120 minutes, the significant greater removal efficiency was observed in the case of entrapped nZVI as compared to bare nZVI. Moreover, when this entrapped nZVI is applied to the groundwater where As(V) is present in low concentration (50–100 μ/L), results indicated that Ca-alginate entrapped nZVI can remove arsenic effectively. Moreover, the presence of Na^+, Ca^{2+}, Cl^-, and HCO_3^- did not affect the removal by entrapped nZVI. CS-nZVI-CMb-CD beads have been successfully tested (Sikder et al. 2014) for the removal of As(III) and As(V) by performing batch adsorption experiments. The presence of the highly functional CS-CMb-CD provided fast kinetics to remove both As(III) and As(V) in water. The data were found to be well fitted with the Langmuir model, and the maximum adsorption capacity of As(III) and As(V) on CS-nZVI-CMb-CD beads was found to be 18.51 and 13.51 mg/g, respectively, at pH 6 and temperature 298 K. The study also reported that the adsorbent used can be magnetically removed; thus, it can be reused again successfully for arsenic removal in water. Yadav et al. (2016) synthesized Fuller's earth immobilized nanoscale zerovalent iron (F-nZVI 1-8) and studied its potential to adsorb As(III) and As(V) from aqueous solution at different pH values from 2 to 11. The maximum adsorption capacity as calculated by Langmuir isotherm was 50.08 mg/g for As(III) and 91.42 mg/g for As(V), at optimum pH 8 and 3, respectively, while stirring for 24 hours at 150 rpm and 25±2 °C. Shaikh et al. (2020) developed nZVI-BC by precipitating ZVI onto biochar synthesized from Cassia fistula's pods via pyrolysis and tested its performance for the removal of Arsenic from water. Langmuir adsorption capacity was 1.04 and 1.40 mg/g for As(III) and As(V), respectively. Residual As concentration thus formed was less than the prescribed detection limit of 10 μg/L by the WHO. At the optimal initial concentration of 1 and 1.25 mg/L for As(III) and As(V), respectively, it depicted 99.1 and 96.1% efficiency removal of As(III) and As(V), respectively. By maintaining the same optimal conditions, when this adsorbent was applied in the groundwater of Sahibganj district of Jharkhand, results showed about 91.5–93.2% of As removal. In another experiment, GB/nZVI NPs were synthesized which showed a maximum adsorption capacity of As(III) removal efficiency of 181.5 mg/g at a pH 4. The addition of GB with nZVI enhanced the arsenic removal by enhancing the conversion of As(III) to As(V) (Liu et al. 2020). It was applied to treat irrigation water and removed 100% of As from irrigation water.

6.5 Merits and Demerits of Using nZVI

The current advancement in nanotechnology has resulted in the synthesis of many new and efficient nanomaterials suitable for water treatment. Having the properties like high specificity, high reactivity, high surface area, high adsorption capacity, and high mobility, these NPs have been considered by many researchers as novel adsorbents for various environmental pollutants. Among these, iron-based NPs, specifically, nZVI NPs have proven to be suitable nanomaterials for treating As-contaminated water. These NPs removed arsenic by substituting the toxic contaminant with iron, enhancing their applicability in the remediation of contaminants from the environment. These adsorbents showed excellent arsenic removal efficiency in the pH range of 4–9. However, as compared to other iron-based oxides, nZVI is more favorable because:

1) They have high reduction potential compared to other NPs, making them to more readily react with contaminants like arsenic, which has an electron acceptance nature.
2) Due to their minimal size, they offer more surface area and sites for the adsorption of arsenic compared to others.
3) They can be directly injected into the groundwater.
4) Some of these are comparatively more cost-effective than other NPs.
5) When capped or supported with some other materials like activated carbon, chitosan, bio char, and montmorillonite, their mobility, effectiveness to remove arsenic, and reusability increases which makes them more preferable over others.

Besides offering specific merits over other NPs, certain limitations are also associated with nZVI. Since extensive research has not been done in this prospect, it can be a challenge to the applicability of nZVI in arsenic removal. Some of the limitations include:

1) Though various methods of nZVI synthesis have been developed, none offer a cost-effective approach to the large-scale production of these NPs.
2) One of the significant limitations of these NPs is that they readily get aggregated when used to remediate groundwater. The groundwater possesses relatively high ionic strength, which causes a decrease in electrostatic repulsion among these particles, thus, resulting in their aggregation. Hence, using bare nZVI for large-scale production does not seem to possess excellent and efficient results.
3) The effectiveness of most of the modified nZVI was assessed in lab conditions. Significantly less work has been done on these particles in field conditions. Therefore, their cost-effectiveness and applicability in field conditions are questionable.
4) To date, no studies have been done from an eco-toxicological point of view. Therefore, the impact of their long-term use on living organisms is still unknown.
5) Various studies indicated the possibility of migration of these particles to a few inches or feet from their injection point. Therefore, the effect of their mobility to a noncontaminated site is unknown and yet to be investigated.

6.6 Conclusions and Future Perspectives

From the past few decades, arsenic contamination in groundwater has increased at a very alarming rate throughout the globe, especially in countries such as India, Bangladesh, Pakistan, and Vietnam. The mobility of arsenic in groundwater has been governed by various oxidative and reductive dissolution processes. Some iron-reducing and sulfate-reducing microbes also play an important role in its mobilization in aquifers. Iron-based NPs, especially nZVI, exhibit good characteristics for their applicability in groundwater treatment and environmental remediation. However, it has been observed that when applied over a large area, these NPs are usually accompanied by many new potential environmental risks such as the tendency to aggregate, and high reactivity with water and oxygen, which ultimately limit their application to a certain extent. Intensive research has been carried out to overcome these limitations by modifying these NPs to increase their stability, reactivity, adsorption capacity, transportability, and mobility in an aqueous medium. To meet this purpose, specific capping and support materials, such as activated carbon, bio char, and montmorillonite, have been introduced. Both iron core (present in nZVI) and modifiers work well together to enhance its adsorption capacity. There also seems to be a fine balance between decreased reactivity and enhanced stability and mobility. Therefore, attempts have been made to produce stable and modified nZVI NPs that can meet specific application requirements and can be applied on a larger scale with cost-efficient methods without any potential risks. Few studies are available on the direct field application of nZVI for arsenic removal; therefore, the long-term toxicity effect of nZVI cannot be predicted based on these studies. Hence, the necessary aspects of future research should be to produce such nZVI particles that can be directly applied to the contaminated areas in an efficient and economical way, and to assess the probable long-term toxicity effects that can be associated with these. Attempts should also be made in future to develop biodegradable, renewable, and environmentally friendly support materials so that potential risks of nZVI on the environment can be controlled. Though certain drawbacks are associated with nZVI and detailed research is required in this aspect, in the present time, nZVI with properties like high adsorption capacity, high surface area, and high reactivity seems to be the best candidate to meet the purpose of remediation of arsenic-contaminated water. A very promising future seems to be associated with the application of nZVI in groundwater remediation. Studies are expected to increase in the coming time at both laboratory and field scales to modify and improve the properties of nZVI NPs and gain further knowledge about its working mechanism and its associated long-term toxic effects.

References

Adusei-Gyamfi, J., & Acha, V. (2016). Carriers for nano zerovalent iron (nZVI): Synthesis, application and efficiency. *RSC Advances*, 6, 91025–91044. https://doi.org/10.1039/c6ra16657a

Alam, M. O., Shaikh, W. A., Chakraborty, S., Avishek, K., & Bhattacharya, T. (2016). Groundwater arsenic contamination and potential health risk assessment of gangetic plains of Jharkhand, India. *Exposure and Health*, 8(1), 125–142. https://doi.org/10.1007/s12403-015-0188-0

Ali, I., Asim, M., & Khan, T. A. (2013). Arsenite removal from water by electrocoagulation on zinc-zinc and copper-copper electrodes. *International Journal of Environmental Science and Technology*, 10(2), 377–384. https://doi.org/10.1007/s13762-012-0113-z

Andjelkovic, I., Nesic, J., Stankovic, D., Manojlovic, D., Pavlovic, M. B., Jovalekic, C., & Roglic, G. (2014). Investigation of Sorbents synthesised by mechanical-chemical reaction for sorption of As(III) and As(V) from aqueous medium. *Clean Technologies and Environmental Policy*, 16(2), 395–403. https://doi.org/10.1007/s10098-013-0635-1

Attia, T.M.S., Hu, X.L., & Qiang (2014). Synthesized magnetic nanoparticles coated zeolite (MNCZ) for the removal of arsenic (As) from aqueous solution. *Journal of Experimental Nanoscience*, 9(6), 551–556.

Babaee, Y., Mulligan, C. N., & Rahaman, M. S. (2018). Removal of arsenic (III) and arsenic (V) from aqueous solutions through adsorption by Fe/Cu nanoparticles. *Journal of Chemical Technology and Biotechnology*, 93(1), 63–71. https://doi.org/10.1002/jctb.5320

Babu, C. M., Vinodh, R., Sundaravel, B., Abidov, A., Peng, M. M., Cha, W. S., & Jang, H. T. (2016). Characterization of reduced graphene oxide supported mesoporous Fe_2O_3/TiO_2

nanoparticles and adsorption of As(III) and As(V) from potable water. *Journal of the Taiwan Institute of Chemical Engineers*, 62, 199–208. https://doi.org/10.1016/j.jtice.2016.02.005

Baikousi, M., Georgiou, Y., Daikopoulos, C., Bourlinos, A. B., Filip, J., Zbořil, R., Deligiannakis, Y., & Karakassides, M. A. (2015). Synthesis and characterization of robust zero valent iron/ mesoporous carbon composites and their applications in arsenic removal. *Carbon*, 93, 636–647. https://doi.org/10.1016/j.carbon.2015.05.081

Basu, T., & Ghosh, U. C. (2013). Applied Surface Science Nano-structured iron (III)– cerium (IV) mixed oxide: Synthesis, characterization and arsenic sorption kinetics in the presence of co-existing ions aiming to apply for high arsenic groundwater treatment. *Applied Surface Science*, 283, 471–481. https://doi.org/10.1016/j.apsusc.2013.06.132

Bentahar, Y., Hurel, C., Draoui, K., Khairoun, S., & Marmier, N. (2016). Adsorptive properties of Moroccan clays for the removal of arsenic(V) from aqueous solution. *Applied Clay Science*, 119, 385–392. https://doi.org/10.1016/j.clay.2015.11.008

Bezbaruah, A. N., Kalita, H., Almeelbi, T., Capecchi, C. L., Jacob, D. L., Ugrinov, A. G., & Payne, S. A. (2014). Ca-alginate-entrapped nanoscale iron: Arsenic treatability and mechanism studies. *Journal of Nanoparticle Research*, 16(1). https://doi.org/10.1007/s11051-013-2175-3

Bhowmick, S., Chakraborty, S., Mondal, P., & Renterghem, W. Van. (2014). Montmorillonite-supported nanoscale zero-valent iron for removal of arsenic from aqueous solution: Kinetics and mechanism. *Chemical Engineering Journal*, 243, 14–23.

Bhowmick, S., Pramanik, S., Singh, P., Mondal, P., Chatterjee, D., & Nriagu, J. (2018). Arsenic in groundwater of West Bengal, India: A review of human health risks and assessment of possible intervention options. *Science of the Total Environment*, 612, 148–169. https://doi.org/10.1016/j.scitotenv.2017.08.216

Blowes, D. W., Ptacek, C. J., Jambor, J. L., Weisener, C. G., Paktunc, D., Gould, W. D., & Johnson, D. B. (2013). The geochemistry of acid mine drainage. In Treatise on Geochemistry: Second Edition (11th ed., Vol. 11). Elsevier Ltd. https://doi.org/10.1016/B978-0-08-095975-7.00905-0

Bodek, I., Lyman, W. J., Reehl, W. F., & Rosenblatt, D. H. (1988). Environmental Inorganic Chemistry: Properties, Processes, and Estimation Methods. Pergamon press, New York.

Brame, J., Li, Q., & Alvarez, P. J. J. (2011). Nanotechnology-enabled water treatment and reuse: Emerging opportunities and challenges for developing countries. *Trends in Food Science and Technology*, 22(11), 618–624. https://doi.org/10.1016/j.tifs.2011.01.004

Bundschuh, J., & Maity, J. P. (2015). Geothermal arsenic: Occurrence, mobility and environmental implications. *Renewable and Sustainable Energy Reviews*, 42, 1214–1222. https://doi.org/10.1016/j.rser.2014.10.092

Butaciu, S., Senila, M., Sarbu, C., Ponta, M., Tanasella, C., Cadar, O., Roman, M., Radu, E., Sima, M., & Frentiu, T. (2017). Chemical modeling of groundwater in the Banat Plain, southwestern Romania, with elevated As content and co-occurring species by combining diagrams and unsupervised multivariate statistical approaches. *Chemosphere*, 172, 127–137. https://doi.org/10.1016/j.chemosphere.2016.12.130

Cantu, J., Gonzalez, L. E., Goodship, J., Contreras, M., Joseph, M., Garza, C., Eubanks, T. M., & Parsons, J. G. (2016). Removal of arsenic from water using synthetic Fe_7S_8 nanoparticles. *Chemical Engineering Journal*, 290, 428–437. https://doi.org/10.1016/j.cej.2016.01.053

Chakraborti, D., Das, B., Rahman, M. M., Chowdhury, U. K., Biswas, B., Goswami, A. B., Nayak, B., Pal, A., Sengupta, M. K., Ahamed, S., Hossain, A., Basu, G., Roychowdhury, T., &

Das, D. (2009). Status of groundwater arsenic contamination in the state of West Bengal, India: A 20-year study report. *Molecular Nutrition and Food Research*, 53(5), 542–551. https://doi.org/10.1002/mnfr.200700517

Chakraborti, D., Rahman, M. M., & Ahamed, S. (2016). Arsenic contamination of groundwater and its induced health effects in Shahpur block, Bhojpur district, Bihar state, India: Risk evaluation. *Environmental Science and Pollution Research*, 9492–9504. https://doi.org/10.1007/s11356-016-6149-8

Chauhan, V. S., Nickson, R. T., Chauhan, D., Iyengar, L., & Sankararamakrishnan, N. (2009). Ground water geochemistry of Ballia district, Uttar Pradesh, India and mechanism of arsenic release. *Chemosphere*, 75(1), 83–91. https://doi.org/10.1016/j.chemosphere.2008.11.065

Chen, G. (2004). Electrochemical technologies in wastewater treatment. *Separation and Purification Technology*, 38(1), 11–41.

Das, T. K., & Bezbaruah, A. N. (2021). Comparative study of arsenic removal by iron-based nanomaterials: Potential candidates for field applications. *Science of the Total Environment*, 764, 142914. https://doi.org/10.1016/j.scitotenv.2020.142914

Das, A., Das, S. S., Chowdhury, N. R., Joardar, M., Ghosh, B., & Roychowdhury, T. (2020). Quality and health risk evaluation for groundwater in Nadia district, West Bengal: An approach on its suitability for drinking and domestic purpose. *Groundwater for Sustainable Development*, 10, 100351. https://doi.org/10.1016/j.gsd.2020.100351

Daus, B., Wennrich, R., & Weiss, H. (2004). Sorption materials for arsenic removal from water: A comparative study. *Water Research*, 38(12), 2948–2954. https://doi.org/10.1016/j.watres.2004.04.003

Dong, H., Guan, X., & Lo, I. M. C. (2012). Fate of As(V)-treated nano zerovalent iron: Determination of arsenic desorption potential under varying environmental conditions by phosphate extraction. *Water Research*, 46(13), 4071–4080. https://doi.org/10.1016/j.watres.2012.05.015

Du, Q., Zhang, S., Pan, B., Lv, L., Zhang, W., & Zhang, Q. (2013). Bifunctional resin-ZVI composites for effective removal of arsenite through simultaneous adsorption and oxidation. *Water Research*, 47(16), 6064–6074. https://doi.org/10.1016/j.watres.2013.07.020

Erbs, J. J., Berquó, T. S., Reinsch, B. C., Lowry, G. V., Banerjee, S. K., & Penn, R. L. (2010). Reductive dissolution of arsenic-bearing ferrihydrite. *Geochimica et Cosmochimica Acta*, 74(12), 3382–3395. https://doi.org/10.1016/j.gca.2010.01.033

Faria, M. C. S., Rosemberg, R. S., Bomfeti, C. A., Monteiro, D. S., Barbosa, F., Oliveira, L. C. A., Rodriguez, M., Pereira, M. C., & Rodrigues, J. L. (2014). Arsenic removal from contaminated water by ultrafine δ-FeOOH adsorbents. *Chemical Engineering Journal*, 237, 47–54. https://doi.org/10.1016/J.CEJ.2013.10.006

Feng, L., Cao, M., Ma, X., Zhu, Y., & Hu, C. (2012). Superparamagnetic high-surface-area Fe_3O_4 nanoparticles as adsorbents for arsenic removal. *Journal of Hazardous Materials*, 218, 439–446.

Fox, D. I., Stebbins, D. M., & Alcantar, N. A. (2016). Combining ferric salt and cactus mucilage for arsenic removal from water. *Environmental Science and Technology*, 50(5), 2507–2513. https://doi.org/10.1021/acs.est.5b04145

Ghosh, M. K., Eddy, G., Poinern, J., Issa, T. B., & Singh, P. (2012). Arsenic adsorption on goethite nanoparticles produced through hydrazine sulfate assisted synthesis method. *Korean Journal of Chemical Engineering* 29(1), 95–102. https://doi.org/10.1007/s11814-011-0137-y

Gillham, R. W., & O'Hannesin, S. F. (1994). Enhanced degradation of halogenated aliphatics. In Leonard F. Konikow, Groundwater (Vol. 32, Issue 6, pp. 958–967). Wiley-Blackwell.

Gonzalez Rodriguez, B., Rietveld, L. C., Longley, A. J., & van Halem, D. (2019). Arsenic contamination of rural community wells in Nicaragua: A review of two decades of experience. *Science of the Total Environment*, 657, 1441–1449. https://doi.org/10.1016/j.scitotenv.2018.12.168

Greenwood, N. N., & Earnshaw, A. (1984). Chemistry of the Elements. Pergamon Press, Oxford. Chap, 10, 435.

Greenwood, N. N., & Earnshaw, A. (2012). Chemistry of the Elements. Elsevier.

Guo, H., Wen, D., Liu, Z., Jia, Y., & Guo, Q. (2014). A review of high arsenic groundwater in Mainland and Taiwan, China: Distribution, characteristics and geochemical processes. *Applied Geochemistry*, 41, 196–217. https://doi.org/10.1016/j.apgeochem.2013.12.016

Guo, H., Zhang, D., Ni, P., Cao, Y., & Li, F. (2017). Hydrogeological and geochemical comparison of high Arsenic groundwaters in Inland Basins, P.R. China. *Procedia Earth and Planetary Science*, 17, 416–419. https://doi.org/10.1016/j.proeps.2016.12.105

Gupta, A., Yunus, M., & Sankararamakrishnan, N. (2012). Zerovalent iron encapsulated chitosan nanospheres – A novel adsorbent for the removal of total inorganic Arsenic from aqueous systems. *Chemosphere*, 86(2), 150–155. https://doi.org/10.1016/j.chemosphere.2011.10.003

Halim, M. A., Majumder, R. K., Nessa, S. A., Hiroshiro, Y., Uddin, M. J., Shimada, J., & Jinno, K. (2009). Hydrogeochemistry and arsenic contamination of groundwater in the Ganges Delta Plain, Bangladesh. *Journal of Hazardous Materials*, 164(2–3), 1335–1345. https://doi.org/10.1016/j.jhazmat.2008.09.046

He, F., & Zhao, D. (2005). Preparation and characterization of a new class of starch-stabilized bimetallic nanoparticles for degradation of chlorinated hydrocarbons in water. *Environmental Science & Technology*, 39, 3314–3320.

Herath, I., Vithanage, M., Bundschuh, J., Maity, J. P., & Bhattacharya, P. (2016). Natural Arsenic in global groundwaters: Distribution and geochemical triggers for mobilization. *Current Pollution Reports*, 2(1), 68–89. https://doi.org/10.1007/s40726-016-0028-2

Hossain, S., Hosono, T., Ide, K., & Matsunaga, M. (2016). Redox processes and occurrence of arsenic in a volcanic aquifer system of Kumamoto Area, Japan. *Environmental Earth Sciences*, 75(9), 1–19. https://doi.org/10.1007/s12665-016-5557-x

Huber, D. L. (2005). Synthesis, properties, and applications of iron nanoparticles. *Small*, 1(5), 482–501. https://doi.org/10.1002/smll.200500006

Huq, M. E., Su, C., Fahad, S., Li, J., Sarven, M. S., & Liu, R. (2018a). Distribution and hydrogeochemical behavior of arsenic enriched groundwater in the sedimentary aquifer comparison between Datong Basin (China) and Kushtia District (Bangladesh). *Environmental Science and Pollution Research*, 25(16), 15830–15843. https://doi.org/10.1007/s11356-018-1756-1

Huq, M. E., Su, C., Li, J., & Sarven, M. S. (2018b). Arsenic enrichment and mobilization in the Holocene alluvial aquifers of Prayagpur of Southwestern Bangladesh. *International Biodeterioration and Biodegradation*, 128(January), 186–194. https://doi.org/10.1016/j.ibiod.2018.01.008

Huq, M. E., Fahad, S., Shao, Z., Sarven, M. S., Al-Huqail, A. A., Siddiqui, M. H., Habib ur Rahman, M., Khan, I. A., Alam, M., Saeed, M., Rauf, A., Basir, A., Jamal, Y., & Khan,

S. U. (2019). High arsenic contamination and presence of other trace metals in drinking water of Kushtia district, Bangladesh. *Journal of Environmental Management*, 242(April 2018), 199–209. https://doi.org/10.1016/j.jenvman.2019.04.086

Huq, M. E., Fahad, S., Shao, Z., Sarven, M. S., Khan, I. A., Alam, M., Saeed, M., Ullah, H., Adnan, M., Saud, S., Cheng, Q., Ali, S., Wahid, F., Zamin, M., Raza, M. A., Saeed, B., Riaz, M., & Khan, W. U. (2020). Arsenic in a groundwater environment in Bangladesh: Occurrence and mobilization. *Journal of Environmental Management*, 262(January 2019), 110318. https://doi.org/10.1016/j.jenvman.2020.110318

Islam, A. B. M. R., Maity, J. P., Bundschuh, J., Chen, C. Y., Bhowmik, B. K., & Tazaki, K. (2013). Arsenic mineral dissolution and possible mobilization in mineral-microbe-groundwater environment. *Journal of Hazardous Materials*, 262, 989–996. https://doi.org/10.1016/j.jhazmat.2012.07.022

Jain, C. K., Sharma, S. K., & Singh, S. (2018). Physico-chemical characteristics and hydrogeological mechanisms in groundwater with special reference to arsenic contamination in Barpeta District, Assam (India). *Environmental Monitoring and Assessment*, 190(7), 1–16.

Janković, M. M. (2020). Arsenic Contamination Status in North America. In Sudhakar Srivastava, Arsenic in Drinking Water and Food (pp. 41–69). Springer Singapore. https://doi.org/10.1007/978-981-13-8587-2_3

Jha, P. K., & Tripathi, P. (2021). Arsenic and fluoride contamination in groundwater: A review of global scenarios with special reference to India. *Groundwater for Sustainable Development*, 13(April), 100576. https://doi.org/10.1016/j.gsd.2021.100576

Jiang, Z., Li, P., Wang, Y., Liu, H., Wei, D., Yuan, C., & Wang, H. (2019). Arsenic mobilization in a high arsenic groundwater revealed by metagenomic and Geochip analyses. *Scientific Reports*, 9(1), 1–10. https://doi.org/10.1038/s41598-019-49365-w

Jin, X., She, Q., Ang, X., & Tang, C. Y. (2012). Removal of boron and arsenic by forward osmosis membrane: Influence of membrane orientation and organic fouling. *Journal of Membrane Science*, 389, 182–187. https://doi.org/10.1016/j.memsci.2011.10.028

Kanel, S. R., Manning, B., Charlet, L., & Choi, H. (2005). Removal of arsenic(III) from groundwater by nanoscale zero-valent iron. *Environmental Science and Technology*, 39(5), 1291–1298. https://doi.org/10.1021/es048991u

Kanel, S. R., Greneche, J. M., & Choi, H. (2006). Arsenic(V) removal from groundwater using nano scale zero-valent iron as a colloidal reactive barrier material. *Environmental Science and Technology*, 40(6), 2045–2050. https://doi.org/10.1021/es0520924

Kaur, T., Bhardwaj, R., & Arora, S. (2017). Assessment of groundwater quality for drinking and irrigation purposes using hydrochemical studies in Malwa region, southwestern part of Punjab, India. *Applied Water Science*, 7(6), 3301–3316. https://doi.org/10.1007/s13201-016-0476-2

Kazmierczak, J., Postma, D., Dang, T., Hoang, H. Van, Larsen, F., Elmelund, A., Hvam, A., Fensholt, R., Quy, N., & Jakobsen, R. (2022). Science of the Total Environment Groundwater arsenic content related to the sedimentology and stratigraphy of the Red River delta, Vietnam. *Science of the Total Environment*, 814, 152641. https://doi.org/10.1016/j.scitotenv.2021.152641

Kilianová, M., Prucek, R., Filip, J., Kolařík, J., Kvítek, L., Panáček, A., Tuček, J., & Zbořil, R. (2013). Remarkable efficiency of ultrafine superparamagnetic iron(III) oxide nanoparticles toward arsenate removal from aqueous environment. *Chemosphere*, 93(11), 2690–2697. https://doi.org/10.1016/j.chemosphere.2013.08.071

Kumar, A., & Singh, C. K. (2019). Arsenic enrichment in groundwater and associated health risk in Bari doab region of Indus basin, Punjab, India. *Environmental Pollution*, 256, 113324. https://doi.org/10.1016/j.envpol.2019.113324

Kumar, P. R., Chaudhari, S., Khilar, K. C., & Mahajan, S. P. (2004). Removal of arsenic from water by electrocoagulation. *Chemosphere*, 55(9), 1245–1252. https://doi.org/10.1016/j.chemosphere.2003.12.025

Kumar, S., Nair, R. R., Pillai, P. B., Gupta, S. N., Iyengar, M. A. R., & Sood, A. K. (2014). Graphene oxide-$MnFe_2O_4$ magnetic nanohybrids for efficient removal of lead and arsenic from water. *ACS Applied Materials and Interfaces*, 6(20), 17426–17436. https://doi.org/10.1021/am504826q

Kumar, R., Patel, M., Singh, P., Bundschuh, J., Pittman, C. U., Trakal, L., & Mohan, D. (2019). Emerging technologies for arsenic removal from drinking water in rural and peri-urban areas: Methods, experience from, and options for Latin America. *Science of the Total Environment*, 694. https://doi.org/10.1016/j.scitotenv.2019.07.233

Kumar, A., Ali, M., Kumar, R., Rahman, M. S., Srivastava, A., Chayal, N. K., Sagar, V., Kumari, R., Parween, S., Kumar, R., Niraj, P. K., Anand, G., Singh, S. K., & Ghosh, A. K. (2020a). High arsenic concentration in blood samples of people of village Gyaspur Mahaji, Patna, Bihar Drinking Arsenic-contaminated water. *Exposure and Health*, 12(2), 131–140. https://doi.org/10.1007/s12403-018-00294-5

Kumar, R., Mittal, S., Peechat, S., & Kumar, P. (2020b). Quantification of groundwater – agricultural soil quality and associated health risks in the agri-intensive Sutlej River Basin of Punjab, India. *Environmental Geochemistry and Health* 42(3), 4245–4268.

Lalhmunsiama, L., Nautiyal, B. P., Tiwari, D., Choi, S. I., Kong, S. H., & Lee, S. M. (2016). Silane grafted chitosan for the efficient remediation of aquatic environment contaminated with arsenic(V). *Journal of Colloid and Interface Science*, 467, 203–212. https://doi.org/10.1016/j.jcis.2016.01.019

Lee, S. M., Lalhmunsiama, T., & Tiwari, D. (2015). Porous hybrid materials in the remediation of water contaminated with As(III) and As(V). *Chemical Engineering Journal*, 270, 496–507. https://doi.org/10.1016/j.cej.2015.02.053

Li, X. Q., Elliott, D. W., & Zhang, W. X. (2006). Zero-valent iron nanoparticles for abatement of environmental pollutants: Materials and engineering aspects. *Critical Reviews in Solid State and Materials Sciences*, 31(4), 111–122. https://doi.org/10.1080/10408430601057611

Li, Z., Deng, S., Yu, G., Huang, J., & Lim, V. C. (2010). As(V) and As(III) removal from water by a Ce-Ti oxide adsorbent: Behavior and mechanism. *Chemical Engineering Journal*, 161(1–2), 106–113. https://doi.org/10.1016/j.cej.2010.04.039

Li, H., De Boer, C. V., Buchau, A., Klaas, N., Rucker, W. M., & Hermes, H. (2012). Development of an inductive concentration measurement sensor of nano sized zero valent iron. *International Multi-Conference on Systems, Signals and Devices, SSD 2012 - Summary Proceedings*. https://doi.org/10.1109/SSD.2012.6198042

Lin, S., Lu, D., & Liu, Z. (2012). Removal of arsenic contaminants with magnetic γ-Fe_2O_3 nanoparticles. *Chemical Engineering Journal*, 211–212, 46–52. https://doi.org/10.1016/J.CEJ.2012.09.018

Lin, Y., Jin, X., Khan, N. I., Owens, G., & Chen, Z. (2021). Efficient removal of As (III) by calcined green synthesized bimetallic Fe/Pd nanoparticles based on adsorption and oxidation. *Journal of Cleaner Production*, 286, 124987. https://doi.org/10.1016/j.jclepro.2020.124987

Litter, M. I., Alarcón-Herrera, M. T., Arenas, M. J., Armienta, M. A., Avilés, M., Cáceres, R. E., Cipriani, H. N., Cornejo, L., Dias, L. E., Cirelli, A. F., Farfán, E. M., Garrido, S., Lorenzo, L., Morgada, M. E., Olmos-Márquez, M. A., & Pérez-Carrera, A. (2012). Small-scale and household methods to remove arsenic from water for drinking purposes in Latin America. *Science of the Total Environment*, 429, 107–122. https://doi.org/10.1016/j.scitotenv.2011.05.004

Liu, S., Kang, S., Wang, G., Zhao, H., & Cai, W. (2015). Micro/nanostructured porous Fe-Ni binary oxide and its enhanced arsenic adsorption performances. *Journal of Colloid and Interface Science*, 458, 94–102. https://doi.org/10.1016/j.jcis.2015.07.038

Liu, J., Kong, L., Huang, X., Liu, M., & Li, L. (2018). Removal of arsenic(v) from aqueous solutions using sulfur-doped Fe3O4 nanoparticles. *RSC Advances*, 8(71), 40804–40812. https://doi.org/10.1039/c8ra08699k

Liu, H., Li, P., Yu, H., Zhang, T., & Qiu, F. (2019). Controlled fabrication of functionalized nanoscale zero-valent iron/celluloses composite with silicon as protective layer for arsenic removal. *Chemical Engineering Research and Design*, 151, 242–251. https://doi.org/10.1016/j.cherd.2019.09.020

Liu, K., Li, F., Cui, J., Yang, S., & Fang, L. (2020). Simultaneous removal of Cd(II) and As(III) by graphene-like biochar-supported zero-valent iron from irrigation waters under aerobic conditions: Synergistic effects and mechanisms. *Journal of Hazardous Materials*, 395(March), 122623. https://doi.org/10.1016/j.jhazmat.2020.122623

Lord, G., Kim, N., & Ward, N. I. (2012). Arsenic speciation of geothermal waters in New Zealand. *Journal of Environmental Monitoring*, 14(12), 3192–3201. https://doi.org/10.1039/c2em30486d

Lu, H. J., Wang, J. K., Ferguson, S., Wang, T., Bao, Y., & Hao, H. X. (2016). Mechanism, synthesis and modification of nano zerovalent iron in water treatment. *Nanoscale*, 8(19), 9962–9975. https://doi.org/10.1039/c6nr00740f

Lunge, S., Singh, S., & Sinha, A. (2014). Magnetic iron oxide (Fe_3O_4) nanoparticles from tea waste for arsenic removal. *Journal of Magnetism and Magnetic Materials*, 356, 21–31. https://doi.org/10.1016/j.jmmm.2013.12.008

Luther, S., Borgfeld, N., Kim, J., & Parsons, J. G. (2012). Removal of arsenic from aqueous solution: A study of the effects of pH and interfering ions using iron oxide nanomaterials. *Microchemical Journal*, 101, 30–36. https://doi.org/10.1016/j.microc.2011.10.001

Maity, J. P., Ho, P. R., Huang, Y. H., Sun, A. C., Chen, C. C., & Chen, C. Y. (2019). The removal of arsenic from arsenic-bearing groundwater in In-situ and Ex-situ environment using novel natural magnetic rock material and synthesized magnetic material as adsorbent: A comparative assessment. *Environmental Pollution*, 253, 768–778. https://doi.org/10.1016/j.envpol.2019.07.048

Mak, M. S. H., Rao, P., & Lo, I. M. C. (2009). Effects of hardness and alkalinity on the removal of arsenic(V) from humic acid-deficient and humic acid-rich groundwater by zero-valent iron. *Water Research*, 43(17), 4296–4304. https://doi.org/10.1016/j.watres.2009.06.022

Mandal, B. K., & Suzuki, K. T. (2002). Arsenic round the world : a review. Talanta, 58, 201–235.

McGrory, E. R., Brown, C., Bargary, N., Williams, N. H., Mannix, A., Zhang, C., Henry, T., Daly, E., Nicholas, S., Petrunic, B. M., Lee, M., & Morrison, L. (2017). Arsenic contamination of drinking water in Ireland: A spatial analysis of occurrence and potential risk. *Science of the Total Environment*, 579, 1863–1875. https://doi.org/10.1016/j.scitotenv.2016.11.171

Medunić, G., Fiket, Ž., & Ivanić, M. (2020). Arsenic contamination status in Europe, Australia, and Other Parts of the World. In Sudhakar Srivastava, Arsenic in Drinking Water and Food (pp. 183–233). Springer Nature Singapore Pte Ltd. https://doi.org/10.1007/978-981-13-8587-2_6

Mohan, D., & Pittman, C. U. (2007). Arsenic removal from water/wastewater using adsorbents – A critical review. *Journal of Hazardous Materials*, 142(1–2), 1–53. https://doi.org/10.1016/j.jhazmat.2007.01.006

Morgada, M. E., Levy, I. K., Salomone, V., Farías, S. S., López, G., & Litter, M. I. (2009). Arsenic (V) removal with nanoparticulate zerovalent iron: Effect of UV light and humic acids. *Catalysis Today*, 143(3–4), 261–268. https://doi.org/10.1016/j.cattod.2008.09.038

Muehe, E. M., Morin, G., Scheer, L., Le Pape, P., Esteve, I., Daus, B., & Kappler, A. (2016). Arsenic(V) incorporation in vivianite during microbial reduction of Arsenic(V)-bearing biogenic Fe(III) (Oxyhydr)oxides. *Environmental Science and Technology*, 50(5), 2281–2291. https://doi.org/10.1021/acs.est.5b04625

Murcott, S. (2012). Arsenic contamination in the World: An international sourcebook 2012. In Susan Murcott, Water Intelligence Online (Vol. 11). IWA Publishing. https://doi.org/10.2166/9781780400396

Neil, C. W., Yang, Y. J., Schupp, D., & Jun, Y. S. (2014). Water chemistry impacts on arsenic mobilization from arsenopyrite dissolution and secondary mineral precipitation: Implications for managed aquifer recharge. *Environmental Science and Technology*, 48(8), 4395–4405. https://doi.org/10.1021/es405119q

Nieto-delgado, C., & Rangel-mendez, J. R. (2012). Anchorage of iron hydro(oxide) nanoparticles onto activated carbon to remove As(V) from water. *Water Research*, 46(9), 2973–2982. https://doi.org/10.1016/j.watres.2012.03.026

Nordstrom, D. K. (2011). Hydrogeochemical processes governing the origin, transport and fate of major and trace elements from mine wastes and mineralized rock to surface waters. *Applied Geochemistry*, 26(11), 1777–1791. https://doi.org/10.1016/j.apgeochem.2011.06.002

Onishi, H. (1969). Chapter 33: Arsenic. In Wedepohl, K.H., Handbook of Geochemistry. New York: Springer-Verlag. us DEPARTMENT oF Legacy.

Park, H., Myung, N.V., Jung, H., Choi, H. (2009). As (V) remediation using electrochemically synthesized maghemite nanoparticles. *Journal of Nanoparticle Research*, 1981–1989. https://doi.org/10.1007/s11051-008-9558-x

Pasinszki, T., & Krebsz, M. (2020). Synthesis and application of zero-valent iron nanoparticles in water treatment, environmental remediation, catalysis, and their biological effects. *Nanomaterials*, 10(5). https://doi.org/10.3390/nano10050917

Patel, K. S., Sahu, B. L., Dahariya, N. S., Bhatia, A., Patel, R. K., Matini, L., Sracek, O., & Bhattacharya, P. (2017). Groundwater arsenic and fluoride in Rajnandgaon District, Chhattisgarh, northeastern India. *Applied Water Science*, 7(4), 1817–1826. https://doi.org/10.1007/s13201-015-0355-2

Phenrat, T., & Lowry, G. V. (2019). Nanoscale Zerovalent Iron particles for environmental restoration: From fundamental science to field scale engineering applications. In Tanapon Phenrat and Gregory V. Lowry, Nanoscale Zerovalent Iron Particles for Environmental Restoration: From Fundamental Science to Field Scale Engineering Applications. Springer Cham. https://doi.org/10.1007/978-3-319-95340-3

Pillewan, P., Mukherjee, S., Roychowdhury, T., Das, S., Bansiwal, A., & Rayalu, S. (2011). Removal of As(III) and As(V) from water by copper oxide incorporated mesoporous alumina. *Journal of Hazardous Materials*, 186(1), 367–375. https://doi.org/10.1016/j.jhazmat.2010.11.008

Poguberović, S. S., Krčmar, D. M., Maletić, S. P., Kónya, Z., Pilipović, D. D. T., Kerkez, D. V., & Rončević, S. D. (2016). Removal of As (III) and Cr (VI) from aqueous solutions using "green" zero-valent iron nanoparticles produced by oak, mulberry and cherry leaf extracts. *Ecological Engineering*, 90, 42–49.

Prabhakar, R., & Samadder, S. R. (2018). Low cost and easy synthesis of aluminium oxide nanoparticles for arsenite removal from groundwater: A complete batch study. *Journal of Molecular Liquids*, 250, 192–201. https://doi.org/10.1016/j.molliq.2017.11.173

Qu, X., Brame, J., Li, Q., & Alvarez, P. J. J. (2013). Nanotechnology for a safe and sustainable water supply: Enabling integrated water treatment and reuse. *Accounts of Chemical Research*, 46(3), 834–843. https://doi.org/10.1021/ar300029v

Rahnamarad, J., Derakhshani, R., & Abbasnejad, A. (2020). Data on arsenic contamination in groundwater of Rafsanjan plain, Iran. *Data in Brief*, 31, 105772. https://doi.org/10.1016/j.dib.2020.105772

Rasheed, H., Iqbal, N., Ashraf, M., & Hasan, F. (2022). Groundwater quality and availability assessment: A case study of District Jhelum in the Upper Indus, Pakistan. *Environmental Advances*, 7, 100148. https://doi.org/10.1016/j.envadv.2021.100148

Rasool, A., Farooqi, A., Xiao, T., Masood, S., Kamran, M. A., & bibi, S. (2016). Elevated levels of arsenic and trace metals in drinking water of Tehsil Mailsi, Punjab, Pakistan. *Journal of Geochemical Exploration*, 169, 89–99. https://doi.org/10.1016/j.gexplo.2016.07.013

Reynolds, G. W., Hoff, J. T., & Gillham, R. W. (1990). Sampling bias caused by materials used to monitor halocarbons in groundwater. *Environmental Science and Technology*, 24, 135.

Robins, R. G. (2006). Some chemical aspects relating to arsenic remedial technologies. *Proceedings of the US EPA Workshop on Managing Arsenic Risks to the EnvironmentAvailable Online* at: http://www.Epa.Gov/Ttbnrmrl/ArsenicPres/78.Pdf Accessed April, 1–3.

Rowland, H. A. L., Omoregie, E. O., Millot, R., Jimenez, C., Mertens, J., Baciu, C., Hug, S. J., & Berg, M. (2011). Geochemistry and arsenic behaviour in groundwater resources of the Pannonian Basin (Hungary and Romania). *Applied Geochemistry*, 26(1), 1–17. https://doi.org/10.1016/j.apgeochem.2010.10.006

Saba, N., & Umar, R. (2016). Hydrogeochemical assessment of Moradabad city, an important industrial town of Uttar Pradesh, India. *Sustainable Water Resources Management*, 2(3), 217–236. https://doi.org/10.1007/s40899-016-0053-8

Saif, S., Tahir, A., Asim, T., Chen, Y., & Adil, S. F. (2019). Polymeric nanocomposites of iron – oxide nanoparticles (IONPs) synthesized using *Terminalia chebula* leaf extract for enhanced adsorption of arsenic (V) from water. *Colloids Interfaces*, 3, 17. https://doi.org/10.3390/colloids3010017

Sarkar, A., & Paul, B. (2016). The global menace of arsenic and its conventional remediation – A critical review. In Chemosphere (Vol. 158, pp. 37–49). Elsevier Ltd. https://doi.org/10.1016/j.chemosphere.2016.05.043

Sarkar, S., Greenleaf, J. E., Gupta, A., Uy, D., & Sengupta, A. K. (2012). Sustainable engineered processes to mitigate the global arsenic crisis in drinking water: Challenges and progress.

Annual Review of Chemical and Biomolecular Engineering, 3, 497–517. https://doi.org/10.1146/annurev-chembioeng-062011-081101

Savina, I. N., English, C. J., Whitby, R. L. D., Zheng, Y., Leistner, A., Mikhalovsky, S. V., & Cundy, A. B. (2011). High efficiency removal of dissolved As (III) using iron nanoparticle-embedded macroporous polymer composites. *Journal of Hazardous Materials*, 192(3), 1002–1008.

Shahid, M., Niazi, N. K., Dumat, C., Naidu, R., Khalid, S., Rahman, M. M., & Bibi, I. (2018). A meta-analysis of the distribution, sources and health risks of arsenic-contaminated groundwater in Pakistan. *Environmental Pollution*, 242, 307–319. https://doi.org/10.1016/j.envpol.2018.06.083

Shaikh, W. A., Alam, M. A., Alam, M. O., Chakraborty, S., Owens, G., Bhattacharya, T., & Mondal, N. K. (2020). Enhanced aqueous phase arsenic removal by a biochar based iron nanocomposite. *Environmental Technology and Innovation*, 19. https://doi.org/10.1016/j.eti.2020.100936

Shankar, S., Shanker, U., & Shikha. (2014). Arsenic contamination of groundwater: A review of sources, prevalence, health risks, and strategies for mitigation. *Scientific World Journal*, 2014. https://doi.org/10.1155/2014/304524

Sharma, S., Kaur, I., & Nagpal, A. K. (2018). Estimation of arsenic, manganese and iron in mustard seeds, maize grains, groundwater and associated human health risks in Ropar wetland, Punjab, India, and its adjoining areas. *Environmental Monitoring and Assessment*, 190(7). https://doi.org/10.1007/s10661-018-6763-7

Sherlala, A. I. A., Raman, A. A. A., Bello, M. M., & Buthiyappan, A. (2019). Adsorption of arsenic using chitosan magnetic graphene oxide nanocomposite. *Journal of Environmental Management*, 246(January), 547–556. https://doi.org/10.1016/j.jenvman.2019.05.117

Sierra, L., Cacciabue, L., Dietrich, S., Weinzettel, P. A., & Bea, S. A. (2016). Arsenic in groundwater and sediments in a loessic aquifer, Argentina. *Arsenic Research and Global Sustainability - Proceedings of the 6th International Congress on Arsenic in the Environment, AS 2016*, April 2018, 92–93. https://doi.org/10.1201/b20466-45

Sikder, T. M., Tanaka, S., Saito, T., & Kurasaki, M. (2014). Application of zerovalent iron impregnated chitosan-caboxymethyl-β-cyclodextrin composite beads as arsenic sorbent. *Journal of Environmental Chemical Engineering*, 2(1), 370–376. https://doi.org/10.1016/j.jece.2014.01.009

Singh, S. K. (2017). Conceptual framework of a cloud-based decision support system for arsenic health risk assessment. *Environment Systems and Decisions*, 37(4), 435–450. https://doi.org/10.1007/s10669-017-9641-x

Singh, A. L., & Singh, V. K. (2018). Assessment of groundwater quality of Ballia district, Uttar Pradesh, India, with reference to arsenic contamination using multivariate statistical analysis. *Applied Water Science*, 8(3), 1–18. https://doi.org/10.1007/s13201-018-0737-3

Singh, S. K., Brachfeld, S. A., & Taylor, R. W. (2016). Emerging issues in groundwater resources. In Ali Fares Emerging Issues in Groundwater Resources. Springer. https://doi.org/10.1007/978-3-319-32008-3

Singh, C. K., Kumar, A., & Bindal, S. (2018). Arsenic contamination in Rapti River Basin, Terai region of India. *Journal of Geochemical Exploration*, 192(June), 120–131. https://doi.org/10.1016/j.gexplo.2018.06.010

Smedley, P. L., & Kinniburgh, D. G. (2001). Source and behaviour of arsenic in natural waters. United Nations Synthesis Report on Arsenic in Drinking Water. *World Health Organization*, Geneva, Switzerland, 1–61. http://www.who.int/water_sanitation_health/dwq/arsenicun1.pdf.

Sorg, T. J., Chen, A. S. C., & Wang, L. (2014). Arsenic species in drinking water wells in the USA with high arsenic concentrations. *Water Research*, 48(1), 156–169. https://doi.org/10.1016/j.watres.2013.09.016

Sridharan, M., & Nathan, D. S. (2018). Chemometric tool to study the mechanism of arsenic contamination in groundwater of Puducherry region, South East coast of India. *Chemosphere*, 208, 303–315. https://doi.org/10.1016/j.chemosphere.2018.05.083

Stumm, W., & Morgan, J. (1996). Aquatic Chemistry: Chemical Equilibria and Rates in Natural Waters, 3rd edition. Wiley.

Tang, W., Li, Q., Gao, S., & Ku, J. (2011a). Arsenic (III,V) removal from aqueous solution by ultrafine α-Fe_2O_3 nanoparticles synthesized from solvent thermal method. *Journal of Hazardous Materials*, 192, 131–138. https://doi.org/10.1016/j.jhazmat.2011.04.111

Tang, W., Li, Q., Li, C., & Gao, S. (2011b). Ultrafine a –Fe_2 O_3 nanoparticles grown in confinement of in situ self-formed 'cage' and their superior adsorption performance on arsenic (III) 2641–2651. https://doi.org/10.1007/s11051-010-0157-2

Thanh, N.D., Singh, M., Ulbrich, P., Strnadova, N., & Štěpánek, F. (2011). Perlite incorporating γ-Fe_2O_3 and α-MnO_2 nanomaterials: Preparation and evaluation of a new adsorbent for As(V) removal. *Separation and Purification Technology*, 82(1), 93–101. https://doi.org/10.1016/j.seppur.2011.08.030

Thy, L. T. M., Thuong, N. H., Tu, T. H., My, N. H. T., Tuong, H. H. P., Nam, H. M., Phong, M. T., & Hieu, N. H. (2020). Fabrication and adsorption properties of magnetic graphene oxide nanocomposites for removal of arsenic (V) from water. *Adsorption Science and Technology*, 38(7–8), 240–253. https://doi.org/10.1177/0263617420942710

Viraraghavan, T., Subramanian, K. S., & Aruldoss, J. A. (1999). Arsenic in drinking water – Problems and solutions. *Water Science and Technology*, 40(2), 69–76. https://doi.org/10.1016/S0273-1223(99)00432-1

Virk, H. S. (2019). Groundwater contamination due to heavy metals and other pollutants in Amritsar District of Punjab. *Research and Review: A Journal of Toxicology*, 9(January), 18–27.

Vu, K. B., Kaminski, M. D., & Nunez, L. (2003). Review of arsenic removal technologies for contaminated groundwaters. *Other Information: PBD: 2 May 2003*, Medium: ED; Size: 43 pages. http://www.osti.gov/scitech//servlets/purl/815660-2c93gd/native/

Vunain, E., Mishra, A. K., & Krause, R. W. (2013). Fabrication, characterization and application of polymer nanocomposites for Arsenic(III) removal from water. *Journal of Inorganic and Organometallic Polymers and Materials*, 23(2), 293–305. https://doi.org/10.1007/s10904-012-9775-8

Wang, L., Chen, A., & Fields, K. (2000). Arsenic Removal from Drinking Water by Ion Exchange and Activated Alumina Plants. U.S. Environmental Protection Agency, Washington, DC, EPA/600/R-00/088 (NTIS PB2001-104343). October, 1–133.

Wang, C., Luo, H., Zhang, Z., Wu, Y., Zhang, J., & Chen, S. (2014). Removal of As(III) and As(V) from aqueous solutions using nanoscale zero valent iron reduced graphite oxide modified composites. *Journal of Hazardous Materials*, 268, 124–131. https://doi.org/10.1016/j.jhazmat.2014.01.009

Weerasundara, L., Ok, Y. S., & Bundschuh, J. (2021). Selective removal of arsenic in water: A critical review. *Environmental Pollution*, 268. https://doi.org/10.1016/j.envpol.2020.115668

Wegelin, M., Gechter, D., Hug, S., Mahmud, A., & Motaleb, A. (2000). *SORAS-a simple arsenic removal process*. In *Water, Sanitation and Hygiene—Challenges of the Millennium, Proceedings of the 26th WEDC Conference, Dhaka, Bangladesh* (pp. 255-258).

World Health Organization (WHO).(February 2018). Arsenic. http://www.who.int/news-room/fact-sheets/detail/arsenic (accessed 16 December 2021).

Wu, C., Tu, J., Liu, W., Zhang, J., Chu, S., Lu, G., Lin, Z., & Dang, Z. (2017). The double influence mechanism of pH on arsenic removal by nano zerovalent iron: Electrostatic interactions and the corrosion of Fe0. *Environmental Science: Nano*, 4(7), 1544–1552. https://doi.org/10.1039/c7en00240h

Wu, K., Jing, C., Zhang, J., Liu, T., Yang, S., & Wang, W. (2019). Magnetic Fe_3O_4@CuO nanocomposite assembled on graphene oxide sheets for the enhanced removal of arsenic(III/V) from water. *Applied Surface Science*, 466, 746–756. https://doi.org/10.1016/j.apsusc.2018.10.091

Wu, Z., Su, X., Lin, Z., Islam, N., Owens, G., & Chen, Z. (2020). Science of the Total Environment Removal of As (V) by iron-based nanoparticles synthesized via the complexation of biomolecules in green tea extracts and an iron salt. *Science of the Total Environment*, 764, 142883. https://doi.org/10.1016/j.scitotenv.2020.142883

Yadav, R., Kumari, A., & Babu, J. N. (2016). Sorptive removal of arsenite [As (III)] and arsenate [As (V)] by fuller's earth immobilized nanoscale zero-valent iron nanoparticles (F-nZVI): Effect of Fe 0 loading on adsorption activity. *Journal of Environmental Chemical Engineering*, 4, 681–694.

Yadav, M. K., Saidulu, D., Gupta, A. K., Ghosal, P. S., & Mukherjee, A. (2021). Status and management of arsenic pollution in groundwater: A comprehensive appraisal of recent global scenario, human health impacts, sustainable field-scale treatment technologies. *Journal of Environmental Chemical Engineering*, 9(3), 105203. https://doi.org/10.1016/j.jece.2021.105203

Yan, W., Lien, H. L., Koel, B. E., & Zhang, W. X. (2013). Iron nanoparticles for environmental clean-up: Recent developments and future outlook. *Environmental Sciences: Processes and Impacts*, 15(1), 63–77. https://doi.org/10.1039/c2em30691c

Yantasee, W., Warner, C. L., Sangvanich, T., Addleman, R. S., Carter, T. G., Wiacek, R. J., Fryxell, G. E., Timchalk, C., & Warner, M. G. (2007). Removal of heavy metals from aqueous systems with thiol functionalized superparamagnetic nanoparticles. *Environmental Science and Technology*, 41(14), 5114–5119. https://doi.org/10.1021/es0705238

Yasunori, Y., Kenichi, I., Akihiko, K., Koichiro, S., Shigeki, T., Mitsuhiro, S., & Hiroshi, Y. (2012). Arsenic polluted groundwater and its countermeasures in the middle basin of the Ganges, Uttar Pradesh State, India. *Journal of Environmental Protection*, 3, 21787.

Zeng, J., Qi, P., Shi, J., Pichler, T., Wang, F., Wang, Y., & Sui, K. (2020). Chitosan functionalized iron nanosheet for enhanced removal of As(III) and Sb(III): Synergistic effect and mechanism. *Chemical Engineering Journal*, 382(September), 122999. https://doi.org/10.1016/j.cej.2019.122999

Zhang, G., Khorshed, A., & Paul Chen, J. (2013). Simultaneous removal of arsenate and arsenite by a nanostructured zirconium–manganese binary hydrous oxide: Behavior and mechanism. *Journal of Colloid and Interface Science*, 397, 137–143. https://doi.org/10.1016/J.JCIS.2012.11.056

Zhou, Y., Zeng, Y., Zhou, J., Guo, H., Li, Q., Jia, R., Chen, Y., & Zhao, J. (2017). Distribution of groundwater arsenic in Xinjiang, P.R. China. In Applied Geochemistry (Vol. 77). Elsevier Ltd. https://doi.org/10.1016/j.apgeochem.2016.09.005

Zhu, H., Jia, Y., Wu, X., & Wang, H. (2009). Removal of arsenic from water by supported nano zerovalent iron on activated carbon. *Journal of Hazardous Materials*, 172(2–3), 1591–1596. https://doi.org/10.1016/j.jhazmat.2009.08.031

Zhu, H., Shi, M., Zhang, X., Liu, B., & Yao, D. (2020). Adsorption kinetics of arsenic (V) on nanoscale zero-valent iron supported by activated carbon. *Nanomaterials*, 10(9), 1–12. https://doi.org/10.3390/nano10091791

Zkeri, E., Aloupi, M., & Gaganis, P. (2018). Seasonal and spatial variation of arsenic in groundwater in a rhyolithic volcanic area of Lesvos Island, Greece. *Environmental Monitoring and Assessment*, 190(1). https://doi.org/10.1007/s10661-017-6395-3

7

Chemical Speciation of Chromium and Arsenic and Biogeochemical Cycle in the Aquatic System

Nitin Verma[1], *Neha Kanojia*[1], *Sourav Kalra*[1], *and Kamal Dua*[2]

[1] *Chitkara University School of Pharmacy, Chitkara University, Baddi, Himachal Pradesh, India*
[2] *Discipline of Pharmacy, Graduate School of Health, University of Technology Sydney, Broadway, NSW, Australia*

7.1 Introduction

The significant issue of heavy metal contamination pollutes the climate and makes genuine harm to the environment (Ali et al. 2013; Ali and Khan 2017; Hashem et al. 2017). As a result of rapid industrial growth and urbanization, heavy metal pollution has increased dramatically since 1940s (Merian 1984; Khan et al. 2004). Iron metal in the environment and volcanic ejections are normal sources of contamination, while anthropogenic sources incorporate modern extraction, mining, refining, and cultivating techniques that incorporate the utilization of pesticides and the use of phosphate manures. Consuming residue fuel likewise delivers heavy metals into the climate (Spiegel 2002). Heavy metals, due to their poisonous nature, endure in the climate, taint natural pecking orders, and cause an assortment of medical issues. Unending natural openness to heavy metals represents a genuine danger to animal categories (Wieczorek-Dąbrowska et al. 2013). Metal concentrations beyond the threshold levels have an impact on soil microbiology and can impair fertility (Barbieri 2016). Toxic heavy metal bioaccumulation in riverine ecosystem biota may have negative consequences on animals and humans (Malik and Maurya 2014). Heavy metal concentrations in biota can negatively affect the ability of species of sea creatures to survive in their habitat and contribute to population decline (Luo et al. 2014). Heavy metals and synthetics can interfere with fish's capacity to communicate with their surroundings (Baatrup 1991), exposing fish species to potent neurotoxins. Both in natural residents and in the research laboratory, these have been associated to fish pathologies. Such malformations, in general, have a negative impact on fish populations as they influence survival, evolution, development, well-being, and external appearance. These fish deformities can be used as biomarkers for heavy metal effluence in the atmosphere (Sfakianakis et al. 2015). "Metals, whether natural or anthropogenic in origin, are pervasive in the aquatic environment and so understanding their behavior and interaction with aquatic organisms, particularly fishes, a major source of protein for human consumption," states Hartl (2013). Various

Hydrogeochemistry of Aquatic Ecosystems, First Edition. Edited by Sughosh Madhav, Virendra Bahadur Singh, Manoj Kumar, and Sandeep Singh.

facets of heavy metals were examined in this article, including their persistence in the environment, toxicity to life forms, and bioaccumulative potential. An emphasis is placed on aquatic life and how these elements accumulate in the body, as well as the consequences of their accumulation on human health.

7.1.1 Heavy Metals

Chemically, heavy metals are "elements that conduct electricity, exhibit a metallic luster, are malleable and ductile, form cations, and have basic oxides" (Atkins and Jones 1997). Steel, metalloid, semimetal, mild steel, heavy metallic, vital steel, beneficent steel, poisonous metal, abundant metallic, to be had metal, hint metallic, and micronutrient are commonplace terms utilized in organic and environmental research when it comes to metals (Duffus 2002). Metals have a huge scope of bundles and expect a sizable component in enterprise networks. A few metals assume a significant part in physiology and biochemical cycles in organic frameworks, and their need or excess can prompt metabolic issues, therefore, a development of issues. Metals and metalloids are required for natural endurance. They assume substantial physiochemical parts in the system as they are essential for the biomolecules and proteins, advancing biochemical compound responses in the structure. As indicated by Csuros and Csuros (2002), heavy metallic is characterized as "a metal with a thickness of more than 5 g/cm^3 (that is, a particular gravity of more than 5)." According to Duffus (2002), "time span" of heavy metals "is frequently utilized as a gathering wire for metals and semimetals (metalloids) connected with disease and poisonousness potential or ecotoxicity."

7.1.2 Sources

Heavy metals can emerge from individual normal/geogenic/lithogenic and anthropogenic sources inside the general climate. Environmental rock bluffs and volcanic ejections are the regular or topographical properties of heavy metals inside the general climate. Global forms in contemporary events and worldwide urbanization have contributed to an increase in the anthropogenic contribution of heavy metals to the climate (Nagajyoti et al. 2010). Mining, business, and agrarian exercises are largely anthropogenic sources of heavy metals in the environment. Heavy metals added to the environment due to mining, liquefication, and other commercial enterprise approaches eventually determine the floor place as wet and dry. The heavy metals were likewise introduced into the environment even as wastewater carries cutting-edge waste and circle of relative's wastewater. The amount of gasoline and compost used each year influences the amount of heavy metal pollution in the atmosphere. Phosphate compost is vital on account of focus of heavy metals in business compound manures. Acidulation is utilized to shape phosphate manures from phosphate rock (PR). Sulfuric corrosive is applied within the acidulation of 1 superphosphate (SSP), whereas phosphoric corrosive is employed within the acidulation of triple superphosphate (TSP) (Dissanayake and Chandrajith 2009). The end result joins generally accessible heavy metals as parts of phosphate stone (Mortvedt 1996). Business composts, particularly phosphate manures, can assume a part in the transportation of heavy metals around the world. Heavy metals that are widely used in agricultural soils in inorganic fertilizers have the potential to contaminate groundwater (Dissanayake and Chandrajith 2009). Heavy metals are heavy in phosphate fertilizers. Two main ways for heavy metals to reach the human body from phosphate fertilizers are given as follows (Dissanayake and Chandrajith 2009):

1) Phosphate rock⟶fertilizer⟶soil⟶plant⟶food⟶human body
2) Phosphate rock⟶fertilizer⟶water⟶human body

Anthropogenic properties of heavy metals include the burning of mineral products in industries, homes, and transportation. The anthropogenic causes of heavy metals including Zn, Cr, Cd, and Pb include motor vehicle tourists (Ferretti et al. 1995). In metropolitan regions, high centralizations of heavy metals influencing the climate have been seen in soil and street vegetation. Extras from coal consuming and other ignition processes are significant anthropogenic sources of heavy metals (Merian 1984). Cd, Pb, and As fluctuate slightly during coal burning, while Hg is entirely dangerous. Coal burning is one of the main reasons for harmful trace detail emissions in China's ecosystem (Tian et al. 2013). Cr (30.37 ppm), Ni (17.44 ppm), Cd (0.61 ppm), Pb (23.04 ppm), Hg (0.20 ppm), As (5.78 ppm), Se (3.66 ppm), Sb (2.01 ppm) show some critical environmental statistics for eight essential hazardous trace elements (HTEs) in Chinese coals that are of primary environmental concern.

Electroplating industries, textile industries, leather tanneries, and steel industries are all anthropogenic sources of Cr (Palaniappan and Karthikeyan 2009). Coal combustion, wood combustion, and trash incineration may emit roughly 50,000 t/year of Cr globally (Merian 1984). Fertilizers frequently contain high levels of Cr (Krüger et al. 2017). Coal burning may yield roughly 60,000 t/year of Ni globally, with the majority of Ni remaining in the ash (Merian 1984). Although nonferrous metal mining, particularly the management of Pb-Zn minerals, is a human source of Cd, volcanic interest and releasing of molten rock are natural assets of the metal within the environment (Hutton 1983). Coal ignition might discharge around 7000 t/year of Cd worldwide, and sewage slime cremation is likewise a source of Cd (Merian 1984). Unreasonable usage of compound manures can likewise set off anthropogenic increment in Cd focuses (Wang et al. 2015). P-containing manures contain Cd as a foreign component at concentrations ranging from trace amounts up to 300 parts per million (ppm) on a dry weight basis, and can therefore be a major source of this metal in agricultural designs (award and Grant and Shepherd 2008). Pb is released into the atmosphere from a variety of sources, including corroded batteries, significant piping systems, and piston-engine aircraft operating on leaded aviation fuel. Pb is also released into the atmosphere when leaded gasoline is burned. Although tetraethyl lead is no longer used as an antiknock ingredient in gasoline, it is still utilized in some underdeveloped countries.

7.1.3 Toxic Effects

Despite the fact that several heavy metals work with vital skills in organic structures, they are often harmful to their living organisms, relying on quantity and exposure time. In toxicology, it is stated that "an excessive amount of it is dangerous." Even in the most reduced parts, insignificant metals (like Pb, Cd, and Hg) and metalloids (like As and so on) can be perilous. Heavy metals are essential in modest amounts in the body;

notwithstanding, they wind up being perilous when the center crosses specific limit or limits. Significant window and toxicity in certain regards are restricted. Heavy metals are linked to diseases, mutagenesis, and teratogenic impacts. They result in production of responsive oxygen species (ROS) and the oxidative kind subsequently. Numerous infections and changes are brought about by a sort of oxidation in living life forms. Heavy metals can likewise be hurtful to outline digestion. The primary means by which heavy metal poisons are propagated is through their association with sulfhydryl (SH) protein frameworks and the inhibition of molecules involved in the energy cycle (Csuros and Csuros 2002). Figure 7.1 shows the portrait of heavy metals (M) containing glutathione (GSH), a significant cell reinforcement inside the human body. At this time, iron replaces H-iotas from SH cells into two contiguous glutathione atoms. The co-operation of two glutathione particles in the development of a strong bond with iron makes them dormant for additional response:

$$2\mathrm{GSH} + \mathrm{M}^{2+} \rightarrow \mathrm{M}(\mathrm{glutathione})_2 + 2\mathrm{H}^+$$

Heavy metals such as Cd, Pb, Hg, and As lower the main antioxidants in cells, especially the antioxidants of the thiol group and the enzymes (-SH). Types of active oxygen (ROS) such as hydroxyl radicals (HO^-), superoxide radicals (O_2^-), and hydrogen peroxide can be supplemented with such metals (H_2O_2). Increased ROS production may cause damage to the cell's natural defenses, leading to a condition known as "oxidative stress" (Ercal et al. 2001). Heavy metals, such as cadmium, lead, and mercury, are nephrotoxic, especially in the renal cortex (Wilk et al. 2017). In terms of toxicity, the chemical composition of heavy metals is important. Mercury toxin is determined primarily by Hg

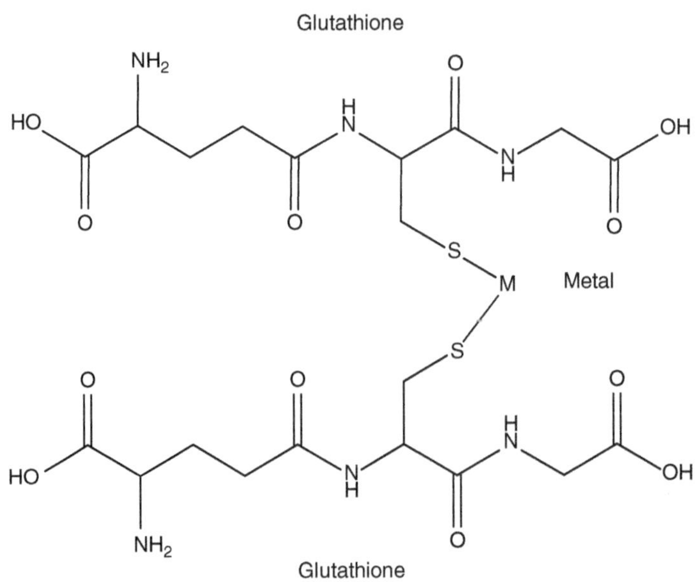

Figure 7.1 Metal reacts with glutathione to form heavy metal glutathione complex in the body.

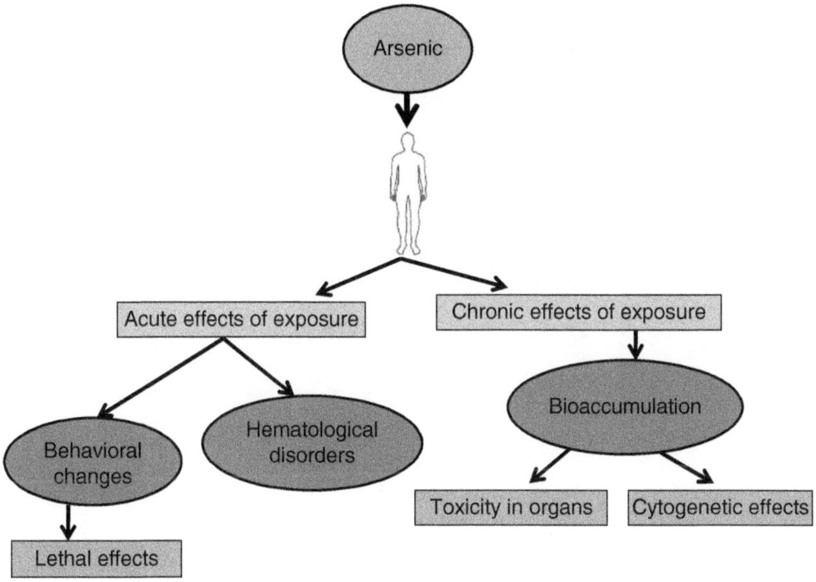

Figure 7.2 Toxicity of As in human body.

speciation (Ebrahimpour et al. 2010). Patients with cancer and diabetes have higher levels of harmful metals, such as Cr, Cd, and Pb, as well as lower concentrations of the antioxidant element Se, compared to healthy people in the city of Lahore, Pakistan (Salman et al. 2011).

Poisons enter the human body in an assortment of ways, including ingestion, scent, and assimilation. In helpless nations, individuals are progressively being presented to perilous metals (Eqani et al. 2016). People, particularly in less evolved nations, are regularly ignorant about the risks of exposure to heavy metal and its effect on the well-being of humans (Afrin et al. 2015). Heavy metals are present in working place and even in the environment to which individuals are exposed. Exposure to work pollutants in the workplace is known as work exposure, whereas exposure to work or environmental exposure is known as environmental exposure. Miners and industrialists are exposed to heavy metals from dust and particles with metal particles. Those who obtain gold by mixing it inhale snuff of mercury smoke. Welders exposed to long-term reproductive waste exhibited higher blood levels of heavy metals such Ni, Cr, Pb, and Cd than the reference group and oxidative pressure-actuated stress (Mahmood et al. 2015). Smoking is an extensive purpose for openness to Cd and furthermore hurtful heavy metals observed in tobacco leaves (Järup 2003). Heavy metals are consumed via meals and by ingesting water, and this is a large source of public exposure. Modern turn of events, metropolitan turn of events, and quick monetary development have prompted an expansion in modern and agrarian exercises around the world. Because of these effects, heavy toxic metals can be released into the water, air, and soil. Growing human food in an area contaminated with heavy metals results in an abundance of biodiversity of these nutrients in human chains, which then reach the human body (Figure 7.3).

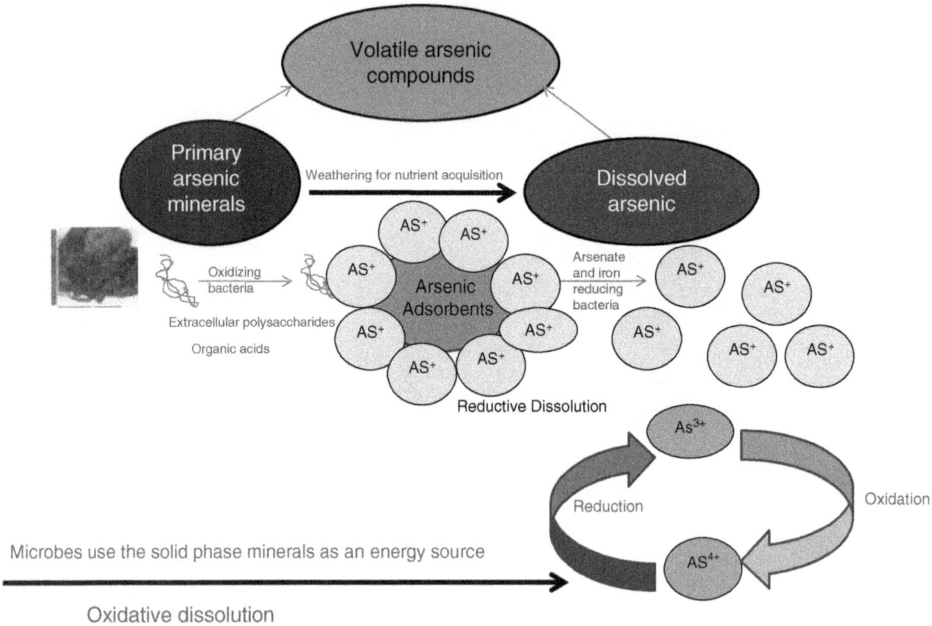

Figure 7.3 Oxidative dissolution of AS (Drewniak and Sklodowska 2013).

7.2 Chemistry of Arsenic in Aquatic System

Arsenic (As) is the most plentiful metalloid found inside the amphibian climate because of its therapeutic and anthropogenic cycles. It is a significant and far and wide ecological contaminant, which is a danger to humankind and is a worldwide social issue (Chowdhury et al. 1999; Rossman 2003). Its harmfulness is demonstrated in numerous organic entities, and insights recommend that non-poisonous types of As show the best degree of toxicity, though organo-arsenicals normally have almost no harmfulness (Duker et al. 2005). Among them, trivalent inorganic As has all the earmarks of being a definitive non-poisonous type of semimetal with harmful impacts (Kovendan et al. 2013). The organs primarily affected by ingestion, aggregation, and/or discharge are the gastrointestinal tract, circulatory system, skin, liver, and kidneys. It influences the hematological, biochemical, and ion regulatory qualities of sea-dwelling species, particularly fish, and changes in these boundaries might be useful in As defiled climate biomonitoring (Lavanya et al. 2011). Gernhofer et al. (2001) and Gaim et al. (2015) accept that fish is the favored strategy for estimating the impacts of contamination on oceanic life. They are continually exposed to pollutants through their pores, skin, and As-rich food varieties. Fish have for some time been utilized as biomonitoring specialists for marine contamination and are strong markers of As harmfulness (Tisler and Zagorc-Koncan 2002). Because of the incredible capability of fish models to foster biomarkers, they are naturally suitable for toxicogenomic studies its openness (Das et al. 2012). Fish seem to have fostered an assortment of systems into which limited quantities of poisons can be effectively separated (Bears et al. 2006). The existence of inordinate degrees of As in the sea-dwelling atmosphere adds to various organic cycles in fish, including generation, development, particle control, quality articulation, smoltification, catalyst action, immunological movement, and histopathology (Datta et al. 2009) (Figure 7.2). Utilization of As-tainted fish could expose people to the metal and cause medical issues. Therefore, iron homeostasis, oxidative pressure, lipid digestion issue, and carcinogenesis could be in every way set off (Xu et al. 2013). Arsenic has a wide

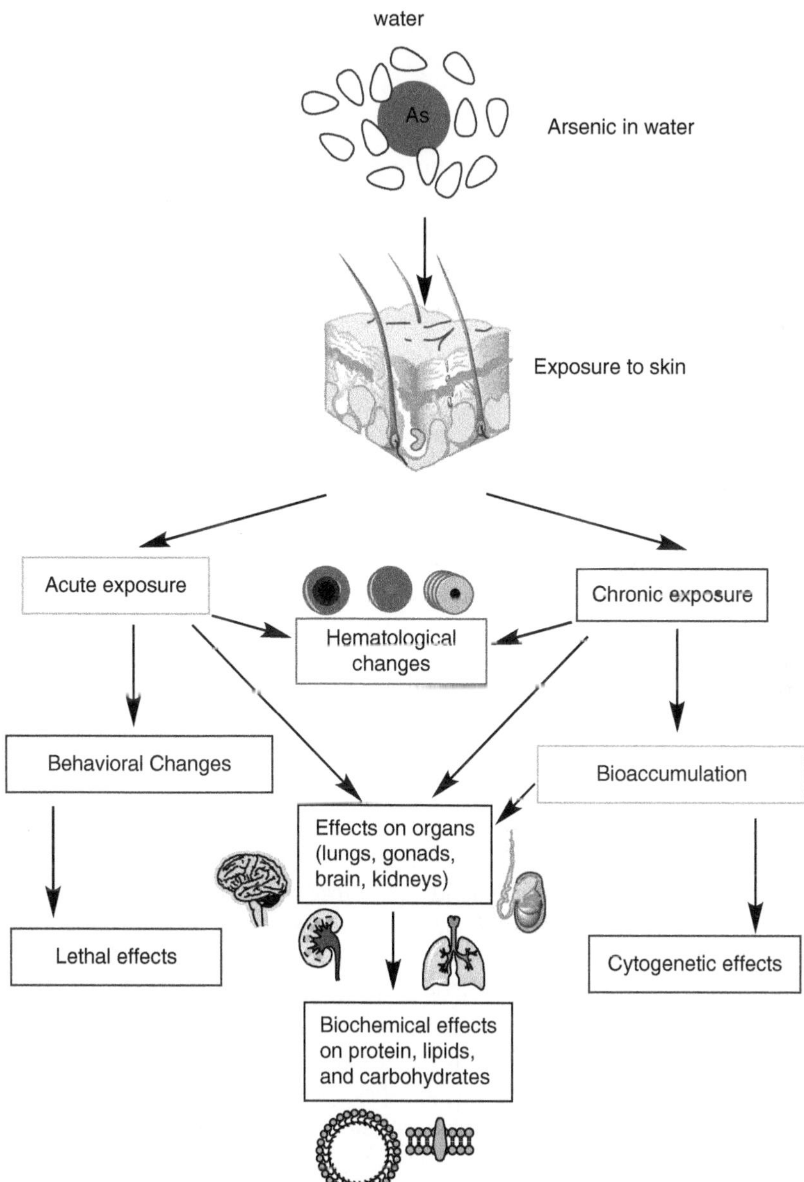

Figure 7.4 Arsenic in water with acute and chronic exposure through skin/gills and the toxicological pathway.

Table 7.1 Physical and chemical characteristics of arsenic chemicals **(IARC 1987)**.

Sr No.	Scientific name	Molecular mass	Physical state	Oxidation state	Solubility in water
1.	Arsenic	74.92	Solid	0	Insoluble
2.	Arsenic acid	141.95	Solid	+5	Freely soluble
3.	Arsenic trioxide	197.82	Solid	+3	Slightly soluble
4.	Arsenic pentaoxide	229.84	Amorphous Solid	+5	Freely soluble
5.	Sodium arsenate	85.91	Solid	+5	Very soluble
6.	Arsine	85.91	Solid	+3	Freely soluble
		77.93	Gas		Slightly soluble
7.	Dimethylarsenic acid	138.01	Solid	+5	Soluble
8.	Sodium dimethyl arsenate	159.98	Solid	+5	Readily soluble
9.	Sodium methane arsonate	161.96	Solid	+5	Soluble
10.	Methanearsonic acid	139.98	Solid	+5	Soluble

assortment of compound responses and may deliver combinations and covalent co-operations with different components. The inorganic As mixtures are solids at typical temperatures and are not liable to be unstable. These synthetic substances' water solvency goes from extremely solvent to basically insoluble (Table 7.1). At room temperature, a few natural As mixtures are gases or fluids (because of their low limits).

They are not promptly water solvent, except for natural arsenic acids. Artificially, there are two kinds of As mixtures: inorganic and natural As (Table 7.2). There are two sorts of inorganic As: trivalent (As III) and pentavalent (As V) (As V). Arsenic oxide is the most critical synthetic. While As can be found naturally, it is often obtained from the mineral arsenopyrite. As efficiently promotes oxidation-decrease, methylation-demethylation, and corrosive base cycles. The most well-known oxidation states of As are 3 in the arsenate III form, 5 in the arsenate V form, the really stable inorganic asoxy compounds, and under reducing circumstances (Table 7.3). Generic Compounds Like Intermetallic Compounds, where III is the main pathway and 3 is an arsenide (De Carnelo et al. 1997).

Both in vivo and in vitro arsenate structures are less poisonous than arsenite (Cervantes et al. 1994). Arsenate or arsenite is the most widely recognized type of As in water. As III oxide, the main business synthetic, is delivered as a side project of copper and lead metal purifying agent. The most widely accepted cause of a strong fixation on the climate is anthropogenic activity. Mining activities have to a great extent added to soil and water contamination. In any case, the different anthropogenic activities utilizing As, which incorporate agribusiness, ranger service, and industry, harm soil and water less significantly (Smith et al. 2003). Arsenic can be conveyed to the ground involving air or water as a stream, or it can infiltrate through rocks containing arsenic, mining tail, vegetation, and modern waste. The spring includes numerous springs that can be delivered into the ionic state under oxidizing conditions. Water or dissolvable ice can change solvent structures into groundwater, and a

Table 7.2 Arsenic mineral groups and examples of common minerals **(IARC 1987)**.

Arsenic oxides	Arsenolite	As_2O_3
	Claudetite	As_2O_3
	Trippkeite	Cu_2 $3As_2O_4(OH)$
	Leiteite	$ZnAs_2O_4$ Mn_2+$3As_2O_4(OH)_4$
	Manganarsenite	$Zn_3(AsO_{3)2}$
	Reinerite	Pb_2Mn_2+$(AsO_3)_2$•$2H_2O$
	Rouseite	$Ca_8(Ti,Fe^{2+,3+},Mn)_{6-7}$
	Cafarsite $(AsO_3)12$•$4H_2O$ Gebhardite	$Pb_8(As_2O_5)2OCl$
Arsenic sulfides	Arsenopyrite	FeAsS
	Cobaltite	CoAsS
	Gersdorffite	NiAsS
	Orpiment	As_2S_3
	Realgar	AsS or As_4S_4
Arsenic sulfosalts	Enargite	Cu_3AsS_4
	Proustite	Ag_3AsS_3
Arsenides	Domeykite	Cu_3As
	Löllingite	$FeAs_2$
	Nickeline (aka Niccolite)	NiAs
	Nickel-skutterudite	$(Ni,Co)As_{2-3}$
	Rammelsbergite	$NiAs_2$
	Safflorite	$CoAs_2$
	Skutterudite	$CoAs_{2-3}$
	Sperrylitc	$PtAs_2$

little part can be changed into a temperamental organization that consolidates arsine with the assistance of microorganisms (Turpeinen et al. 1999). As could be available in much depleted soils like H_2AsO_4 under acidic conditions or as $HAsO_4$ 2 under soluble conditions. Decrease oxidation, adsorption, dissolving, volatilization, and precipitation, i.c. variances of As typically happen in soil. Under regular conditions, water is the principal transporting agent of As. In a relative investigation of the design of water resources, straightforward arsenate and arsenite have been accounted for in various dosages (Hughes et al. 2011). When provided water, the concentration of arsenate/arsenite isn't always very low, but this is also due to the redox conditions in surface water, including the ocean, which is sometimes considered to "clean" because of As, for example, As the arrangement is destroyed through residue precipitation. In anoxic or decreasing frameworks, for example, V, the oxidation-stable condition of the liquid can be diminished to As III (Oremland et al. 2000). In any case, the degree of As III in those fluids, just as water containing bless interstitial, is low; a total drop from one As to another III ended up being imperceptible. As the III is delivered into the oxygenated ocean, and it very well may be reinfused into As V within specific days.

Alternatively, water consolidates four types of As mixtures. Arsenite, arsenate, monomethylarsonic corrosive, and dimethylarsinic corrosive are the four types. Arsenate is changed due to methylation of synthetic substances in clean water and salt water.

Table 7.3 Arsenic compounds of environmental significance **(IARC 1987)**.

Sr. No.	Name	Synonyms	Molecular formula
1.	**Inorganic arsenic, trivalent**		
	As (III) oxide	As trioxide, arsenous oxide, white As	As_2O_3 (or As_4O_6)
	Arsenenous acid	Arsenious acid	$HAsO_2$
	As (III) chloride	As trichloride, arsenous trichloride	$AsCl_3$
	As (III) sulfide	As trisulfide orpiment, auripigment	As_2S_3
2.	**Inorganic arsenic, pentavalent**		
	As (V) oxide	As pentoxide	As_2O_5
	Arsenic acid	*ortho*-Arsenic acid	H_3AsO_4
	Arsenenic acid	*meta*-Arsenic acid	$HAsO_3$
		Arsenates, salts of *ortho*-arsenic acid	$H_2AsO_4^-$, $HAsO_4^{2-}$, AsO_4^{3-}
3.	**Organic As**		
	Methylarsine	Organic AS	CH_3AsH_2
	Dimethylarsine	Organic AS	$(CH_3)_2AsH$
	Trimethylarsine	Organic AS	$(CH_3)_3As$
	Dialkylchloroarsine	Organic AS	R_2AsCl
	Alkyldichloroarsine	Organic AS	$RasCl_2$

Unadulterated water and salt water both produce mono-, di-, and trimethyl arsenicals. As in open salt water is generally under 2 lg/L (Ng 2005), despite the fact that it fluctuates in the range of 0.15 and 0.45 lg/L in new water (Bissen and Frimmel 2003). Nonetheless, because of the liquid corrosive substance and dregs load, there have been a couple of reports of higher than typical bits in saline waters close to the headwaters. In regions with sulfide mineralization and mining, high volumes of up to 5000 lg/L can be found in groundwater (Smedley and Kinniburgh 2002). Inorganic As seen in groundwater is utilized as savoring water in various regions (Ahmed et al. 2011) and is utilized for hydroponic processes for an assortment of purposes. In districts connected with aqueous amusement in Japan, 35 and 25.7 mg As/L have been accounted for (Tanaka 1990).

The best-known source of As disease in groundwater is an amalgamation of what occurs as silt. Normally, As-carrying minerals are oxidized and distributed from the water. Therefore, speculation of "pyrite oxidation" is not a major pathway, and the "lower oxyhydroxide" principle (Acharyya et al. 1999) may be the main cause of the apparent infection of groundwater. According to Acharyya et al. (1999), combined microbiological and chemical cycles may increase As, and increased phosphate fixation may be required to advance the growth of dregs biota and the fusion of As from removes.

7.2.1 Aqueous Arsenic Distribution and Speciation

Arsenic is subtle to changes in water under pH and redox energy. Arsenic may be found with the increase in oxidation conditions in the environment (−3, 0, +3, and +5), but it is usually found in the inorganic materials of herbal water, as well as arsenite (As (III)) and/or arsenate (As (V)). Organic bureaucracy is created in shallow waters using organic matter, but they have never been so abundant. As adsorption of As decreases as pH increases (Cui et al. 2018). As anions may be present in the solution in excessive quantities, even near neutral pH, and may be more abundant in groundwater under certain conditions. The oxyanions are the most intricate environmental trace because of their comparable movement over a broad range of pH and redox conditions (Shakoor et al. 2017). In environments rich in sulfur, shrinking, a few trace elements, including As, may form soluble sulfides.

At low pH (pH 6.9), H_2AsO_4 reigns under oxidizing conditions, but at extreme pH, $HAsO_4$ takes over. Non-rechargeable strains of As (III) rule as H_3AsO_3 in reducing frequency with a pH below 9.2. In fact, a high amount of research within the literature supply records in speciation without considering the protonation diploma. The soluble forms of As-sulfide may be the largest in the presence of highly concentrated components of reduced sulfur. Orpiment (As_2S_3), realgar ($-As_4S_4$), and various sulfide minerals include co-induced As they are most likely to form under reduced, acidic conditions. As a result, large amounts of uncultivated sulfide are no longer effective at high water levels. At a neutral and alkaline pH, thioarsenate varieties are very important in concentrating high sulfide concentrations (Niazi and Burton 2016).

7.2.2 River Water

Awareness on As presence in river water is limited (between 0.1 and 0.8 g/L). It differs mainly based on the formation of the upper complement, the lower motion contribution, and the lithology of the sediments. Within the Krka region of Yugoslavia, where the karst limestone rock is found, the average flow rate was 0.13 g/L, while the overflow of tired rivers with the help of Norwegian rivers became 0.25 g/L. In some regions, high environmental values of As may also be due to inputs from geothermal properties or groundwater levels (Smedley and Kinniburgh 2002). Presence of As in water in the river in tropical areas was found to be 10–70 g/L. As attention to other water-sensitive geothermal fluctuations varies significantly with the seasons. The Madison River has an excellent concentration with low waft prices.

That increase will be due to a decrease in temperature controlled by small insects As (V) and As (III), which improve the flow of As (III) (McLaren and Kim 1995). The concentration of arsenic is further recorded to be thrown up in a few rivers in arid areas where the rivers dominate the lower waters, the water flowing from the surrounding rocks to the water. The pH and alkalinity of the surface water have the result that is usually high. For example, within the Loa River Valley in the waters off northern Chile. Caceres et al. (2005) pointed out that natural phenomena As concentrations ranged from 190 to 800 g/L. Overwhelmingly, since concentration is related to river water salinity, which is governed by groundwater, and may be associated with attention to river water evaporation. Items contained such as seven Bangladeshi river water samples were found to be between 0.5 and a pair of 0.7 g/L, at the same time one sample had a value of up to 29 g/L. The value of As in 79% of the samples (n1 / four 727) is below the acquisition limit of 10 g/L, with a high value of 1100 g/L found below the commercial Park River in Sugar Creek, South Carolina. Excess As concentration of surface water (usually about 200–300 g/L) has been shown to be exaggerated by tin and gold mining, according to Ko et al. (2015).

7.2.3 Lake Water

As level in seawater is generally the same or lesser than that of river water. The basic awareness of 1 g/L was modified in Canada. Due to the limited evaporation and/or geothermal input, large amounts of As were found in ponds in a few alkaline-sealed containers. For example, dissolved content in Lake Mono, California, is 10–20 g/L. The pH is in the range of 9.5–10 due to geothermal penetration and volcanic climates, which are detected by evaporation. There is little indication that As level of certain waters in the lake are separated. Azcue and Nriagu (1995) found that the amount of As (III) to As (V) increased dramatically (up to 10 m) in lake waters from Ontario, due to the incorporation of pore fluid into the interface and water surface, increasing the rate of As (III) to As (V) in depth. During the summer, when the As (III) section was large, the concentration was high. The high concentration of As in deep lake water is most likely due to the reduction of oxygen levels in the backwater of the lake due to biological production during the summer months.

7.2.4 Seawater and Estuaries

Recommendation of As sensitivity to open seawater fluctuates slightly; it is about 1.5 g/L. Due to fluctuations in flow rates, salts, and redox gradients, the concentration of oral water varies greatly, although it may usually decrease under natural conditions, usually less than 4 g/L. When rivers are fed through industrial or mining waste (e.g. Tamar, Scheldt, Loire estuaries) or hot water, the concentration is usually large. In contrast to various minerals such as boron, seawater infiltration into saline waters is less likely to cause significant upward pressure such as underwater groundwater (Smedley and Kinniburgh 2002). Several studies have revealed that As behaves differently during certain estuarine interactions. Sequential behaviors have been described using multiple researchers. Yao et al. (2006) presented a line that usually rises as the salinity increases in freshwater, from 0.13 g/L in glossy water to 1.8 g/L overseas (i.e. the cost of seawater). However, consistent behavior (from simple mixing) in wetlands has been documented in various studies due to mechanisms involving the distribution of liquids from sediment pore fluids, rainfall associated with iron oxides, or anthropogenic inputs (Smedley and Kinniburgh 2002).

7.2.5 Groundwater

The past's history of As is prevalent in many nations. Since groundwater usually gradually increases in content and is less than 10 g/L, however, the books charge a wide variety of degrees, ranging from 0.5 to 5000 g/L (i.e. four orders of value). In many cases, most As categories can be found in groundwater. This consists of oxidizing (excess pH) and depleted underground areas, in addition to affected areas with geothermal support, mining, and commercial activities (Niazi et al. 2018). There are many cases of infection caused by mines in the literature, although they are usually limited. Local industrial ills such as pollution (similar to those from agriculture) can be extreme, though not uncommon.

7.2.6 Sediment Pore Waters

Excessive concentration of As was expressed in the fluid retrieval in uninsulated vessels, which is usually examined more closely by the values found within groundwater

(Rahman et al. 2014). Yan et al. (2000) found it to be concentrated in a concentrated liquid from clay soils in Saskatchewan, Canada, from 3.2 to 99 g/L. An even greater quantity was found in the liquid fluid from the mineral-infected structures (tail, mineral-rich poles). Anawar et al. (2004) detected significant quantities of As, 0.001–0.005 g/L in saltwater collected from Ontario lakes. Elevated inland water because of the high concentration may be related to strong redox gradients that occur beneath the visible sediment from water to sediment usually above the depth of the inch scale. The slight distribution of oxygen in the sedimentary, next to the recent ecological burial, results in low-level conditions without delay under sedimentary water interactions. This promotes the reduction of As (V) and desorption from Fe and Mn oxides, in addition to the reduced degradation of those minerals (la force et al. 2000; Fakhreddine et al. 2015; Wang et al. 2017). In response to the temporary changes in redox conditions, there is ample evidence that like cycling between shallow pore sediment liquids and surface water (Yan et al. 2000, El-Naggar et al. 2018).

7.2.7 Oilfield and Different Brines

Although there may be limited records of As focusing on the oil field and the different oil fields, some published studies suggest that awareness may be far more extreme. In the sodium bicarbonate water from the 1000 m depth of oil field at Ellis Pool, Alberta, Canada, White et al. (1963) refers to a molten substance such as 230 g/L contents. In sodium-chloride-rich brine from Tisak € urt, Hungary, a concentration of 5800 g/L was obtained. Most As speciation experiments in herbal water aim to classify inorganic species into (III) and As (V) chromatographic subdivisions or a relatively lazy disc (As) V with sodium borohydride (Edwards et al. 1998). To avoid the assignment of keeping species for later testing in the laboratory, the classification of species may be eliminated within the area (Smedley and Kinniburgh 2002). Storing HCl and ascorbic acid, too, has been successful. Oxidation conditions for rainwater may vary depending on As supply. While arsine (As (III) H3) can be found in dumping and decontamination sites, it is formed by solvents, coal burning, and volcanic springs. Natural species may be taken up by soil. Soils such as peat and As (IV) can be found in sea aerosols. Decreased bureaucracy may involve O_2 in the environment, and interaction with SO_2 or O_3 in the ecosystem is predictable (He et al. 2010). In oxygen-rich seawater, As (V) tends to dominate, at the same time as As (III) persists and becomes important in anointing backwater. In open seawater, the values of As (V)/As (III) are usually between 10 and 100°. Within the pH range of salt water, arsenate needs to be present mainly as $HAsO_4^{2-}$ and $H_2AsO_4^-$, whereas As (III) needs to be present primarily because the H_3AsO_3 type is impartial (Smedley and Kinniburgh 2002). Ecological expansion As animal species were found in these areas due to phytoplankton methylation reaction (Yang et al. 2018). Due to the variety of redox and salts, in addition to groundwater, the related components of As species in river water are very diverse (Yang et al. 2018, Ma et al. 2016, Yang and Rosen 2016). However, they are governed as As (V). Adjustment of input resources, redox conditions, and biological activity affects concentration with moderate values of As (V) and As (III). The presence of As (III) in oxidized water can be maintained in the form of natural degradation of As (V), especially throughout the summer months. The length of the river adjacent to the input of As (III) filled with commercial wastewater (Acharya et al. 2018) and streams with geothermal water feature has high relative affinity for As (III). In stagnant pools, where redox gradients may be wider and wider at certain times of the year, the concentrations of As (III) and As (V) vary in particular. The driving force behind the redox variety is to change the varieties in that as a specific feature of the pool, done like they do in the estuary. For example, there is a strong concentration of As (V) in the upper layer of oak and As (III) inside the hypersaline and hyperalkaline Mono Lake (California, United States). Dimethylarsinic (DMAA; (CH_3) 2AsO (OH)) and monomethylarsinic (MMAA; CH_3AsO (OH) 2) are the best known species, as well as in the world of five-sided oxidation in both cases. Because of expanded bacterial movement, appraisals of those two species were seen to develop throughout the late spring. The proportion of As (III) to As (V) in groundwater can change significantly because of enormous varieties in spring redox conditions, redox slopes, and history. As (III) frequently happens in underground regions that are extremely low. As (III)/AsT gauges in Bangladesh lower groundwater As-rich fluctuate from 0.1 to 0.9, however, as a rule around 0.5–0.6 (Smedley and Kinniburgh 2002).

7.2.8 Biogeochemical Cycle of Arsenic

Inside the floor and underground habitats, herbal geochemistry and biological sports are vital in figuring out the biogeochemical cycle of As (Islam et al. 2013). As mobilization happens in groundwater in oxidizing and decreasing instances with pH ranging from 6.5 to 8.5 (Kim et al. 2011). In humid and dry situations, biotic and abiotic sports resource natural redox transformation and mobilization of As.

Although there may be limited records of As focusing on the oil field and the different oil fields, some published studies suggest that awareness may be far more extreme. In the sodium bicarbonate water from the 1000 m depth of oil field at Alberta, Ellis Pool, Canada, White et al. (1963) refers to a molten substance such as 230 g/L contents. In a sodium-chloride-rich brine from Tisak € urt, Hungary, a concentration of 5800 g/L was obtained. Most As speciation experiments in herbal water aim to classify inorganic species into (III) and As (V) chromatographic subdivisions or a relatively lazy disc (As) V with sodium borohydride (Edwards et al. 1998). To evade the assignment of keeping species for later testing in the laboratory, the classification of species may be eliminated within the area (Smedley and Kinniburgh 2002). Storing mineral acid like HCl and ascorbic acid, too, has been efficacious. Oxidation conditions for rainwater may vary depending on As supply. Arsine (As (III) H3) is a naturally occurring substance that can be discovered in dumping and decontamination sites. It is produced by solvents, coal burning, and volcanic springs, and is about as distant from (III) 2O3 as is conceivable. Soils like peat and As (IV) can be found in sea aerosols. Decreased bureaucracy may involve O_2 in the environment, and interaction with SO_2 or O_3 in the ecosystem is predictable (He et al. 2010). In oxygen-rich seawater, As (V) tends to dominate, at the same time as As (III) persists and becomes important in anointing backwater. In open seawater, As aggregation under natural conditions is often associated with delta and alluvial deposits (Sracek et al. 2004). Where there are natural organisms in which fossils are deposited, extreme conditions may arise. The arrival of As from the indistinct Fe oxides usually found in spring residue is helped by these profoundly restricting conditions. Feeling of As from spring soil in groundwater is worked with by the decrease of As-containing nebulous Fe (III) oxide. Also, broke up natural dissolved organic carbon (DOC) and anaerobic iron-diminishing microorganisms found in vessels, for example, Shewanella green growth, *Shewanella oneidensis*, and *Geobacter metallireducens*, may assume a part in compounding As (Islam et al. 2004, Eiche et al. 2017). Under high-impact conditions, type As (III) is not thermodynamically steady and is quickly decreased to a subtype As (V) cell (Figure 7.1). At the point when oxygen is the main oxidant, nonetheless, the cycle dials back. Other redox-sensitive species, such as Fe (III) oxides, manganese (MnO_2) oxides, soil minerals, and some microorganisms, can accelerate the oxidation of As (III) to the Ace (III) type (III) is less harmful (V) (van Geen et al. 2008). Manganese oxides have been singled out as the primary oxidant for the oxidation of As (III) to As (V), and the reaction is thermodynamically successful over a wide pH range (Eiche et al. 2008). The co-operation of As (III) and As (V) with the produced birnessite (MnO_2) was tested. When (III) becomes MnO_2, the reaction object is (V) that promoted for the duration of MnO_2. Likewise, several bacteria, for example, *Pseudomonas aeruginosa*, *Alcaligenes faecalis*, *Legionella pneumophila*, *Thermus thermophilus* and *Agrobacterium tumefaciens*, have been exposed to use oxygen to provide As (III) oxidation. As displayed in Figure 7.4, iron hydroxide Fe (OH) three is possibly the most well-known stage associated with spring shops (Valenzuela et al. 2009). For As (III) the convergence of groundwater is firmly connected with the dissolving way of As from Fe (OH) 3 (Islam et al. 2013). Decrease of dissimilatory Fe bio-intervened responses that can deliver As from Fe (OH) 3 where natural matter and residing carbon made by regular atoms are combined. Fe (III) is decreased to Fe (II) by the scattering of Fe to lessen microscopic organisms and green growth to work with the arrival of As (V) into translucent mineral scorodite (Frommer et al. 2011).

7.3 Chemistry of Chromium in Aquatic System

With an absolute grouping of 64 mg/kg in waterway residue, chromium is tenth of the most plentiful in the world's hull (Fendorf 1995) and is viewed as a minor component (De Caritat and Reimann 2012). The principal sources of chromium pollution are modern and metropolitan sources, causing the vast majority of the Cr outside layer in residue. Because of its toxicity and/or consumption inhibitor properties, Cr (VI) compounds are generally utilized in numerous modern applications. Decrease of Cr (VI) to Cr (III) makes Cr (III) precipitation layers, which are steadier under different substance conditions. Cr is frequently utilized consequently, for instance, wood conservation, skin care regime, concrete and paint, metal drenching, and electroplating (Reid 2011). The principal sources of Cr (VI) in the sea-dwelling environment are unplanned spillage, inappropriate removal, and modern contamination (Kotas and Stasicka 2000; Kumar and Riyazuddin 2011). Because of the antagonistic consequences for amphibian life, chromium contamination remains significant for ecological concern. Since it effectively infiltrates into the cell layer through backhanded anion channels, Cr (VI) is viewed as hazardous for animal groups. Cr (III) is consumed by phagocytosis or inactive dispersion. Types of dynamic oxygen (ROS) like superoxide, hydroxyl revolutionaries, and hydrogen peroxide are significant reasons for harm to intracellular Cr (VI) (IPCS 2009). Cr (III) structures complex coordination buildings containing nucleic corrosive and proteins as an end product of Cr (VI) digestion. Therefore, Cr (VI) decrease is an inception occasion answerable for the advancement of genotoxic harm and different types of poisons inside the cell (Salnikow and Zhitkovich 2008). Also, Cr (VI) is profoundly intrusive in amphibian frameworks, while Cr (III) is frequently exhausted and/or presented to minerals, particularly aluminosilicates, prompting slow silt development (Rai et al. 1989). The distinguishing proof of redox processes that might add to the chromium oxidation state in water frameworks is especially significant due to these distinctions between Cr (VI) and Cr (III).

Surface dregs are adaptable and complex. In any case, this lattice comprises liquid pores, strong inorganic particles, organic entities, and living beings (particularly microbes that structure biofilms in strong surfaces) that can change many stages (Berner 1980). Since the sedimentary region might change geologically, creating ideal signs in a huge area of exploration and/or observing individual site arrangement over the long run can be tricky. Second, solid redox slopes can create in structures, particularly in the earth, where bacterial action is associated with the absorption of rotting natural minerals, including the utilization of broken-down oxygen and different solvents (like nitrates, iron hydroxides, and sulfates) (Froelich et al. 1979). This cycle, likewise called early diagenesis, upsets the harmony between the various stages. Nonetheless, it is feasible to appear as a nearby gauge. While analyzing the particularity of the redox factor and the elements (worldwide impulse, synthetic, and organic) that influence their structures, spatial elements are significant (Borch et al. 2009). Therefore, deciding the detail of Cr in the upper silt is significant, particularly on the off chance that the metal is available in little amounts. Subsequent to estimating the all-out chromium content, most speciation strategies center chiefly around computing the quantity of Cr (III) or Cr (VI) species, and afterward deciding the quantity of different species by difference (Marques et al. 2000). As it ignores the type of polymeric Cr (III) (the intermediary step in the continuous progression from free particles to Cr III (OH) 3 accelerates) and/or Cr (III) - broke up living things, this cycle is fragmented. Subsequently, almost certainly, the convergences of chromium and toxicity will drop (Hu et al. 2016). Because of the first-rate recovery of Cr species and the presence of tricky debris and combinations like carbonate, chloride, and humic corrosive, which can basically disable chromatographic division of Cr (III), all of that at the same time as envisioned by way of the two sorts of Cr, Cr is but difficult (Xing and Beauchemin 2010). Relating techniques have been endeavored to represent the trouble of straightforwardly characterizing Cr in sullied water tests by tannery (Burbridge et al. 2012). Thus, there are various references in the writing on the disclosure of instruments that impact Cr redox speciation in water settings. Different examples and in situ estimations, then again, are relied upon to work on the ecological examination of chromium soon (Devillers et al. 2016; Suárez et al. 2016).

Chromium species exist in an assortment of oxidation states, from +II to +VI, with trivalent [Cr(III)] and hexavalent [Cr(VI)] species predominant in amphibian settings. CrIII(OH)three, $HCrVIO_4$ –, and $CrVIO_4$ 2- are the vital loose fluid Cr species inside the surrounding pH variety (for instance 5 pH nine) (Kotas and Stasicka 2000). Since their synthesis requires high chromium concentrations and/or high temperatures, polynuclear complexes (CrIII 2(OH)2 4+, CrIII 3(OH)4 5+, CrIII 4(OH)6 6+, [CrIII(OH)4]n n-, and $CrVI_2O_7^{2-}$) and mixed ligand complexes (CrIII(OH)Cl+, CrIII(SO4) +) never have significant contents in solution (Torapava et al. 2009; Rakhunde et al. 2012). Cr (III) may be complexed with an assortment of everyday substance ligands, expanding the trivalent chromium's restrained solvency (Eq. 7.1).

$$\mathrm{Cr3++3OH-} \leftrightarrow \mathrm{Cr}\left(\mathrm{OH}\right)\mathrm{three\ with\,Ks} = 6.710-31 \tag{7.1}$$

Various complexation constants among Cr(III) and engineered chelating professionals are reachable in the writing (Carbonaro et al. 2008; Carbonaro and Stone 2015); however, associations with normal natural member are difficult to understand (Gustafsson et al. 2014). This can make displaying Cr(III) motion in the oceanic climate tough. Cr(VI) is discovered as oxyanions [$HCrVIO_4$ – and $CrVIO_4{}^{2-}$], just as an exceedingly dissolvable nonpartisan inorganic species that is probably introduced in the sight of the good sized cations found in silt pore liquids (as an example Ca^{2+}, K^+, Na^+, and NH^{4+}) (Kimbrough et al. 1999).

7.3.1 Biogeochemical Cycle of Chromium

7.3.1.1 Oxidation of Cr (III)

Numerous oxidants are capable of oxidizing Cr (III) into Cr (VI) in sediments, particularly at the water-sediment interface. Due to its delayed oxidative kinetics, dissolved oxygen has been shown to be an inactive (indirect) oxidant (Schroeder and Lee 1975). Although redox reactions appear to be thermodynamically plausible for these two species, nitrates and nitrites are never mentioned as potential oxidants. Thermodynamically, the reduction of sulfates by Cr(III) is not possible. Finally, only manganese (III, IV) (hydr)-oxides (Fendorf 1995) are capable of oxidizing the Cr(III) species present in both the liquid and solid phases appreciably. The biological oxidation of Cr has received little attention (III). Mn (II)-oxidizing bacteria, such as Bacillus sp. strain SG-1 (Murray and Tebo 2007) and *Pseudomonas putida* (Murray et al. 2005), have been demonstrated to indirectly increase the oxidation of Cr (III) to Cr(VI) in aerobic to partly anaerobic environments, albeit the reaction pathways remain unknown. Bacterial oxidation of Mn (II) to Mn(IV) (hydr)-oxides proceeds as a series of two one-electron transfer events driven by enzymes (Webb et al. 2005), with the enzyme-Mn(III) intermediate serving as a possible Cr(III) oxidant (Murray and Tebo 2007).

7.3.1.2 Reduction of Cr(VI)

In the presence of Fe(II), HS-, natural organic matter, and bacteria, the reduction of Cr(VI) to Cr(III) is most effective (Cheung and Gu 2007). Because Cr(VI) is more effective in soluble forms, its solubility can be a limiting factor in redox reactions in sediments, as its dissolved concentrations are influenced by a variety of elements in pore fluids, including Ba(II), Ca(II), Fe(III), K(I), and Sr(II) (Kimbrough et al. 1999).

Biotic reduction of Cr(VI) to Cr(III) occurs either as a result of detoxification or as a result of dissimilatory reduction (Barrera-Diaz et al. (2012). Several Cr(VI)-reducing bacteria have been identified in aerobic and/or anaerobic environments, including *Achromobacter* Ch1 (Ma et al. 2007), *Agrobacterium radiobacter* EPS-916 (Llovera et al. 1993), *Enterococcus gallinarum* (Sayel et al. 2012), *Escherichia coli* ATCC 33456 (Shen and Wang 1993), and *Leucobacter* (DeLeo and Ehrlich 1994). Both insoluble CrIII(OH)3 precipitates and soluble organo-Cr(III) complexes are often produced by microbial reduction of Cr(VI) (Cheng et al. 2010). Due to sluggish ligand exchange mechanisms, these complexes are very stable over time, conferring more stability to Cr(III) in pore fluids than can be expected thermodynamically (Beck et al. 2008). The presence of inhibitor, biomass concentration, starting Cr(VI) level, carbon supply, pH, temperature, redox potential, and other factors all impact Cr(VI)-reducing bacteria (Chen and Hao 1998). In aerobic conditions, the rate of Cr(VI) reduction by Fe(II), S(-II), and bacteria implies that biotic reductive processes predominate. In an anaerobic environment, bacteria contribute to the decrease of Cr(VI). For pH > 5, Fe(II) is the major reducer in anaerobic settings, whereas sulfide reduction of Cr(VI) dominates for pH 5.

7.4 Conclusion

Heavy metals and metalloids are common contaminants found in aquatic habitats. The danger of natural chemicals is determined by their persistence in environment, toxicity, and bioaccumulative power. Persistent and bioaccumulative pollutants are highly toxic. There are three characteristics of heavy metals that make it vulnerable to persistence, bioaccumulation, and toxicity (PBT). Cr, Ni, Cu, Zn, Cd, Pb, Hg, and As are some of the heavy metals that damage the environment and metalloids. Transmission of these nutrients into the food chains of the web has important effects on animal health and human health. The concentration of potentially harmful metals and metalloids in various parts of the environment, as well as biodiversity, should be evaluated.

References

Acharya J, Kumar U, Rafi PM (2018) Removal of heavy metal ions from wastewater by chemically modified agricultural waste material as potential adsorbent – a review. *Int J Curr EngTechnol* 8(3):526–530.

Acharyya SK, Chakraborty P, Lahiri S, Raymahashay BC, Guha S, Bhowmik A (1999) Arsenic poisoning in the Ganges delta. *Nature* 401(6753):545–547.

Afrin R, Mia MY, Ahsan MA, Akbor A (2015) Concentration of heavy metals in available fish species (bain, *Mastacembelus armatus*; taki, *Channa punctatus* and bele, *Glossogobius giuris*) in the Turag river, Bangladesh. *Pakistan Journal of Scientific and Industrial Research Series B: Biological Sciences* 58(2):104–110.

Ahmed AAM, Alam MJB, Ahmed AAM (2011) Evaluation of socioeconomic impact of arsenic contamination in Bangladesh. *J Toxicol Environ Heal Sci* 3(10):298–307.

Ali, H, Khan E (2017) Environmental chemistry in the twenty-first century. *Environmental Chemistry Letters* 15(2):329–346.

Ali H, Khan E, Sajad MA (2013) Phytoremediation of heavymetals-concepts and applications. *Chemosphere* 91(7):869–881.

Anawar HM, Akai J, Sakugawa H (2004) Mobilization of arsenic from subsurface sedimentsby effect of bicarbonate ions in groundwater. *Chemosphere* 54:753–762.

Atkins P, Jones L (1997) *Chemistry – Molecules, Matter and Change*, 3rd edition, W. H. Freeman, New York, NY, USA.

Azcue JM, Nriagu JO (1995) Impact of abandoned mine tailings on the arsenic concentrations in Moira Lake, Ontario, *J Geochem Explor* 52:81–89.

Baatrup E (1991) Structural and functional effects of heavy metals on the nervous system, including sense organs, of fish. *Comparative Biochemistry and Physiology Part C: Comparative Pharmacology* 100(1–2):253–257.

Barbieri M (2016) The importance of enrichment factor (EF) and geoaccumulation index (Igeo) to evaluate the soil contamination. *J Geol Geophys* 5(1):237.

Barrera-Diaz CE, Lugo-Lugo V, Bilyeu B (2012). A review of chemical, electrochemical and biological methods for aqueous Cr(VI) reduction. *Journal of Hazardous Materials* 223–224:1–12.

Bears H, Richards JG, Schulte PM (2006) Arsenic exposure alters hepatic arsenic species composition and stress-mediated gene expression in the common killifish (*Fundulus heteroclitus*). *Aquat Toxicol* 77:257–266.

Beck M, Dellwig O, Schnetger B, Brumsack H-J (2008). Cycling of trace metals (Mn, Fe, Mo, U, V,Cr) in deep pore waters of intertidal flat sediments. *Geochimica et Cosmochimica Acta*; 72: 2822–2840.

Berner RA (1980) Early Diagenesis: A Theoretical Approach. Princeton University Press.

Bissen M, Frimmel FH (2003) Arsenic a review. Part I: occurrence, toxicity, speciation, and mobility. *Acta Hydrochim Hydrobiol* 31:9–18.

Borch T, Kretzschmar R, Kappler A, Cappellen PV, Ginder-Vogel M, Voegelin A, Campbell K (2009). Biogeochemical redox processes and their impact on contaminant dynamics. *Environ Sci Technol* 44:15–23.

Burbridge DJ, Koch I, Zhang J, Reimer KJ (2012) Chromium speciation in river sediment pore water contaminated by tannery effluent. *Chemosphere* 89:838–843.

Caceres DD, Pino P, Montesinos N, Atalah E, Amigo H, Loomis D (2005) Exposure to inorganic arsenic in drinking water and total urinary arsenic concentration in a Chilean population. *Environ Res* 98:151–159.

Carbonaro RF, Stone AT (2015) Oxidation of Cr(III) aminocarboxylate complexes by hydrous manganese oxide: products and time course behaviour. *Environ Chem* 12:33–51.

Carbonaro RF, Gray BN, Whitehead CF, Stone AT (2008) Carboxylate-containing chelating agent interactions with amorphous chromium hydroxide: adsorption and dissolution *Geochimica et Cosmochimica Acta* 72:3241–3257.

Carnelo LG de L, Miguez SR de, Marbán L (1997) Heavy metals input with phosphate fertilizers used in Argentina. *Science of the Total Environment* 204(3):245–250.

Cervantes C, Ji G, Ramirez JL (1994) Silver resistance to arsenic compounds in microorganisms. *FEMS Microbiol Rev* 15:355–367.

Chen JM, Hao OJ (1998) Microbial chromium (VI) reduction. *Crit Rev Environ Sci Technol* 28:219–251.

Cheng Y, Yan F, Huang F, Chu W, Pan D, Chen Z, Zheng J, Yu M, Lin Z, Wu Z (2010) Bioremediation of Cr (VI) and immobilization as Cr (III) by *Ochrobactrum anthropi*. *Environ Sci Technol* 44:6357–6363.

Cheung KH, Gu J-D (2007) Mechanism of hexavalent chromium detoxification by microorganisms and bioremediation application potential: a review. *International Biodeterioration & Biodegradation* 59:8–15.

Chowdhury TR, Basu GK, Mandel BK, Biswass BK, Samanta G, Chowdhury UK, Chandra CR, Lodh D, Roy SL, Saha KC, Roy S, Kabir S, Quamruzzaman Q, Chakraborti D (1999) Arsenic poisoning in the Ganges delta. *Nature* 401:545–546.

Csuros M, Csuros C (2002) *Environmental Sampling and Analysis for Metals*, Lewis Publishers, Boca Raton, FL, USA.

Cui J-L, Zhao Y-P, Li J-S, Beiyuan J-Z, Tsang DCW, Poon C-S, Chan T-S, Wang W-X, Li X-D (2018) Speciation, mobilization, and bioaccessibility of arsenic in geogenic soil profile from Hong Kong. *Environmental Pollution* 232:375–384.

Das S, Unni B, Bhattacharjee M, Wann SB, Rao PG (2012) Toxicological effects of arsenic exposure in a freshwater teleost fish, *Channa punctatus. African J Biotech* 11(19):4447–4454.

Datta S, Ghosh D, Saha DR, Bhattacharaya S, Mazumder S (2009) Chronic exposure to low concentration of arsenic is immunotoxic to fish: role of head kidney macrophages arsenic biomarkers of arsenic toxicity to *Clarias batrachus. Aquat Toxicol* 92:86–94.

De Caritat P, Reimann C (2012) Comparing results from two continental geochemical surveys to world soil composition and deriving Predicted Empirical Global Soil (PEGS2) reference values. *Earth & Planet Sci Lett* 319–320:269–276.

DeLeo PC, Ehrlich HL (1994). Reduction of hexavalent chromium by *Pseudomonas fluorescens* LB300 in batch and continuous cultures. *Appl. Microbiol Biotechnol* 40:756–759.

Devillers D, Buzier R, Simon S, Charriau A, Guibaud G (2016) Simultaneous measurement of Cr(III) and Cr(VI) in freshwaters with a single diffusive gradients in thin films device. *Talanta* 154:533–538.

Dissanayake CB, Chandrajith R (2009) Phosphate mineral fertilizers, trace metals and human health. *Journal of the National Science Foundation of Sri Lanka* 37(3):153–165.

Drewniak L, Sklodowska A (2013) Arsenic-transforming microbes and their role in biomining processes. *Environmental Science and Pollution Research* 20(11):7728–7239.

Duffus JH (2002) "Heavy metals" a meaningless term? (IUPAC Technical Report). *Pure and Applied Chemistry* 74(5):793–807.

Duker AA, Carranza EJM, Hale M (2005) Arsenic geochemistry and health. *Environ Int* 31(5):631–641.

Ebrahimpour M, Mosavisefat M, Mohabbati R (2010) Acute toxicity bioassay of mercuric chloride: an alien fish from a river. *Toxicological & Environmental Chemistry* 92(1):169–173.

Edwards M, Patel S, McNeill L, Chen Hw, Frey M, Eaton AD, Antweiler RC, Taylor HE (1998) Considerations in as analysis and speciation. *J Am Water Works Assoc* 90:103–113.

Eiche E, Neumann T, Berg M, Weinman B, van Geen A, Norra S, Berner Z, Trang PTK, Viet PH, Stüben D (2008) Geochemical processes underlying a sharp contrast in groundwater arsenic concentrations in a village on the red river delta, Vietnam. *Appl Geochem* 23:3143–3154.

Eiche E, Berg M, Hönig S-M, Neumann T, Lan VM, Pham TKT, Pham HV (2017) Origin and availability of organic matter leading to arsenic mobilisation in aquifers of the red river delta, Vietnam. *Appl Geochem* 77:184–193.

El-Naggar A, Shaheen SM, Ok YS, Rinklebe J (2018) Biochar affects the dissolved and colloidal concentrations of Cd, Cu, Ni, and Zn and their phytoavailability and potential mobility in a mining soil under dynamic redox-conditions, *Sci Total Environ* 624:1059–1071.

Eqani SAMAS, Khalid R, Bostan N, Saqib Z, Mohmand J, Rehan M, Ali N, Katsoyiannis IA, Shen H (2016) Human lead (Pb) exposure via dust from different land use settings of Pakistan: a case study from two urban mountainous cities. *Chemosphere* 155:259–265.

Ercal N, Gurer-Orhan H, Aykin-Burns N (2001) Toxic metals and oxidative stress Part I: mechanisms involved in metal-induced oxidative damage. *Current Topics in Medicinal Chemistry* 1(6):529–539.

Fakhreddine S, Dittmar J, Phipps D, Dadakis J, Fendorf S (2015) Geochemical triggers of arsenic mobilization during managed aquifer recharge. *Environ Sci Technol* 49:7802–7809.

Fendorf SE (1995) Surface reactions of chromium in soils and waters. *Geoderma* 67:55–71.

Ferretti M, Cenni E, Bussotti F, Batistoni P (1995) Vehicle induced lead and cadmium contamination of roadside soil and plants in Italy. *Chemistry and Ecology* 11(4):213–228.

Froelich PN, Klinkhammer GP, Bender ML, Luedtke NA, Heath GR, Cullen D, Dauphin P, Hammond D, Hartman B, Maynard V (1979) Early oxidation of organic matter in pelagic sediments of the eastern equatorial Atlantic: suboxic diagenesis. *Geochimica et Cosmochimica Acta* 43:1075–1090.

Frommer J, Voegelin A, Dittmar J, Marcus MA, Kretzschmar R (2011) Biogeochemical processes and arsenic enrichment around rice roots in paddy soil: results from microfocused x-ray spectroscopy. *Eur J Soil Sci* 62:305–317.

Gaim K, Gebru G, Abba S (2015) The effect of arsenic on liver tissue of experimental animals (fishes and mice) – a review article. *Int J SciRes Publ* 5(5):1–9.

van Geen A, Radloff K, Aziz Z, Cheng Z, Huq M, Ahmed K, Weinman B, Goodbred S, Jung H, Zheng Y (2008) Comparison of arsenic concentrations in simultaneously-collected groundwater and aquifer particles from Bangladesh, India, Vietnam, and Nepal. *Appl Geochem* 23:3244–3251.

Gernhofer M, Pawert M, Scharmm M, Muller E, Triebskom R (2001) Ultrastructural biomarkers arsenic tool to characterize the health status of fish in contaminated streams. *J Aquat Ecosys Stress Reco* 8:241–260.

Grant CA, Sheppard SC (2008) Fertilizer impacts on cadmium availability in agricultural soils and crops. *Human and Ecological Risk Assessment: An International Journal* 14(2):210–228.

Gustafsson JP, Persson I, Oromieh AG, van Schaik JWJ, Sjostedt C, Kleja DB (2014). Chromium(III) complexation to natural organic matter: mechanisms and modeling. *Environ Sci Technol* 48:1753–1761.

Hartl MGJ (2013) Book review homeostasis and toxicology of non-essential metals. Edited by C. M. Wood, A. P. Farrell & C. J. Brauner. 528 pp. Published by Elsevier, London, U.K.,2012. Price £66.99. ISBN: 978-0-12-378636-0. *Journal of Fish Biology* 83:1476–1477.

Hashem MA, Nur-A-Tomal MS, Mondal NR, Rahman MA (2017) Hair burning and liming in tanneries is a source of pollution by arsenic, lead, zinc, manganese and iron. *Environ Chem Lett* 15(3):501–506.

He YT, Fitzmaurice AG, Bilgin A, Choi S, O'Day P, Horst J, Harrington J, Reisinger HJ, Burris DR, Hering JG (2010) Geochemical processes controlling arsenic mobility in groundwater: a case study of arsenic mobilization and natural attenuation. *Appl Geochem* 25:69–80.

Hu L, Cai Y, Jiang G (2016) Occurrence and speciation of polymeric chromium (III), monomeric chromium (III) and chromium (VI) in environmental samples. *Chemosphere* 156:14–20.

Hughes MF, Beck BD, Chen Y, Lewis AS, Thomas DJ (2011) Arsenic exposure and toxicology: a historical perspective. *Toxicol Sci* 123(2):305–332.

Hutton M (1983) Sources of cadmium in the environment. *Ecotoxicology and Environmental Safety* 7(1):9–24.

IARC (1987). Overall evaluations of carcinogenicity: an updating of IARC Monographs volumes 1 to 42. *IARC Monogr Eval Carcinog Risks Hum Suppl.* 7:1–440. PMID: 3482203.

IPCS (2009) Inorganic Chromium(III) compounds. World Health Organization, International Programme on Chemical Safety (Concise International Chemical Assessment Document) 76, http://www.who.int/ipcs/publications/cicad/cicad76.pdf (accessed 15 June 2016).

Islam FS, Gault AG, Boothman C, Polya DA, Charnock JM, Chatterjee D, Lloyd JR (2004) Role of metal-reducing bacteria in arsenic release from Bengal delta sediments. *Nature* 430:68.

Islam A, Maity JP, Bundschuh J, Chen C-Y, Bhowmik BK, Tazaki K (2013). Arsenic mineral dissolution and possible mobilization in mineral–microbe–groundwater environment, *J Hazard Mater* 262:989–996.

Järup L (2003) Hazards of heavy metal contamination. *British Medical Bulletin* 68(1):167–182.

Khan FU, Rahman AU, Jan A, Riaz M (2004) Toxic and trace metals (Pb, Cd, Zn, Cu, Mn, Ni, Co and Cr) in dust, dustfall/soil. *J Chem Soc Pakistan* 26(4):453–456.

Kim K-W, Chanpiwat P, Hanh HT, Phan K, Sthiannopkao S (2011) Arsenic geochemistry of groundwater in Southeast Asia. *Front Med* 5:420–433.

Kimbrough DE, Cohen Y, Winer AM, Creelman L, Mabuni C (1999) A critical assessment of chromium in the environment. *Crit Rev Environ Sci Technol* 29:1–46.

Ko M-S, Kim J-Y, Park H-S, Kim K-W (2015) Field assessment of arsenic immobilization in soil amended with iron rich acid mine drainage sludge. *J Clean Prod* 108:1073–1080.

Kotas J, Stasicka Z (2000) Chromium occurrence in the environment and methods of its speciation. *Environ Pollut* 107:263–283.

Kovendan K, Vincent S, Janarthanan S, Saravanan M (2013) Expression of metallothionein in liver and kidney of freshwater fish Cyprinus carpio var. communis (Linn) exposed to arsenic trioxide. *Am J Sci Ind Res* 4(1):1–10.

Krüger O, Fiedler F, Adam C, Vogel C, Senz R (2017) Determination of chromium (VI) in primary and secondary fertilizer and their respective precursors. *Chemosphere* 182:48–53.

Kumar A, Riyazuddin P (2011) Chromium speciation in a contaminated groundwater: redox processes and temporal variability. *Environmental Monitoring and Assessment* 176:647–662.

La Force MJ, Hansel CM, Fendorf S (2000) Arsenic speciation, seasonal transformations,and co-distribution with iron in a mine waste-influenced palustrine emergent wetland. *Environ Sci Technol* 34:3937–3943.

Lavanya S, Mathan R, Chokkalingam K, Annamalai M (2011) Hematological, biochemical and ionoregulatory responses of Indian major carp Catla catla during chronic sub lethal exposure to inorganic arsenic. *Chemosphere* 82:977–985.

Llovera S, Bonet R, Simon-Pujol MD, Congregado F (1993). Chromate reduction by resting cells of Agrobacterium radiobacter EPS. *Appl Environ Microbiol* 59:3516–3518.

Luo J, Ye Y, Gao Z, Wang W (2014) Essential and nonessential elements in the red-crowned craneGrus japonensisof Zhalong Wetland, northeastern China. *Toxicol Environ Chem* 96(7):1096–1105.

Ma Z, Zhu W, Long H, Chai L, Wang Q (2007) Chromate reduction by resting cells of Achromobacter sp. Ch-1 under aerobic conditions. *Process Biochemistry* 42:1028–1032.

Ma J, Mi Y, Li Q, Chen L, Du L, He L, Lei M (2016) Reduction, methylation, and translocation of arsenic in Panax notoginseng grown under field conditions in arsenic-contaminated soils. *Sci Total Environ* 550:893–899.

Mahmood Q, Wang J, Pervez A, Meryem SS, Waseem M, Ullah Z (2015) Health risk assessment and oxidative stress in workers exposed to welding fumes. *Toxicological & Environmental Chemistry* 97(5):634–639.

Malik DS, Maurya PK (2014) Heavy metal concentration in water, sediment, and tissues of fish species (*Heteropneustis fossilis* and *Puntius ticto*) from Kali River, India. *Toxicol Environ Chem* 96(8):1195–1206.

Marques MJ, Salvador A, Morales-Rubio AE, De la Guardia M (2000) Chromium speciation in liquid matrices: a survey of the literature. *Fresenius' Journal of Analytical Chemistry* 367:601–613.

McLaren SJ, Kim ND (1995) Evidence for a seasonal fluctuation of arsenic in New Zealand's longest river and the effect of treatment on concentrations in drinking water, *Environ Pollut* 90:67–73.

Merian E (1984) Introduction on environmental chemistry and global cycles of chromium, nickel, cobalt beryllium, arsenic, cadmium and selenium, and their derivatives. *Toxicological & Environmental Chemistry* 8(1):9–38.

Mortvedt JJ (1996) Heavy metal contaminants in inorganic and organic fertilizers. *Fertilizer Research* 43(1–3):55–61.

Murray KJ, Tebo BM (2007) Cr(III) is indirectly oxidized by the Mn(II)-oxidizing *Bacterium Bacillus* sp. strain SG-1. *Environ Sci Technol* 41:528–533.

Murray KJ, Mozafarzadeh ML, Tebo BM (2005) Cr (III) oxidation and Cr toxicity in cultures of the manganese (II)-oxidizing *Pseudomonas putida* strain GB-1. *Geomicrobiol J* 22:151–159.

Nagajyoti PC, Lee KD, Sreekanth TVM (2010) Heavy metals, occurrence and toxicity for plants: a review. *Environmental Chemistry Letters* 8(3):199–216.

Ng JC (2005) Environmental contamination of arsenic and its toxicological impact on humans. *Environ Chem* 2:146–160.

Niazi NK, Burton ED (2016) Arsenic sorption to nanoparticulate mackinawite (Fes): an examination of phosphate competition, *Environ Pollut* 2018:111–117.

Niazi, N.K., Bibi, I., Shahid, M., Ok, Y.S., Burton, E.D., Wang, H., Shaheen, S.M., Rinklebe, J. and Lüttge, A., (2018) Arsenic removal by perilla leaf biochar in aqueous solutions and groundwater: an integrated spectroscopic and microscopic examination. Environmental Pollution, 232, pp.31–41.

Oremland RS, Dowdle PR, Hoeft S, Sharp JO, Schaefer JK, Miller LG, Blum JS, Smith RL, Bloom NS, Wallschlaeger D (2000) Bacterial dissimilatory reduction of arsenate and sulfate in meromictic Mono Lake, California. *Geochim Cosmochim Acta* 64(18):3073–3084.

Palaniappan PR, Karthikeyan S (2009) Bioaccumulation and depuration of chromium in the selected organs and whole body tissues of freshwater fish *Cirrhinus mrigala* individually and in binary solutions with nickel. *Journal of Environmental Sciences* 21(2):229–236.

Rahman MM, Mondal D, Das B, Sengupta MK, Ahamed S, Hossain MA, Samal AC, Saha KC, Mukherjee SC, Dutta RN (2014) Status of groundwater arsenic contamination in all 17 blocks of Nadia district in the state of West Bengal, India: a 23-year study report. *J Hydrol* 518:363–372.

Rai D, Eary LE, Zachara JM (1989) Environmental chemistry of chromium. *Science of The Total Environment* 86:15–23.

Rakhunde R, Deshpande L, Juneja H (2012). Chemical speciation of chromium in water: a review. *Crit Rev Environ Sci Technol* 42:776–810.

Reid SD (2011) Molybdenum and chromium. In: Chris M. Wood APF, Colin JB, editors. Fish Physiology. Volume 31, Part A. Academic Press, pp. 375–415.

Rossman TG (2003) Mechanism of arsenic carcinogenesis: an integrated approach. *Mutat Res* 533:37–65.

Salman M, Rehman R, Mahmud T et al (2011) Assessment of concentration of lead, cadmium, chromium and selenium in blood serum of cancer and diabetic patients of Pakistan. *Journal of the Chemical Society of Pakistan* 33(6):869–873.

Salnikow K, Zhitkovich A (2008). Genetic and epigenetic mechanisms in metal carcinogenesis and cocarcinogenesis: nickel, arsenic and chromium. *Chem Res Toxicol* 21(1):28–44.

Sayel H, Bahafid W, Joutey NT, Derraz K, Benbrahim KF, Koraichi SI, El Ghachtouli N (2012) Cr (VI) reduction by Enterococcus gallinarum isolated from tannery waste-contaminated soil. *Annals of microbiology* 62:1269–1277.

Schroeder D, Lee GF (1975). Potential transformations of chromium in natural waters. *Water, Air, and Soil Pollution* 4:355–365.

Sfakianakis DG, Renieri E, Kentouri M, Tsatsakis AM (2015) Effect of heavy metals on fish larvae deformities:a review. *Environ Res* 137:246–255.

Shakoor MB, Nawaz R, Hussain F, Raza M, Ali S, Rizwan M, Oh S-E, Ahmad S (2017) Human health implications, risk assessment and remediation of as-contaminated water: a critical review. *Sci Total Environ* 601–602:756–769.

Shen H, Wang Y (1993). Characterization of enzymatic reduction of hexavalent chromium by *Escherichia coli* ATCC 33456. *Appl. Environ Microbiol* 59:3771–3777.

Smedley P, Kinniburgh D (2002) A review of the source, behaviour and distribution of arsenic in natural waters, *Appl Geochem* 17:517–568.

Smith E, Smith J, Smith L, Biswas T, Correll R, Naidu R (2003) Arsenic in Australian environment: an overview. *J Environ Sci Heal A* 38(1):223–239.

Spiegel H (2002) Trace element accumulation in selected bioindicators exposed to emissions along the industrial facilities of Danube Lowland. *Turkish Journal of Chemistry* 26(6):815–823.

Sracek O, Bhattacharya P, Jacks G, Gustafsson J-P, Von Brömssen M (2004) Behavior of arsenic and geochemical modeling of arsenic enrichment in aqueous environments, *Appl Geochem* 19:169–180.

Suárez CA, de Simone TV, Menegario AA, Rolisola AMCM, Luko KS, Gastmans D, et al. (2016) In-situ redox speciation analysis of chromium in water by diffusive gradients in thin films using a DE81 anion exchange membrane. *Talanta* 154:299–303.

Tanaka T (1990) Arsenic in the natural environment. Part II: arsenic concentrations in thermal waters from Japan. *Appl Organomet Chem* 4:197–203.

Tian HZ, Lu L, Hao JM, Gao JJ, Cheng K, Liu KY, Qiu PP, Zhu CY (2013) A review of key hazardous trace elements in Chinese coals: abundance, occurrence, behavior during coal combustion and their environmental impacts. *Energy & Fuels* 27(2):601–614.

Tisler T, Zagorc-Koncan J (2002) Acute and chronic toxicity of arsenic to some aquatic organisms. *Bull Environ Contam Toxicol* 69:421–429.

Torapava N, Radkevich A, Davydov D, Titov A, Persson I (2009) Composition and structure of polynuclear chromium(III) hydroxo complexes. *Inorg. Chem.* 48:10383–10388.

Turpeinen R, Pantsar-Kallio M, Haggblom M, Kairesalo T (1999) Influence of microbes on the mobilization, toxicity and Biomethylation of arsenic in soil. *Sci Total Environ* 236:173–180

Valenzuela C, Campos V, Yañez J, Zaror C, Mondaca M (2009) Isolation of arsenite oxidizing bacteria from arsenic-enriched sediments from Camarones river, Northern Chile. *Bull. Environ. Contam. Toxicol* 82:593–596.

Wang D, Dang Z, Feng H, Wang R (2015) Distribution of anthropogenic cadmium and arsenic in arable land soils of Hainan, China. *Toxicological & Environmental Chemistry* 97(3–4):402–408.

Wang Y, Pi K, Fendorf S, Deng Y, Xie X (2017) Sedimentogenesis and hydrobiogeochemistry of high arsenic late pleistocene-holocene aquifer systems. *Earth Sci Rev* 189:78–98.

Webb SM, Dick GJ, Bargar JR, Tebo BM (2005) Evidence for the presence of Mn (III) intermediates in the bacterial oxidation of Mn (II). *Proceedings of the National Academy of Sciences* 102(15):5558–5563.

White DE, Hem JD, Waring GA (1963) Chemical composition of sub-surface waters. In: Fleischer M (Ed.), Data of Geochemistry, 6th ed., Geological Survey, Washington, pp. 1–67.

Wieczorek-Dąbrowska M, Tomza-Marciniak A, Pilarczyk B, Balicka-Ramisz A (2013) Roe and red deer as bioindicators of heavy metals contamination in north-western Poland. *Chem Ecol* 29(2):100–110.

Wilk A, Kalisińska E, Kosik-Bogacka DI, Romanowski M, Różański J, Ciechanowski K, Słojewski M, Łanocha-Arendarczyk N (2017) Cadmium, lead and mercury concentrations in pathologically altered human kidneys. *Environmental Geochemistry and Health* 39(4):889–899.

Xing, L. and Beauchemin, D., (2010) Chromium speciation at trace level in potable water using hyphenated ion exchange chromatography and inductively coupled plasma mass spectrometry with collision/reaction interface. Journal of Analytical Atomic Spectrometry, 25(7), pp.1046–1055.

Xu H, Lam SH, Shen Y, Gong Z (2013) Genome-wide identification of molecular pathways and biomarkers in response to arsenic exposure in zebrafish liver. *PLoS ONE* 8(7):e68737.

Yan X-P, Kerrich R, Hendry MJ (2000) Distribution of arsenic (III), arsenic (V) and total inorganic arsenic in porewaters from a thick till and clay-rich aquitard sequence, Saskatchewan, Canada, *Geochim Cosmochim Acta* 64:2637–2648.

Yang H-C, Rosen BP (2016) New mechanisms of bacterial arsenic resistance. *Biom J* 39:5–13.

Yang Y-P, Zhang H-M, Yuan H-Y, Duan G-L, Jin D-C, Zhao F-J, Zhu Y-G (2018) Microbe mediated arsenic release from iron minerals and arsenic methylation in rhizosphere controls arsenic fate in soil-rice system after straw incorporation. *Environ Pollut* 236:598–608.

Yao Q-Z, Zhang J, Qin X-G, Xiong H, Dong L-X (2006) The behavior of selenium and arsenic in the zhujiang (pearl river) estuary, South China Sea, *Estuar Coast Shelf Sci* 67:170–180.

8

Occurrences and Mobility of Uranium in Soil Profile Due to Groundwater–Soil Interaction

Somenath Ganguly and Uday Bhan

Department of Petroleum Engineering & Earth Sciences, University of Petroleum and Energy Studies, Dehradun, Uttarakhand, India

8.1 Introduction

Uranium is a naturally occurring radioactive mineral found in Earth's crust in association with radium and other radioactive elements. Average uranium content for supracrustal rocks of outer crust (including acidic, basic, and ultrabasic igneous rocks as well as sedimentary rocks) is approximately 1 g/ton. Considering all the rock types, granite has maximum concentration of uranium. To be more specific, uranium enrichment has been reported maximum for end-stage magmatic products of granitic magma particularly of pegmatite and mineral veins. The reason behind this is uranium does not form stable compounds with common rock forming elements like silicon or aluminum. In the process of magma generation, uranium only combines with rare earth elements like thorium, niobium, tantalum, and other similar elements. In the magma chamber, these constitute the volatile fraction of the magma. Therefore, they tend to concentrate at the upper part of the magma chamber with maximum enrichment in the form of pegmatite dykes or hydrothermal veins and emplace as intrusive igneous bodies at outer continental crust (Tomkeieff 1946; Bali 2012).

Radioactive elements have significant contribution in heat generation at shallow crustal level. Among them, the most significant radioactive elements are uranium, thorium, and potassium (^{40}K). Uranium and thorium present in the upper crust of the Earth in very small quantities and quantifiable only in parts per million (ppm). However, in the upper mantle, their abundance is two orders of magnitude less than the abundance in upper crust. ^{238}U, ^{235}U, ^{232}Th, and ^{40}K are the most common radioactive isotopes for heat generation at shallow crustal level. For shallow crustal level, the common radioactive isotopes of uranium are ^{238}U, ^{235}U, and ^{234}U with present relative abundance of 99.275, 0.720, and 0.0054%, respectively. Average amount of uranium present in granitic rock is 4.6 ppm contributing to Earth's internal heat at shallow crustal level of 9.8 10^{-5}W/kg (Lowrie 2007). Table 8.1 gives uranium, thorium, and ^{40}K abundance and heat generation of some common crustal rock types. It is evident from Table 8.1 that, uranium and thorium on average contribute more

Hydrogeochemistry of Aquatic Ecosystems, First Edition. Edited by Sughosh Madhav, Virendra Bahadur Singh, Manoj Kumar, and Sandeep Singh.

Table 8.1 Average abundance of U, Th, and ^{40}K and its contribution to heat production of some common rock types and undepleted mantle.

Concentration by weight (ppm)	Granite	Tholeiitic basalt	Alkali basalt	Peridotite	Average continental upper crust	Average continental crust	Average oceanic crust	Undepleted mantle
U	4	0.1	0.8	0.006	2.8	11.1	0.9	0.02
Th	15	0.4	2.5	0.04	10.7	4.2	2.7	0.10
^{40}K	3.5	0.2	1.2	0.01	3.4	1.3	0.4	0.04
Heat generation ($\mu W/m^3$)	2.5	0.08	0.5	0.006	1.8	0.7	0.5	0.02

Source: Fowler (2005)/Cambridge University Press.

radiogenic heat at shallow crustal level in comparison to potassium (^{40}K). On average, the contribution of granite to internal radiogenic heat generation at shallow crustal level is more than that of mafic igneous rocks, whereas the contribution of undepleted mantle to radioactive heat generation at shallow crustal level is very low.

From Table 8.1, it is clear that contribution of crust to the radioactive heat generation is about twice than that of undepleted mantle. However, considering the whole Earth, mantle contributes significantly to the heat production rate since the volume of the mantle is way larger than that of total volume of crust. Radiogenic heat contributes one fifth of crustal heat production. Combining crust and mantle, the mean abundances of ^{40}K, thorium, and uranium are in the ranges of 150–260 ppm, 80–100 ppb, and 15–25 ppb, respectively. As a whole for crust and mantle, total radioactive heat generation is in the order of $1.4–2.7 \times 10^{13}$ W, with best-guess approximation of 2.1×10^{13} W (Fowler 2005; p. 275).

In the igneous rock category, some uncommon igneous rocks like certain lamprophyric rocks and kimberlite are generally enriched in uranium (average abundance of 5 ppm) and thorium (average abundance of 13.5 ppm). Porphyritic granite and allied rocks and certain alkali-rich syenite may contain up to 100 ppm of thorium and 100 ppm of uranium. In sedimentary rocks, like saprolites, or pyritic carbonaceous shale or oil shale, may contain up to 20 ppm of thorium and 1200 ppm of uranium. Elevated concentration of uranium and thorium has also been reported in lignite and various other types of coal. Chemogenic sedimentary rocks like phosphorite are generally enriched in both uranium and thorium with concentration up to 12 ppm thorium and 300 ppm uranium (IAEA, Vienna 1988; Cuney 2009; Fayek et al. 2011; Alloway 2013).

Subsequent weathering of igneous rocks lead to uranium enrichment in soil. The process of chemical weathering generally does not change the uranium concentration in the residual soil with respect to parent bed rock. However, in igneous rocks, uranium is associated with heavy minerals which are resistant to chemical weathering. Thus, mechanical weathering is the process by which uranium is transported by rivers and streams and may form placer deposits in the perspective of ore mineralization. In some region, uranium and thorium enrichment may happen due to the accumulation of resistant minerals like zircon or mobilization of uranium by metal complex formation or by gels of vanadate, phosphate, silicate, and carbonate (Bourdon et al. 2003). However, local uranium depletion has reported due to the presence of active leaching agents like sulfuric carbonic or organic acids (Idiz et al. 1986). This will be covered in detail in the following sections of this chapter.

8.2 Uranium Mineralogy and Mode of Occurrences

Uranium mineralogy is a vast topic and beyond the scope of this chapter. Uranium is classified under lithophile element. Uranium can have all the valence states from 2+ to 6+. However, in the mineralogical point of view only tetravalent (4+) and hexavalent (6+) oxidation states are important, though the presence of 5+ oxidation state is also reported as solution species and for some of the uranium oxides (Mehta et al. 2014). Uranium minerals can be divided into two main group namely "primary" and "secondary" minerals. The basis of this classification is all uranium was first deposited as uraninite considering it as primary mineral. Further oxidation of primary mineral leads to the formation of other minerals which comes under the heading of secondary minerals. However, recent studies reported that several uranyl minerals can be formed directly from source solutions and without oxidizing primary uraninite. Therefore, in the mineralogical point of view, the better way to classify uranium minerals is based on their valance states. As per this mode of classification, U^{4+} comes under "reduced" species and U^{6+} comes under "oxidized" species. Uraninite comes under reduced minerals with or without associated with rare earth elements and other actinide elements. Common primary uranium minerals are niobate, tantalate, and titanate minerals (Hazen et al. 2009). Oxidized minerals are usually occurred in the form of hydrated oxides, silicates, and phosphates. Table 8.2 captures common uranium minerals of reduced (4+) species, whereas Table 8.3 captures common uranium minerals of oxidized (6+) species. However, uraninite is the commonest reduced mineral for

Table 8.2 Common uranium minerals of reduced (4+) state.

Mineral	Formula	Crystal system
Brannerite	$(U, Ca, Ce)(Ti, Fe)_2O_6$	Monoclinic
Coffinite	$U(SiO_4)_{,1-x}(OH)_{4x}$	Tetragonal
Ishikawaite	$(U, Fe, Y, Ca)(Nb, Ta)O_4$	–
Lermontovite	$(U, Ca, Ce)PO_4(OH)\cdot H_2O$	–
Mourite	$U\,Mo_5\,O_{12}(OH)_{10}$	Monoclinic
Ningyoite	$(U, Ca, Ce)_2(PO_4)_2.1–2H_2O$	Orthorhombic
Petscheckite	$U\,Fe(Nb, Ta)_2O_8$	Hexagonal
Sedovite	$U\,(MoO_4)_2$	Orthorhombic
Uraninite	$UO_{2+X}\,(0.0 < X < 0.25)$	Cubic
Uranmicrolite	$(U, Ca, Ce)_2(Nb, Ta)_2O_6(OH,F)$	Cubic
Uranpyrochlore	$(U, Ca, Ce)_2(Ta, Nb)_2O_6(OH,F)$	Cubic

Source: De Vivo et al. (1984)/Springer Nature.

Table 8.3 Common Uranium minerals of oxidized (6+) state.

Mineral	Formula	Crystal system
Ianthinite	$U(UO_2)_5(OH)_{14}.3H_2O$	Orthorhombic
Liandratite	$U(Nb, Ta)_2O_8$	Hexagonal
Moluranite	$H_4U(UO_2)_3(MoO_4)_7.18H_2O$	Amorphous
Orthobrannerite	$UUTi_4O_{12}(OH)_2$	Orthorhombic
Wyartite	$Ca_3U(UO_2)_6(CO_3)_2(OH)_{18}.3\text{-}5H_2O$	Orthorhombic

Source: De Vivo et al. (1984)/Springer Nature.

uranium. Uraninite is found in all uranium deposits. The rare exception is for a uranium-bearing ore body where the oxidizing zone has eliminated the reduced mineralization zone (De Vivo et al. 1984; p. 44).

The primary uranium deposits are of pegmatite origin and/or in sedimentary host rocks. Uranium deposits of pegmatite origin are directly related to igneous activity. Weathered residue or soil on granitic, gneissic, and alkali igneous rock terranes shows uranium enrichment in comparison to pelitic terranes formed due to weathering of basic, ultrabasic, and carbonate rocks (Murakami et al. 1997). However, sedimentary uranium deposits are considered as weathered residue of previously dispersed uranium, may be all of igneous origin. Uranium deposits of sedimentary origin can be grouped/classified as following materials such as weathered residuum, soil, gossan, silcrete, calcrete, ferricrete and various glacial depositional products like till, gravel, and sand (IAEA 2009). For weathered residuum, the concentration of uranium remains same as that of bedrock. However, for gossan, limonitic zones, limonitic cemented soil, sand gravel, and other similar materials concentration of uranium is highly variable. Also uranium content of soil is dependent on soil provenance and soil–water interaction for localized uranium enrichment. Concentration of uranium in soil may range from less than 1 ppm to 8 ppm, with an average abundance of about 1 ppm. However, in the uraniferous areas, uranium concentration may be as high as 50 ppm, and in some areas, it may range up to as high as 100 ppm (IAEA 1988).

In the soil profile, "A" horizon is likely to be the richest in uranium concentration. This may be attributed to uranium sequestration by organic humic materials/acids, clay-humic complexes, humic-silicate, and humic-iron complexes of uncertain origin. However, in and around uraniferous deposits, the "B" horizon may have higher uranium enrichment compared to "A" horizon. This may be attributed to the adsorption and/or sequestration of uranium by the hydroxides and hydrous oxides of iron, manganese, aluminum, or other metal ions which is highly dependent on oxidation-reduction (Eh) condition and pH of the soil profile (Beazley et al. 2011). However, for uraniferous deposits, the "C" horizon of the soil profile is likely to be the richest in uranium concentration due to uranium precipitation in anoxic condition (IAEA 1988).

The uranium dissolution, migration, and precipitation in weathered residue, soil profile, gossan, and calcretes, are dependent on the following factors (Susan et al. 2016).

1) Type of bed rock and its weathering product.
2) Eh and pH of the water in the soil profile.

3) The chemical condition for coprecipitating VO_4, PO_4, AsO_4, CO_3, as well as soil colloids, hydrous silicate, alumina and titania, and hydrous ferric and manganese oxides which influence adsorption of uranium ions as well as facilitate the formation of metal complexes.
4) The various complex forming reactions of uranium with organic humic acid in the soil profile.
5) Microbiological oxidation leading to uranium enrichment in the form of metal complexes.

The process of uranium fixation in soil, gossan, and weathered residuum is highly variable. Uranium may present in soil and allied materials as resistant heavy minerals like zircon, monazite, epidote, biotite allanite, apatite, sphene, and leucoxene. Uranium may bind with soil in the form of colloidal silicate components, hydrous ferric and manganese oxides, humic-clay complexes, humic-iron-manganese complexes as well as by the process of adsorption on clay in the presence of humic acids (Nakashima et al. 1984). However, uranium binding in gossan and limonitic, aluminous and siliceous residue overlying uraniferous deposits are highly variable. For uraniferous deposits, uranium mobilization from oxidized and leached zone of the ore body is influenced by adsorption, ion exchange, and coprecipitation processes on the wad, limonite, jarosite, colloidal silica/chert, clay minerals, and other components of the gossan (IAEA 1988).

8.3 Uranium Mobilization Due to Soil–Water Interaction

Soil–water interaction releases uranium and thorium from soil to groundwater. Then it is transported in the form of emulsoids, and/or through diffusion in the groundwater. Uranium form hydrated cations of UO^{2+}. These hydrated species are responsible for uranium solubility over a wide pH range (Echevarria et al. 2001). Several organic acids like humic and fulvic acid in soil profile have influence on increasing the solubility of uranium and thorium in groundwater (Kantar 2007). However, uranium and thorium mobility in soil profile is limited due to adsorption on clays and organic matter and also influenced by the presence of less soluble compounds of oxides and phosphates. Generally, uranium in soil profile is transported as uranyl ion. Stability of uranyl ion depends largely on Eh-pH condition of soil and groundwater along with the absence or presence of iron oxides, iron sulfides, and carbonates (Newsome et al. 2014). In general, thorium has more adsorption affinity onto clays or organic matter than uranium. Thorium generally forms ThO_2 under most naturally occurring pH-Eh conditions. Thorium may present in the form of $Th(OH)_4$ initially but eventually it age to ThO_2. For pH below 3 and in the presence of sulfur, ThO_2 dissolve to form $ThSO^{4+}$. However, thorium carbonates are not common (Kelepertsis 1981). Both pH and Eh are main factors in the process of uranium and thorium transportation by groundwater movement. Under reducing conditions, uranium and thorium are considered to migrate from groundwater and adsorbed by organic matters present in soil/rocks, as well as on clays (Tedesco 1995; p. 78).

Adsorption and precipitation are two main processes identified for uranium isotope enrichment in groundwater. Dissolved UO_2^{2+} is prone to adsorption, especially in the presence of

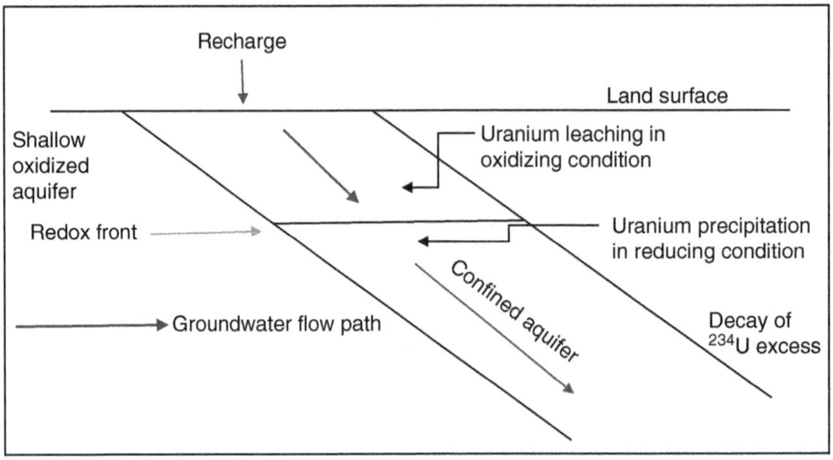

Figure 8.1 Schematic diagram of evolution of uranium isotope along a confined aquifer system. Uppermost part is characterized by preferential leaching of uranium in oxidizing condition, followed by precipitation in the reducing condition. Deepest part is characterized by recoil ejection and radioactive decay of ^{234}U in aqueous phase.

organic matter such as humic acids in soil, iron oxyhydroxides, and micas. The retention of adsorbed uranium mainly depends on the pH of the solution present in the soil profile. Silicate mineral surfaces generally show high uranium retention at pH range of 5–10. However, for pH less than 4, the uranium retention is low. UO_2^{2+} ion shows a very close affinity with iron for adsorption (Payne et al. 1996). Therefore, iron minerals can retain a significant fraction of adsorbed uranium as surface coatings for long-term precipitation condition (IAEA 2013).

Three specific zones have been identified with studies on uranium isotope composition integrated with the groundwater flow from the outcrop to confined aquifers. At the uppermost part, there exists an oxidizing zone which often coincides with the vadose zone of regional aquifer system. In this zone, weathering takes place and uranium mobilizes into the groundwater. This zone is marked by high uranium concentration and moderate $^{234}U/^{238}U$ ratio. This zone is followed by a redox front characterized by decreasing uranium concentration. In this zone, hexavalent (6+) uranium reduces to tetravalent (4+) uranium which is rather an insoluble phase. In addition, this zone is marked by high $^{234}U/^{238}U$ ratio. This has been attributed to enhanced recoil of ^{234}U from precipitated phases. The deepest zone is a reduced zone characterized by uranium precipitation and varying $^{234}U/^{238}U$ ratio as shown in Figure 8.1 (Osmond and Cowart 1982).

There are three mechanisms identified for disequilibrium between ^{234}U and ^{238}U as mentioned earlier. They are described as follows (Zielinski et al. 1981):

1) Preferential release of ^{234}U which is more loosely bound at damaged lattice sites of mineral.
2) Preferential release of ^{234}U adsorbed on surfaces of mineral grain when it is in contact with the aqueous phase.
3) Direct recoil of ^{234}Th into the surrounding aqueous phase from mineral surface which is the immediate short-lived daughter of ^{238}U.

The first effect is attributed to the nuclear recoil due to energetic alpha decay of ^{238}U and ^{234}Th. This recoil process induces breaking of bond and causes lattice damage. This displaces the decay products of radio nuclides into weakly bound interstitial or surficial sites of the mineral grain. However, different minerals have different susceptibility to alpha decay induced lattice damage, as well as differing ability to anneal lattice damage considering lattice defect recovery process is a slow process at normal shallow sub-surface groundwater temperature ($\approx$30°C).

However, high $^{234}U/^{238}U$ ratios more than three cannot be exclusively explained by above-mentioned mechanisms. Proper justification is required combining with additional fractionation mechanism/process based on direct alpha recoil ejection of ^{234}Th from the solid to aqueous phase. At the transition from oxic to anoxic condition in a confined aquifer, high $^{234}U/^{238}U$ ratio is attributed to direct recoil effects and preserved by the anoxic groundwater condition. Therefore, in the anoxic zone, uranium is concentrated as secondary phases by the process of precipitation. As a consequence, due to recoiling $^{234}U/^{238}U$ ratios increases in these phases. However, the recoil effect depends on the uranium concentration in the mineral and grain size, the recoil range of ^{234}Th, and also time. Considering long residence time of ^{234}Th in the aqueous phase compared with its short half-life (24.1 days), ^{234}U can be produced directly due to the decay of recoil ejected ^{234}Th within the aqueous phase. Since ^{234}Th is quickly adsorbed by mineral surfaces, therefore ^{234}U produced at the mineral surface sites can easily get into the surrounding groundwater. Considering a slow-moving groundwater into the aquifer with long residence time (>100 ka) and low leaching rate, the radioactive decay of ^{234}U may overcompensate due to direct recoiling, thus increasing $^{234}U/^{238}U$ ratio along the groundwater flow path of the aquifer (IAEA 2013).

8.4 Hydro-Geochemistry of Uranium

Lot of experimental works have been done to understand uranium dissolution and precipitation in natural condition which includes hydrothermal systems as well as normal aquifer temperature condition. It is a general agreement that uranium transported in aqueous phase as U^{6+} and U^{4+} is considered insoluble. It has also been reported that uranium is soluble in acid oxidizing environments and reducing environment is favorable for uranium precipitation. Transport mechanism of uranium in aqueous solution is mostly dependent on various complexes that are formed by uranium in aqueous phase. Considering hydrothermal systems (temperature $\approx$ 300°C), the mobilization of uranium and allied elements is studied by fluid inclusion study (Nakashima et al. 1999). The study revealed that uranium enrichment in hydrothermal system is not only restricted to derive from differentiating silicate magmas but also attributed to alteration and leaching of uranium from the rocks through which the hydrothermal solution percolated. This process involves release of metallic components including uranium from the crystal lattice by hydrothermal solution and leading to subsequent precipitation with various secondary oxides and/or adsorbed on these secondary metal oxides or secondary clay minerals. Uranium mobilization for uranium-carbonate complexes may happen at pH range of neutral to alkaline solution. The source of carbonate may be the host rock through which the solution percolated.

Uranium transported by solution involves evaluation of various complexes formed by it. Experimental works on various uranyl complexes reveal that with different complexing ions like F^-, HPO_4^{2-}, SO_4^{2-}, and CO_3^{2-}, stability of uranyl complexes is dependent on pH and concentration of complexing components at constant temperature and with no influence of total dissolved uranium present in the given complex (Romberger 1984). Partial pressure of carbon dioxide (PCO_2) also plays an important role in stability of uranyl complex when the complexing ion is carbonate (Garrels and Christ 1975). Salinity of the solution particularly in the presence of NaCl has significant influence on stability of uranyl complex in the solution phase, considering a constant temperature (Huang and Cuentas 1989). Therefore, from the above discussion, it is evident that factors that can influence the solubility of uranium in a hydrothermal system are temperature, pressure, Eh-pH of the solution, activity of complexing anions, and also partial pressure of volatile components in the system such as CO_2. Precipitation of uranium oxide may happen due to change in oxidation state of the transporting hydrothermal solution that is from oxidizing to reducing condition thus increasing pH. In addition, variety of uranium complexes may occur as per the chemical composition of fluid which may lead to uranium precipitation due to change in any governing conditions of their stability like pH, oxidation state, and temperature (De Vivo et al. 1984; pp. 12–17).

Considering shallow sub-surface groundwater–surface water condition (temperature ≈ 25°C), only considering U-O-H_2O system, the predominant dissolved species of uranium are function of pH and Eh of the system. Distribution and stability of different oxide, hydroxide, and carbonate complexes of uranium are shown in Figure 8.2 in the presence of CO_2. It has also been observed that the presence of carbonate and phosphate complexes makes total dissolved uranium concentration more in the solution in comparison to carbonate and phosphate free water (Drever 2002).

From Figure 8.2, it is evident that uranium oxides and hydroxides are stable in acidic condition, whereas with increasing pH, uranium carbonate species are more stable. Not only that, uranium hydroxides are more stable in reducing environment, whereas oxidizing

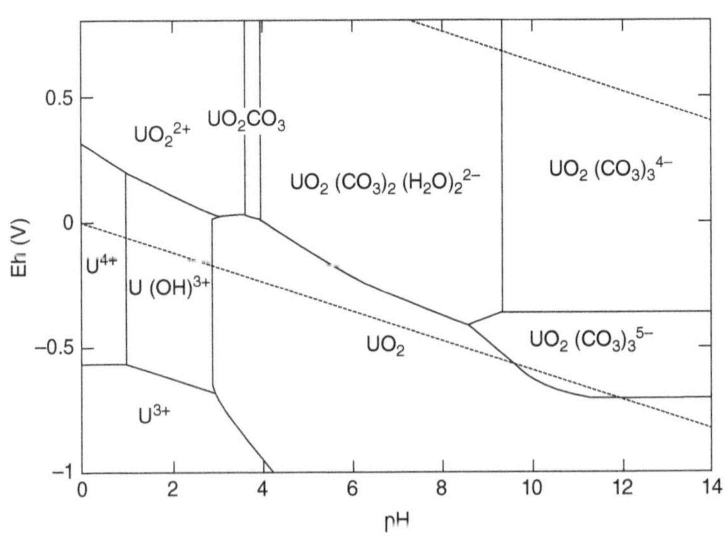

Figure 8.2 Eh-pH phase diagram of U-CO_2-H_2O system with log∑CO_2 = -1. *Source:* Adapted from Huang (2016).

environment favors uranium oxides. At neutral pH (pH $\approx$ 7), dissolved uranium aquatic species are mainly in the form of stable uranyl dicarbonate and tricarbonate complexes in the presence of CO_2.

8.5 Applicability of Uranium Mobility in Groundwater Management, Geochemical Hydrocarbon Exploration, and Earthquake Prediction

Effort of uranium disequilibrium isotope dating method for very old groundwater is not new in scientific literatures (Cherdyntsev et al. 1955; Osmond et al. 1968). Measuring uranium isotopes ^{238}U and ^{234}U requires alpha spectrometry combining isotope dilution methods using ^{236}U or ^{232}U as analytical tracers. However, detail discussion of this analytical method is beyond the scope of this chapter. Dating very old groundwater by uranium isotope disequilibrium requires following prerequisites to fulfill.

1) The initial state of disequilibrium should be known.
2) The concentrations of both isotopes (^{238}U and ^{234}U) in the aquifer condition should not change by any chemical or physical processes other than radioactive decay.

First prerequisite can be satisfied empirically by extrapolation of measured uranium isotope concentration along the regional groundwater flow direction. However, second prerequisite is difficult to satisfy since various geochemical processes like dissolution, precipitation, and sorption/desorption are involved in the deeper part of the aquifer which may change the concentration of ^{238}U and ^{234}U. Not only that, water and soil/rock interaction may affect the uranium concentration/evolution in groundwater. This dating method is also dependent on analytical accuracy and on the retardation factor. However, constant effort is going on by various researchers to improve uranium disequilibrium dating method for very old groundwater (Fröhlich et al. 1984).

Governing equation of the uranium disequilibrium dating method is described by **Eq.** 8.1. Detailed derivation of this uranium disequilibrium dating equation is beyond the scope this chapter.

$$\frac{\partial N}{dt} = D\frac{\partial^2 N_m}{\partial x^2} - V\frac{\partial N_m}{\partial x} - \lambda N + \lambda_{-1}N_{-1} - p\left(N_m + N_{im}\right) + eN_r + \gamma\lambda_{-1}N_{-1r} \tag{8.1}$$

Equation 8.1: Governing equation of dating very old groundwater by uranium isotope (^{238}U and ^{234}U) disequilibrium method.

where $\frac{\partial N}{dt}$: time (t) derivative of the given amount of radionuclide in the liquid and adsorbed phases (N) of the unit volume; $D\frac{\partial^2 N_m}{\partial x^2}$: dispersive/diffusive flux change, assuming that dispersion/diffusion coefficient (D) is constant; $V\frac{\partial N_m}{\partial x}$: advective flux change, assuming constant flow velocity (v), when N = total number of radionuclide atoms in the mobile (N_m), immobile (N_{im}), and adsorbed (N_a) phases and $N = N_m + N_{im} + N_a$; λN: radioactive decay in absorbed and liquid phases; $\lambda_{-1}N_{-1}$: radioactive decay of parent nuclide in absorbed and liquid phases; p = precipitation rate constant; e: dissolution (leaching) rate constant; eN_r: dissolution from rock phase; γ: recoil efficiency; $\gamma\lambda_{-1}N_{-1r}$: recoil ejection from rock phase.

However, a simplified version of the equation of uranium disequilibrium method for dating very old groundwater is reported by Osmond and Cowart (2000) given by **Eq. 8.2.**

$$t = \frac{1}{\lambda} \ln\left[\frac{C-1}{C_0-1}\right] \tag{8.2}$$

Equation 8.2: Simplified groundwater age determination equation by uranium isotope disequilibrium method.

where λ: ^{234}U decay constant, C: the $^{234}U/^{238}U$ ratio of a deep groundwater sample, and C_0: $^{234}U/^{238}U$ ratio in the upflow direction of the groundwater near the redox front to calculate initial $^{234}U/^{238}U$ ratio.

Though there are lots of disagreements for uranium disequilibrium groundwater age determination as well as satisfying all the conditions in field, this method is used to calculate residence time of very old groundwater in confined aquifer. This method of dating old groundwater has implication in groundwater management in terms of classifying water in the aquifer as young, old, and very old in absolute age. Not only that, usefulness of dating groundwater may extend up to characterizing renewability/recharge of the aquifers and their susceptibility to pollution. Age determination of recharge, transmission, and discharge area/zone of the aquifers may be helpful for making isochrone map (groundwater map of same age) of the aquifer and its farther application in groundwater management. However, there is significant amount of scope of improvement in this field.

Considering geochemical hydrocarbon exploration, the relation of radioactive anomaly and presence of sub-surface hydrocarbon is not so straightforward. There is no such empirical mathematical model exists that can correlate radioactive anomaly in soil profile particularly uranium and thorium with sub-surface hydrocarbon reserve. However, radioactive geophysical logging is a common practice to gather petrophysical data of sub-surface horizons at drilling as well as production stage of a hydrocarbon reserve. One of the most common radiometric logging is Natural Gamma Radiation log due to natural radioactivity produced in the formation by uranium and thorium in borehole environment. It is measured in American Petroleum Institute (API) unit which can be defined as 1/200 of the difference between the rate of count recorded in the middle of the radioactive bed by a logging tool and that recorded in the middle of the nonradioactive bed. Shale usually contains elevated amount of radioactive elements. Therefore, Natural Gamma Radiation log identifies shale horizons and can provide clay content estimation of sedimentary rock succession which is an important information at hydrocarbon production stage and also for hydrocarbon reservoir simulation (Kilmer 1985).

Detecting surface radioactive anomaly and correlating it with sub-surface hydrocarbon reserve are constant efforts in geochemical hydrocarbon exploration (Armstrong and Heemstra 1972). However, there is very limited success to correlate sub-surface hydrocarbon reserve with surface radioactive anomaly directly. Expectation of radioactive anomaly particularly of uranium and thorium just above the sub-surface hydrocarbon reserve is considered as an effect of hydrocarbon micro-seepage and its interaction with radioactive elements to form organometallic compounds (Bloch and Key 1981). Uranium removal from solution to insoluble phase occurs on the surface layer in the presence of insoluble organic phase like wood, cellulose, peat, oil shale, and kerogen. Thus these organic phases are likely to have high radioactivity (Murphy et al. 1999). However, most uranium-organic associations are thought to be unstable, but the uranium complexes with ethylene, propylene, and other organics are considered stable (Marks 1982). Hydrocarbon micro-seepage removes uranium from groundwater which is in the form of colloidal suspension, resulting in the formation of insoluble organometallic complexes. Therefore, non-uraniferous area may show very small uranium positive anomaly or enrichment due to above-mentioned process. Under reducing condition, uranium migrates out from groundwater and gets absorbed by hydrocarbon micro-seepage moving up through the rock/soil. Migrating hydrocarbon makes a reducing environment in the soil profile and, as a consequence, facilitates the precipitation of uranium. Generally, compared to uranium, thorium is adsorbed more effectively on clay in the form of ThO_2 for most of the naturally occurring Eh-pH conditions. Positive radioactive anomalies in and around hydrocarbon deposits are common expectation and considered as important hydrocarbon prospecting clue (Morse and Alewine 1983).

Asphalt deposits generally show high concentration of uranium and thorium. This happens due to microbial oxidation of vertically migrating hydrocarbon micro-seepage resulting in the formation of solid bitumen (Lovley et al. 1991). Eventually bitumen makes a reducing environment in and around soil profile and becomes enriched in uranium, thorium, and other metals. Hydrogen sulfide (H_2S) present in hydrocarbon micro-seepage also creates reducing environment thus facilitating uranium precipitation from aqueous solution, along with other metal compounds (Duff et al. 2002). Along with uranium and thorium, other elements like radon, helium, argon, and radium are found to be present in elevated amount around a vertically migrating hydrocarbon plume.

However, an alternative observation that is surface negative radioactive anomaly has also been reported in and around sub-surface hydrocarbon deposits. Uranium and other radioactive elements (including parent and daughter element) present in the soil in the area of petroleum micro seepage generate a reducing environment thus facilitating uranium enrichment. Change in oxidation/reduction condition, that is from reducing to oxidizing condition, may be due to erosion of the outcrop can take uranium and thorium in solution thus migrating with groundwater as contaminant. In a high radioactive zone, well-defined negative radiometric anomaly or lows are indicative of old oil fields.

Radioactive decay of uranium, thorium, and potassium (^{40}K) isotopes produces several inert gases. These chemically inert gases are physically immobilized below water table in the formation. Thus, those inert gases dissolve into the groundwater. As the deep seated groundwater reaches shallow sub-surface water table, these gases tend to leave the water, go into a vapor phase, and ultimately released to the pores of the rock and soil. Along with radiometric anomaly, the presence of different inert gases particularly ^4He which is essentially radiogenic in origin is a strong indicative of the presence of sub-surface hydrocarbon (Durrance 1986).

Regional geology, topography, and groundwater condition have strong influence on radiometric anomaly data collection in an oilfield. Changes in soil moisture content affect gamma ray emissions. High soil moisture in monsoon will suppress radioactive radiation emissions also. Topography has strong influence for radiometric data collection particularly radiometric survey in hydrocarbon exploration. Uranium removed from solution may be redeposited downhill due to groundwater movement and mislead an explorer to pinpoint sub-surface hydrocarbon reserve. Not only that, barometric pressure variation due to changing weather condition can suppress or enhance radioactive radiation emissions from the soil profile (Tedesco 1995; pp. 73–93).

At the time of geochemical exploration, it has also been observed that radioactive emission of natural decay of uranium and thorium has correlation with short-term stress release event like earthquake. Radon which is a natural decay product of uranium and thorium (half-life of 3.82 days) shows strong correlation with earthquake. It has been observed that high emission of helium and radon with respect to regional threshold value in the earthquake-prone areas indicates a short-time stress release event like earthquake. This empirical correlation can be further supported by Radon anomaly found in association with faults (Lang 1950; Fyfe and Brown 1979).

8.6 Uranium Health Hazard

Uranium radioactivity and toxicity in very minute quantity also have huge health effect on human being. Uranium intake in human body can happen due to uranium-contaminated food or drinking water. As per World Health Organization (WHO 2011), the recommended standard of maximum uranium tolerable intake (TI) is 0.0006 mg per kg of body weight per day and for drinking water 0.002 mg/l. On average, approximately 90 μg (micrograms) of uranium exist in an adult human body. In that approximately 66% of uranium is found in the skeleton, 16% in the liver, 8% in the kidneys, and 10% in other tissues. Uranium contamination in human body can take place due to inhalation, ingestion, or dermal contact. More than 95% of the uranium getting into human body is not absorbed. Approximately 67% of the uranium is filtered by the kidney and excreted by urine within 24 hours of intake. Approximately 2% of the uranium getting into a human body by food or water is absorbed by the gastrointestinal tract. Therefore, high intake of uranium by food or drinking water due to localized uranium enrichment in soil or groundwater can have adverse effect on community health. The main isotopes identified for uranium toxicity are ^{238}U and ^{235}U (WHO 2011).

Uranium radioactive health hazard can be grouped under two broad categories; one is chemical toxicity and another is radioactive toxicity (Moss et al. 1983). Two most commonly targeted human organs due to radioactive chemical toxicity are kidneys and the lungs. It has been observed in clinical research that long-term uranium chemical toxicity can lead to kidney impairment (Kurttio et al. 2006). Insoluble uranium particles of 1–10 μm in size if inhaled can retain in the lungs. Long-term exposure may lead to irradiation damage of the lung and may even lead to lung cancer. As per WHO recommendation, maximum inhalation of soluble or insoluble uranium compounds in the respirable fraction should not exceed 1 $\mu g/m^3$. Direct skin contacts of uranium-bearing minerals may cause superficial inflammation of skin (WHO 2017).

Considering radiation toxicity for naturally occurring uranium enrichment, the probability of community contamination is very low. However, as per NHMCR recommendation, the effective radiation dose should not exceed 20 millisieverts (mSv) per year averaged over 5 consecutive years or an effective dose of 50 mSv for a single year for occupational exposure. Not only that, the equivalent radiation toxicity dose to the extremities like hands, skin, and feet should not exceed 500 mSv for a single year (NHMRC 2011).

From the above discussion, it should be clear that uranium mobility and transportation by groundwater in soil profile have huge effect on community health. Therefore, localized enrichment of uranium needs to be taken under serious concern in terms of public health when any geochemical survey is going on, for example geochemical hydrocarbon exploration. Uranium mineralized zone when going for mining needs to take high concern about groundwater contamination of radioactive elements.

8.7 Conclusions

Uranium the naturally occurring radioactive element found on Earth has huge implication in the field of geochemistry. Geochemistry of uranium, its mobilization, and deposits are constantly getting attraction due to its ability to serve as radioactive fuel for mitigation of global warming due to greenhouse gas emission. Uranium is spread all over the crust in the form of igneous or sedimentary deposits. Subsequent weathering of host rock makes uranium liberate to weathered product keeping the concentration same with host rock. Soil–water interaction liberates uranium from weathered product and spreads it to hydrosphere. The process of uranium liberation from source is identified as leaching and subsequent transportation of uranium by groundwater. It has been observed that uranium (6+) is most soluble in natural groundwater condition as well as hydrothermal systems. Detailed study of different uraniferous zones revealed that oxidizing environment is more favorable for uranium leaching and anoxic condition favors uranium precipitation. In aqueous phase, uranium may present in the form of metal complex or may be transported as colloids. Uranium may form organometallic compounds depending on pH and Eh of the solution as well as availability of organic matter like humic or fulvic acid. At the time of precipitation/adsorption, uranium complexes show strong affinity with clay minerals as well as carbonate rocks and different metal oxides. In addition, naturally occurring organic matters like pit and kerogen generally show elevated uranium concentration. This preferential concentration of uranium leads to the development of a strong petrophysical tool to identify clay content of a formation in the hydrocarbon industry known as natural radioactive logging. Applicability of uranium radioactivity has also been extended to geochemical hydrocarbon exploration to identify radiation anomalies associated with uprising hydrocarbon plume due to micro-seepage. Extending the domain to groundwater management uranium disequilibrium dating of very old groundwater has significant application in groundwater management particularly delineating isochrone map of the aquifer. Effort has also been made to correlate radioactive radiation (helium and radon) of uranium and thorium with earthquake event. Localized uranium concentration though in minute quantity has significant impact on community health as chemical and radiation toxicity. Therefore, it is advised that when dealing with naturally occurring radioactive elements like uranium and thorium, one needs to be very careful about its long-term health effect for individual as well as for community.

8.8 Case Studies

In this section, an effort has been made to illustrate uranium pollution of groundwater and geochemical behavior of uranium with published research. This will help reader to better conceptualize the geochemical behavior of uranium in soil profile, its interaction with groundwater, and mobilization as well as behavior of uranium series decay products like radon in soil profile. Both the case studies are purposefully selected from different parts of globe, one is from Grand Pré region, Nova Scotia, Canada (**Case study, a**) and other is from Icheon, Middle Korea (**Case study, b**), to illustrate the influence of geology and hydrogeology in the uranium mobilization process.

a) **Uranium Mobility in Groundwater – The Grand Pré Region, Nova Scotia, Canada**

The study area is located at southern shore of Minas Basin, Kings County, Nova Scotia, Canada. The study area is well known for its local high anomalous uranium concentration. Lithological distribution of the area is extended from Cambrian to Jurassic in geological timescale. Among the litho units, Horton Group which is essentially composed of sandstone and shale and of Carboniferous in age is best reported for its uranium occurrence. The two most significant aquifers in the study area are of the Wolfville Formation which is essentially porous sandstone and conglomerate of Triassic Age and carbonaceous siltstone and sandstone of the Horton Bluff Formation (HBF) of Carboniferous Age. Geochemical problem identified in the study area is elevated uranium concentration in drilled groundwater wells. In those wells, uranium concentration is way higher than the level of uranium for drinking water standard as per recommendation.

Systematic groundwater sampling and chemical analysis revealed that major cations are either Ca^{2+} or Na^{+} and dominant anion is Cl^{-}. Piper trilinear plot of water sample demonstrates three primary types of groundwater that is of calcium carbonate water, sodium chloride water, and calcium chloride water. pH of the water samples ranges from 7.45 to 6.45 with pE range of 3.12–0.84. pE-pH diagram of water samples demonstrates that the dominant uranium species are uraninite and uranium-hydroxide at 8°C. These uranium speciation in aquatic phase indicates that some of the water was supposed to be saturated with uraninite. It is also possible that the water may be derived from an aquifer containing uraninite. Iron has been identified as tracer to understand the redox condition of the aquifer because of its availability in aquifer water of the study area. The rule of thumb was the presence of excess Fe^{2+} in the water indicates reducing environment, whereas drastically low Fe^{2+} concentration indicates oxidizing environment. Elemental analysis of dissolved ions of water demonstrated that groundwater of the Horton Group possesses reducing condition, whereas for the Wolfville Formation, it is essentially oxidizing.

Considering regional geology and coupled with regional hydrogeology, it turned out that weathered granite of South Mountain Batholith (Late Devonian) and arkosic arenite and shale from the Horton Group are the main source of uranium into groundwater in the study area. Regional groundwater study indicates groundwater flow from the Horton Group to Wolfville Group. Field survey indicates that the contact between the Horton Group and the Wolfville Formation is unconformable. Interbedded carbon-rich shale and siltstone of the HBF is forming a reduced groundwater environment, whereas the clean, porous sandstone mainly of mature quartz arenite of the Wolfville Formation produces an oxidized groundwater environment. Oxidized groundwater of the Wolfville Formation facilitates uranium enrichment in groundwater compared to reducing condition of Horton Group which is characterized by low uranium concentration in the groundwater. Special distribution of uranium in the groundwater of the study area also support this conclusion. Source: Adapted from Samolczyk et al. (2012).

b) **Uranium and Radon Geochemistry in Groundwater of Jurassic Granite, Icheon, Middle Korea**

The study area is located at Icheon Province of Middle Korea, where variable concentrations of uranium and radon have been reported from groundwater samples. Geologically, the study area is dominated by biotite granite, two mica granite, and porphyritic granite of Jurassic age. Petrogenetically this granite is considered as S-type granite, with two mica granite as zonal stocks and associated pegmatite dikes. Water samples of the study area indicates mostly $CaHCO_3$ and Ca-Na-HCO_3 type, but some of the water samples are of Ca-SO_4 and Na-SO_4 type. Ca-Na-HCO_3 groundwater is considered as product of mineral water interaction with granite. Whereas Ca-SO_4 and Na-SO_4 type water originates from discharge areas or intermediate to lower zones in the aquifer. Uranium concentration in the study area shows wide range of concentration from 0.02 to 1640 μg/L with a mean of 56.77 μg/L, and radon concentration of 1.48–865.8 Bq/L with mean of 211.29 Bq/L. Interesting finding of the study area was that uranium has no statistical correlation with radon which is a radioactive gas originates from radioactive decay of radium. However, significant statistical correlation were found between uranium and redox potential (Eh), HCO_3^-, dissolved oxygen, SiO_2, and Sr. It has also been observed that uranium concentration increases with increasing concentration of certain cations and anions like HCO_3^-, SO_4^{2-}, F^-, and Sr. Uranium enrichment in groundwater has been considered due to water–rock interaction of Jurassic granite. Uranium concentration in the groundwater shows wide special variability. However, general trend is northern and central part of the granitic aquifer has higher uranium enrichment than the rest of it. Spatial distribution and variability of uranium enrichment in the aquifer fail to draw any direct statistical correlation with Jurassic granite as a whole. This observation suggests that special distribution of uranium in the granitic aquifer is heterogeneous and higher values are attributed to localized enrichment of uranium in the aquifer water. Radon concentration in the groundwater also shows special variability. However, radon concentration of groundwater in the study area shows higher value than the rest of the Jurassic granite aquifer of Korea. Water analysis data revealed that uranium came to groundwater in oxidizing environment and precipitated in reducing environment. Water analysis data could not make any direct statistical correlation between uranium enrichment in groundwater and radon concentration. Localized radon concentration has been attributed to localized precipitation of uranium-bearing carbonates as coatings in the fractures of the granite releasing radioactive radon gas. Source: Adapted from Cho and Choo (2019).

References

Alloway, B.J., (2013) "Uranium". In: Alloway, B.J. (Ed.), *Heavy Metals in Soils*. Springer, p. 565.

Armstrong, F.E., Heemstra, R.J., (1972) "Radiometries proposed for exploration". *Oil Gas J.* 12, 152–161.

Bali, E., (2012) "Economic geology: uranium-ore giants". *Nat. Geosci.* 5 (2), 96–97.

Beazley, M.J., Martinez, R.J., Webb, S.M., Sobecky, P.A., Taillefert, M., (2011) "The effect of pH and natural microbial phosphatase activity on the speciation of uranium in subsurface soils". *Geochim. Cosmochim. Acta* 75 (19), 5648–5663.

Bloch, S., Key, R.M., (1981) "Modes of formation of anomalous high radioactivity in oil-field brines". *Am. Assoc. Pet. Geol. Bull.* 65, 154.

Bourdon, B., Henderson, G.M., Lundstrom, C.C., Turner, S.P., (2003) *Uranium-Series Geochemistry*. Washington, DC: Mineralogical Society of America. Reviews of Mineralogy and Geochemistry, Vol. 52, 656pp.

Cherdyntsev, V.V., Chalov, P.I., Khasdarov, G.Z., (1955) "On isotopic composition of radioelements in natural objects, and problems of geochronology". 2012 Trans. 3rd Session of the Commission on Determination of Absolute Age of Geological Formations, Izv. Acad. Nauk SSR (1955).

Cho, B.W., Choo, C.O., (2019) "Geochemical behavior of uranium and radon in groundwater of Jurassic Granite Area, Icheon, Middle Korea". *Water* 11, 1278, doi:https://doi.org/10.3390/w11061278

Cuney, M., (2009) "The extreme diversity of uranium deposits". *Mineral. Deposita* 44 (1), 3–9.

De Vivo, B., Ippolito, F.; Capaldi, G., Simpson, P.R. (Eds.) (1984) "*Uranium Geochemistry, Mineralogy, Geology, Exploration and Resources*". The Institution of Mining and Metallurgy. Springer Dordrecht, London. e-ISBN-13: 978-94-009-6060-2.

Drever, J.I., (2002) *The Geochemistry of Natural Water*. Third Edition. Prentice Hall (NJ). ISBN: 0-13-272790-0.

Duff, M.C., Coughlin, J.U., Hunter, D.B., (2002) "Uranium co-precipitation with iron oxide minerals". *Geochim. Cosmochim. Acta* 66 (20), 3533–3547.

Durrance, E.M., (1986) "*Radioactivity in Geology: Principles and Applications*". Ellis Hoorwood, West Sussex, England.

Echevarria, G., Sheppard, M.I., Morel, J., (2001) "Effect of pH on the sorption of uranium in soils". *J. Environ. Radioact.* 53 (2), 257–264.

Fayek, M., Horita, J., Ripley, E.M., (2011) "The oxygen isotopic composition of uranium minerals: a review". *Ore Geol. Rev.* 41 (1), 1–21.

Fowler, C.M.R., (2005) "*The Solid Earth- An Introduction to Global Geophysics*". Cambridge University Press. ISBN 0 521 89307 0.

Fröhlich, K., Gellermann, R., Hebert, D., (1984) "Uranium isotopes in a Sandstone Aquifer. Interpretation of data and implications for groundwater dating", Isotope Hydrology 1983 (Proc. Vienna, 1983), IAEA, Vienna.

Fyfe, W.S., Brown, G.C., (1979) "The geochemical cycle of uranium and discussion". *Philos. Trans. R. Soc. Lond. A* 291 (1381), 433–445.

Garrels, R.M.; Christ, C.L. (1975) "*Eh-pH. In Solutions, Minerals, and Equilibria*". Freeman, Cooper & Company, New York, NY, USA; Chapter 7; pp. 172–277.

Hazen, R.M., Ewing, R.C., Sverjensky, D.A., (2009) "Evolution of uranium and thorium minerals". *Am. Mineral.* 94, 1293–1311.

Huang, H.H., (2016) "The Eh-pH diagram and its advances". *Metals*, 6, 23 https://doi.org/10.3390/met6010023.

Huang, H.H.; Cuentas, L. (1989) "Construction of Eh-pH and other stability diagrams of uranium in a multicomponent system with a microcomputer – I. Domains of predominance diagram". *Can. Metall. Q.* 28, 225–234.

Idiz, E.F., Carlisle, D., Kaplan, I.R., (1986) "Interaction between organic matter and trace metals in a uranium rich bog, Kern County, California, U.S.A.". *Appl. Geochem.* 1 (5), 573–590.

International Atomic Energy Agency, Vienna, (1988) "Geochemical Exploration for Uranium". Technical Reports Series No. 284.

International Atomic Energy Agency, Vienna, (2009) "World Distribution of Uranium Deposits (UDEPO) with Uranium Deposit Classification". ISBN 978–92–0–110509–7

International Atomic Energy Agency, Vienna, (2013) "Isotope Methods for Dating Old Groundwater". ISBN 978–92–0–137210–9

Kantar, C., (2007) "Heterogeneous processes affecting metal ion transport in the presence of organic ligands: reactive transport modeling". *Earth Sci. Rev.* 81 (3), 175–198.

Kelepertsis, A.E., (1981) "The geochemistry of uranium and thorium in some Lower Carboniferous sedimentary rocks (Great Britain)". *Chem. Geol.* 34 (3–4), 275–288.

Kilmer, C. (1985) "Shale normalized radiation surveying for petroleum exploration in geochemical prospecting". IGM/Inside Oil & Gas Proceedings, Houston, TX.

Kurttio, P., Harmoinen, A., Saha, H., Salonen, L., Karpas, Z., Komulainen, H., Auvinen, A., (2006) "Kidney toxicity of ingested uranium from drinking water". *Am. J. Kidney Dis.* 47 6, 972–982, ISSN 0272-6386. https://doi.org/10.1053/j.ajkd.2006.03.002

Lang, B. (1950). Gammatron surveys, *World Oil* 131 (6), 86–88.

Lovley, D.R., Phillips, E.J.P., Gorby, Y.A., Landa, E.R., (1991) "Microbial reduction of uranium". *Nature* 350 (6317), 413–416.

Lowrie, W., (2007) "*Fundamentals of Geophysics*." Second Edition. Cambridge University Press. ISBN-13: 978-0-521-27038-0.

Marks, T.J., (1982) "Actinide organometallic chemistry". *Science* 217(4564), 989–997. https://doi.org/10.1126/science.217.4564.989

Mehta, V.S., Maillot, F.,Wang, Z., Catalano, J.G., Giammar, D.E., 2014. "Effect of co-solutes on the products and solubility of uranium (VI) precipitated with phosphate". *Chem. Geol.* 364, 66–75.

Morse. J.G., Alewine. J.W., (1983). "Airborne radiometrics: case study of Patrick Draw Field, Wyoming". *Oil Gas J.* 28, 145–147.

Moss, M.A., McCurdy, R.F., Dooley, K.C., Givner, M.L., Dymond, L.C., Slayter, J.M., Courneya, M.M., (1983) "Uranium in drinking water –report on clinical studies in Nova Scotia". In: Brown, S.S., Savory, J. (Eds.), *Chemical Toxicology and Clinical Chemistry of Metals.* Academic Press, London, pp. 149–152.

Murakami, T., Ohnuki, T., Isobe, H., Sato, T., (1997) "Mobility of uranium during weathering". *Am. Mineral.* 82 (9), 888–899.

Murphy, R.J., Lenhart, J.J., Honeyman, B.D., (1999) "The sorption of thorium (IV) and uranium (VI) to hematite in the presence of natural organic matter". *Colloids Surf. A Physicochem. Eng. Asp.* 157 (1–3), 47–62.

Nakashima, S., Disnar, J.R., Perruchot, A., Trichet, J., (1984) "Experimental-study of mechanisms of fixation and reduction of uranium by sedimentary organic-matter under diagenetic or hydrothermal conditions". *Geochim. Cosmochim. Acta* 48 (11), 2321–2329.

Nakashima, S., Disnar, J.R., Perruchot, A., (1999) "Precipitation kinetics of uranium by sedimentary organic matter under diagenetic and hydrothermal conditions". *Econ. Geol. Bull. Soc. Econ. Geol.* 94 (7), 993–1006.

Newsome, L., Morris, K., Lloyd, J.R., 2014. "The biogeochemistry and bioremediation of uranium and other priority radionuclides". *Chem. Geol.* 363, 164–184.

NHMRC, (2011) Australian Drinking Water Guidelines, Paper 6, National Water Quality Management Strategy. National Health and Medical Research Council, National Resource Management Ministerial Council, Commonwealth of Australia, Canberra Version 3.1.

Osmond, J.K., Cowart, J.B., (1982) "Groundwater". In: Ivanovic, M., Harmon, R.S. (Eds.), *Uranium Series Disequilibrium: Applications to Environmental Problems*, Oxford University Press, Oxford.

Osmond, J.K., Cowart, J.B. (2000) "U-Series nuclides as tracers in groundwater hydrology". In: Cook, P.G., Herczeg, A. (Eds.) *Environmental Tracers in Subsurface Hydrology*. Kluwer Academic Publishers, Boston, p. 145.

Osmond, J.K., Rydell, H.S., Kaufman, M.I., (1968) "Uranium disequilibrium in groundwater: an isotope dilution approach in hydrologic investigations". *Science* 162 (1968), 997–999.

Payne, T., Davis, J., Waite, T., (1996) "Uranium adsorption on ferrihydrite-effects of phosphate and humic acid". *Radiochim. Acta* 74 (Supplement), pp 239–244.

Romberger, S.B., (1984) "Transport and deposition of uranium in hydrothermal systems at temperatures up to 300°C: geological implications". In: De Vivob, B. et al. *Uranium Geochemistry, Mineralogy, Geology, Exploration and Resources*. Springer Dordrecht, London ISBN-13: 978-94-009-6062-6.

Samolczyk, M.A., Spooner, I.S., Stanley, C.R., (2012) "A model for uranium mobility in groundwater in the Grand Pré region, Nova Scotia, Canada". *Atlantic Geology* 48, 1–13. doi: https://doi.org/10.4138/atlgeol.2012.001

Susan, A., Cumberland, G.D., Grice, K., Moreau, J.W., (2016) "Uranium mobility in organic matter-rich sediments: a review of geological and geochemical processes", *Earth Sci. Rev.* 159, 160–185.

Tedesco, S.A., (1995) "*Surface Geochemistry in Petroleum Exploration*" Springer-Science-Business Media, BV, New York, NY ISBN 978-1-4615-2660-5.

Tomkeieff, S.I., (1946) "The Geochemistry of Uranium" Science Progress (1933-1946), October 1946, Vol. 34, No. 136 (October 1946), pp. 696-712. Sage Publications, Ltd.

WHO. (2011) Uranium in Drinking Water, Background Document for Development of WHO Guidelines for Drinking Water Quality. World Health Organization WHO/SDE/WSH/03.04/118/Rev/1.

WHO, (2017) *Guidelines for Drinking-Water Quality*, (4th ed). Incorporating the first addendum World Health Organization, Geneva.

Zielinski, R.A., Peterman, Z.E., Stuckless, J.S., Rosholt, J.N., Nkomo, I.T., (1981) "The chemical and isotopic record of rock-water interaction in the Sherman granite, Wyoming and Colorado". *Contrib. Mineral. Petrol.* 78, 209–219.

9

Study of the Rate of CO_2 Consumption with Silicate and Carbonate Weathering in Aquatic System

Uday Bhan[1], *Vamsi Krishna Kudapa*[2], *and Ranjit Kumar*[3]

[1] *Department of Petroleum Engineering and Earth Sciences, UPES, Dehradun, Uttarakhand, India*
[2] *Department of Chemical Engineering, UPES, Dehradun, Uttarakhand, India*
[3] *Department of Chemistry, UPES, Dehradun, Uttarakhand, India*

9.1 Basic Information

The movement of chemical species from zones to oceans is facilitated by chemical weathering (Mackenzie and Garrels 1966; Berner et al. 1983). For a wide range of timescales, various findings have demonstrated that continental weathering is complex to ecological characteristics and is influenced by these elements (Calmels et al. 2014; Norton et al. 2014). This process involves the carbonate minerals weathering, which are one of the utmost common minerals found on the Earth's surface (Hartmann and Moosdorf 2012). According to the most common lithological classes, evaporites have the highest rate of weathering in comparison (Moosdorf et al. 2011). Over long periods of time, carbonate weathering does not serve as a sink for CO_2 emissions from the atmosphere (Walther 2007). It has an impact on the distribution of carbon between the ocean and the atmosphere, but only on timeframes shorter than 105 years, which are shorter than the mixing time of the ocean (Martin 2017). Around 10–14% of the Earth's surfaces is occupied by sedimentary carbonate rocks and weathering of these carbonate rocks generates 50–60% of dissolved products (excluding ice shields) (Gaillardet et al. 1999; Hartmann and Moosdorf 2012). Additionally, rocks other than carbonate like siliciclastic, metamorphic, and felsic intrusion rocks contribute to carbonate weathering (Hartmann et al. 2014). Carbonate weathering is also thought to be responsible for anywhere from 34 to 50% of worldwide CO_2 consumption on shorter time periods, according to some estimates (Gaillardet et al. 1999). In carbonate-dominated lithologies, CO_2 partial pressure (P_{CO_2}) is the principal source of weathering flux and is one of the most significant processes in carbonate management (Calmels et al. 2014). Weathering processes are aided by organisms, which concentrate CO_2 and organic acids and provide more protons for the reactions to take place (Kuzyakov 2006). Similar CO_2 utilization is also applicable in oil and gas reservoirs like shale and coal bed methane for enhancing the gas production (Kudapa et al. 2017a, 2017b, 2018).

Hydrogeochemistry of Aquatic Ecosystems, First Edition. Edited by Sughosh Madhav, Virendra Bahadur Singh, Manoj Kumar, and Sandeep Singh.

The global rate of CO_2 uptake by chemical ageing was the primary focus of previous investigations on land-ocean carbonate weathering fluxes (Hartmann et al. 2009), utilizing empirically based functions with respect to temperature and soil properties (Hartmann et al. 2014). A number of researchers have used ecosystem numerical models that incorporate kinetic theory and chemical equilibrium, as well as hydrological models, to create mechanistic models (Goddéris et al. 2013). Model methods to weathering flux and CO_2 consumption by chemical processes are very different, despite their similar goals of quantifying the overall weathering flux to the ocean. Phenomenological models, which are based on observable relationships, are the most straightforward and easy to use for global computations, such as Earth system models (Goll et al. 2014). The primary flaw of existing models is that they assume persistent alkalinity regardless of temperature or soil-rock CO_2 concentration, illustrating the impact of dilution in heavy runoff (Bluth and Kump 1994). As a result of calcite's rapid dissolution, annual carbonate weathering calculations are constrained by the need to reduce the interval for each calculation, which greatly increases the iteration time required for each calculation (Roland et al. 2013).

9.2 Silicate Weathering

The amount of CO_2 absorbed and radiated back to Earth terrestrial radiation has a significant impact on world temperature. Geochemical and geophysical processes influence the partial pressure of CO_2 (P_{CO_2}) in the atmosphere. During subduction and regional metamorphism, carbonates metamorphize, releasing carbon dioxide into the atmosphere.

$$CaCO_3 + SiO_2 \rightarrow CO_2 + CaSiO_3 \tag{9.1}$$

Weathering of silicate minerals primarily removes CO_2 from the atmosphere.

$$2CO_2 + 4H_2O + CaAl_2Si_2O_8 \rightarrow Ca^{++} + 2Al\left(OH\right)_3 + 2SiO^{aq}{}_2 + 2HCO_3^{-} \tag{9.2}$$

Carbon dioxide (CO_2) is emitted into the atmosphere when marine carbonates are generated through silicate weathering. By dissolving calcium silicates and magnesium silicates rather than carbonates and potassium feldspars, the weathering process consumes CO_2. CO_2 is spent in the process of dissolving limestone, but it is released again when the carbonate is deposited in the ocean and begins to corrode. Also, feldspars like K and Na consume less CO_2 than silicates like Ca and Mg (Berner et al. 1983).

Walker et al. (1981) proposed a feedback mechanism involving silicate weathering regulated world temperatures. The greenhouse effect causes temperatures to rise as a result of high CO_2 levels. At higher temperatures, silicates decompose more rapidly, requiring more CO_2. CO_2 levels and the world temperature T are reduced as a result. CO_2 accumulates in the atmosphere at low temperatures because silicate weathering and CO_2 consumption are negligible. As a result, the feedback loop between weathering and CO_2 prevents drastic changes in global temperature. The efficiency of this weathering–climate coupling is obviously dependent on the weathering rates to temperature and P_{CO_2} variations. The feedback is strong and the coupling is tight if a slight change in T and P_{CO_2} causes a substantial increase in weathering. Temperatures can vary over a large range if the rate of change in

T and P_{CO_2} is very small, and the feedback is minimal. It has been shown that the silicate control on CO_2 and T in the atmosphere has been overestimated in prior models.

Models that focus on mineral weathering are those of Walker et al. (1981), termed as WHAK, Berner et al. (1983), termed as BLAG, and Volk (1987). According to WHAK, a link between silicate weathering rates, temperature, and CO_2 concentration was utilized to suggest that silicate dissolution reactions may have regulated Earth surface temperatures over geologic time. BLAG calculated P_{CO_2} and the mean global surface temperature between now and 100 million years ago using a carbon mass balance model that took into account measurements of seafloor spreading rates and continental surface areas. When it comes to silicate weathering rates, temperature, and P_{CO_2}, BLAG employed the same method as WHAK. The organic carbon reservoir was included in the BLAG model in later revisions. Variations in the rate of seafloor spreading and the size of the continents have a significant impact on CO_2 levels and global temperatures, according to the BLAG model (Volk 1987). It was shown that feedback processes involving organic productivity could mitigate some of the rise in worldwide temperatures due to changes in CO_2 levels in the soil.

BLAG model from Volk (1987), representing a steady-state behavior is likely on interval scales stretched over 105-year CO_2 residence time in the ocean-atmosphere system. Weathering and volcanic activities counteract each other in terms of carbon dioxide production and consumption. After that, the CO_2 mass balance is computed.

$$f_{wr}f_a = f_{sr} \tag{9.3}$$

where f_{wr}, f_a, and f_{sr} are the fractions of the continental weathering rate, the overall continental area, and the seabed creation rate to the current values in terms of area. In Eq. 9.3, limestones sensitivity to metamorphic decarbonation has been overlooked over time. Equation 9.3 represents the CO_2 production from the metamorphism of marine carbonates, and the degassing of the mantle at seafloor spreading centers is assumed to be closely correlated with spreading rates. Volk (1987) states that in BLAG model, which is computationally more difficult, yields the same key findings when Eq. 9.3 is combined with the other BLAG identities for f_{wr} as a function of P_{CO_2}, temperature (T) in degrees kelvin, runoff (R), and Pc percent (Pc%) as a function of time (t). As suggested by Volk (1987), in this section, the authors employed Eq. 9.3 as an analytical method to measure the impact of silicate dissolution on global carbon cycling.

River discharge per unit area (R) and P_{CO_2} (T) are shown in the following BLAG relationships:.

$$\frac{R}{R_0} = 1 + 0.038\left(T - T_0\right) \tag{9.4}$$

$$T - T_0 = 2.88 \ln\left(\frac{P_{CO_2}}{P_{CO_{2,0}}}\right) \tag{9.5}$$

Present-day values are indicated by the 0 subscript. For continental weathering, the BLAG formula connects bicarbonate flow with the following expression:

$$f_{wr} = \frac{R}{R_0}\frac{C}{C_0} \tag{9.6}$$

Water in a river has a bicarbonate content of C. Mineral weathering converts CO_2 to bicarbonate, and the C/C_0 model captures this transformation, with an assumption that the amount of bicarbonate in continental runoff is related to the rate at which minerals weather. This means that when the temperature rises, for example, the amount of C in the water also rises.

$$\frac{C}{C_0} = \frac{r}{r_0} \tag{9.7}$$

As an example, the ratio of mineral weathering rate to current value is r/r_0, which may be expressed as Eq. 9.7. In the carbon cycle models of WHAK and Volk (1987), experimental weathering rates were employed to measure C/C_0 using Eq. 9.7 along with the proportionality between weathering rate and bicarbonate generation.

9.3 Weathering of Ca and Mg Silicates

pH and temperature can have a significant impact on the rate at which silicate dissolves. Because adsorbed protons catalyze mineral dissolution at the mineral surface at 25 °C and below pH = 5, rates rise with decreasing pH (Stumm and Wollast 1990). Adsorbed hydroxyls catalyze silicate dissolution reactions above pH = 8, increasing rates and CO_2 consumption. CO_2 does not seem to be able to control the rate at which silicates dissolve because aqueous carbonate species (e.g. H_2CO_3, HCO_3^-, and CO_3^{2-}) do not adsorb liquefying silicates at soil pH or have a direct effect on their rates (Brady and Walther 1989).

P_{CO_2} increases rates indirectly by raising temperatures. P_{CO_2} increases may also indirectly affect rates by decreasing pH to a level where proton catalysis increases silicate-dissolving rates (pH = 5). Due to the fact that noncalcareous soils' pH is generally (but not always) higher than this, the effect of pH on weathering feedback is omitted below. When you add a pH term to the statement about global weathering, it makes the feedback mechanism even stronger. If CO_2 levels rise, the pH of the soil will go below five. This will cause rates to rise because of the lower pH and the more heat that is in the soil. Consequently, the temperature dependence of the weathering process was found to be on the low side, while the atmospheric CO_2 and T values were found to be the highest values.

As CO_2 levels in the atmosphere rise, it is reasonable to predict that organic production and the concentration of organic acids in groundwater will rise as a result. Volk (1987) increases the weathering-temperature feedback loop by increasing the productivity of the biotic system. An experimental effort has been made to determine how temperature affects the weathering of organically mediated materials.

9.4 Carbonate Weathering

Earth's surface is dotted with carbonate minerals, which are among the most prevalent on the planet, and this process involves their weathering (Hartmann and Moosdorf 2012). Evaporites, followed by quartzites, are the most weathered of the most common lithological types (Moosdorf et al. 2011). Over geological timeframes, carbonate weathering does not operate as a sink for CO_2 emissions from the atmosphere (Berner et al. 1983). It has an impact on the distribution of carbon between the ocean and the atmosphere, but only on timeframes shorter than 105 years, which are shorter than the mixing time of the ocean (Martin 2017). Despite the fact that just 10–14% of the Earth's surface is covered by sedimentary carbonate rocks, they play a critical role in our understanding of the past (Hartmann and Moosdorf 2012); carbonate rock weathering accounts for approximately 50–60% of the dissolved products from rock weathering in the environment (Gaillardet et al. 1999). This lithology, as well as others dominated by carbonate, including metamorphic rocks and felsic intrusive rocks, all contribute significantly to weathering fluxes (Hartmann et al. 2014). Carbonate weathering is also thought to be responsible for anywhere from 34 to 50% of global CO_2 consumption on short timescales, according to some estimates (Hartmann et al. 2009). When the lithology is dominated by carbonates, the partial pressure of CO_2 in the soil is an important mechanism affecting carbonate weathering flux (Calmels et al. 2014). As part of this system, organisms play a significant role in weathering by concentrating CO_2 and organic acids in the soil profile, as well as increasing the amount of proton available for weathering processes (Kuzyakov 2006).

Furthermore, it was found out that how much carbonate could be dissolved by using simple equilibrium equations and a soil-rock P_{CO_2} function based on the amount of water that evaporates from the groundwater table each year (Brook et al. 1983). The generalization of weathering dynamics driven by ecosystem respiration in the crucial zone is made possible by the use of a soil-rock CO_2 partial pressure representation for averaged conditions. Using equilibrium equations instead of kinetics has the advantage of reducing computational time, which is comparable to the time savings achieved through the use of phenomenological techniques in global weathering models. According to annual evapotranspiration data, the P_{CO_2} equation is implemented (Brook et al. 1983) in Gombert (2002) has certain drawbacks for low-temperature regions and may not adequately address the role of soil water content in CO_2 generation in the soil system (Romero-Mujalli et al. 2019b). There are recorded changes in carbonate dissolution within timescales of hours and rapid precipitation, both of which are dependent on the amount of CO_2 in the aquatic system (Pu et al. 2013). Carbonate dissolution response kinetics can also reach equilibrium in three hours in laboratory investigations, implying that the system can easily approach equilibrium with a given P_{CO_2} in the critical zone (soil and rock). Equilibrium principles and data on soil-rock P_{CO_2} concentrations can be used to determine the amount of carbonate weathering products present in water moving out from the soil-rock system (Romero-Mujalli et al. 2019b). Pu et al. (2013) discovered that groundwater in Karstic locations had consistently high calcite concentrations and that soil CO_2 generation was responsible for seasonal hydro chemical fluctuations in those waters. During the warm season, the concentration of P_{CO_2} in water was at its maximum. Moreover, Calmels et al. (2014) have revealed that variations in CO_2 levels in soils across the Jura Mountains explain the chemical weathering gradient. CO_2 is produced in soils through microbial activity, root respiration, and acidic carbonate dissolution (Wang et al. 2015).

In most cases, only calcite dissolution is taken into account when modelling global carbonate weathering; this leaves out other minerals that have an impact on fluxes during weathering (Gombert 2002). When it comes to carbonate weathering, worldwide methodologies do not take into account acidic conditions, such as H_2SO_4 (Calmels et al. 2014). Soil-rock P_{CO_2} and temperature differences are not taken into consideration by runoff-based functions (Hartmann et al. 2014); and, when it comes to global carbonate weathering models, anthropogenic perturbations such as the use of fertilizers are not taken into account (Romero-Mujalli et al. 2019a). Because of this, a number of methodologies are compared and assessed utilizing calcite-dominated river chemistry data to understand the function of carbonate weathering in the Earth's climate and ecosystem. A yearly representative P_{CO_2} in the soil-rock system was calculated by Hartmann and colleagues using the river alkalinity parameterization and equilibrium equations (mechanistic approach). Both phenomenological and mechanical assumptions underlie these methods. Additionally, the existence of other minerals (such as calcium evaporites) can have a significant impact on alkalinity fluxes; this is due to the presence of open, semi-closed, or closed system conditions in relation to soil CO_2 (Romero-Mujalli et al. 2019a).

9.5 Rate of Silicate Weathering and CO_2 Consumption

9.5.1 Rate of Silicate Weathering

This study used the forward model to calculate the silicate weathering rates of the study area in 2016 and 2017 (Wu et al. 2008), output per unit area, and key element concentrations combined and represented as Eq. 9.8.

$$SWR = Q\left(\Sigma\left(Na + K + Mg + Ca\right)_{sil} + SiO_2\right) \tag{9.8}$$

where, $(Na+K+Mg+Ca)_{sil}$ = dissolved cations derived from silicate weathering. Q = water discharge per unit area.

In atmospheric decomposition, it is considered that the elements Na^+ and K^+ are extracted from the mineralogy of feldspar, whereas Ca^{2+} and Mg^{2+} are derived from the pyroxene minerals, according to the literature. After considering weather-related factors and human activity, the only cations left are those resulting from the atmospheric decomposition of key rocks in the basin. The following formulae (Eqs. 9.9 and 9.10) were used to compute the silicate-derived Ca^{2+} and Mg^{2+} from the gneissic and charnockite rock samples.

$$\text{Ca from silicate rock} = \frac{Na^*}{\left(Ca/Na\right)_{bedrock}} \tag{9.9}$$

$$\text{Mg from silicate rock} = \frac{Na^*}{\left(Mg/Na\right)_{bedrock}} \tag{9.10}$$

where Na^* is the corrected value. The mean concentration of sodium derived from atmospheric inputs. Here, basement rock had Ca/Na and Mg/Na ratios of 0.41 and 0.325 because it was fully gneissic and charnockite in composition.

Changing ion compositions at Adoor and Kottody sites are thought to be the cause of system of rivers Payaswini and Chandragiri silicate weathering rate (Figure 9.1). There was a silicate weathering rate on a yearly basis, which is 42 tonnes/km^2/year in the Payaswini and

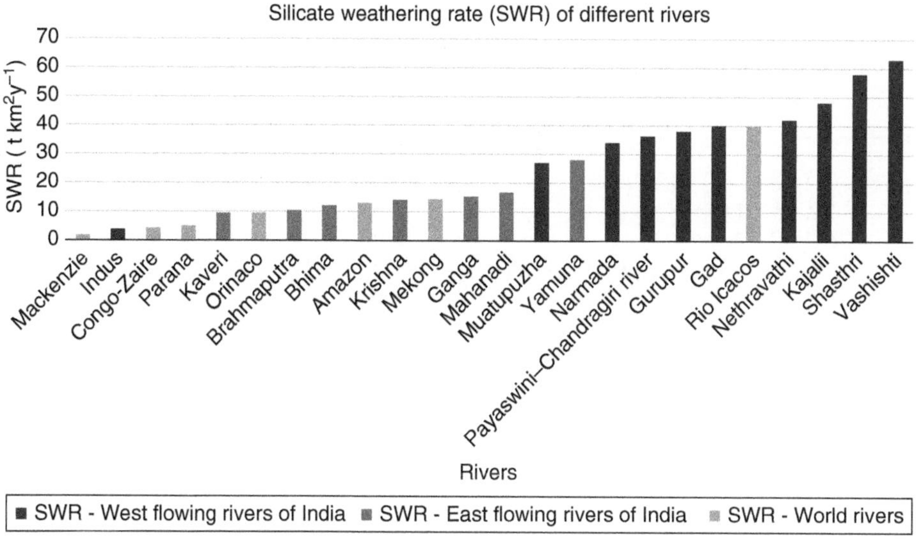

Figure 9.1 Comparative of yearly silicate weathering rates of different rivers across world with respect to Payaswini–Chandragiri river.

Chandragiri river basin in 2016, compared to 36 tonnes/km^2 in 2017. A greater silicate weathering rate (SWR) is observed in the river system during the monsoon season (37 t/km^2 in 2016 and 26 t/km^2 in 2017), accounting for 82% of total discharge. When compared to the rivers draining basaltic rock rivers, the southwest-flowing rivers have lower SWR; however, they are still greater than the other Himalayan rivers in terms of SWR. The resulting weathering variety could be attributable to the different lithologies of the Indian river basin as well as the climatic differences between the tropical and temperate regions of the world. High SWR in southwest-flowing rivers is primarily controlled by increased runoff, temperature, and basin development, which are all elements that influence SWR. In the Deccan Traps, the Krishna Basin, and the Western Ghats, higher SWR in the monsoon season than the east-flowing rivers is observed. Rivers that move from west to east tend to have a higher volume of precipitation and runoff (Pattanaik et al. 2013). Strong physical erosion from heavy rains (4000 mm/year) contributed significantly to the basin's high silicate-weathering rates. Runoff, granitic terrain, and the topology of the study region were the primary factors influencing the rate of silicate weathering in the area under investigation. The weathering of silicate minerals and the storage of CO_2 in the Western Ghats are accelerated by the tropical climate, whereas runoff into the Arabian Sea is increased by the high runoff (Kumar Reddy et al. 2019).

9.5.2 Rate of Carbon Consumption (CCR) During Silicate Rock Weathering

Atmospheric CO_2 levels are decreasing as a result of extensive weathering by silicates. The following equation can be used to compute the real-world carbon dioxide consumption rates associated with weathering of silicate rocks:

$$\mathrm{CCR} = {}^{Q}\!/_{A} \cdot \Sigma\left(\mathrm{Na}^{+} + \mathrm{K}^{+} + \mathrm{Mg}^{2+} + \mathrm{Ca}^{2+}\right) \tag{9.11}$$

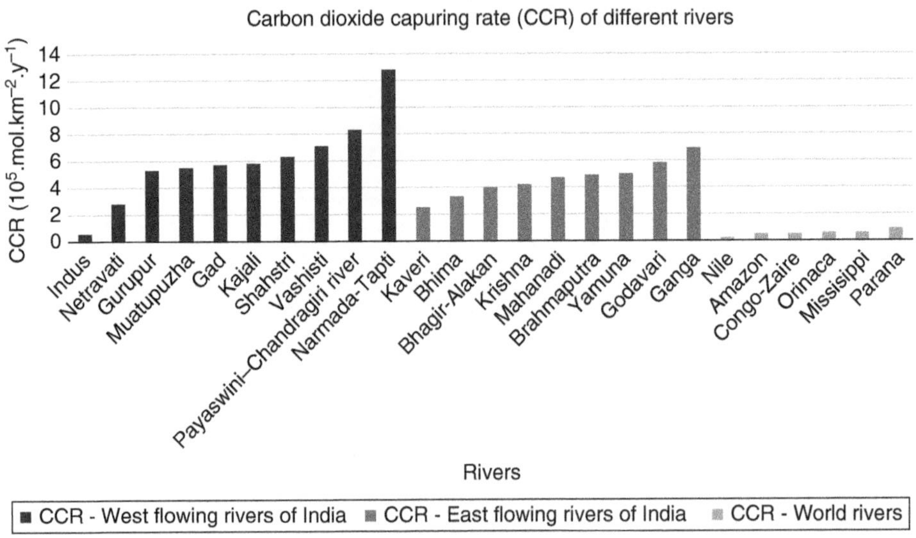

Figure 9.2 Comparative of yearly carbon dioxide capturing rates of different rivers across world with respect to Payaswini–Chandragiri river.

The discharge in cubic meter per second is denoted by Q, the surface area of the watershed is denoted by A, and the silicate-derived cations are denoted by $(Na^+ + K^+ + Mg^{2+} + Ca^{2+})_{Sil}$. The average annual carbon consumption (CCR) due to weathering of silicate rock in the Payaswini–Chandragiri river basin was reported to be 9.6 105 mol/km^2/year in 2016 and 8.3105 mol/km^2/year in 2017, according to the findings (Figure 9.2). The CCR of the Netravathi was lower (2.9 105 mol/km^2/year, compared to the other species) (Gurumurthy et al. 2012) compared to the findings of the study by Nisha et al. (2021). When silicate weathering is performed, the calculated CCR from Nisha et al. (2021) was, on the other hand, higher than the world river watersheds. As a result of the heavy monsoon rain and abundant silicate minerals in the catchment area, CO_2 consumption is likely to be higher during the monsoon season (8 × 105 mol/km^2 in 2016 and 2017).

9.6 Importance of Weathering as a Carbon Sequestration Mechanism

Each year, it is estimated that rock weathering removes about a gigatonne of CO_2 from the atmosphere. When compared to the 30 gigatonnes of CO_2 emitted each year by humans burning fossil fuels, this appears bleak. However, over geological timescales, weathering removes a massive amount of CO_2 from the atmosphere, at around a million gigatonnes per million years. Compared to the 720 gigatonnes of carbon dioxide in the atmosphere and the 38,400 gigatonnes of carbon dioxide in the oceans, limestones and other carbonate-based sedimentary rocks constitute a phenomenally important and relatively stable carbon sink.

As a result, limestones have a low carbon sequestration efficiency. Some of the most important mechanisms for letting go of it are metamorphism, subduction, and industrialization.

Metamorphism and volcanism occur across geological timescales, whereas human industry occurs in the present day. Carbon dioxide is released back into the atmosphere as a result of the process of weathering and the formation of limestones. The weathering process, even without geological replenishment and even with ocean outgassing, is projected to eliminate all atmospheric carbon dioxide in less than half a million years, even without geological replacement. Losing it all at once would be just as dreadful as having a rapidly building surplus of gas (Taylor et al. 2009).

9.7 Case Study – Himalayan Glaciers

The Himalayas are the world's tallest and youngest mountain range, combining Mediterranean, tropical, and polar influences (Pandey et al. 1999; Borgaonkar et al. 2018). As a result of the large amount of glacier ice and snow that accumulates in the Himalayas, they are often referred to as the "third pole" (Qiu 2008; Online 2010; Kulkarni and Karyakarte 2014). The Indian Himalayan region covers an area of about 0.35×10^6 km^2 and is home to 9575 glaciers (Shanker and Srivastava 2001) (Raina and Srivastava 2008). In response to rising global temperatures, lower Himalayan glaciers are retreating faster than high-altitude Himalayan glaciers (Kulkarni et al. 2007). The Indian summer monsoon provides moisture for eastern and central Himalayan glaciers, while the mid-latitude westerlies provide moisture for western Himalayan glaciers (Sharma et al. 2016). CO_2 emissions from the atmosphere have been reduced since the Cenozoic era because of an increase in silicate weathering on continental areas due to Himalayan uplift (Raymo et al. 1988). Glacier hydrogeochemistry in the Western Himalayas has seen a great deal of investigation (Sharma et al. 2013; Singh et al. 2017; Kumar et al. 2019). However, glacier CO_2 consumption rates and chemical weathering in this region, on the other hand, have received comparatively little study. (Shukla et al. 2018).

Shukla et al. (2018) looked at how CO_2 was used in the global carbon cycle because of chemical deforestation in heavily glaciated Himalayan catchments. They looked at the data from previous studies of solute concentrations in glacier-fed rivers, which were reprocessed to figure out how much CO_2 was used and how much solute hydrolysis took place across the country. About 90% of the SO_4^{-2} is derived through crustal sulfide oxidation, and the remaining 10% is derived from aerosols and sea salts, according to the analysis. While sulfide oxidation contributes around 20% to carbonate dissolution, the HCO_3 flux estimate shows that silicate dissolution and simple hydrolysis provide nearly the same percentage (both about 20%). According to CO_2 consumption estimates in the atmosphere, 10.6 million tonnes of CO_2 are absorbed each year through silicate weathering, 15.7 million tonnes of CO_2 are absorbed through simple hydrolysis, 9.6 million tonnes of CO_2 are consumed through the sulfate oxide and carbon dissolution reaction, and 5.9 million tonnes of CO_2 are absorbed through the carbonate carbonation reaction (SO-CD). These are the CO_2 consumption estimates in the atmosphere. Solute provenance calculations show that silicate weathering reactions produce CO_2 and HCO_3, but simple hydrolysis and the SO-CD process also produce CO_2. This shows that a counter mechanism is at work in subglacial areas of the Himalayas. Instead of being a sink for CO_2, these areas are a source of CO_2.

Chemical weathering and glacial stream dissolved inorganic carbon (DIC) fluxes in the Himalayas have imposed new restrictions on the amount of CO_2 that may be removed from the atmosphere. Between 1989 and 2017, Shukla et al. looked again at the ionic composition of meltwater from five Himalayan glacierized basins. As you can see, silicate weathering can remove a lot of CO_2 from the atmosphere. For Na-bearing minerals, the amount of CO_2 that is removed each year is about 17%. For K-bearing minerals, it is about 22%, and it is about 0.5105 mol/km^2/year. In the same way, carbonate weathering draws CO_2 out of the air, which accounts for about 57% of the minerals that are made up of calcium and magnesium (Ca – Mg carbonate). The amount of DIC that flows from the Himalayan glacierized basins to the ocean is thought to be about 0.02 gm C/km^2/year. This is made up of 0.09 gm C/km^2/year of silicate weathering (Na, K, and Ca postured minerals) and 0.11 gm C/km^2/year of carbonate weathering (Mg and Ca postured minerals). We found that carbonate weathering is the main reason why CO_2 levels are dropping in the atmosphere and releasing DIC in glacierized basins in the Himalayas. To better understand the effects of climate change on people, our research shows the importance of carbonate weathering from glacial terrains (Ganguly et al. 2015, 2019a, 2019b; Sundriyal et al. 2018, 2021; Shukla et al. 2020).

9.8 Conclusion

This chapter uses weathering rates from experiments to show that global carbon cycle models should be more sensitive to temperature. On this basis, CO_2 and temperature levels are less different from preindustrial levels than the BLAG model. There are a lot of things that could speed up the process of weathering that have been left out for simplicity and because we are not clear about their functioning. The Payaswini–Chandragiri river basin, on India's southwest coast, was studied for its geochemical characteristics and atmospheric CO_2 drawdown rate during silicate weathering. Seasonal changes in runoff, climate, and temperature have a smaller impact on river water's hydrochemical features, according to the findings. Anions dominate major ions in the Payaswini–Chandragiri river system, while major cations dominate anions in the order $Na^+ > Ca^{2+} > Mg^{2+} > K^+$. The Na^+ normalized Ca^{2+} versus Mg^{2+} and HCO_3 plots indicated major ion contribution via silicate minerals. In 2016, the Payaswini–Chandragiri river catchment weathered 42 t/km^2/year, and 36 t/km^2/year in 2017, which is 0.9 times the neighboring west-flowing river Nethravathi.

References

Berner, R. A., Lasaga, A. C., & Garrels, R. M. (1983). Berner_83_Mg-Ca-model. *American Journal of Science*, 283, 641–683).

Bluth, G. J. S., & Kump, L. R. (1994). Lithologic and climatologic controls of river chemistry. *Geochimica et Cosmochimica Acta*, 58(10), 2341–2359. https://doi.org/10.1016/0016-7037(94)90015-9

Borgaonkar, H. P., Gandhi, N., Ram, S., & Krishnan, R. (2018). Tree-ring reconstruction of late summer temperatures in northern Sikkim (eastern Himalayas). *Palaeogeography, Palaeoclimatology, Palaeoecology*, 504, 125–135. https://doi.org/10.1016/j.palaeo.2018.05.018

Brady, P. V., & Walther, J. V. (1989). Controls on silicate dissolution rates in neutral and basic pH solutions at 25 °C. *Geochimica et Cosmochimica Acta*, 53(11), 2823–2830. https://doi.org/10.1016/0016-7037(89)90160-9

Brook, G. A., Folkoff, M. E., & Box, E. O. (1983). A world model of soil carbon dioxide. *Earth Surface Processes and Landforms*, 8(1), 79–88. https://doi.org/10.1002/esp.3290080108

Calmels, D., Gaillardet, J., & François, L. (2014). Sensitivity of carbonate weathering to soil CO_2 production by biological activity along a temperate climate transect. *Chemical Geology*, 390, 74–86. https://doi.org/10.1016/j.chemgeo.2014.10.010

Gaillardet, J., Dupre, B., Louvat, P., & Allegre, C. J. (1999). Global silicate weathering and CO_2 consumption rates deduced from the chemistry of large rivers. *Chemical Geology*, 159(8), 3–30.

Ganguly, S., Tiwari, S., Bhan, U., Mittal, S., Rai, S., Osta, E. F. (2015). Melting of sea ice inexplicable for recent global eustatic sea level rise. *Journal of Earth Science & Climatic Change*, 6(1), 245. https://doi.org/10.4172/2157-7617.1000245

Ganguly, S., Bhan, U., Rai, S. K., Mittal, S., Ahluwalia, R. S., & Verma, A. (2019a). An experimental approach to estimate groundwater temperature from ^{18}O fractionation. *Groundwater for Sustainable Development*, 9(June), 100257. https://doi.org/10.1016/j.gsd.2019.100257

Ganguly, S., Rai, S. K., Mittal, S., Bhan, U., & Ahluwalia, R. S. (2019b). Estimation of groundwater temperature from ^{18}O fractionation – A deterministic analytical model. *Groundwater for Sustainable Development*, 9, 100234. https://doi.org/10.1016/j.gsd.2019.100234

Goddéris, Y., Brantley, S. L., François, L. M., Schott, J., Pollard, D., Déqué, M., & Dury, M. (2013). Rates of consumption of atmospheric CO_2 through the weathering of loess during the next 100 yr of climate change. *Biogeosciences*, 10(1), 135–148. https://doi.org/10.5194/bg-10-135-2013

Goll, D. S., Moosdorf, N., Hartmann, J., & Brovkin, V. (2014). Climate-driven changes in chemical weathering and associated phosphorus release since 1850: Implications for the land carbon balance. *Geophysical Research Letters*, 41(10), 3553–3558. https://doi.org/10.1002/2014GL059471

Gombert, P. (2002). Role of karstic dissolution in global carbon cycle. *Global and Planetary Change*, 33(1–2), 177–184. https://doi.org/10.1016/S0921-8181(02)00069-3

Gurumurthy, G. P., Balakrishna, K., Riotte, J., Braun, J. J., Audry, S., Shankar, U. H. N., & Manjunatha, B. R. (2012). Controls on intense silicate weathering in a tropical river, southwestern India. *Chemical Geology*, 300–301, 61–69. https://doi.org/10.1016/j.chemgeo.2012.01.016

Hartmann, J., & Moosdorf, N. (2012). The new global lithological map database GLiM: A representation of rock properties at the Earth surface. *Geochemistry, Geophysics, Geosystems*, 13(12), 1–37. https://doi.org/10.1029/2012GC004370

Hartmann, J., Jansen, N., Dürr, H. H., Kempe, S., & Köhler, P. (2009). Global CO_2-consumption by chemical weathering: What is the contribution of highly active weathering regions? *Global and Planetary Change*, 69(4), 185–194. https://doi.org/10.1016/j.gloplacha.2009.07.007

Hartmann, J., Moosdorf, N., Lauerwald, R., Hinderer, M., & West, A. J. (2014). Global chemical weathering and associated p-release – The role of lithology, temperature and soil properties. *Chemical Geology*, 363, 145–163. https://doi.org/10.1016/j.chemgeo.2013.10.025

Kudapa, V.K., Kumar, S., Gupta, D. K., Sharma, P., & KunhAbdulla, R. (2017a). Modelling of gas production from shale matrix to fracture network. *Society of Petroleum Engineers – SPE Intelligent Oil and Gas Symposium*, 2017. https://doi.org/10.2118/187473-ms

Kudapa, V. K., Sharma, P., Kunal, V., & Gupta, D. K. (2017b). Modeling and simulation of gas flow behavior in shale reservoirs. *Journal of Petroleum Exploration and Production Technology*, 7(4), 1095–1112. https://doi.org/10.1007/s13202-017-0324-4

Kudapa, V. K., Gupta, D. K., & Sharma, P. (2018). Modeling of gas flow within the shale fracture network. *Journal of Civil and Environmental Engineering*, 1, 21–43. https://doi.org/10.1007/978-981-10-7281-9_3

Kulkarni, A. V., & Karyakarte, Y. (2014). Observed changes in Himalayan glaciers. *Current Science*, 106(2), 237–244. https://doi.org/10.18520/cs/v106/i2/237-244

Kulkarni, A. V., Bahuguna, I. M., Rathore, B. P., Singh, S. K., Randhawa, S. S., Sood, R. K., & Dhar, S. (2007). Glacial retreat in Himalaya using Indian Remote Sensing satellite data. *Current Science*, 92(1), 69–74. https://doi.org/10.1117/12.694004

Kumar Reddy, S. K., Gupta, H., & Reddy, D. V. (2019). Dissolved inorganic carbon export by mountainous tropical rivers of the Western Ghats, India. *Chemical Geology*, 530(September), 119316. https://doi.org/10.1016/j.chemgeo.2019.119316

Kumar, R., Kumar, R., Singh, A., Singh, S., Bhardwaj, A., Kumari, A., Sinha, R. K., & Gupta, A. (2019). Hydro-geochemical analysis of meltwater draining from Bilare Banga glacier, Western Himalaya. *Acta Geophysica*, 67(2), 651–660. https://doi.org/10.1007/s11600-019-00262-w

Kuzyakov, Y. (2006). Sources of CO_2 efflux from soil and review of partitioning methods. *Soil Biology and Biochemistry*, 38(3), 425–448. https://doi.org/10.1016/j.soilbio.2005.08.020

Mackenzie, F. T., & Garrels, R. M. (1966). Chemical mass balance between rivers and oceans. *American Journal of Science*, 264, 7, 507–525. https://doi.org/10.2475/ajs.264.7.507

Martin, J. B. (2017). Carbonate minerals in the global carbon cycle. *Chemical Geology*, 449, 58–72. https://doi.org/10.1016/j.chemgeo.2016.11.029

Moosdorf, N., Hartmann, J., Lauerwald, R., Hagedorn, B., & Kempe, S. (2011). Atmospheric CO_2 consumption by chemical weathering in North America. *Geochimica et Cosmochimica Acta*, 75(24), 7829–7854. https://doi.org/10.1016/j.gca.2011.10.007

Nisha, B. K., Balakrishna, K., Udayashankar, H. N., & Manjunatha, B. R. (2021). Chemical weathering and carbon dioxide consumption in a small tropical river catchment, southwestern India. *Aquatic Geochemistry*, 27(3). Springer Netherlands. https://doi.org/10.1007/s10498-021-09394-2

Norton, K. P., Molnar, P., & Schlunegger, F. (2014). The role of climate-driven chemical weathering on soil production. *Geomorphology*, 204, 510–517. https://doi.org/10.1016/j.geomorph.2013.08.030

Online, I. B. (2010). Status of the Glacier Research in the HKH region By Dr. S. I. Hasnain School of Environmental Sciences Jawahar Lal Nehru University. 1–6.

Pandey, S. K., Singh, A. K., & Hasnain, S. I. (1999). Weathering and geochemical processes controlling solute acquisition in Ganga Headwater-Bhagirathi River, Garhwal Himalaya, India. *Aquatic Geochemistry*, 5(4), 357–379. https://doi.org/10.1023/A:1009698016548

Pattanaik, J. K., Balakrishnan, S., Bhutani, R., & Singh, P. (2013). Estimation of weathering rates and CO_2 drawdown based on solute load: Significance of granulites and gneisses dominated weathering in the Kaveri River basin, Southern India. *Geochimica et Cosmochimica Acta*, 121, 611–636. https://doi.org/10.1016/j.gca.2013.08.002

Pu, T., He, Y., Zhang, T., Wu, J., Zhu, G., & Chang, L. (2013). Isotopic and geochemical evolution of ground and river waters in a karst dominated geological setting: A case study from Lijiang basin, South-Asia monsoon region. *Applied Geochemistry*, 33, 199–212. https://doi.org/10.1016/j.apgeochem.2013.02.013

Qiu, J. (2008). China news feature. *Nature*, 454(7203), 393–396. https://doi.org/10.1038/454393a

Raina, V. K. and Srivastava, D.. (2008). Glacier Atlas of India. *Geological Society of India*, 315, 315.

Raymo, M. E., Ruddiman, W. F., & Froelich, P. N. (1988). Influence of late Cenozoic mountain building on ocean geochemical cycles. *Geology*, 16, 649–635. https://doi.org/10.1130/0091-7613(1988)016<0649:IOLCMB>2.3.CO;2

Roland, M., Serrano-Ortiz, P., Kowalski, A. S., Goddéris, Y., Sánchez-Cañete, E. P., Ciais, P., Domingo, F., Cuezva, S., Sanchez-Moral, S., Longdoz, B., Yakir, D., Van Grieken, R., Schott, J., Cardell, C., & Janssens, I. A. (2013). Atmospheric turbulence triggers pronounced diel pattern in karst carbonate geochemistry. *Biogeosciences*, 10(7), 5009–5017. https://doi.org/10.5194/bg-10-5009-2013

Romero-Mujalli, G., Hartmann, J., & Börker, J. (2019a). Temperature and CO_2 dependency of global carbonate weathering fluxes – Implications for future carbonate weathering research. *Chemical Geology*, 527, 118874. https://doi.org/10.1016/j.chemgeo.2018.08.010

Romero-Mujalli, G., Hartmann, J., Börker, J., Gaillardet, J., & Calmels, D. (2019b). Ecosystem controlled soil-rock pCO_2 and carbonate weathering – Constraints by temperature and soil water content. *Chemical Geology*, 527(January), 1–11. https://doi.org/10.1016/j.chemgeo.2018.01.030

Shanker, R., & Srivastava, D. (2001). Glaciated regime environmental interaction and Himalayan ecosystem. *Snow Ice*, 17–21.

Sharma, P., Ramanathan, A. L., & Pottakkal, J. (2013). Etude des sources de soluté et de l'volution des processus hydrogochimiques des eaux de la fonte du glacier Chhota Shigri, Himachal Himalaya, Inde. *Hydrological Sciences Journal*, 58(5), 1128–1143. https://doi.org/10.1080/02626667.2013.802092

Sharma, S., Chand, P., Bisht, P., Shukla, A. D., Bartarya, S. K., Sundriyal, Y. P., & Juyal, N. (2016). Factors responsible for driving the glaciation in the Sarchu Plain, eastern Zanskar Himalaya, during the late Quaternary. *Journal of Quaternary Science*, 31(5), 495–511. https://doi.org/10.1002/jqs.2874

Shukla, T., Sundriyal, S., Stachnik, L., & Mehta, M. (2018). Carbonate and silicate weathering in glacial environments and its relation to atmospheric CO_2 cycling in the Himalaya. *Annals of Glaciology*, 59(77), 159–170. https://doi.org/10.1017/aog.2019.5

Shukla, T., Sundriyal, S., & Sen, I. S. (2020). Contemporary inorganic carbon fluxes from rapidly changing glacierized watersheds of the Himalaya. *Journal of Hydrology*, 587, 124972. https://doi.org/10.1016/j.jhydrol.2020.124972

Singh, A. T., Laluraj, C. M., Sharma, P., Patel, L. K., & Thamban, M. (2017). Export fluxes of geochemical solutes in the meltwater stream of Sutri Dhaka Glacier, Chandra basin, Western Himalaya. *Environmental Monitoring and Assessment*, 189(11). https://doi.org/10.1007/s10661-017-6268-9

Stumm, W., & Wollast, R. (1990). Coordination chemistry of weathering: Kinetics of the surface-controlled dissolution of oxide minerals. *Reviews of Geophysics*, 28(1), 53–69. https://doi.org/10.1029/RG028i001p00053

Sundriyal, S., Shukla, T., Tripathee, L., Dobhal, D. P., Tiwari, S. K., & Bhan, U. (2018). Deposition of atmospheric pollutant and their chemical characterization in snow pit profile at Dokriani Glacier, Central Himalaya. *Journal of Mountain Science*, 15(10), 2236–2246. https://doi.org/10.1007/s11629-017-4817-x

Sundriyal, S., Bhan, U., Selvakumar, S., Singh, R., & Dobhal, D. P. (2021). Two decadal changes in the major ions chemistry of melt water draining from Dokriani Glacier, Central Himalaya, India. *Journal of the Geological Society of India*, 97(3), 308–314. https://doi.org/10.1007/s12594-021-1682-6

Taylor, L. L., Leake, J. R., Quirk, J., Hardy, K., Banwart, S. A., & Beerling, D. J. (2009). Biological weathering and the long-term carbon cycle: Integrating mycorrhizal evolution and function into the current paradigm. *Geobiology*, 7(2), 171–191. https://doi.org/10.1111/j.1472-4669.2009.00194.x

Volk, T. (1987). Feedbacks between weathering and atmospheric CO_2 over the last 100 million years. *America Journal of Science*, 287(8), 763–779. https://doi.org/10.2475/ajs.287.8.763

Walker, J. C. G., Hays, P. B., & Kasting, J. F. (1981). Weathering rates in 106 gm/km^2/yr of dissolved Si02. Parentheses signify. *Journal of Geophysical Research*, 86(1), 9776–9782.

Walther, J. V. (2007). MAGic: A phanerozoic model for the geochemical cycling of major rock-forming components. *American Journal of Science*, 307(5), 856–857. https://doi.org/10.2475/05.2007.04

Wang, S., Yeager, K. M., Wan, G., Liu, C. Q., Liu, F., & Lü, Y. (2015). Dynamics of CO_2 in a karst catchment in the southwestern plateau, China. *Environmental Earth Sciences*, 73(5), 2415–2427. https://doi.org/10.1007/s12665-014-3591-0

Wu, W., Yang, J., Xu, S., & Yin, H. (2008). Geochemistry of the headwaters of the Yangtze River, Tongtian He and Jinsha Jiang: Silicate weathering and CO_2 consumption. *Applied Geochemistry*, 23(12), 3712–3727. https://doi.org/10.1016/j.apgeochem.2008.09.005

10

Carbonate Chemistry, Carbon Cycle, and Its Sequestration in Aquatic System

Mallika Vashist, Harshit Chawla, and S. K. Singh

Department of Environmental Engineering, Delhi Technological University, Delhi, India

10.1 Introduction

Every element in the ecosystem has a specific pathway by which it moves through biotic and abiotic aspects of the Earth and consequently accelerates the relocation of matters from one form to another and one location to another (Zalewski 2015). Carbon is one such critical element of the environment; it is the universal currency that is fundamental for all the life-forms on the Earth to store and expand energy. Carbon contributes to ~50% dry weight of the living matter, which indicates the biosphere's health on the basis of the organic matter production and its breakdown. The exchange of carbon is highly controlled between the land (terrestrial), oceanic (aquatic system), and atmosphere, which are the reservoir for the carbon, out of which ocean being the most dominating sink for carbon by storing almost 30% of the atmospheric carbon dioxide (Gruber et al. 2019). The influx of CO_2 from the ocean to the atmosphere during the preindustrial era was 70.6 Gt C/yr and the opposite direction flux was 70 Gt C/yr (Friedlingstein et al. 2019). Over the past 200 years, due to industrialization, the natural flux has been superimposed by the anthropogenic flux due to which the global carbon cycle has been altered significantly. During the preindustrial period, the concentration of carbon dioxide ranged between 172 and 300 parts per million by volume (ppmv) for over past 800,000 years (Lüthi et al. 2008), but during the industrial era, it reached 387 ppmv in 2009 and presently it is around ~420 ppmv. This very evidently suggests that the rate of increase was about 1.0% per year during the 1990s and it accelerated to 3.4% between 2000 and 2008 (Le Quéré et al. 2009). This trend suggests that socioeconomic factors are the major drivers of the future levels of atmospheric carbon and might reach 1071 ppmv in 2100 (Plattner et al. 2001), which since 1750 has been approximately a fourfold increase. The CO_2 which is emitted from anthropogenic sources has three fates, about ~29% gets absorbed by the terrestrial biosphere, around 30% by the ocean, and the rest sustains in the atmosphere (Le Quéré et al. 2009). The accretion of CO_2, in the atmosphere, leads to increasing natural greenhouse effect and subsequently changes in climate (IPCC 2007). The ocean has been serving as the major sink for

Hydrogeochemistry of Aquatic Ecosystems, First Edition. Edited by Sughosh Madhav, Virendra Bahadur Singh, Manoj Kumar, and Sandeep Singh.

anthropogenic carbon emissions as it has taken up about one-quarter (118 Pg C, or about 25%) of anthropogenic carbon generated by humans (Canadell et al. 2007). This ocean sink of CO_2 has been advantageous for humankind by considerably curtailing the elevated level of CO_2 in the atmosphere thereby plummeting the global warming apprehend till date because the level of atmospheric CO_2 would have reached to approximately 450 ppmv today, if these aquatic bodies were not present, and this high concentration of CO_2 would have led to even greater climate change than witnessed today. However, the uptake of CO_2 by the aquatic systems, majorly oceans, is not benign; as it causes alteration in seawater carbon dioxide – carbonic acid systems, including a decrease in pH and the saturation state concerning $CaCO_3$ mineral phases, which is collectively known as ocean acidification (Gattuso and Hansson 2011), and in addition to this, it has several biological, ecological, and biogeochemical implications as well (Kroeker et al. 2013). This suggests the significance of studies on thermodynamics of carbonate processes and the carbon cycle in aquatic systems to understand climate change and to control the permissible emission levels of carbon. Thus, the aim of this chapter is to understand the reactivity of carbonate system, its pathways and formation in aquatic systems, the present-day alterations in its concentration due to anthropogenic activities, and its impacts on biodiversity subsequently suggesting the mitigating strategies.

10.2 Carbon Cycle in Aquatic Systems

The dissolution of atmospheric carbon at the interface of air–water is one of the most important steps in the carbon cycle. After dissolution, CO_2 forms three sorts of inorganic carbon (DIC) species upon reaction, and the concentration of these inorganic species is dependent on the pH of the system. For instance, at pH 8.17 (alkaline solution), bicarbonate will be dominantly present which is responsible for about 90% of the total DIC, 9% as carbonates, and the remaining 1% as aqueous carbon dioxide (Falkowski et al. 1998; Neven et al. 2011). The succeeding step of carbon cycle is the formation of organic matter by photosynthesis which is known as the primary productivity of any aquatic system. During photosynthesis, the major sources of carbon for the formation of organic matter are taken by the phytoplankton from the bicarbonate ions (HCO_3^-) and the aqueous CO_2 (Thornton 2014). These phytoplankton cells formed during photosynthesis are taken up by the zooplankton available in the water column. Thus, these zooplanktons are the primary consumer of the aquatic carbon cycle which is later devoured by the consumers of the higher trophic level. The biomass of the remaining phytoplankton, which are either dead or living, sinks to the deeper oceans and gets deposited in their sediments, which leads to another important step of the carbon cycle, i.e. organic matter decomposition which is known as re-mineralization (Agrawal and Gopal 2013). The decomposition of the organic matter is majorly performed by the bacteria present in the water, which revert the DIC to the aquatic system. Other than this, two other significant processes are highly pH dependent, which are dissolution and precipitation of carbonate. The carbonate dissolution adds DIC to the system, hence increasing the level of carbon. On the other hand, the precipitation of carbonate removes DIC from the system which eventually removes the carbon from the system (Marx et al. 2017).

10.3 Reactivity of Carbonate System

The difference in partial pressure of the carbon dioxide in the atmosphere and the surface water regulates the flux of carbon dioxide across the interface of air and sea. Once the CO_2 is in the solution, it equilibrates between the ions of carbonate and bicarbonate where the precipitation and dissolution rates of $CaCO_3$ are regulated by the concentration of carbonate in the ocean (Renforth and Henderson 2017). The characterization of carbonate can be done by the quantification of any two out of the four parameters: pH, the fugacity of COP (FCS), the total carbonate (TCO_2), and the total alkalinity (TA).

The following reaction thermodynamically controls the amount of CO_2 in the solution (Millero 1995):

$$CO_2(g) = 2CO_2(aq); \quad [K_0] \tag{10.1}$$

$$CO_2(aq) + H_2O = H_2CO_3; \quad [K_H] \tag{10.2}$$

$$CO_2 + H_2O = H^+ + HCO^{-3}; \quad [K_1] \tag{10.3}$$

$$HCO^-{}_3 = H^+ + CO^{2-}{}_3 \quad [K_2] \tag{10.4}$$

In the above reaction, K_0 refers to the solubility coefficient of the CO_2 in seawater. In Eqn. 2, the dissolved CO_2 is in the aqueous form. While the $K_H \sim 10^{-3}$, most of the undissociated CO_2 is dissolved in the aqueous form. The carbonate system dissociation constant (Eqs. 10.3 and 10.4) does not differentiate between CO_2 (aq) and H_2CO_3; consequently, the sum of their concentration is used and is denoted by $[CO_2]$. Thereby the fugacity of CO_2 is defined by the relationship:

$$f_{CO_2} = [CO_2]/K_0 \tag{10.5}$$

The first dissociation constant of carbonic acid is denoted as:

$$K_1 = [H^*][HCO_3]/[CO_2] \tag{10.6}$$

And second dissociation constant is denoted as:

$$K_2 = [H^*][CO_3{}^{2-}]/[HCO^-{}_3] \tag{10.7}$$

The TA is the sum of all the bases that can accept protons at the endpoint of carbonic acid (Dickson 1981). And it helps in the quantification of the constituents of the CO_2 system which can accept protons. It is defined by:

$$TA = [HCO_3] + 2[CO_3{}^{2-}] + [B(OH)_4{}^-] + [OH^-] - [H] + \text{Bases}, \tag{10.8}$$

where $[OH^-]$ consists of the OH^- ions which are bound to Mg^{2+} and $[H^+]$ includes the H^+ ions which are bound to $SO_2{}^{-4}$ and F^-. TA includes the minor concentration of other bases as well (Dickson 1981; Bradshaw and Brewer 1988) which can accept a proton.

$$\text{Bases} = [SiO(OH)_3{}^-] + [HPO_4{}^{2-}] + 2[PO_4{}^{3-}] + [\text{organics}], \tag{10.9}$$

But only the alkalinity contributed by the components of the CO_2, i.e. the carbonate alkalinity (CA), is required to characterize the system. The CA is defined by:

$$CA = \left[HCO_3^-\right] + 2\left[CO^{2-}{}_3\right] \tag{10.10}$$

The TA measured can be used to calculate the CA, by taking into account the contribution by $[B(OH)_4^-]$, $[OH^-]$, $[H^+]$, and other proton-accepting bases. To measure CA, one needs to consider the following reaction:

$$B(OH)_3 + H_2OK_B = B(OH) + H^*, \tag{10.11}$$

and,

$$H_2O = H^+ + OH^-;\ \left[K_w\right] \tag{10.12}$$

where K_B is the dissociation constant of boric acid and K_W is the dissociation constant of water

$$K_B = \left[H^*\right]\left[B(OH)_4{}^-\right] / \left[B(OH)_3\right] \tag{10.13}$$

$$K^-{}_w = \left[H^+\right]\left[OH^-\right] \tag{10.14}$$

In terms of deep water, it is also important to take into account the contribution of the phosphate as well as the silicate in alkalinity (Dickson 1981):

$$H_3PO_4 + H^+ + H_2PO_4{}^-; \tag{10.15}$$

$$H_2PO^-{}_4 + H^+ + HPO_4{}^{2-}; \tag{10.16}$$

$$HPO^{2-}{}_4 = H^* + PO_4{}^{3-}; \tag{10.17}$$

$$SiO_2 + 2H_2O = H^+ + Si(OH)_3O^-, \tag{10.18}$$

where the dissociation constants are given by:

$$Kp_1 = \left[H^+\right]\left[H_2PO^4\right] / \left[H.PO.\right] \tag{10.19}$$

$$Kp_2 = \left[H^+\right]\left[HPO1 / \left[H_2PO.\right]\right. \tag{10.20}$$

$$Kp_3 = \left[H^+\right]\left[PO\right] / \left[HPO\right] \tag{10.21}$$

$$Ksi = \left[H^+\right] / \left[Si(OH),O - 1\left[SiO_2\right]\right.. \tag{10.22}$$

Additionally, factors like solubility and precipitation of calcium carbonates like calcite and aragonite also cause alterations in the carbon dioxide system, and it is given as:

$$CaCO_{3(S)} = Ca^{2+} + \left[CO_3^{2-}\right]\ \left[K_{sp}\right] \tag{10.23}$$

And the solubility product for calcite and aragonite will be:

$$K_{sp} = \left[Ca^{2+}\right]\left[CO_3^{2-}\right] \tag{10.24}$$

In the above equations, the bracketed species are stoichiometric concentrations (molarity/mol/kg) regardless of any complex formation or ion pairs, which suggest that it is important to define not only the pH scale (Dickson 1984) used in the evaluation of the dissociation constant but also their correlation with ionic strength, temperature, and pressure.

10.3.1 The Contemporary Marine $CaCO_3$ Cycle

The carbonate chemistry in seawater is controlled by a series of chemical reactions as follows:

The first and the prime step in the carbonate cycle is the dissolution of CO_2 into seawater:

$$CO_2(atm) \leftrightharpoons CO_2(aq) \tag{10.25}$$

This exchange of CO_2 is rapid and occurs within a few months, as a result of which the concentration of CO_2 in the surface water of the ocean is increased proportionally with the increase in the CO_2 levels in the atmosphere from year to year.

The subsequent set of reaction formation of carbonic acid by the hydration of CO_2, which further dissociates into acid-base reactions:

$$CO_2(aq) + H_2O \leftrightharpoons H_2C0_3 \leftrightharpoons H^+ + HCO^-{}_3 \leftrightharpoons 2H^+ + CO_3^{2-} \tag{10.26}$$

These reactions are relatively very rapid, on the timescale of translate of seconds for the hydration of CO_2 and microseconds for next acid-base dissociation reactions (Zeebe and Wolf-Gladrow 2001; Dickson et al. 2007). This partitioning of inorganic carbonate species is considered to be in a steady state of equilibrium in most cases. In usual conditions, bicarbonate is the most dominating dissolved inorganic species followed by carbonate ions and dissolved CO_2 (aq) and H_2CO_3 in the least amount.

The next step is associated with formation as well as dissolution of solid calcium carbonate minerals ($CaCO_3$ (s)):

$$Ca^{2+} + CO_3^{2-} \leftrightharpoons CaCO_3(s) \tag{10.27}$$

By the elimination of calcium in the form of $CaCO_3$, its intake from terrestrial sources such as a river, land runoff, and groundwater input and hydrothermal sources by the oceans is balanced through burial reefs and sediments. In terms of chemical reactions, it can be said that at equilibrium, the calcium concentration times the carbonate concentration gives us a constant which is called a solubility product, and it is given as:

$$\left[Ca2+\right] \times \left[CO_3^{2-}\right] = K_{sp} \tag{10.28}$$

The saturation state of calcium carbonate is defined by:

$$\Omega = \left[Ca^{2+}\right]\left[CO_3^{\ 2-}\right]/K_{sp} \tag{10.29}$$

where the solubility product depends on the temperature, salinity, and pressure, and fluctuates among different calcium carbonate minerals like calcite and aragonite (Mucci 1983). The last reaction is that of photosynthesis and respiration/decomposition which is:

$$CO_2 + H_2O \leftrightharpoons CH_2O + O_2 \tag{10.30}$$

In the present time, the increase in the concentration of CO_2 leads to an imbalance in these reactions of the carbonate cycle. This is because the CO_2 gas reacts with water to form H_2CO_3 which subsequently dissociates into H^+ and HCO_3^-, where the H^+ forms HCO_3^- ions on reaction with CO_3. Consequently, this dissolution in the ocean decreases the pH as well as the concentration of CO_3.

As a result of a decrease in CO_3^{2-} ions, the saturation rate of $CaCO_3$ also reduces, which primarily impacts the $CaCO_3$ secreting organisms like planktonic foraminifera, coccolithophores and pteropods, and invertebrates such as mollusks and corals, to produce their shells or skeletons (Doney et al. 2009). The saturation of $CaCO_3$ reduces even though most of the surface water of the ocean is in the supersaturated state for $CaCO_3$, due to exceedingly slow abiotic precipitation rates for minerals of $CaCO_3$. The saturation level of seawater exceeds above 1 (saturation level > 1) which favors calcification more than the dissolution. However, many organisms can optimally undergo calcium carbonate precipitation at supersaturated states for the minerals which were present in the preindustrial oceans, consequently decreasing the concentration of the CO_3^{2-} which results in a decrease in calcification rates for a number of these species.

Over a long period, the capacity of the ocean for the absorption of more atmospheric CO_2 depends on the extent of interactions of marine carbonate with CO_2 via the dissolution reaction:

$$CO_2 + CaCO_3 + H_2O \leftrightharpoons 2HCO^-_3 + Ca^{2+} \tag{10.31}$$

In the above reaction, if the TA increases, it will also enhance the absorption of CO_2 by the ocean from the atmosphere (Zeebe and Wolf-Gladrow 2001). The quantification of TA is done by measuring the excess of cations such as Ca^{2+} and Mg^{2+}which are balanced by the anions that are formed by the dissociation of weak acids like carbonic acid, that are present in the seawater (Clark 2015). Carbonate shells formed in the euphotic zone are the main contributors to the dissolution process. When calcifying organisms perish, their carbonate shells are either deposited in sediments or dissolved in water. The carbonate mineral saturated state decreases naturally with depth, due to an increase in the concentration of dissolved oxygen because of the biological respiration and cold temperature in deep seawater (Doney et al. 2009). Approximately below 1500 m, the depth of saturation controls the pressure effect on the solubility of $CaCO_3$. When the mineral component's degree of saturation is less than one, the unprotected shells and skeletons of the $CaCO_3$ phases begin to dissolve, as a result of which some of the shells and skeletons dissolve even before reaching to the deep sediments which are below the saturation horizons, or we can say the depth where seawater becomes undersaturated with respect to $CaCO_3$. The point of saturation horizon, where $\Omega = 1$, varies in different carbonate minerals and leads to differences in the response of materials that are composed of aragonite, calcite, or high-magnesium carbonate (Williamson and Turley 2012). With the enrichment of anthropogenic CO_2 in the ocean, the sites in the ocean where dissolution takes place will proliferate as a function of the decrease in $CaCO_3$ saturation state (Johnson et al. 2012).

10.3.2 Formation and Pathway of $CaCO_3$

Presently the range of production of $CaCO_3$ ranges from 0.5 to 1.6 Pg C/yr, and the global export of $CaCO_3$ ranges from 0.4 to 1.8 Pg C/yr. Coccolithophorids are the largest contributor in the production of $CaCO_3$ on a global scale and account for almost 60% of the total production (Schiebel 2002). At the higher latitudes, export flux is dominated by pteropods (Fabry 1989; Honjo 2004), and in global terms, they contribute 10% of the total $CaCO_3$ production (Fabry 1990). Other than these planktic calcifiers, there are other benthic foraminifera as well such as corals, coralline algae, sea urchins, and many others which are most abundant in shallow-water environments, nevertheless many are also found in the deep ocean beneath the photic zone including azooxanthellae corals that forms extensive $CaCO_3$ (Guinotte and Fabry 2008). The maximum depth at which different calcareous structures can occur in the ocean is the point at which dissolution exceeds the organisms' ability to maintain its $CaCO_3$ structures (Lebrato et al. 2016). The shallow waters like a reef, banks, and estuaries also produce $CaCO_3$ which range from 0.25 to 0.35 Pg C/yr (Milliman 1993; Wollast 1994; Milliman and Droxler 1996; Mackenzie et al. 2005), and this production is credited by coral and various other calcifying algae (Nelson 2009); the coral reef environment alone can accumulate as much as 0.08–0.1 Pg C/yr (Rees et al. 2007), but some calcifiers like benthic foraminifera, mollusks, and bryozoans can significantly contribute at local and regional scales as well. The sediments present in the tropical and subtropical shallow-water comprise 63% aragonite, 24% high Mg-calcite, and 13% calcite/low Mg-calcite (Land 1967), which is mainly contributed by the corals and caducean green algae, coralline red algae, echinoderms, bryozoans, benthic foraminifera (high Mg-calcite), and mollusks (calcite/low Mg-calcite). However, as such, there is no published quantitative estimate of the global budget of $CaCO_3$, but other than the marine calcifying organisms, the abiotic factors also contribute to $CaCO_3$ formation but relatively in a small fraction. The large fraction abiotically precipitated $CaCO_3$ is headed by the dissolution of biologically produced $CaCO_3$ followed by the reprecipitation by the abiotic factors, as a result of which the new production of $CaCO_3$ is less than the actual observed quantity. Methane seeps in conjunction with cement, carbonate muds, and crystalline carbonate deposits from the source of abiotic $CaCO_3$. The abiotic $CaCO_3$ contributes to important processes, consolidation, and strengthening of reef structures, and also lithification of sediments. The $CaCO_3$ cement is produced in the form of particles not only within the void of reef structure, and in the sediments, but also on the exterior of carbonate particles (Morse and Mucci 1984; Morse and Mackenzie 1990). Additionally, some microbial communities and algae are also related to the formation of cement (Morse and Mackenzie 1990). The hypothesis of many studies suggested that the whitings which give a milky white appearance are suspended fine grain particles which are originated from the resuspension of sediments from the seafloor by different mechanisms like feeding fish. The carbonate muds which are sometimes originated by the breakdown or recrystallization of biogenic calcareous hard parts are one of the major constituents of shallow marine carbonate sediments. Some studies have even suggested that aragonite needles which are precipitated by the green algae also contribute a fraction of mud to the sediments (Neumann and Land 1975; Reid and Macintyre 1998).

Other than this, the extensive amount of abiotic carbonate deposits is mostly allied with the methane seeps, which are formed as a result of anaerobic oxidation of methane mediated by the methanotrophic archaea and sulfate-reducing bacteria, which consequently elevate the TA and saturation state of water in context with $CaCO_3$ and hence stimulate precipitation (Peckmann and Thiel 2004):

$$CH_4 + SO_4 = HCO_3 + HS + H_2O \tag{10.32}$$

The extent of the carbonate deposited by methane seeps ranges from a few centimeters to a few meters, but their contribution to the widespread $CaCO_3$ budget is still unknown.

10.3.3 Destruction and Dissolution of $CaCO_3$

The reverse process of calcification is referred to as dissolution of $CaCO_3$ as it leads to disintegration and breakdown of solid $CaCO_3$ into different components like Ca^{2+}, CO_3^{2-}, and Mg^{2+} (if present), which dissolve in the solution (Alkhatib 2016). As suggested by the principles of thermodynamics, at $\Omega < 1$ the net dissolution is expected, nevertheless $CaCO_3$ dissolution being the continuous process is a ubiquitous process that occurs in almost all environments, even in the conditions where the sweater saturation state is well above the thermodynamic equilibrium threshold, and this is because of the various factors like (Morse et al. 2007):

- Microbial decomposition of organic material which will result into corrosive seawater conditions in the microenvironments and the reefs and sediment pore water
- Bio-eroders that vigorously chew or penetrate the $CaCO_3$, through different approaches and mechanisms.

However, in terms of $CaCO_3$, majority of the deep sea is undersaturated, therefore it dissolves and progressively disappears from the sediments as a function of depth unless there is a balance between the removal and supply rate of $CaCO_3$. The microbial decomposition and subsequent $CaCO_3$ dissolution are referred to as metabolic dissolution (Burdige et al. 2008), the organic matter microbial decomposition results in the production of CO_2, due to which the sea become undersaturated with respect to $CaCO_3$ mineral phases, which eventually start dissolving. The magnitude of dissolution is primarily related to the quantity of organic material which is remineralized (Sekula-Wood et al. 2012). And it is expressed as:

$$CaCO_3 + CH_2O + O_2 = Ca^{2+} + 2HCO_3^- \tag{10.33}$$

The property of dissolved inorganic carbon can be used to demonstrate the amount of dissolution under the aerobic condition as a function of TA (Morse and Mackenzie 1990).

10.3.3.1 Bio-erosion

Bio-erosion refers to the breakdown of $CaCO_3$ by different organisms which can rasp, chew, secrete acids, and mechanically erode or dissolve $CaCO_3$ into smaller fragments (Alexandersson 1975). The first step that facilitates dissolution is a mechanical breakdown to smaller fragments, by sea urchins or grazing fish; this broadens the surface area to volume ratio of particles and hence increases the vulnerability to dissolution (Walter and Morse 1984). Along with large bioeroders, there is a multitude of microorganisms called endoliths, which penetrate $CaCO_3$ grains and structures by secretion of CO_2 and other acids. These endoliths not only dissolve $CaCO_3$ but also, due to their photosynthetic nature, significantly contribute to the production of organic material in coral reefs and other environments (Aline et al. 2006).

10.4 Present-Day Changes in the Seawater CO_2: Carbonic Acid System Due to Anthropogenic Activities

In current times, the natural cycle of carbon in the seawater has been significantly altered because of the superfluous influx of carbon due to unsustainable anthropogenic activities. Presently the atmospheric CO_2 has been increased to ~402 ppm, which is 40% from 589 Pg C in 1750 to 849 ± 5 Pg C in 2015 with an average increase of ~ 0.98 Pg C/yr (Ciais et al. 2014; Le Quéré et al. 2015). As already discussed, oceans act as the sink for ~30% of all the anthropogenic carbon emissions, which reduces its concentration in the atmosphere and consequently slow down global climate change. Even though the contribution of inland waters contributes ~1% of Earth's surface, its collective contribution to the global carbon cycle is very significant. Around 2.7 Pg carbon is mineralized, transported, and buried annually by inland waters, which is analogous to the sink for anthropogenic CO_2 ~ 2.8 Pg C/yr (Battin et al. 2009; Tranvik et al. 2009). The exchange of CO_2 between the ocean (or any aquatic body) and the atmosphere is driven by three different factors which are chemical effect which regulates the CO_2 solubility by the temperature-salinity variations; biological effect from the water surface which owe to the fixation, organic carbon export, and precipitation or dissolution of biogenic carbonate; and advection and mixing (Roy et al. 2011; Ciais et al. 2014). For the year 2015, the net decadal uptake of CO_2 by ocean has been estimated to be 3.0 ± 0.5 Pg C/yr. But after the beginning of postindustrial era, because of elevated concentration of carbon dioxide and methane, the surface temperature of land and ocean has been increased by ~ 0.85 ± 0.21 °C globally over the timescale of 1880–2014 (IPCC 2014).

Based on various reports, it has been suggested that raised levels of atmospheric CO_2 lead to the CO_2 fertilization effect which is defined as the increase in photosynthesis in leaf (Zak et al. 2011; De Kauwe et al. 2013). Another most important and severe impact of rising levels of CO_2 is ocean acidification, which is a decline in pH and reduction in CO_3^{2-} mineral saturation state. In the past 200 years, the ocean has absorbed ~ 40% of atmospheric CO_2 which has led to a pH drop of ~ 0.1 of the world's oceans which is the same as the 30% increase in ocean acidity (Caldeira and Wickett 2003; Zeebe et al. 2008; Johnson and White 2014). Moreover, anthropogenic CO_2-induced global warming has also led to an increase in water column stratification (Manabe and Stouffer 1993). The investigation on several estuaries and coastal ecosystem of India has been done in context of CO_2, and they suggested that these ecosystems are source of CO_2, i.e. the flux of CO_2 is from water to atmosphere (Sarma et al. 2001; Bouillon et al. 2003; Biswas et al. 2004). Very high bacterial respiration (20.7 ± 0.7 μmol C/L/d) has been observed in the Godavari estuary leading to very high partial pressure of CO_2 (pCO_2) of ~ 30,000 μatm during peak discharge period (Sarma et al. 2011). Not only estuaries, but coastal zones also have pCO_2 more than atmospheric value suggesting heterotrophic activity, especially during peak discharge time (Prasad et al. 2013). Additionally, Arabian Sea, which is a part of the Indian ocean, is also a perennial source of CO_2 as reported by many authors (Goyet et al. 1998; Sarma et al. 2001).

10.5 Impact on Marine $CaCO_3$ Cycle Owing to Anthropogenic Ocean Acidification

10.5.1 Formation of $CaCO_3$

Based on thermodynamics point of view, it is expected that there will be a decline in the production rate of $CaCO_3$ with the decreasing seawater Ω. And this has also been proven by most of the experiments that there is a decrease in the calcification rate of calcifying marine biogenics with escalating CO_2 levels and decreasing Ω (Doney et al. 2009; Erez et al. 2011; Andersson and Mackenzie 2011). Other than this, recent studies have also shown that ocean acidification also significantly affects the biogenic calcification of organisms (Kroeker et al. 2010). Also, it is not very much evident that which factor of the CO_2 carbonic system has the major role in controlling the calcification rates in organisms, whether it is $[CO_3^{2-}]$, O, $[H^+]$, or $[HCO_3^-]$ (Allemand et al. 2011). But, most likely ocean acidification and decreasing Ω are the major factors that reduce marine calcification.

Additionally, considering the significant role and importance of coccolithophorids for both organic as well as inorganic carbon production, various studies have been carried out on them, the majority of which have concluded that the rate of calcification in coccolithophorids decreases in response to increasing CO_2 level and decreasing pH and Ω, even though the extent of the response can vary (Feng et al. 2008; Gao et al. 2009). Thus, by extrapolation of these results in the natural environment, it can be said that pelagic $CaCO_3$ production could decrease by 10–70% by the end of the twenty-first century (Kleypas et al. 2005).

Similar to the pelagic calcifiers, several studies have shown that the ocean acidification also have a negative effect on the rate of calcification of benthic calcifiers (Doney et al. 2009; Andersson and Mackenzie 2011). Mostly the studies are carried out on tropical and subtropical corals, but large experiments have also been conducted with bivalves, gastropods, echinoderms, and coralline algae, and their results have shown consistent outcomes indicating a drop in calcification rate under the elevated levels of CO_2. But some studies contradicted these results by indicating no effect or even enhanced calcification in eminent CO_2 conditions (Ries et al. 2009); however, this enhanced calcification to compensate for ocean acidification takes place at the cost of a lot of energy loss, which consequently impacts the essential processes like protein synthesis, growth, and/or reproduction

(Wood et al. 2008; Findlay et al. 2011). Hence regardless of the opposing responses from certain calcifiers, it can be concluded that the ocean acidification has consistently the negative impact, and extrapolation of observed result suggests that by the end of the century, individual coral colony calcification can drop by 10–90% if the same CO_2 emission scenario persists.

10.5.2 Dissolution of $CaCO_3$

Another major effect of ocean acidification is its potential to change the ocean biogeochemical dynamic for both organic carbon and nutrients via indirect pathways. For example, increased dissolution of $CaCO_3$ reduces the influence of $CaCO_3$ on the ballasting of organic carbon to the deep sea, as it causes more remineralization of organic carbon in shallow water (Armstrong et al. 2001; Klaas and Archer 2002). As a result, the carbonate dissolution process is crucial because it decreases the ocean's CO_2 uptake efficiency and hence acts as a negative feedback to increasing CO_2 and lowering pH. However, this also produces alkalinity, which further increases the capacity of the water to absorb CO_2 (Wang et al. 2016); therefore, it is important to have an understanding of the process of dissolution, as it tells about the rate and the extent of atmospheric CO_2 absorbed by the oceans, the rate at which absorption of the anthropogenic CO_2 lead to changes in the surface seawater chemistry, whether the dissolution of CO_2 can act as a buffer to alleviate changes in the pH of the seawater and saturation state of carbonate imposed as a result of rising CO_2 level, and whether the ocean acidification will adversely affect the carbonate dominant communities and ecosystem or not.

With the penetration of anthropogenic CO_2 deeper into the oceans, the carbonate minerals saturation horizon will move upward in the water column, increasing the area of the seafloor and quantities of $CaCO_3$ which are immersed in the seawater undersaturated with respect to these minerals in the seawater. Based on different studies it has been suggested that in high-latitude environment, the surface seawater could become undersaturated in context with aragonite by the middle of the twenty-first century (Orr 2011). In temperate and tropical latitudes, it is expected that the seawater could already be near the equilibrium with Mg-calcite of the bulk composition in the shallow-water sediments (Andersson et al. 2008). As a result of which, the dissolution of seafloor carbonate sediments is expected to escalate and the lysocline and the carbonate compensation depth will get shallow as a result of ocean acidification. In the coming thousands of years, the dissolution of carbonate minerals will be the ultimate sink for CO_2, the buffer, and the antacid of the world (Archer et al. 1998; Broecker 2003).

The area which remains supersaturated with the carbonate minerals leads to an increase in the dissolution of $CaCO_3$ due to ocean acidification and decreasing seawater Ω. However, it is not just the only $CaCO_3$ sediments and substrates that may undergo dissolution because of ocean acidification, but if the seawater Ω and the pH drop significantly, some organisms may also find it hard to sustain their calcareous structures. For instance, the living marine gastropods show the visual representation of shell dissolution in the environment with high seawater CO_2 and low pH because of volcanic venting (Hall-Spencer et al. 2008). Although there are organisms that have protective coating or tissues which serve as the protective layer from dissolution (Morse and Mackenzie 1990), other physiological processes such as metabolism, health, and behavior were negatively affected under the high levels of CO_2, perhaps as an outcome of upregulation of the calcification process (Findlay et al. 2011).

Additionally, it has been proposed that to counteract the anthropogenic changes in seawater pH and increase in the ocean's capacity to take up CO_2, the dissolution of carbonate minerals in the shallow water could act as a buffer by the generation of TA (Garrels and Mackenzie 1981). The dissolution of Ca, CO_2 minerals in the global ocean surface needs to be balanced with the net invasion of anthropogenic CO_2 (~170 × 10^{12} mol C/yr), so that the surface water pH and the carbonate saturation state against the atmospheric CO_2 can be buffered. Even if the dissolution of $CaCO_3$ in the layer of sediment was equal to the current rate of $CaCO_3$ production (~95 × 10^{12} mol C/yr) (Wollast 1994), the surface water would still only be partially buffered. But the dissolution of this magnitude is not very possible, as most of the $CaCO_3$ which are produced in pelagic environments mostly absorb and dissolve at the deeper depths due to the oversaturation of the surface water with the carbonate minerals (Milliman et al. 1999). Therefore, due to the fast and the short residence time (2–3 years) of the physical exchange of water between the shallow-water ocean environment and the open sea (Chavez 1995), it is necessary for the carbonate minerals to undergo substantial dissolution, so that it can produce sufficient alkalinity to counteract any alterations in the surface water chemistry owing to escalated atmospheric CO_2.

10.6 Biological, Ecological, and Biogeochemical Impacts

The successive decline in the surface water carbonate saturation state and elevated atmospheric CO_2 (causing ocean acidification) have a detrimental impact on calcifying organisms. This is due to the fact that the ability of these organisms to calcify is dependent on the carbonate saturation state. It has been very evidently demonstrated by many studies that the rate of calcification and the seawater carbonate saturation state are directly related to each other for various organisms like coccolithophorids (Riebesell et al. 2000; Zondervan et al. 2001), foramnifera (Spero et al. 1997; Bijma et al. 1999), coralline algae (Smith and Roth 1979; Borowitzka 1981), and scleractinian corals (Marubini and Thake 1999; Reynaud et al. 2003). Other than this, there have been various studies conducted on the calcareous communities, which sometimes showed substantial disparities between species and communities, but majority of studies have concluded similar results that is the rate of calcification decreases as a result of rising pCO_2 or decreasing carbonate saturation state. Based on these results and the expected projections of surface water carbonate saturation, it is expected that the calcification of marine biogenics may decline by 11–44%because of doubling of the atmospheric CO_2 levels relative to preindustrial conditions (Langdon 2002). But it is also important to understand whether this lower carbonate saturation state and decreased rate of calcification will have adverse effect on the calcifying organisms and corals ecologically or not. For this, one need to have understanding of mechanism of calcification and its function in calcifying organism. The obvious advantages of calcification include functions like providing structural support, protection from predators and desiccation, and increase in surface area; other than this, several studies have also suggested that calcification can promote nutrient and /or bicarbonate uptake (McConnaughey and Whelan 1997). Due to decline in surface water carbonate saturation state, the physical strength, and the structure of skeletons, shells and tests which accumulated marine calcifiers may become more fragile and vulnerable to environmental stress, both due to natural and anthropogenic reasons (Kleypas et al. 1999; Buddemeier et al. 2004). Hence, it can be said that calcareous organism may not thrive in the world with high CO_2 levels and low carbonate saturation state and may also be at a competitive disadvantage with the non-calcifying organisms in their environment that occupy same habitat. For instance, it is anticipated that habitat which is dominant by the coral reef might undergo successive transition from being dominated by the coral and coralline algae to being dominated by seaweed and fleshy macroalgae. According to Hamilton et al. (2017), coral reef habitat reduced from 3528 to 1941 ha in Solomon islands, and other numerous studies have suggested similar significant decline in coral reef health along with future climate projections which can lead to a global coral reef crisis.

10.7 Risk Mitigation and Adaptation Strategies

The threats which are associated with ocean acidification to marine organism, ecosystem, and human society include not just the probability of ocean to affect fundamental, physiological, and ecological processes of organisms but also the magnitude of resulting impacts on ecosystem and services it provide to the society. For example, it is well known that ocean acidification impacts formation as well as maintenance of coral reefs and consequently the goods and services they provide like fisheries and tourism, and coastal protection (Hoegh-Guldberg et al. 2007). Based on the quantitative analysis of ocean acidification in monetary terms, it has been estimated that the annual economic damage induced by ocean acidification by 2100 is around US$100 billion (Andersson et al. 2008), which represents the very large GDP loss for the economies of many coastal regions or small islands which are dependent on the ecological goods and service of coral reefs. All these impacts associated with ocean acidification point out the urgency of implementation of some mitigation measures which can minimize these losses and adversity

There are two general approaches for the successful management of the impact of ocean acidification: reducing the anthropogenic CO_2 emissions and/or adaption by reducing the impacts of past and future ocean acidification (Billé et al. 2013). Reduction of atmospheric CO_2 to mitigate ocean acidification is the most effective and least risky method to limit ocean acidification and its impacts (Kleypas et al. 2005). Solar radiation management-based climate geoengineering techniques are not very effective in abatement of ocean acidification and might result in increase in ocean acidification. These techniques are very costly which may limit their functionality due to lack of CO_2 storage capacity. In addition to this, techniques like iron fertilization can also be considered, but they only have the ability to relocate the ocean acidification from the upper region of the ocean to its interior, with probable ramification on deep water oxygen levels (Board 2015). Other than this, approaches like low-regret approach can be used which focus on reducing the number of drivers other than diminishing CO_2 levels (Kleypas et al. 2005). At local level, mitigation of ocean acidification involves reducing nutrients and organic matter in oceans due to anthropogenic sources. Additionally, some adaptation strategies which include drawing water from local watershed from aquaculture at right pH range and selecting less-sensitive species or relocation of industries to some other places can also be considered (Eknath and Hulata 2009).

10.8 Conclusion

According to the current study, uptake of anthropogenic carbon dioxide lowers the pH as well as CO_3^{2-} concentration. It also lowers the saturation state of aragonite and calcite in the upper ocean, causing ocean acidification which decreases the pH of ocean to 0.3–0.4 units relative to preindustrial values (Cao and Caldeira 2010). However, based on the best available data and model projections, as dissolved CO_2 levels in seawater rise, skeletal growth of organisms which secrete calcium will slow down, making the oceans less conducive to the production and preservation of $CaCO_3$ minerals. High CO_2 concentration along with low water pH will indeed negatively affect $CaCO_3$ production and preservation, as discussed in this chapter. And, because anthropogenic CO_2 emissions show no signs of slowing (for example, CO_2 emissions increased by 5% in 2010), CO_2 concentrations in the ocean will increase leading to them becoming increasingly acidic. Based on the findings reviewed in this chapter, it can be predicted that $CaCO_3$ production will decrease in both shallow and open-ocean conditions, $CaCO_3$ dissolution will rise, and $CaCO_3$ preservation in seabed sediments would decrease. Marine communities dominated by calcifiers in both open-ocean and shallow-water environments could change in favor of noncalcifying organisms and coral reefs, for example, could become increasingly dominated by non-calcifying macroalgae. Hence, to adapt to the long-term changes associated with altering carbon dioxide and carbonate concentration in oceans, proper temporal and spatial studies of atmospheric and surface water pCO_2 should be done. Furthermore, long-term studies of pH, alkalinity of oceans, carbonate system perturbation and calcification experiments following standardized methodologies and data reporting criteria should also be carried out. Consequently, methods for investigating and addressing long-term subtle changes that far more closely imitate natural circumstances are required to aid policymakers in formulating management plans to adapt, mitigate, and minimize the negative effects of ocean acidification.

References

Agrawal, A. and Gopal, K., 2013. Biomonitoring of Water and Waste Water. Springer Science & Business Media.

Alexandersson, E.T., 1975. Etch patterns on calcareous sediment grains: petrographic evidence of marine dissolution of carbonate minerals. *Science*, 189(4196), pp.47–48.

Aline, T., Atkinson, M.J. and Christopher, L., 2006. Effects of elevated pCO2 on epilithic and endolithic metabolism of reef carbonates. *Global Change Biology*, 12(11), pp.2200–2208.

Alkhatib, M., 2016. Trace element and isotope measurement on artificially precipitated calcium carbonate (Doctoral dissertation, Christian-Albrechts-Universität).

Allemand, D., Tambutté, É., Zoccola, D. and Tambutté, S., 2011. Coral calcification, cells to reefs. In: Dubinsky, Z., Stambler, N. (eds) Coral Reefs: An Ecosystem in Transition (pp.119–150). Springer, Dordrecht.

Andersson, A.J. and Mackenzie, F.T., 2011. Ocean acidification: setting the record straight. *Biogeosciences Discussions*, 8(3), pp.6161–6190.

Andersson, A.J., Mackenzie, F.T. and Bates, N.R., 2008. Life on the margin: implications of ocean acidification on Mg-calcite, high latitude and cold-water marine calcifiers. *Marine Ecology Progress Series*, 373, pp.265–273.

Archer, D., Kheshgi, H. and Maier-Reimer, E., 1998. Dynamics of fossil fuel CO_2 neutralization by marine $CaCO_3$. *Global Biogeochemical Cycles*, 12(2), pp.259–276.

Armstrong, R.A., Lee, C., Hedges, J.I., Honjo, S. and Wakeham, S.G., 2001. A new, mechanistic model for organic carbon fluxes in the ocean based on the quantitative association of POC with ballast minerals. *Deep Sea Research Part II: Topical Studies in Oceanography*, 49(1–3), pp.219–236.

Battin, T.J., Luyssaert, S., Kaplan, L.A., Aufdenkampe, A.K., Richter, A. and Tranvik, L.J., 2009. The boundless carbon cycle. *Nature Geoscience*, 2(9), pp.598–600.

Bijma, J., Spero, H.J. and Lea, D.W., 1999. Reassessing foraminiferal stable isotope geochemistry: Impact of the oceanic carbonate system (experimental results). In: Fischer, G., Wefer, G. (eds) Use of Proxies in Paleoceanography (pp. 489–512). Springer, Berlin, Heidelberg.

Billé, R., Kelly, R., Biastoch, A., Harrould-Kolieb, E., Herr, D., Joos, F., Kroeker, K., Laffoley, D., Oschlies, A. and Gattuso, J.P., 2013. Taking action against ocean acidification: a review of management and policy options. *Environmental management*, 52(4), pp.761–779.

Biswas, H., Mukhopadhyay, S.K., De, T.K., Sen, S. and Jana, T.K., 2004. Biogenic controls on the air – water carbon dioxide exchange in the Sundarban mangrove environment, northeast coast of Bay of Bengal, India. *Limnology and Oceanography*, 49(1), pp.95–101.

Board, O.S. and National Research Council, 2015. Climate Intervention: Carbon Dioxide Removal and Reliable Sequestration. National Academies Press.

Borowitzka, M.A., 1981. Photosynthesis and calcification in the articulated coralline red algae *Amphiroa anceps* and *A. Foliacea*. *Marine Biology*, 62(1), pp.17–23.

Bouillon, S., Frankignoulle, M., Dehairs, F., Velimirov, B., Eiler, A., Abril, G., Etcheber, H. and Borges, A.V., 2003. Inorganic and organic carbon biogeochemistry in the Gautami Godavari estuary (Andhra Pradesh, India) during pre-monsoon: the local impact of extensive mangrove forests. *Global Biogeochemical Cycles*, 17(4), p.1114.

Bradshaw, A.L. and Brewer, P.G., 1988. High precision measurements of alkalinity and total carbon dioxide in seawater by potentiometric titration—1. Presence of unknown protolyte (s)?. *Marine Chemistry*, 23(1–2), pp. 69–86.

Broecker, W.S., 2003. Does the trigger for abrupt climate change reside in the ocean or in the atmosphere?. *Science*, 300(5625), pp.1519–1522.

Buddemeier, R.W., Kleypas, J.A. and Aronson, R.B., 2004. Potential Contributions of Climate Change to Stresses on Coral Reef Ecosystems. Coral Reefs and Global Climate Change. Pew Center on Global Climate Change, Virginia, USA.

Burdige, D.J., Zimmerman, R.C. and Hu, X., 2008. Rates of carbonate dissolution in permeable sediments estimated from pore-water profiles: the role of sea grasses. *Limnology and Oceanography*, 53(2), pp.549–565.

Caldeira, K. and Wickett, M.E., 2003. Anthropogenic carbon and ocean pH. *Nature*, 425(6956), pp.365–365.

Canadell, J.G., Le Quéré, C., Raupach, M.R., Field, C.B., Buitenhuis, E.T., Ciais, P., Conway, T.J., Gillett, N.P., Houghton, R.A. and Marland, G., 2007. Contributions to accelerating atmospheric CO_2 growth from economic activity, carbon intensity, and efficiency of natural sinks. *Proceedings of the National Academy of Sciences*, 104(47), pp.18866–18870.

Cao, L. and Caldeira, K., 2010. Atmospheric carbon dioxide removal: long-term consequences and commitment. *Environmental Research Letters*, 5(2), p.024011.

Chavez, F.P., 1995. Physical estimates of global new production: the upwelling contribution. *Upwelling in the Ocean: Modern Processes and Ancient Records.*

Ciais, P., Sabine, C., Bala, G., Bopp, L., Brovkin, V., Canadell, J., Chhabra, A., DeFries, R., Galloway, J., Heimann, M. and Jones, C., 2014. Carbon and other biogeochemical cycles. In Climate change 2013: the physical science basis. Contribution of Working Group I to the Fifth Assessment Report of the Intergovernmental Panel on Climate Change (pp. 465–570). Cambridge University Press.

Clark, I., 2015. Groundwater Geochemistry and Isotopes. CRC Press.

De Kauwe, M.G., Medlyn, B.E., Zaehle, S., Walker, A.P., Dietze, M.C., Hickler, T., Jain, A.K., Luo, Y., Parton, W.J., Prentice, I.C. and Smith, B., 2013. Forest water use and water use efficiency at elevated CO_2: a model-data intercomparison at two contrasting temperate forest FACE sites. *Global Change Biology*, 19(6), pp.1759–1779.

Dickson, A.G., 1981. An exact definition of total alkalinity and a procedure for the estimation of alkalinity and total inorganic carbon from titration data. *Deep Sea Research Part A. Oceanographic Research Papers*, 28, 609. https://doi.org/10.1016/0198-0149(81)90121-7.

Dickson, A.G., 1984. pH scales and proton-transfer reactions in saline media such as sea water. *Geochimica et Cosmochimica Acta*, 48(11), pp.2299–2308.

Dickson, A.G., Sabine, C.L. and Christian, J.R., 2007. Guide to Best Practices for Ocean CO_2 Measurements. North Pacific Marine Science Organization.

Doney, S.C., Fabry, V.J., Feely, R.A. and Kleypas, J.A., 2009. Ocean acidification: the other CO_2 problem. *Annual Review of Marine Science*, 1, pp.169–192.

Eknath, A.E. and Hulata, G., 2009. Use and exchange of genetic resources of Nile tilapia (Oreochromis niloticus). *Reviews in Aquaculture*, 1(3-4), pp.197–213.

Erez, J., Reynaud, S., Silverman, J., Schneider, K. and Allemand, D., 2011. Coral calcification under ocean acidification and global change. In: Dubinsky, Z., Stambler, N. (eds) Coral Reefs: An Ecosystem in Transition (pp. 151–176). Springer, Dordrecht.

Fabry, V.J., 1989. Aragonite production by pteropod molluscs in the subarctic Pacific. *Deep Sea Research Part A: Oceanographic Research Papers*, 36(11), pp.1735–1751.

Fabry, V.J., 1990. Shell growth rates of pteropod and heteropod molluscs and aragonite production in the open ocean: implications for the marine carbonate system. *Journal of Marine Research*, 48(1), pp.209–222.

Falkowski, P.G., Barber, R.T. and Smetacek, V., 1998. Biogeochemical controls and feedbacks on ocean primary production. *Science*, 281(5374), 200–206.

Feng, Y., Warner, M.E., Zhang, Y., Sun, J., Fu, F.X., Rose, J.M. and Hutchins, D.A., 2008. Interactive effects of increased pCO_2, temperature and irradiance on the marine coccolithophore *Emiliania huxleyi* (Prymnesiophyceae). *European Journal of Phycology*, 43(1), pp.87–98.

Findlay, H.S., Wood, H.L., Kendall, M.A., Spicer, J.I., Twitchett, R.J. and Widdicombe, S., 2011. Comparing the impact of high CO_2 on calcium carbonate structures in different marine organisms. *Marine Biology Research*, 7(6), pp.565–575.

Friedlingstein, P., Jones, M.W., O'sullivan, M., Andrew, R.M., Hauck, J., Peters, G.P., Peters, W., Pongratz, J., Sitch, S., Le Quéré, C. and Bakker, D.C., 2019. Global carbon budget 2019. *Earth System Science Data*, 11(4), pp.1783–1838.

Gao, K., Ruan, Z., Villafane, V.E., Gattuso, J.P. and Helbling, E.W., 2009. Ocean acidification exacerbates the effect of UV radiation on the calcifying phytoplankter *Emiliania huxleyi*. *Limnology and Oceanography*, 54(6), pp.1855–1862.

Garrels, R.M. and Mackenzie, F.T., 1981. Carbon dioxide effects research and assessment program. Some aspects of the role of the shallow ocean in global carbon dioxide uptake (No. CONF-8003115-). Northwestern Univ., Evanston, IL (USA).

Gattuso, J.P. and Hansson, L., 2011. Ocean acidification: background and history. In: Gattuso, J.P., Hansson, L. (eds.) Ocean Acidification (pp.1–20). Oxford University Press, New York.

Goyet, C., Millero, F.J., O'Sullivan, D.W., Eischeid, G., McCue, S.J. and Bellerby, R.G.J., 1998. Temporal variations of pCO_2 in surface seawater of the Arabian Sea in 1995. *Deep Sea Research Part I: Oceanographic Research Papers*, 45(4–5), pp.609–623.

Gruber, N., Clement, D., Carter, B.R., Feely, R.A., Van Heuven, S., Hoppema, M., Ishii, M., Key, R.M., Kozyr, A., Lauvset, S.K. and Lo Monaco, C., 2019. The oceanic sink for anthropogenic CO_2 from 1994 to 2007. *Science*, 363(6432), pp.1193–1199.

Guinotte, J.M. and Fabry, V.J., 2008. Ocean acidification and its potential effects on marine ecosystems. *Annals of the New York Academy of Sciences*, 1134(1), 320–342.

Hall-Spencer, J.M., Rodolfo-Metalpa, R., Martin, S., Ransome, E., Fine, M., Turner, S.M., Rowley, S.J., Tedesco, D. and Buia, M.C., 2008. Volcanic carbon dioxide vents show ecosystem effects of ocean acidification. *Nature*, 454(7200), pp.96–99.

Hamilton, R.J., Almany, G.R., Brown, C.J., Pita, J., Peterson, N.A. and Howard Choat, J., 2017. Logging degrades nursery habitat for an iconic coral reef fish. *Biological Conservation*, 210, 273–280. https://doi.org/10.1016/j.biocon.2017.04.024

Hoegh-Guldberg, O., Mumby, P.J., Hooten, A.J., Steneck, R.S., Greenfield, P., Gomez, E., Harvell, C.D., Sale, P.F., Edwards, A.J., Caldeira, K. and Knowlton, N., 2007. Coral reefs under rapid climate change and ocean acidification. *Science*, 318(5857), pp.1737–1742.

Honjo, S., 2004. Particle export and the biological pump in the Southern Ocean. *Antarctic Science*, 16(4), pp.501–516.

IPCC, 2007. In: Solomon, S., Qin, D., Manning, M., Chen, Z., Marquis, M., Averyt, K.B., Tignor, M., Miller H.L. (eds.) Climate Change 2007: The Physical Science Basis. Contribution of Working Group I to the Fourth Assessment Report of the Intergovernmental Panel on Climate Change. Cambridge University Press, Cambridge.

IPCC, 2014. Climate change 2014: synthesis report. Contribution of working groups I, II and III to the fifth assessment report of the intergovernmental panel on climate change. Intergovernmental Panel on Climate Change, Geneva, Switzerland.

Johnson, A. and White, N.D., 2014. Ocean acidification: the other climate change issue. *American Scientist*, 102(1), pp.60–63.

Johnson, V.R., Russell, B.D., Fabricius, K.E., Brownlee, C. and Hall-Spencer, J.M., 2012. Temperate and tropical brown macroalgae thrive, despite decalcification, along natural CO_2 gradients. *Global Change Biology*, 18(9), 2792–2803.

Klaas, C. and Archer, D.E., 2002. Association of sinking organic matter with various types of mineral ballast in the deep sea: implications for the rain ratio. *Global Biogeochemical Cycles*, 16(4), pp.63-1–63-14.

Kleypas, J.A., Buddemeier, R.W., Archer, D., Gattuso, J.P., Langdon, C. and Opdyke, B.N., 1999. Geochemical consequences of increased atmospheric carbon dioxide on coral reefs. *Science*, 284(5411), pp.118–120.

Kleypas, J.A., Feely, R.A., Fabry, V.J., Langdon, C., Sabine, C.L. and Robbins, L.L., 2005. Impacts of ocean acidification on coral reefs and other marine calcifiers: a guide for future research. In Report of a workshop held (Vol. 18, No. 2005, p. 20).

Kroeker, K.J., Kordas, R.L., Crim, R.N. and Singh, G.G., 2010. Meta-analysis reveals negative yet variable effects of ocean acidification on marine organisms. *Ecology Letters*, 13(11), pp.1419–1434.

Kroeker, K.J., Kordas, R.L., Crim, R., Hendriks, I.E., Ramajo, L., Singh, G.S., Duarte, C.M. and Gattuso, J.P., 2013. Impacts of ocean acidification on marine organisms: quantifying sensitivities and interaction with warming. *Global Change Biology*, 19(6), pp.1884–1896.

Land, L.S., 1967. Diagenesis of skeletal carbonates. *Journal of Sedimentary Research*, 37(3), pp.914–930.

Langdon, C., 2002. Review of experimental evidence for effects of CO_2 on calcification of reef builders. In Proceedings of 9th International Coral Reef Symposium (Vol. 2, No. 2002, pp. 1091–1098).

Le Quéré, C., Raupach, M.R., Canadell, J.G., Marland, G., Bopp, L., Ciais, P., Conway, T.J., Doney, S.C., Feely, R.A., Foster, P. and Friedlingstein, P., 2009. Trends in the sources and sinks of carbon dioxide. *Nature Geoscience*, 2(12), pp.831–836.

Le Quéré, C., Moriarty, R., Andrew, R.M., Canadell, J.G., Sitch, S., Korsbakken, J.I., Friedlingstein, P., Peters, G.P., Andres, R.J., Boden, T.A. and Houghton, R.A., 2015. Global carbon budget 2015. *Earth System Science Data*, 7(2), pp.349–396.

Lebrato, M., Andersson, A.J., Ries, J.B., Aronson, R.B., Lamare, M.D., Koeve, W., . . . & McClintock, J. B., (2016). Benthic marine calcifiers coexist with $CaCO_3$-undersaturated seawater worldwide. *Global Biogeochemical Cycles*, 30(7), 1038–1053.

Lüthi, D., Le Floch, M., Bereiter, B., Blunier, T., Barnola, J.M., Siegenthaler, U., Raynaud, D., Jouzel, J., Fischer, H., Kawamura, K. and Stocker, T.F., 2008. High-resolution carbon dioxide concentration record 650,000–800,000 years before present. *Nature*, 453(7193), pp.379–382.

Mackenzie, F.T., Andersson, A.J., Lerman, A. and Ver, L.M., 2005. Boundary exchanges in the global coastal margin: implications for the organic and inorganic carbon cycles. *The Sea*, 13, pp.193–225.

Manabe, S. and Stouffer, R.J., 1993. Century-scale effects of increased atmospheric CO_2 on the ocean–atmosphere system. *Nature*, 364(6434), pp.215–218.

Marubini, F. and Thake, B., 1999. Bicarbonate addition promotes coral growth. *Limnology and Oceanography*, 44(3), pp.716–720.

Marx, A., Dusek, J., Jankovec, J., Sanda, M., Vogel, T., Van Geldern, R., . . . & Barth, J.A.C., (2017). A review of CO_2 and associated carbon dynamics in headwater streams: a global perspective. *Reviews of Geophysics*, 55(2), 560–585.

McConnaughey, T.A. and Whelan, J.F., 1997. Calcification generates protons for nutrient and bicarbonate uptake. *Earth-Science Reviews*, 42(1–2), pp.95–117.

Millero, F.J., 1995. Thermodynamics of the carbon dioxide system in the oceans. *Geochimica et Cosmochimica Acta*, 59(4), 661–677.

Milliman, J.D., 1993. Production and accumulation of calcium carbonate in the ocean: budget of a nonsteady state. *Global Biogeochemical Cycles*, 7(4), pp.927–957.

Milliman, J.D. and Droxler, A.W., 1996. Neritic and pelagic carbonate sedimentation in the marine environment: ignorance is not bliss. *Geologische Rundschau*, 85(3), pp.496–504.

Milliman, J.D., Troy, P.J., Balch, W.M., Adams, A.K., Li, Y.H. and Mackenzie, F.T., 1999. Biologically mediated dissolution of calcium carbonate above the chemical lysocline?. *Deep Sea Research Part I: Oceanographic Research Papers*, 46(10), pp.1653–1669.

Morse, J.W. and Mackenzie, F.T., 1990. Geochemistry of Sedimentary Carbonates. Elsevier.

Morse, J.W. and Mucci, A., 1984. Composition of carbonate overgrowths produced on Iceland spar calcite crystals buried in Bahamian carbonate-rich sediments. *Sedimentary Geology*, 40(4), pp.287–291.

Morse, J.W., Arvidson, R.S. and Lüttge, A. 2007. Calcium carbonate formation and dissolution. *Chemical Reviews*, 107(2), 342–381.

Mucci, A., 1983. The solubility of calcite and aragonite in seawater at various salinities, temperatures, and one atmosphere total pressure. *American Journal of Science*, 283(7), pp.780–799.

Nelson, W.A., 2009. Calcified macroalgae–critical to coastal ecosystems and vulnerable to change: a review. *Marine and Freshwater Research*, 60(8), pp.787–801.

Neumann, A.C. and Land, L.S., 1975. Lime mud deposition and calcareous algae in the Bight of Abaco, Bahamas; a budget. *Journal of Sedimentary Research*, 45(4), pp.763–786.

Neven, I.A., Stefels, J., van Heuven, S.M., de Baar, H.J. and Elzenga, J.T.M., 2011. High plasticity in inorganic carbon uptake by Southern Ocean phytoplankton in response to ambient CO_2. *Deep Sea Research Part II: Topical Studies in Oceanography*, 58(25–26), pp.2636–2646.

Orr, J.C., 2011. Recent and future changes in ocean carbonate chemistry. In: Gattuso, J.-P., Hansson, L. (eds.) Ocean Acidification, 1 (pp.41–66). Oxford University Press, New York.

Peckmann, J. and Thiel, V., 2004. Carbon cycling at ancient methane–seeps. *Chemical Geology*, 205(3–4), pp.443–467.

Plattner, G.K., Joos, F., Stocker, T.F. and Marchal, O., 2001. Feedback mechanisms and sensitivities of ocean carbon uptake under global warming. *Tellus B*, 53, pp. 564–592.

Prasad, M.H.K., Sarma, V.V.S.S., Sarma, V.V., Krishna, M.S. and Reddy, N.P.C., 2013. Carbon dioxide emissions from the tropical Dowleiswaram Reservoir on the Godavari River, Southeast of India. *Journal of Water Resource and Protection*, 5(05), pp.534–545.

Rees, S.A., Opdyke, B.N., Wilson, P.A. and Henstock, T.J., 2007. Significance of Halimeda bioherms to the global carbonate budget based on a geological sediment budget for the Northern Great Barrier Reef, Australia. *Coral Reefs*, 26(1), pp.177–188.

Reid, R.P. and Macintyre, I.G., 1998. Carbonate recrystallization in shallow marine environments: a widespread diagenetic process forming micritized grains. *Journal of Sedimentary Research*, 68(5), pp.928–946.

Renforth, P. and Henderson, G., 2017. Assessing ocean alkalinity for carbon sequestration. *Reviews of Geophysics*, 55(3), 636–674.

Reynaud, S., Leclercq, N., Romaine-Lioud, S., Ferrier-Pagés, C., Jaubert, J. and Gattuso, J.P., 2003. Interacting effects of CO_2 partial pressure and temperature on photosynthesis and calcification in a scleractinian coral. *Global Change Biology*, 9(11), pp.1660–1668.

Riebesell, U., Zondervan, I., Rost, B., Tortell, P.D., Zeebe, R.E. and Morel, F.M., 2000. Reduced calcification of marine plankton in response to increased atmospheric CO_2. *Nature*, 407(6802), pp.364–367.

Ries, J.B., Cohen, A.L. and McCorkle, D.C., 2009. Marine calcifiers exhibit mixed responses to CO2-induced ocean acidification. *Geology*, 37(12), pp.1131–1134.

Roy, T., Bopp, L., Gehlen, M., Schneider, B., Cadule, P., Frölicher, T.L., Segschneider, J., Tjiputra, J., Heinze, C. and Joos, F., 2011. Regional impacts of climate change and atmospheric CO_2 on future ocean carbon uptake: a multimodel linear feedback analysis. *Journal of Climate*, 24(9), pp.2300–2318.

Sarma, V.V., Kumar, M.D. and Manerikar, M., 2001. Emission of carbon dioxide from a tropical estuarine system, Goa, India. *Geophysical Research Letters*, 28(7), pp.1239–1242.

Sarma, V.V.S.S., Kumar, N.A., Prasad, V.R., Venkataramana, V., Appalanaidu, S., Sridevi, B., Kumar, B.S.K., Bharati, M.D., Subbaiah, C.V., Acharyya, T. and Rao, G.D., 2011. High CO_2 emissions from the tropical Godavari estuary (India) associated with monsoon river discharges. *Geophysical Research Letters*, 38(8), p. L08601.

Schiebel, R., 2002. Planktic foraminiferal sedimentation and the marine calcite budget. *Global Biogeochemical Cycles*, 16(4), pp.3–1.

Sekula-Wood, E., Benitez-Nelson, C.R., Bennett, M.A. and Thunell, R., 2012. Magnitude and composition of sinking particulate phosphorus fluxes in Santa Barbara Basin, California. *Global Biogeochemical Cycles*, 26(2), p. GB2023.

Smith, A.D. and Roth, A.A., 1979. Effect of carbon dioxide concentration on calcification in the red coralline alga *Bossiella orbigniana*. *Marine Biology*, 52(3), pp.217–225.

Spero, H.J., Bijma, J., Lea, D.W. and Bemis, B.E., 1997. Effect of seawater carbonate concentration on foraminiferal carbon and oxygen isotopes. *Nature*, 390(6659), pp.497–500.

Thornton, D.C., 2014. Dissolved organic matter (DOM) release by phytoplankton in the contemporary and future ocean. *European Journal of Phycology*, 49(1), 20–46.

Tranvik, L.J., Downing, J.A., Cotner, J.B., Loiselle, S.A., Striegl, R.G., Ballatore, T.J., Dillon, P., Finlay, K., Fortino, K., Knoll, L.B. and Kortelainen, P.L., 2009. Lakes and reservoirs as regulators of carbon cycling and climate. *Limnology and Oceanography*, 54(6 part2), pp.2298–2314.

Walter, L.M. and Morse, J.W., 1984. Reactive surface area of skeletal carbonates during dissolution; effect of grain size. *Journal of Sedimentary Research*, 54(4), pp.1081–1090.

Wang, W., Chen, X., Zhang, Y., Yu, J., Ma, T., Lv, Z., . . . & Zou, H., 2016. Nanodeserts: a conjecture in nanotechnology to enhance quasi-photosynthetic CO_2 absorption. *International Journal of Polymer Science*, 2016, 5027879.

Williamson, P. and Turley, C., 2012. Ocean acidification in a geoengineering context. *Philosophical Transactions of the Royal Society A: Mathematical, Physical and Engineering Sciences*, 370(1974), 4317–4342.

Wollast, R., 1994. The relative importance of biomineralization and dissolution of $CaCO_3$ in the global carbon cycle. *Bulletin de l'Institut océanographique de Monaco*, 13, pp.13–35.

Wood, H.L., Spicer, J.I. and Widdicombe, S., 2008. Ocean acidification may increase calcification rates, but at a cost. *Proceedings of the Royal Society B: Biological Sciences*, 275(1644), pp.1767–1773.

Zak, D.R., Pregitzer, K.S., Kubiske, M.E. and Burton, A.J., 2011. Forest productivity under elevated CO_2 and O_3: positive feedbacks to soil N cycling sustain decade-long net primary productivity enhancement by CO_2. *Ecology Letters*, 14(12), pp.1220–1226.

Zalewski, M., 2015. Ecohydrology and hydrologic engineering: regulation of hydrology-biota interactions for sustainability. *Journal of Hydrologic Engineering*, 20(1), A4014012.

Zeebe, R.E. and Wolf-Gladrow, D., 2001. CO_2 in Seawater: Equilibrium, Kinetics, Isotopes (No. 65). Gulf Professional Publishing.

Zeebe, R.E., Zachos, J.C., Caldeira, K. and Tyrrell, T., 2008. Carbon emissions and acidification. *Science*, 321(5885), pp.51–52.

Zondervan, I., Zeebe, R.E., Rost, B. and Riebesell, U., 2001. Decreasing marine biogenic calcification: a negative feedback on rising atmospheric pCO_2. *Global Biogeochemical Cycles*, 15(2), pp.507–516.

11

Recent Trends in Fate, Transport, and Transformation of Inorganic and Organic Carbon in Freshwater Reservoirs

Sanchit Kumar and Anshumali

Department of Environmental Science and Engineering, Indian Institute of Technology (Indian School of Mines), Dhanbad, Jharkhand, India

11.1 Introduction

Reservoirs are constructed by lying down impoundments across river boundaries, which slows down the water velocity, facilitating the accumulation and settling of large amounts of carbon (C) to the reservoir sediments behind the impoundments (Tranvik et al. 2009; Sobek et al. 2012). Recent studies show the contribution of reservoirs in the global carbon cycle, because they act as both carbon sinks (C burial) and carbon sources (emission of potential greenhouse gases [GHGs] such as CO_2 and CH_4) by undergoing a series of biogeochemical transformations (Phyoe and Wang 2019). Their importance will increase in the future as most developing countries continue to rely on river impoundments for energy and irrigation needs (Zarfl et al. 2015). In recent decades, the primary hotspots of reservoir construction are Amazon, La Plata (South America), Yangtze basin (China), and Ganga-Brahmaputra (India and Nepal) river basins (Maavara et al. 2017; Saklani et al. 2020).

The worldwide annual CO_2 emissions from inland waterways to the atmosphere are roughly equal to the CO_2 intake by the seas (Holgerson and Raymond 2016). The regional and global studies show evidence of high organic carbon (OC) storage in freshwater sediments, especially in reservoirs and lakes than the seabed (Wilkinson and McElroy 2007; Qin et al. 2020) because of ~50 times higher sediment deposition and retention (Einsele et al. 2001) regulating 1–4 times higher OC burial efficiency (Clow et al. 2015). Further, reservoirs have been found to bury carbon faster than natural lakes (Knoll et al. 2013). Cole et al. (2007) estimated that out of the total C received by the inland water bodies, about 42.2% is lost to the atmosphere as gaseous exchange, 47.3% of C discharged to the oceans, and 10.5% is buried as sedimental storage. By multiplying area-specific burial rates, total OC burial in reservoirs is predicted to be 21.6–73.0 Tg C/year (Mendonça et al. 2017). On a three-decadal timescale, the construction of new dams dramatically increased the quantity of carbon retained in reservoirs; in 2000, the carbon buried in reservoirs was about twice that buried in global reservoirs in 1970. The OC burial in all global reservoirs is expected to reach 52 Tg C/year by 2030 (Maavara et al. 2017).

Hydrogeochemistry of Aquatic Ecosystems, First Edition. Edited by Sughosh Madhav, Virendra Bahadur Singh, Manoj Kumar, and Sandeep Singh.

GHGs emissions from reservoir water surfaces are estimated to be 800 (500–1200) Tg CO_2 equivalents per year, with CH_4 accounting for most of this forcing (Deemer et al. 2016). Carbon gases emitted from reservoirs to the atmosphere are recycled in the biosphere on contemporary timescales, while C stored in sediments enters the long-term geological cycle (Chmiel et al. 2016). Further, the burial and emission of C depend on several biotic (vegetation type, indigenous primary production, and microbial degradation) and abiotic (watershed area, hydraulic retention time (HRT), seasonal variability, photodegradation, and sedimentation) factors (Teodoru et al. 2012; Sobek et al. 2012; Mendonça et al. 2016; Åkerman Fulford 2017; Phyoe and Wang 2019; Wang et al. 2019a; Quadra et al. 2020). Despite increasing understanding of different factors influencing the C burial in reservoirs, the magnitude and mechanism of C burial are debatable (Mendonça et al. 2014; Keller et al. 2021). Three basic factors determine the potential of the reservoir system to operate as a source or sink of GHGs: (i) physical, chemical, biological, and biogeochemical processes that transport dissolved organic carbon (DOC) and particulate organic carbon (POC) from the water column to the sediment (e.g. OC flocculation, microbial biomass settlement, and POC sedimentation); (ii) photochemical and microbiological activities degrade and mineralize DOC and POC (resulting in CH_4, N_2O, and CO_2 emissions); and (iii) various carbon fractions are passively outfluxed downstream to reservoirs (Jacinthe et al. 2012; Mendoza-Lera et al. 2012; Cardoso et al. 2013; Åkerman Fulford 2017). Further, the restricted river water velocities and lower turbidity in the reservoir system have improved the photosynthetic conditions (Sun et al. 2010). Increased autochthonous production in reservoirs in the presence of essential nutrients and its decomposition may make reservoirs a more critical source of POC and DOC (Bianchi et al. 2004). Furthermore, growing industrialization, urbanization, and agricultural effluents, in river watershed regions will result in higher N and P supplies, facilitating C burial through improved autotrophy and resulting in significant GHG emissions such as CH_4 and N_2O (Kellerman et al. 2015). The carbon metabolism in the inland waters is likely to change in response to climate change (Del Giorgio et al. 1999; Gudasz et al. 2010) and may contribute more mineralization than carbon burial with increased CO_2 and CH_4 fluxes (St. Louis et al. 2000; Barros et al. 2011). However, higher C burial may be aided by higher global mean temperature due to reduced mineralization rates at the sediment level because of increased duration of thermal stratification (Sobek et al. 2009). Reservoirs currently encompass an area of roughly $3.1 \times 10^5\,km^2$ (Deemer et al. 2016). Their contribution to the global carbon (C) cycle will continue to change in the future as reservoir development, sediment sequestration, and emissions into the atmosphere increase. In this chapter, we will discuss (i) different sources and fractions of C input to the reservoir, (ii) transformation of organic and inorganic C fractions, (iii) their burial and outfluxes from the reservoir system, (iv) source vs sink anticipations, (v) factors affecting the burial and emissions from reservoir, and (vi) climate change impacts on reservoir C dynamics.

11.2 The Influx of Carbon to the Reservoir

Reservoirs act as depositional sites for organic and inorganic carbon carried by the streams and rivers from the watershed area. The watershed area provides a network of streams and rivers that are the source of fluvial carbon to the reservoirs (Ward et al. 2017). The fresh litter OC derived from the terrestrial plants, shrubs, herbs, and the organic and inorganic carbon stored in soils and bedrocks is the major source of fluvial carbon in rivers making its way to the reservoirs (Mayorga et al. 2005). The inorganic carbon in the form of carbonates sourced from the weathering of bedrock limestone influxes into the rivers with precipitation is a prime source of dissolved inorganic carbon (DIC) (Aucour et al. 1999). The OC transported and contributed to the reservoir by external sources is allochthonous. The OC produced in the reservoirs is of autochthonous origin. Generally, riverain systems are dominant with allochthonous OC; as a result, the reservoir systems are also dominated by allochthonous C (Par et al. 2009; Fuentes et al. 2013). However, the magnitude of these two carbon sources to the reservoir depends on several factors such as tropic state, HRT, nutrient input, and eutrophic state of the reservoirs (Wetzel 2001; Barros et al. 2011; Kunz et al. 2011b). The enhanced supply of allochthonous OC and the fixation of in situ produced carbon (autochthonous) resulted in a considerable rise in the quantity of OC trapped behind the reservoirs (approximately 26 Tg C/year in 2000) (Maavara et al. 2017).

Lakes and reservoirs have similarities in terms of input of carbon loads as both have a watershed and catchment area (Hayes et al. 2017). Table 11.1 comprises comparative data for the influx and outflux of different carbon fractions. DIC and DOC are the predominant carbon inputs to most lakes and reservoirs, followed by POC and particulate inorganic carbon (PIC). Most lakes receive carbon as DIC and POC, while reservoirs receive comparatively lesser POC content than lakes. However, there is also a difference in outflux C factions between lakes and reservoirs. In lakes, the C outflux is primarily in DIC, while DOC dominates reservoir C outflux. The relative quantity of different C fractions varies with reservoir location, watershed area, vegetation type, and hydrology (Kim et al. 2000; Yi et al. 2021).

Moreover, significant latitudinal differences in the proportions of POC, DOC, and total inorganic carbon in global rivers (Meybeck 1993) reflect climate, soil texture and geochemistry, and land use (Tranvik et al. 2009). It is also evident from Table 11.1 that the sedimental carbon storage efficiency in reservoirs is far better than in lakes. Sun et al. (2010) observed that the impoundment construction over the rivers might intensify the concentration of DOC. However, DIC concentrations are correlated with the magnitude of precipitation and discharge, while the POC concentrations are well connected with sediment supply (Mendonça et al. 2016; Wang et al. 2019b).

11.2.1 Autochthonous

The impoundment over the rivers slows down the movement of dissolved, particulate organic, and inorganic loads, providing a better condition for photosynthetic activities (Sun et al. 2010). According to Par et al. (2009), most of the carbon sources to some large reservoirs in South Korea are allochthonous (52–88%), while autochthonous sources vary from 12 to 48%. The phytoplankton, hydrophytes, and blue-green algae (BGA) are the main autotrophs in the reservoir system (Mineeva and Metelyeva 2019). The C/N ratios, depth of water column, turbidity, the concentration of POC, chlorophyll-*a* content, ultraviolet absorption by DOC, the influx of nutrients, their availability, light intensity, and mixing depth are the main contributing factors defining the primary production rate in reservoirs and show great spatiotemporal fluctuations for different tropical states (Kuo et al. 2013; Ashour et al. 2021; De Paoli 2021; Yi et al. 2021). Water temperature, phosphorus input,

Table 11.1 Quantitative comparison of carbon fraction influx, outflux, and burial for selected lakes and reservoirs of the world.

S. No.	Study area		Influx (mean) POC (%)	Influx (mean) DOC (%)	Influx (mean) DIC (%)	Outflux (mean) POC (%)	Outflux (mean) DOC (%)	Outflux (mean) DIC (%)	Sediment storage (%)	References
					Lakes					
1	Lake Frisksjo¨in, Sweden		66		34	34[a]		5	19	Sobek et al. (2006)
2	Northern Great Plain lakes, Canada		2	19	77	1	13	56	30	Finlay et al. (2009)
3	Williams Lake, USA		1	3	96	–	10	58	11	Stets et al. (2009)
4	Lake Calado, Brazil		95[a]		5	16[a]		9	4	Melack and Engle (2009)
					Reservoirs					
1	Petit Saut reservoir, French Guiana		12	64	24	11	69	20	–	Abril et al. (2005)
2	Kariba reservoir, Zambia, Zimbabwe		10	90	–	20	80	–	43.4	Zuijdgeest and Wehrli (2017)
3	Agricultural reservoir, Ohio, USA		10	5	85	5	6	55	33	Knoll et al. (2013)
4	Kinder reservoir, UK	2012	10.6	68.5	3.5	66.7	13.8	15	50	Stimson et al. (2017)
		2013	28	84.2	5.2	71.2	18.3	9	90	

Code Key: PC, particulate carbon; PIC, particulate inorganic carbon; POC-particulate organic carbon [PC = POC + PIC]
[a] Sum of POC and DOC.

and seasonal dilutions all play a role in the growth of BGA, impacting the reservoir's overall C dynamics (Kim et al. 2005). Phytoplankton may be flushed out of the reservoir system in the rainy season as water HRT is comparatively low, limiting the autochthonous C production (Williamson et al. 2021). The production of autochthonous carbon is also affected by seasonal variations. Kim et al. (2000) and Par et al. (2009) found that in the rainy season, carbon loading to the reservoir is dominated by the allochthonous sources of C, while autochthonous C loading dominates in the summer season or just after the rainy season. Hence, the autochthonous C production is different in different seasons and may vary from the reservoir to reservoir.

In comparison to worldwide GPP of 100–150 Pg C/year, the projected global gross primary production (GPP) of lakes is 0.65 Pg C/year, and for reservoirs, it is 7.8–14.4 Tg C/year (Randerson et al. 2002; Williamson et al. 2009; Maavara et al. 2017). So, the magnitude of primary production of lakes is far greater than reservoirs which are well connected with their surface area.

11.2.2 Allochthonous

The organic and inorganic carbon washed out of the watershed area during precipitation is received by the streams and rivers that ultimately transport these C fractions to reservoirs. The fresh litter OC derived from the terrestrial plants, shrubs, herbs, and secondarily, the organic and inorganic carbon stored in soils and bedrocks are the significant sources of fluvial carbon in rivers making its way to the reservoirs (Mayorga et al. 2005). The reservoirs act as depositional sites for this allochthonous supply of C. Allochthonous OC accounts for around 47–88 % of the OC accumulating in reservoir sediments (Cardoso et al. 2013; Luo et al. 2016). The worldwide input of allochthonous OC to reservoirs is approximately 79 Tg C/ year. It has increased by 74% from 1970 to 2000 due to the catchment area's doubling combined with anthropogenic OC loading to rivers (Maavara et al. 2017). The OC retention efficiency in sediments can be linked to the frequent and considerable influx of allochthonous organic matter (Sobek et al. 2011). The quantity and quality of allochthonous C entering in a reservoir are highly dependent on its tropical state, watershed area, vegetation type, land use, landcover pattern of the watershed area, seasonal variabilities (Mayorga et al. 2005; Knoll et al. 2014; Grasset et al. 2018; Anderson et al. 2020). The quality and amount of allochthonous C have an impact on the generation of autochthonous C because it serves as a growth medium for certain macro-hydrophytes and releases critical nutrients (C, N, and P) for autotrophs in the reservoir system following decomposition (Kunz et al. 2011a; Wang et al. 2019a; Quadra et al. 2020).

11.3 Mineralization and Biogeochemical Transformation of C in the Reservoir

11.3.1 Organic Carbon

The dam produces an artificial reservoir on flowing rivers resulting in reduced water velocities and acts as a depositional and transformation hotspot for inflowing C (Figure 11.1). Carbon from the organic origin is influxes into the reservoir by a network of streams and rivers. As stated earlier, the influx includes POC and DOC (Figure 11.2). The OC influxes from outside the reservoir are dominated by allochthonous sources, which account for around ~70% of the OC accumulation in reservoirs (Cardoso et al. 2013; Maavara et al. 2017). Furthermore, increased water retention time by dam construction and increased nutrient loadings from residential and agricultural effluents improve photosynthetic conditions, resulting in the reservoirs' second significant source of POC (autochthonous origin) (Sun et al. 2010). Larger watershed areas contribute more allochthonous organic C than a smaller watershed area, and large water surface areas in reservoirs

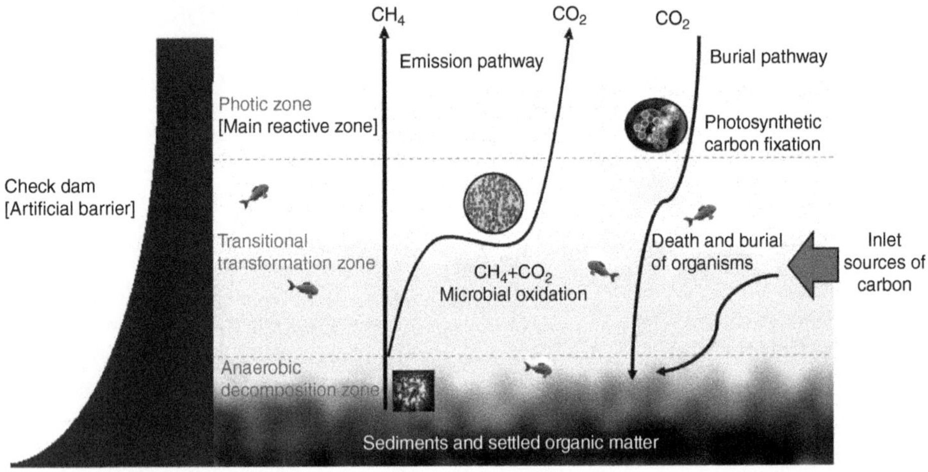

Figure 11.1 Conceptual model of carbon input, transformation zones, emission, and burial pathway of the reservoir system.

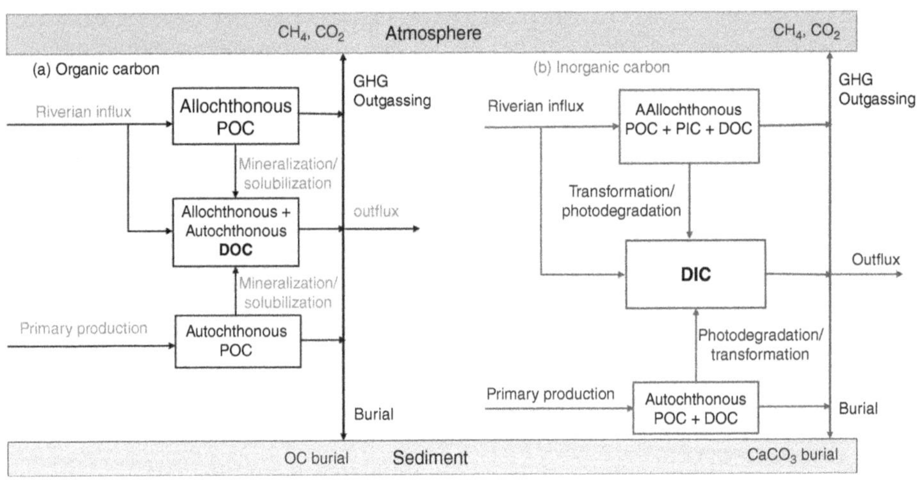

Figure 11.2 Schematic diagram of fate, transport, and transformation of different (a) organic and (b) inorganic carbon fractions in reservoir system.

support more autochthonous C (Urabe et al. 2011). The seasonal vegetation cover growing in the drawdown region of the reservoir is also included in the in situ produced (Autochthonous) OC. Still, it is just a tiny part depending on the extent to which a reservoir draws downed (Figure 11.3a).

When OC reaches a reservoir, it has three primary fates, each of which influences the sink and emission of carbon. It can be (i) resuspended and mineralized immediately in the water column, (ii) decomposed and returned to the atmosphere (from sediments to water column and from the water column to atmosphere) as CO_2 if the oxic conditions persist, or CH_4 if anoxic conditions persist, or (iii) buried in the sediment over long periods (Mendoza-Lera et al. 2012; Jin et al. 2016; Åkerman Fulford 2017).

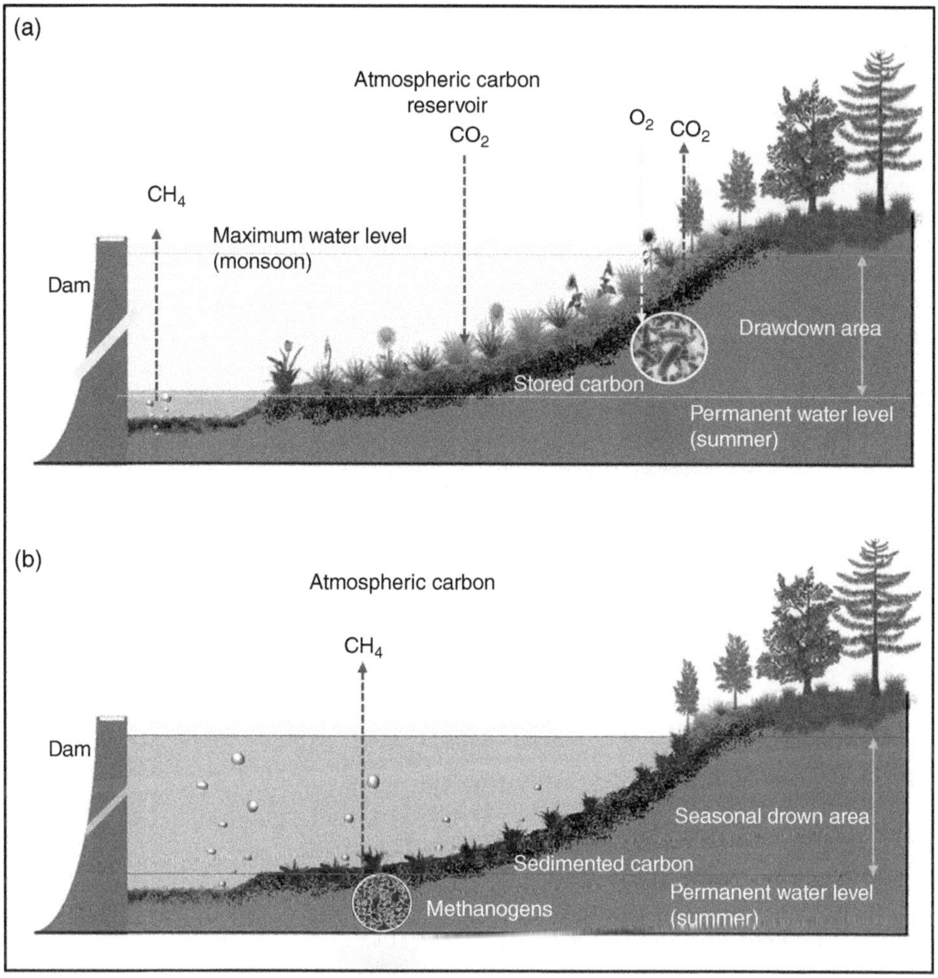

Figure 11.3 (a) C dynamics in drawdown area of reservoir at permanent water level (b) C dynamics of drawdown area at full water level.

POC moves through a lake or reservoir significantly slower than DOC (Johnson et al. 2018). Except for kinder reservoir in the United Kingdom, where DOC values decreased drastically, DOC escapes the reservoir system downstream virtually at the same pace as water is outfluxed from the reservoir, enabling a relatively short timeframe for its transformation, as shown in (Table 11.1) (Stimson et al. 2017). However, some DOC is used by microbes as C energy sources and respired again to the atmospheric C pool as GHGs (Figuer 11.2). The allochthonous DOC pool is generally recalcitrant compared to the autochthonous DOC, and its breakdown rate may be sluggish for a shorter HRT (Evans et al. 2017). Autochthonous organic matter is thought to have more energy and nutritional content per unit weight and be more physiologically labile than allochthonous organic matter. As a result, autochthonous organic matter may be recycled rapidly (Groeger and Kimmel 1984).

Allochthonous POC flows through the reservoir at a significantly slower rate, so it is transformed into DOC and DIC by the action of microbial or photochemical degradation (Figure 11.2). The terrestrial C has a more complex and recalcitrant molecular structure; hence the allochthonous POC can directly be settled to the sediment layer at the bottom of the reservoir (Thomas 1988). Later, the terrestrial allochthonous POC is replaced by the phytoplankton-derived POC (Yi et al. 2021). The reservoir system mainly has three zones of biogeochemical transformation of C, photic zone, transitional transformation zone, and anaerobic decomposition zone (Figure 11.1).

Photic Zone: Because of the high turbulence caused by atmospheric disturbances and aeration opportunities in this zone, it is pretty mobile. This zone also serves as an exchange phase between the reservoir system and the atmosphere (Cole et al. 2002). The DO concentrations in this zone are relatively high, allowing for aerobic OC breakdown (Koehler et al. 2012). A high concentration of POC and DOC is observed in this reaction zone, while the DIC concentrations are low as it is directly consumed by the phytoplankton (Yi et al. 2021). A decrease in DIC concentration can directly be used as an indicator of enhanced primary production in reservoir systems (Wang et al. 2019b). So, overall, a high concentration of C fractions is observed in this zone. The DOC can also be removed from the active reaction chamber by the sorption on sinking heavy inorganic or organic particles in the turbid conditions; however, this process is limited to only a small portion of the upstream region of the reservoir (Hansen et al. 2016). In the presence of sunlight, the colored part of DOC absorbs light and goes through a mineralization phase due to photochemical processes (Yi et al. 2021). Heterotrophs can consume or degrade the POC and DOC carbon fractions to release CO_2 and nutrients (Bastviken et al. 2004). Non-aromatic OC is degraded photochemically, whereas more complex terrestrial aromatic C is oxidized by heterotrophic microbial degradation (Kellerman et al. 2015). Heterotrophic microorganisms may quickly break down POC produced by photoautotrophs inside the reservoir (He et al. 2020a). Gravitational forces cause the more dense and non-decomposed components of OC to descend to the reservoir's bottom (Kunz et al. 2011b). Substantial amounts of CO_2 emission dominate this zone because of C mineralization under oxic conditions.

Transitional Transformation Zone: This zone is placed between the photic and anaerobic decomposition zone. As a result of the limited sunlight penetration in this zone, there are few photosynthetic activities. As we proceed deeper into this layer, DO concentrations drop sharply, and oxygen demand rises (Siriwardana et al. 2019). Lower POC concentrations indicate low primary productivity. On the other hand, allochthonous sources result in a significant quantity of POC. In addition, because of the decreased solar radiation intensity, photodegradation is rare in this region (Romera-Castillo et al. 2011). However, in the presence of inorganic nutrients, heterotrophic microbial mineralization of OC is noticeable, potentially producing significant amounts of DOC and CO_2 (Koehler et al. 2012). Since water is frequently outfluxed from the reservoir at this water level, the features of the outflux water from the reservoir are most likely comparable to the characteristics of this zone.

Anaerobic Decomposition Zone: The heavier, more recalcitrant, and suspended inorganic granules bound C settles to the bottom of the reservoir (Groeger and Kimmel 1984). The two primary phenomena in this zone are POC resuspension caused by deep water currents and gas diffusion from the bottom sediment layer (Yi et al. 2021). The anaerobic decomposition of bottom sediments by heterotrophic benthic microbes resulted in higher concentrations of DIC, DOC along with ebullition of CH_4 in this zone (Sobek et al. 2009; Bottino et al. 2016). By the action of water currents, the DIC from this zone reaches the transition zone and photic zone, boosting photosynthetic activity once again (Wang et al. 2019b). As a result, reservoirs operate as factories, continually converting DIC to OC, which is then either stored in sediments or emitted as GHGs.

11.3.2 Inorganic Carbon

The weathering of bedrock limestone influxes into rivers with precipitation, providing inorganic carbon as carbonates in particulate and dissolved form (Aucour et al. 1999). The dissolved fraction of the inorganic carbon is more labile and reactive (Aarnos 2016). As a result, the DIC plays an essential role in the biogeochemical cycle of C in the reservoir. The source, fate, transport, and transformation of IC are shown in Figure 11.2b. The riverine influx includes terrestrial POC, PIC, DOC, and primary production adds more in situ produced POC (Finlay et al. 2009; Yang et al. 2017). As inorganic carbon is heavier in particulate form, it sinks to the bottom of the reservoir and remains buried for a long time (Li et al. 2021). The POC can be degraded and decomposed by aerobic and anaerobic microbial actions, leading to the production of DOC and DIC (Koehler et al. 2012; Li et al. 2018). The DOC from both the allochthonous and autochthonous sources is further mineralized by photochemical degradation, resulting in DIC production (Figure 11.2b) (Hansen et al. 2016).

The photoautotrophs use the DIC and convert it to OC, completing the cycle (Bardhan et al. 2017). However, photosynthetic activity near the water surface increased pH and, as a result, a decrease in calcite solubility; this mechanism may increase Ca^+ precipitation during daylight hours (De Montety et al. 2011). Precipitation of carbonates in the water column will cause a decrease in Ca^{2+} and DIC concentrations. Ca precipitation in reservoirs is explained briefly in the chemical mechanism described below.

$$\mathbf{CaCO_3 + CO_2 + H_2O \leftrightarrow HCO^{3-} + Ca^{2+} \rightarrow CaCO_3 \downarrow}$$

The three major processes that govern the DIC concentration in the reservoir system are: (i) increased weathering of carbonate rocks leads to greater DIC concentrations in the watershed area, which then enters the reservoir with precipitation. However, the weathering process varies according to the season (Li et al. 2010). (ii) In the summer, thermal stratification of the topwater column ensures a stable environment for OM breakdown, resulting in increased DIC concentrations in the upper water column. In contrast, thermal stratification causes poor DIC exchange from the deeper water layer (Kim 2015). (iii) Flow disturbances: both at the surface and deeper levels, significantly influence decomposition processes (Wang et al. 2019b). Reduced water velocity encourages OC breakdown in deeper layers, resulting in higher DIC concentrations (Yi et al. 2021). The phytoplanktons are highly dependent on water-soluble C for photosynthesis. Hence, a reduction of DIC concentration in photic zone indicates enhanced primary production in the reservoir system (Maberly et al. 2013). With an annual uptake of 200–700 Tg C globally, the biological carbon pump in reservoirs plays a significant role in transporting DIC to aquatic photoautotrophs, forming an unbreakable connection between DIC and autochthonous OC (Passow and Carlson 2012; He et al. 2020b). Various biogeochemical processes influence DIC concentration in reservoirs, including OC breakdown by microorganisms, CO_2 release from the reservoir surface, uptake by phytoplankton, rainfall, and calcium mineral dissolution (Liu et al. 2021; Yi et al. 2021).

11.3.3 Accumulation and Mineralization of C

Microbial decomposition limits the amount of OC stored in the sediment through mineralization processes. The mineralization of C produced in situ, and influxes from the upstream watershed area continue throughout the reservoir's lifecycle. Large amounts of DOC, DIC, CO_2, and methane are formed during the mineralization of OC (Gudasz et al. 2015; Bottino et al. 2016). Further, the mineralization processes greatly depend on the type of OC (allochthonous or autochthonous), oxygen exposure time, tropic state, age, and temperature (Barros et al. 2011; Cardoso et al. 2013; Knoll et al. 2013). Allochthonous OC is more labile than autochthonous OC generated in the reservoir (Sobek et al. 2009). On a global scale, mineralization of allochthonous OC exceeds that of autochthonous OC by a factor of 7–11 as allochthonous OC input is considerably greater than in-reservoir photosynthetic carbon fixation. (Maavara et al. 2017).

The mineralization processes are higher in younger reservoirs associated with the respiration of fresh soil OC influx and flooded biomass (Mendonça et al. 2012a). Reservoirs are projected to have net mineralization to burial ratio of 1.3 globally (Cole et al. 2007; Maavara et al. 2017). The OC accumulation rate for some of the recently studied reservoirs varies from 32.9 to 1113 $gC/m^2/year$, while mineralization rate varies from 34.3 to 325 $gC/m^2/year$ (Kunz et al. 2011a; Sobek et al. 2012; Teodoru et al. 2012; Mendonça et al. 2016; Åkerman Fulford 2017; Wang et al. 2019a; Qin et al. 2020; Quadra et al. 2020).

Three principal processes determine the accumulation and mineralization of C: (i) DOC and POC are transported from the water column to the sediment by different physical, chemical, biological, and biogeochemical processes (i.e. OC flocculation, microbial biomass settlement, and POC sedimentation); (ii) DOC and POC are degraded and mineralized by photochemical and microbial processes (resulting in the emission of CO_2, CH_4, and N_2O); and (iii) outflux of different carbon fractions passively downstream to the reservoirs (Tranvik et al. 2009; Evans et al. 2017).

11.4 Driving Factors Controlling the C Cycle in Reservoirs

11.4.1 Biotic Factors

11.4.1.1 Vegetation

The quality and amount of OC intake to the reservoir are directly proportional to the quality and quantity of litter and humus deposition in the watershed (Ward et al. 2017). The complexity of the structure of terrestrial OC supplied to the reservoir by inflowing streams and rivers affects the deposition and mineralization of OC in the reservoir (Ran et al. 2014).

Microbial, photochemical degradation, and mineralizing of more complex OC are difficult, so these compounds can bypass the degradation process and descend directly to the anoxic bottom sediment layer, where they can be permanently buried over time (Gougoulias et al. 2014). As plants fix carbon in more complex carbohydrates and phenolic compounds in lignin and tannins, plant species play an essential role in determining the breakdown of litter and humus (Rahman et al. 2013). The cropland is the most crucial component regardless of the watershed's natural vegetation pattern. Agricultural methods typically provide a high nitrogen and phosphorus input to the soil, resulting in a higher nutrient supply to the soils (Quadra et al. 2020). High nitrogen loading from terrestrial soils will directly influence reservoir primary production (Anderson et al. 2020). Furthermore, increased nutrition availability will encourage bacterial growth, resulting in faster OC breakdown, greater DOC concentrations, bacterial respiration, and CO_2 emissions to the atmosphere. Furthermore, climatic, geogenic, and anthropogenic variables such as temperature, precipitation, atmospheric CO_2 concentrations, and soil nutrient availability influence the quality and amount of OC.

11.4.1.2 Indigenous Primary Production

Restricted river water velocities and decreased water turbidity due to reservoir construction provide an ideal condition for the photosynthetic activities (Sun et al. 2010). The projected global GPP for reservoirs is 7.8–14.4 Tg C/year (Maavara et al. 2017). The principal autotrophs in the reservoir system are phytoplankton, hydrophytes, and BGA (Brothers et al. 2013). Hydrophytes are mainly limited to the reservoir's littoral zone; phytoplankton and BGA are found throughout the reservoir (Stefanidis and Papastergiadou 2010). Despite microplanktons, reservoir primary production also includes photosynthetic activities that occur in the reservoir drawdown area, as illustrated in Figure 11.3a. Higher primary productivity can fix a large amount of atmospheric CO_2 and, hence, can help accumulate higher OC behind the reservoirs. However, this indigenously produced OC is degraded by microbes that produce a high amount of DOC and CO_2 (Yi et al. 2021). The DOC can leave the reservoir system with the outfluxed water and become a source of GHGs emissions post-discharge (Mendonça et al. 2015). Phytoplanktons are the main driving force in inorganic carbon cycling and biogeochemical transformation (Li et al. 2021). As phytoplankton is mainly dependent on water-soluble C, a drop in DIC concentration is commonly observed in the photic zone. A decrease in DIC concentration can directly be used as an indicator of enhanced primary production in reservoir systems (Wang et al. 2019b).

Water temperature, phosphorus input, and seasonal dilutions all play a role in the primary production, impacting the reservoir's overall C dynamics (Kim et al. 2005). The production of autochthonous carbon shows great spatiotemporal fluctuations (Kim et al. 2000; Par et al. 2009). Phytoplankton may be flushed out of the reservoir system in the rainy season as water HRT is comparatively low, limiting the autochthonous C production (Kim et al. 2000). Increased availability of nutrients such as nitrogen and phosphorus due to agricultural and industrial runoff promotes phytoplankton blooms, pushing the reservoir system toward eutrophic conditions and potentially assisting in burying a considerable quantity of CO_2 from the atmosphere (Downing et al. 2008; Sobek et al. 2012). Furthermore, when the buried C is exposed to oxic conditions, primary production in the drawdown region functions as a source of C, resulting in significant CO_2 emissions from these locations.

11.4.1.3 Microbial Degradation

Microbes in the reservoir absorb allochthonous and autochthonous POC in two ways: under oxic and anoxic environments. The decomposition process results in DOC, DIC, CO_2, CH_4, and other gases (Bottino et al. 2016). Allochthonous OC oxygenates as soon as it enters the reservoir system, and reactive oxygen species form a crucial step in OC mineralization (Groeger and Kimmel 1984). However, this kind of breakdown is restricted to an oxygen-rich zone, and mineralization of allochthonous OC is inhibited in anoxic environments (Bastviken et al. 2004). Furthermore, because allochthonous C has already undergone a significant degree of degradation and biogeochemical process before reaching the reservoir, the decomposition process of allochthonous OC is slower than autochthonous OC (Sobek et al. 2009).

Furthermore, allochthonous C contains more complex resistant aliphatic and phenolic compounds in the form of lignin and tannins, which can only be degraded by aerobic microorganisms, reactive oxygen species, and radicals (Wetzel et al. 1995; Bottino et al. 2016). Since microbial mineralization of more complex OC is difficult, these compounds could skip the degradation process and descend directly to the anoxic bottom sediment layer, where they can be buried permanently over time (Grasset et al. 2018). Autochthonous phytoplankton has a C:N ratio of 6–9, but allochthonous C has a C:N ratio of 10–40, implying that autochthonous C has a higher mineralization rate and autochthonous OC mineralizes at equal rates in both oxic and anoxic environments (Quadra et al. 2020). Microbial decomposition of sediments dominated by autochthonous material is more rapid than allochthonous material (Gudasz et al. 2015).

Oxygen-deficient conditions dominate the microbial decomposition activities in the reservoir's deeper layer, and the anaerobic breakdown of bottom sediments by heterotrophic benthic microorganisms results in greater concentrations of DIC, DOC, as well as CH_4 ebullition (Yi et al. 2021). In the summer season, thermal stratification reduces the mixing of the oxic and anoxic layers, limiting mineralization activities in reservoir sediments sustaining the C burial (Finlay et al. 2009; Minor et al. 2012). As a result, the increasing global mean temperature is thought to aid in the persistence of thermal stratification for extended periods, resulting in higher burial rates and lower mineralization at the sediment level.

11.4.2 Abiotic Factors

11.4.2.1 Watershed Area

The size of the watershed is one of the most important external factors influencing C supply and transformation in reservoirs (Urabe et al. 2011). Reservoirs often have a large watershed to reservoir area ratio providing a substantial amount of terrestrial soil and plant biomass to the reservoirs (Ward et al. 2017). The plant diversity and density in the watershed region directly influence litterfall and humus buildup, affecting the quantity and quality of OC delivered to the reservoir (Sobek et al. 2006). The deposition and mineralization of OC in the reservoir are influenced by the complexity of the structure of terrestrial OC delivered to the reservoir by inflowing streams and rivers (Grasset et al. 2018). Increased primary productivity in the watershed region aids heterotrophic microbe development, respiration, and generation of DOC, DIC, and GHGs in the reservoir (Hosen et al. 2019; Quadra et al. 2020).

Furthermore, the watershed region promotes the influx of nutrients from dry and wet atmospheric deposition, as well as agricultural and industrial operations (Anderson and Downing 2006; Yi et al. 2021). Enhanced soil erosion due to the conversion of forestland to agricultural land increases OC input to reservoir sediments (Anderson et al. 2013). Anderson et al. (2013) also discovered that the prevalence of agriculture in watershed areas is linked to the pace of OC buildup in lakes. When the bean cultivated land is switched to maize crops, Thothong et al. (2011) observed increased soil OC loss in the watershed. As a result, the land use-landcover pattern in the watershed impacts reservoir C dynamics.

11.4.2.2 Photodegradation

The reservoir's extended HRT helps to stabilize the water column, facilitating the biogeochemical transition of allochthonous and autochthonous OC, culminating in the release of significant amounts of DOC (Minor et al. 2012). The DOC in the reservoir is classified into two types: chromophoric (CDOC) and fluorescent (FDOC) (Bittar et al. 2015). In lakes and reservoirs, OC photodegradation is a well-known phenomenon. The photochemical degradation of the FDOC and CDOC components occurs when they absorb sunlight. Depending on the turbidity of the water column, the photodegradation of DOC is confined to a few meters of the photic zone (Aarnos 2016). Light can only penetrate a few centimeters beyond the water surface in the presence of significant quantities of colored DOC, limiting photodegradation to a narrow zone (Bertilsson and Tranvik 2000). The photodegradation of complex and long-chain DOC results in molecular size reduction (Hansen et al. 2016). When DOC is photodegraded, aromatic OC is transformed into a more aliphatic-rich DOC pool (Kellerman et al. 2015). As a result, DOC degradation resulting in direct photolysis to CO_2 might be an essential mechanism for DOC removal and transformation at the reservoir surface.

11.4.2.3 Hydraulic Retention Time

One of the most important influencing factors that govern OC dynamics in reservoirs is water retention time, also known as HRT (Lee et al. 2015). Based on the inflow and outflux rates, the impoundment over a flowing river enhances the reservoir's retention time. A reservoir's water retention duration can be significantly longer than a river, ranging from hours to years (Wang et al. 2018). The longer retention time allows photochemical and biological processes in a stable water column. C retention is determined by HRT behind river impoundments (Groeger and Kimmel 1984). Longer contact time between the OC and microorganisms improves allochthonous and autochthonous POC decomposition conditions. The formation of significant amounts of DOC is facilitated by the longer HRT (Prairie et al. 2018). However, a short water retention period causes allochthonous and autochthonous OC to flush rapidly, leaving less time for transformation and settling. HRT has a significant influence on DOC concentrations because DOC travels through the reservoir at the same rate as water (Groeger and Kimmel 1984). Furthermore, the HRT controls the nutrient and DIC exchange between the water layers by affecting the development of thermal stratification.

OC degradation rate in inland waterways is strongly and adversely connected to HRT, according to Catalán et al. (2016). In addition, the extended contact time caused by higher HRT and OC exposure to mineralization over the inland water continuum resulted in a lower degradation rate with increasing WRT as the composition of OC becomes skewed toward compounds that are degraded more slowly (Kellerman et al. 2014). Researchers claim that when the HRT is larger than 10 days, the reservoir retains 60–75% of the total OC intake, but shorter HRTs have low OC retention (Downing et al. 2008).

Furthermore, through changing nutrient ratios and nutrient buildup in the water column, the HRT can impact the composition and productivity of aquatic systems (Bottino et al. 2016). The rate of nutrient removal is strongly proportional to the HRT. Reservoirs are expected to remove 5% of dissolved and biogenic silica, 7.4% nitrogen, and 12% phosphorous from total nutrient loads transported by rivers (Maavara et al. 2020). Another study on Lake Diefenbaker reservoir having HRT of about 1.1 years was found to eliminate 28% of dissolved silica, 64% of total nitrogen, and 92% of total phosphorous (Donald et al. 2015; North et al. 2015). The mean N:P ratio tends to increase in the water column with the increased HRT as the loss of P takes place via sedimentation burial. So, the HRT is well associated with the primary production, terrestrial OC accumulation, nutrient retention, and biogeochemical transformation of the C.

11.4.2.4 Seasonal Variability

Seasonal variations directly influence the amount of water received and reservoir availability throughout the year (Coursolle et al. 2012). Snowfall and snow cover in the watershed region significantly affect reservoirs fed by glacial waters, which directly impact carbon transport across the watershed (Milner et al. 2017). Furthermore, changes in precipitation patterns due to climate change would most likely affect the organic and inorganic carbon inputs to aquatic systems from the watershed (Urabe et al. 2011).

The availability of the growth limiting nutrients (P & N), which varies greatly throughout seasons, significantly impacts primary productivity (Par et al. 2009). Seasonal hydrological variations affect the terrestrial primary production and mineralization processes in the reservoir (Siriwardana et al. 2019). Seasonal fluctuations in GHG emissions are intimately connected since less runoff during the summer season promotes photochemical breakdown and microbial respiration (Barros et al. 2011). Furthermore, summer stratification causes persistent anoxic conditions in bottom sediments, resulting in a low mineralization rate that promotes C burial (Sobek et al. 2009). During the high flow season (July to September), Korean reservoirs receive 71% river inflow and 69% allochthonous OC, while the total OC load was dominated by autochthonous C in the dry seasons. For example, in April, autochthonous OC accounted for 73% C load in the reservoir (Par et al. 2009). Reduced photosynthetic activity and increased respiration are associated with the winter season. However, increased photosynthetic activity, higher water temperature, increased CO_2 absorption by the water column, and the algal bloom are observed throughout the summer season (Kim et al. 2005; Lee et al. 2015). Furthermore, the reservoir's water release during the summer season will result in CO_2 outgassing, making the reservoir a net C source (Tranvik et al. 2009; Coursolle et al. 2012). As a result of enhanced weathering intensity, Wang et al. (2019b) detected an increased supply of

DIC and Ca^{2+} during the rainy season. As higher water temperatures cause quicker breakdown and respiration, tropical reservoirs decay OC more quickly than temperate reservoirs (St. Louis et al. 2000).

11.4.2.5 Sedimentation

The sedimentation rate in recently studied global reservoirs varies from 0.3 to 7.8 cm/year, as shown in Table 11.2. Except for the Hongfeng reservoir in China, reservoirs with a higher sediment accumulation rate have a higher OC retention efficiency (Table 11.2). The flocculation and sedimentation processes impact the settling of PIC and POC. Higher hydraulic retention leads to increased photosynthetic activity near the water surface, raising pH and lowering calcite solubility; this process might enhance Ca^{2+} flocculation and sedimentation during daylight hours. Ca^{2+} and DIC concentrations will decrease as carbonates precipitate in the water column (Wang et al. 2019b). Particle concentration, turbidity, pH, and reservoir temperature will all impact DOC and POC flocculation (Li et al. 2021).

According to Brothers et al. (2013), very turbid lakes bury approximately twice as much OC than cleaner lakes, implying that turbid lakes have a greater sedimentation rate. Furthermore, there is a lot of variation in sedimentation rates across the reservoir. According to Mendonça et al. (2016), the efficiency of OC burial reduces as the rate of sediment deposit drops closer to the dam head. Reservoirs have a higher OC burial efficiency than lakes because reservoirs have a much faster sedimentation rate than lakes, resulting in a low oxygen concentration in the bottom sediments, preserving earlier carbon settlings (Sobek et al. 2012). High sedimentation rates provide a physical barrier that inhibits oxygen from accessing earlier sediment layers, impeding OC mineralization (Li et al. 2012). Because slower C mineralization is linked with anoxic conditions, reservoirs with a higher sedimentation rate will have a higher C burial efficiency (Fuentes et al. 2013). OC deposited on a sediment's surface is either mineralized into CO_2 or CH_4 by heterotrophic bacteria or buried in the sediments over a geological timescale (Yang et al. 2017). OC burial in sediments is a significant sink in the global carbon cycle, removing carbon from actively cycling carbon reservoirs (Sobek et al. 2009).

11.5 Carbon Cycling in Relation with Other Nutrients (N and P)

Nutrients such as N and P are necessary for autotrophic production in the aquatic system. However, nutrient enrichment can lead to eutrophication in aquatic bodies, limiting light penetration and causing DO depletion in subsurface water layers (Williamson et al. 2021). The limiting availability of nutrients such as nitrogen and phosphorus inhibits phytoplankton production in aquatic environments (Elser et al. 2007). It is predicted that impoundments over river water systems retain 70–80% of incoming nutrient loads, including a reduction of 7.4% nitrogen and 12% phosphorous (Maavara et al. 2020). Another research indicated that the lake Diefenbaker reservoir could remove 64% of total nitrogen and 92% of total phosphorus (Donald et al. 2015; North et al. 2015). Excess nitrogen and phosphorus from agriculture development and land-use change are causing significant eutrophication in European lowland waters (Anderson et al. 2014).

Table 11.2 Age, surface area, mean depth, hydraulic retention time, sediment accumulation rate, OC accumulation rate, OC mineralization rate, and carbon retention efficiency of selected global reservoirs.

S.no	Reservoir name	Location	Age (years)	Surface area (km^2)	Mean depth (m)	Hydraulic retention time (years)	Sediment accumulation rate (cm/year)	OC accumulation rate (g C/m^2/year)	OC mineralization rate (g C/m^2/year)	Carbon retention efficiency (%)	References
1.	Mascarenhasde Moraes	Brazil	64	272	15	0.14	0.72	42.2	34.3	57	Mendonça et al. (2016)
2.	Curuá Una	Amazon	44	72	6	29	0.60	91	325	22	Quadra et al. (2020)
3.	Kariba	Zimbabwe	63	5550	29	3	0.30	61	–	41	Kunz et al. (2011a)
4.	Wohlen	Switzerland	101	2.5	9	0.005	7.8	1113	136	87	Sobek et al. (2012)
5.	Xin'anjiang	China	62	567	37	2	0.5	34	–	–	Phyoe et al. 2020
6.	Camargos	Brazil	61	64	9.6	0.16	1.23	121.3	–	–	Åkerman Fulford (2017)
7.	Eastmain-1	Canada	15	602.9	11	0.4	–	32.9	–	–	Teodoru et al. (2012)
8.	Hongfeng	China	61	57.2	10.5	0.76	0.95	117.5	–	13	Wang et al. (2019a)
9.	Itezhi-Tezhi	Zambia	43	364	15.4	0.7	1.3	43	70	48	Kunz et al. (2011b)
10.	Wujiangdu	China	42	47.5	–	0.14	–	579.2	–	–	Qin et al. (2020)

Note: For studies with a wide range of estimates, mean values are taken.

Further, Anderson et al. (2020) recently demonstrated that nutrient availability is responsible for 70% of the observed increase in worldwide lake C sequestration. The availability of nutrients is also linked to the mineralization and decomposition processes. After mineralization, the P-bound OC releases the available phosphorous into the water column, resulting in the highest autochthonous primary production in the water column near the P-bound OC (Yi et al. 2021). The sewage input from the residential settlements contributes high N loads to the reservoir, which stimulates the aquatic production and mineralization of OC by heterotrophic microbes due to the low C:N ratio (Quadra et al. 2020). Growing industrialization and urbanization in river catchment areas alters the N-deficient bed sediments to N-rich bed sediments, which accelerates the rate of C decomposition resulting in increased GHG emissions (Venkatesh and Anshumali 2020). Globally, eutrophic reservoirs have been discovered to emit a significant amount of CH_4 (Anderson et al. 2014). The high nitrogen loads to the reservoir may also result in substantial release in N_2O emissions due to the denitrification (Fearnside and Pueyo 2012). These regional and global studies reveal greater nutrient loading in reservoirs, which aid C burial through increased autotrophy, and they may also accelerate significant GHG emissions such as CH_4 and N_2O.

11.6 Source vs Sink

Inland waters play an essential part in the global carbon balance, as they act both as carbon sinks (OC burial in sediments) and as carbon sources (emission of potential GHGs such as CO_2 and CH_4) to the atmosphere (Tranvik et al. 2009; Sobek et al. 2012; Phyoe and Wang 2019). This is a common perception that the reservoirs provide a clean energy source. However, the carbon burial in reservoirs, which might help offset emissions, has been widely overlooked and is still poorly defined (Mendonça et al. 2012b). The entire consequences of dam construction on the carbon cycle are yet unknown. According to Barros et al. (2011), the reservoirs might be a significant source of CO_2 and CH_4 and constitute a new carbon source. The scattered or dense vegetation in terrestrial environments slowly accumulates carbon in soils, making them net carbon sinks (Teodoru et al. 2012). But these terrestrial environments have been significantly affected by the construction of dams in areas where the river used to flow freely (Wang et al. 2018). C can be emitted or buried in three ways by the reservoir. First, the dam's construction drowns the region's flora in a thick water column in its early stages, possibly strangling a significant carbon sink. As a result of the microbial breakdown of pre-existing drowned vegetation, significant GHG emissions became evident a few years after the dam was built (Fearnside and Pueyo 2012; Prairie et al. 2018; Phyoe and Wang 2019). Second, the buildup of highly methanogenic sediments under a shallow water column has resulted from influxes of terrestrial and in situ generated biomass, facilitating the conversion of complex organic C to atmospheric CO_2 and CH_4 (Yang 2019). Also, high sedimentation rates in reservoirs feed reactive organic materials to deep, anoxic sediment layers, driving methanogenesis by CH_4 ebullition (Sobek et al. 2012). Further, a third phenomenon occurs just after the outflux of reservoir water through turbines; as methane-rich water passes through turbines, the hydrostatic pressure drops, and a significant quantity of the CH_4 gas escapes into the atmosphere (Kemenes et al. 2007). The estimated GHGs emission from reservoir water surfaces account for ~1.3% of all anthropogenic GHG emissions, with the majority (79%) of this forcing by CH_4 (Deemer et al. 2016). It has been stated that the net GHG consequences of some tropical hydroelectric reservoirs can be greater than those of fossil fuel alternatives; hence downstream degassing of methane from hydroelectric reservoirs has been controversial (Tranvik et al. 2009).

The retention of inflow carbon in reservoirs is greater than in natural lakes, as seen in Table 11.1. On the other hand, reservoirs are distinguished from lakes by their overall construction, hydrology, and flow management. As a result, the reservoir's biogeochemical cycles, burial, and emissions differ from lakes (Dean and Gorham 1998). Lakes typically retain 4–30% of total C intake, whereas reservoirs have a higher retention efficiency (33–90%) (Table 11.1). Furthermore, several recently investigated reservoirs' OC burial efficiency ranges from 13 to 87% (Table 11.2). The rate of OC accumulation varies from 32.9 to 1113 g C/m^2/year, whereas the mineralization rate ranges from 34.3 to 325 g C/m^2/year (Table 11.2). The reservoirs may be expected to be a net C sink since the accumulation rate is larger than the mineralization rate, as evidenced by the findings.

Using varying aerial reservoir surface areas (3.0 to $15 \times 10^5\,km^2$), different researchers estimated different CO_2 and CH_4 emission and burial rates (Table 11.3). CO_2 emissions estimates range from 37 to 273 Tg/year, whereas CH_4 emissions estimates range from 3 to 48 Tg/year. In addition, net C burial rates range from 26 to 290 Tg/year (Table 11.3). The estimate GHG emission from reservoir water surfaces account for ~1.3 % of all anthropogenic GHG emissions, with the majority (79%) of this forcing by CH_4. In some reservoirs, the high autochthonous production rate increases C sequestration in sediments and lowers C emissions (Keaveney et al. 2020; Liu et al. 2021). A study carried out by Sobek et al. (2012) in reservoir Lake Wohlen, Switzerland, found that C burial was three times greater than the CO_2 emissions. Roughly, the carbon burial rates in reservoir sediments are 200 Tg C/year globally (Phyoe and Wang 2019), with C emission rates of 37 Tg C/year as CO_2 and 13 Tg C/year as CH_4 (Deemer et al. 2016). This also supports the notion that GHG emissions from reservoirs are nearly three times lower than those from carbon burial.

Certain reservoirs have a higher mineralization rate than the rate of accumulation that may surpass the C burial, making the system a C emitter (Quadra et al. 2020; Kunj et al. 2011b). Further, the emissions from the drawdown area of global reservoirs are compromised by many researchers. The development of drawdown areas exposes the bottom sediments to the atmosphere, consequently changing the oxygen-deficient conditions to oxygen-rich conditions and resulting in enhanced emission of GHGs (Figure 11.2a). Keller et al. (2021) examined satellite data of 6794 reservoirs and discovered that 15% of the total reservoir surface was exposed to the atmosphere, with the exposure being more prominent in the tropics, and that this exposure is expected to increase GHG emissions by 53% when the drawdown area's emissions are considered. Furthermore, according to Li and Zhang (2014), global reservoir drawdown areas emit an additional 15.0% CO_2 and 12.03% CH_4 compared to total GHG emissions from all global reservoirs. Concluding that reservoirs are C sources or sink remains challenging since emission rates vary widely from reservoir to reservoir, and the quantity of emissions is highly dependent on reservoir location and age (Phyoe and Wang 2019). Tropical reservoirs emit more GHGs, while temporal reservoirs bury carbon at a much faster rate than tropical reservoirs, simultaneously reducing C emissions (Evans et al. 2017). In addition, younger reservoirs have more rapid rates of mineralization and emission than older reservoirs (Sobek et al. 2012). Both emissions and burial occur simultaneously in reservoirs, although their amplitudes vary widely within and among reservoirs (Teodoru et al. 2012).

Table 11.3 Global emission and burial estimates of reservoirs along with emissions from the drawdown areas.

S. No.	Measured system	Area × 10^5 (km^2)	CO_2 emissions (Tg/year)	CH_4 emissions (Tg/year)	References
		Emissions from whole reservoir surface			
1	All reservoirs	15	273	48	St. Louis et al. (2000)
2	Hydroelectric reservoir	3.4	48	3	Barros et al. (2011)
3	Hydroelectric reservoir	3.4	301	18.7	Li and Zhang (2014)
4	All reservoirs	3.1	37	13	Deemer et al. (2016)
5	All reservoirs	3.0	39.5	12.7	Keller et al. (2021)
		Emissions from drawdown areas			
1	Hydroelectric reservoir drawdown	–	45.1	1.6	Li and Zhang (2014)
2	Lakes, reservoirs, and ponds drawdown	–	92[a]		Marcé et al. (2019)
3	Reservoir drawdown	–	26.2[a]		Keller et al. (2021)

S. No.	Measured System	Area × 10^5 (km^2)	OC burial (Tg/year)	References
		Burial from whole reservoir surface		
1	All reservoirs	3.05	26	Maavara et al. (2017)
2	All reservoirs	3.5	60	Mendonça et al. (2017)
3	All reservoirs	4.0	180	Cole et al. (2007)
4	All reservoirs	6.6	290	Stallard (1998)

[a] Carbon equivalent emissions.
All reservoirs = hydroelectric + non-hydroelectric impoundments.

11.7 Water Drawdown Area as a New Emission Hotspot

Generally, large fluctuations in reservoir water levels occur in three phases: (i) the highest-level phase (January to mid-June), (ii) the emptying phase (mid-June to the beginning of September), and (iii) the minimum-level phase (mid-September to the first autumn/winter rain events) (Geraldes and Boavida 2005) due to the seasonal differences in precipitation and the rate of glacier melting (Marcé et al. 2019). As a result, reservoirs experience the formation of ephemeral drawdown zones (Figure 11.3), which has several implications for reservoir limnology and biogeochemical cycling (Thornton et al. 1990). The mapping and calculation of the drawdown area are complex operations. Still, estimates suggest that a considerable portion of the world's inland waters is prone to periodic or seasonal drying, with this phenomenon being especially evident in arid and semiarid reservoirs (Jin et al. 2016). Keller et al. (2021) examined satellite data of 6,794 reservoirs and discovered that 15% of the total globe reservoir surface was exposed to the atmosphere, with the exposure being more prominent in reservoirs near the tropics. Further, he observed that the reservoirs commonly utilized for agricultural irrigation have the largest drawdown areas, whereas deep hydroelectric reservoirs have smaller drawdown areas. Weather extremities (Heatwave, delayed precipitation) further worsen the conditions that happened to the Theewaterskloof Dam, South Africa, culminating in a 73% drawdown area of the maximum reservoir area during the water crisis of 2018 (Pekel et al. 2016). Furthermore, climatic and human factors influenced the global distribution of reservoir drawdown areas.

The anoxic condition at the bottom sediments helps fix the C for a long time. However, anoxic conditions prevail in the thin top layer of a few millimeters of sediments, but the anthropogenic disturbances such as seasonal agricultural activities (a common practice in southeast Asia in small irrigation purpose reservoirs which are mostly drawdown in the summer season) expose deeper sediment layers to the atmosphere (Sobek et al. 2009; Marcé et al. 2019), consequently changing the oxygen-deficient conditions to oxygen-rich conditions (Figure 11.3a). When the reservoir is fully filled, the seasonal vegetation that grew in the drawdown zone is submerged in the water column, permitting anoxic breakdown of this biomass, which results in CH_4 emissions (Figure 11.3b). Methanogenesis fades away when a system becomes dry, and aerobic respiration in oxic conditions stimulates enzymatic activities and overall microbial growth. The major implication of the drawdown area is enhanced emission of GHGs (Sobek et al. 2009; Deemer et al. 2016; Marcé et al. 2019).

In a laboratory incubation experiment of sediments, Jin et al. (2016) showed significant increases in CO_2, CH_4, and N_2O generation due to enhanced activities of phenoloxidase and three hydrolases, indicating a transient stimulation of microbial organic matter decomposition in the drying sediment. In terms of climatic forcing, CH_4 dominates total reservoir C emissions. However, as compared to CO_2 emissions, CH_4 emissions from drawdown regions have been proven to be low, but the supporting data on CH_4 emissions from drawdown areas is rare (Harrison et al. 2017). Assuming that all reservoirs are entirely filled, global GHG emissions from reservoirs are estimated to be around 500–1,200 Tg CO_2 equivalent per year (Deemer et al. 2016). However, the additional global emission of CO_2 from the drawdown area is quantified as 43.2–79.5 TgCO_2-C/year by (Keller et al. 2021). When areas with a non-reduced OC burial are examined, the emission-to-burial ratio climbs to 1.72 (previously estimated as 1.2) (Maavara et al. 2017). The findings imply that drought-induced bursts of GHG emission from exposed sediments can counteract C buildup in reservoir sediments across time periods ranging from years to decades, reversing the downward trend in C emissions from older reservoirs (Mendonça et al. 2017). As a result, including drawdown regions in reservoirs' global C budget is expected to shift the C mass balance toward C emission, with emissions outpacing OC burial, especially if global drawdown areas continue to rise (Jin et al. 2016; Marcé et al. 2019; Keller et al. 2021).

11.8 Climate Change Effects on C Dynamics in Reservoirs

The carbon metabolism in the inland waters is likely to change in response to climate (Tranvik et al. 2009). Gudasz et al. (2010) revealed that increased OC mineralization is associated with the temperature rise. This means that high water temperatures in reservoirs will contribute more mineralization than carbon burial with increased CO_2 and CH_4 fluxes (St. Louis et al. 2000; Barros et al. 2011). Climate change will cause an increase in CO_2 levels and mean global temperature; as a result, water bodies may become warmer, and drought frequency and severity may increase (Jin et al. 2016). Such changes may increase autochthonous primary production in rivers and reservoirs with broad implications for carbon cycling in aquatic ecosystems (Hosen et al. 2019). Autochthonous production in inland waterways appears to have strengthened in response to climate changes and land use (Liu et al. 2021). Climate anomalies may result in excessive precipitation in the watershed region, resulting in significant erosion of terrestrial soil OC, increasing the C input to the reservoir (Kim et al. 2000). Greater allochthonous organic and inorganic carbon inputs will result in increased sedimentation rates, which will create a physical barrier that prevents oxygen from reaching older sediment layers, preventing OC mineralization, and resulting in higher C burial efficiency (Sobek et al. 2012). Global temperature rise can cause increased drought frequency and severity that will ultimately affect the drawdown area that happened to the Theewaterskloof Dam, South Africa, culminating in a 73% drawdown area of the maximum reservoir area during the water crisis of 2018 (Pekel et al. 2016). The development of drawdown areas exposes the bottom sediments to the atmosphere, consequently changing oxygen-deficient to oxygen-rich conditions (Keller et al. 2021). Methanogenesis fades away; aerobic respiration stimulates enzymatic activities and overall microbial growth (Harrison et al. 2017). The major implication of the drawdown area is enhanced emission of GHGs (Marcé et al. 2019). The increased temperature will also cause elongated thermal stratification of the topwater column ensures poor DIC exchange from the deeper water layer, which will help in the burial of inorganic carbon as $CaCO_3$ (Wang et al. 2019b). Further, thermal stratification reduces the mixing of the oxic and anoxic layers, limiting mineralization activities in reservoir sediments sustaining the C burial (Williamson et al. 2009; He et al. 2020a). As a result, the increasing global mean temperature is thought to aid in the persistence of thermal stratification for extended periods, resulting in higher burial rates and lower mineralization at the sediment level (Sobek et al. 2009). Hence, climate change will have positive and negative effects, with positive effects like C burial in sediments, increased C supply, increased autotrophic production, and negative consequences like faster mineralization rates and massive CO_2 and CH_4 emissions. However, many researchers concur that, in the long term, climate change will lead to an increase in C burial in reservoirs.

11.9 Conclusion

Reservoirs are an essential part of the global carbon cycle; however, their contribution may be altered in the future because of climate change, increased impoundment area, and anthropogenic activities. C sequestration in sediments, GHGs emissions to the atmosphere, downstream degassing, and altered C transit to the sea are examples of these alterations. Despite increased anthropogenic OC loading to watersheds, riverine OC export might significantly decrease OC export to the coastal zone as the number of dams continues to rise. According to published research, reservoirs release more GHG and store more carbon than rivers and lakes. Changes in land use and landcover in the watershed region will affect C supply and biogeochemical processes involved in C cycling. Agriculture, a major source of nitrogen and phosphorus for the reservoir system, will also have a considerable influence. Despite being a little-studied portion of the reservoir, the drawdown zone is crucial in reservoir C dynamics. However, more research is needed to determine how the drawdown area influences the overall carbon footprint of a reservoir. Climate change will influence reservoir C mineralization and sequestration; however, evidence of these phenomena is limited, and additional study is needed to support the claims. As a result, global climate change will have positive and negative effects, with positive effects such as increased C supply, higher autotrophic production, and negative consequences such as faster mineralization rates and large CO_2 and CH_4 emissions. Carbon and nutrient availability, temperature distribution, light availability, and the reservoir's capacity to function as both long- and short-term CO_2 sinks are all predicted to be affected by climate change. As a result, reservoirs are essential regulators in the global C cycle, and their contributions have altered drastically due to climate change and human-induced activities and will continue to change as the number of reservoirs increases.

References

Aarnos, H., 2016. Photochemical transformation of dissolved organic matter in aquatic environment: From a boreal lake and Baltic Sea to global coastal ocean.

Abril, G., Guérin, F., Richard, S., Delmas, R., Galy-Lacaux, C., Gosse, P., Tremblay, A., Varfalvy, L., Dos Santos, M.A. and Matvienko, B., 2005. Carbon dioxide and methane emissions and the carbon budget of a 10-year old tropical reservoir (Petit Saut, French Guiana). *Global Biogeochemical Cycles*, 19(4), pp.1–16.

Åkerman Fulford, E., 2017. Organic carbon getting buried deep: A study on a subtropical reservoir and comparison with a chain of reservoirs.

Anderson, K.A. and Downing, J.A., 2006. Dry and wet atmospheric deposition of nitrogen, phosphorus and silicon in an agricultural region. *Water, Air, and Soil Pollution*, 176(1), pp.351–374.

Anderson, N.J., Dietz, R.D. and Engstrom, D.R., 2013. Land-use change, not climate, controls organic carbon burial in lakes. *Proceedings of the Royal Society B: Biological Sciences*, 280(1769), p.20131278.

Anderson, N.J., Bennion, H. and Lotter, A.F., 2014. Lake eutrophication and its implications for organic carbon sequestration in Europe. *Global Change Biology*, 20(9), pp.2741–2751.

Anderson, N.J., Heathcote, A.J., Engstrom, D.R. and Globocarb Data Contributors, 2020. Anthropogenic alteration of nutrient supply increases the global freshwater carbon sink. *Science Advances*, 6(16), p.eaaw2145.

Ashour, M.F., Guan, D. and Zheng, G., 2021. Phytoplankton dynamics at a reservoir in northeast China. In *IOP Conference Series: Earth and Environmental Science* (Vol. 772, No. 1, p. 012092). IOP Publishing.

Aucour, A.M., Sheppard, S.M., Guyomar, O. and Wattelet, J., 1999. Use of ^{13}C to trace origin and cycling of inorganic carbon in the Rhône River system. *Chemical Geology*, 159(1–4), pp.87–105.

Bardhan, P., Naqvi, S.W.A., Karapurkar, S.G., Shenoy, D.M., Kurian, S. and Naik, H., 2017. Isotopic composition of nitrate and particulate organic matter in a pristine dam reservoir of western India: implications for biogeochemical processes. *Biogeosciences*, 14(4), pp.767.

Barros, N., Cole, J.J., Tranvik, L.J., Prairie, Y.T., Bastviken, D., Huszar, V.L., Del Giorgio, P. and Roland, F., 2011. Carbon emission from hydroelectric reservoirs linked to reservoir age and latitude. *Nature Geoscience*, 4(9), pp.593–596.

Bastviken, D., Persson, L., Odham, G. and Tranvik, L., 2004. Degradation of dissolved organic matter in oxic and anoxic lake water. *Limnology and Oceanography*, 49(1), pp.109–116.

Bertilsson, S. and Tranvik, L.J., 2000. Photochemical transformation of dissolved organic matter in lakes. *Limnology and Oceanography*, 45(4), pp.753–762.

Bianchi, T.S., Filley, T., Dria, K. and Hatcher, P.G., 2004. Temporal variability in sources of dissolved organic carbon in the lower Mississippi River. *Geochimica et Cosmochimica Acta*, 68(5), pp.959–967.

Bittar, T.B., Stubbins, A., Vieira, A.A. and Mopper, K., 2015. Characterization and photodegradation of dissolved organic matter (DOM) from a tropical lake and its dominant primary producer, the cyanobacteria Microcystis aeruginosa. *Marine Chemistry*, 177, pp.205–217.

Bottino, F., Cunha-Santino, M.B. and Bianchini, I., 2016. Decomposition of particulate organic carbon from aquatic macrophytes under different nutrient conditions. *Aquatic Geochemistry*, 22(1), pp.17–33.

Brothers, S.M., Hilt, S., Attermeyer, K., Grossart, H.P., Kosten, S., Lischke, B., Mehner, T., Meyer, N., Scharnweber, K. and Köhler, J., 2013. A regime shift from macrophyte to phytoplankton dominance enhances carbon burial in a shallow, eutrophic lake. *Ecosphere*, 4(11), pp.1–17.

Cardoso, S.J., Vidal, L.O.V., Mendonça, R.F., Tranvik, L.J., Sobek, S. and Roland, F., 2013. Spatial variation of sediment mineralization supports differential CO_2 emissions from a tropical hydroelectric reservoir. *Frontiers in Microbiology*, 4, p.101.

Catalán, N., Marcé, R., Kothawala, D.N. and Tranvik, L.J., 2016. Organic carbon decomposition rates controlled by water retention time across inland waters. *Nature Geoscience*, 9(7), pp.501–504.

Chmiel, H.E., Kokic, J., Denfeld, B.A., Einarsdóttir, K., Wallin, M.B., Koehler, B., Isidorova, A., Bastviken, D., Ferland, M.È. and Sobek, S., 2016. The role of sediments in the carbon budget of a small boreal lake. *Limnology and Oceanography*, 61(5), pp.1814–1825.

Clow, D.W., Stackpoole, S.M., Verdin, K.L., Butman, D.E., Zhu, Z., Krabbenhoft, D.P. and Striegl, R.G., 2015. Organic carbon burial in lakes and reservoirs of the conterminous United States. *Environmental Science & Technology*, 49(13), pp.7614–7622.

Cole, J.J., Carpenter, S.R., Kitchell, J.F. and Pace, M.L., 2002. Pathways of organic carbon utilization in small lakes: Results from a whole-lake ^{13}C addition and coupled model. *Limnology and Oceanography*, 47(6), pp.1664–1675.

Cole, J.J., Prairie, Y.T., Caraco, N.F., McDowell, W.H., Tranvik, L.J., Striegl, R.G., Duarte, C.M., Kortelainen, P., Downing, J.A., Middelburg, J.J. and Melack, J., 2007. Plumbing the global carbon cycle: integrating inland waters into the terrestrial carbon budget. *Ecosystems*, 10(1), pp.172–185.

Coursolle, C., Margolis, H.A., Giasson, M.A., Bernier, P.Y., Amiro, B.D., Arain, M.A., Barr, A.G., Black, T.A., Goulden, M.L., McCaughey, J.H. and Chen, J.M., 2012. Influence of stand age on the magnitude and seasonality of carbon fluxes in Canadian forests. *Agricultural and Forest Meteorology*, 165, pp.136–148.

De Montety, V., Martin, J.B., Cohen, M.J., Foster, C. and Kurz, M.J., 2011. Influence of diel biogeochemical cycles on carbonate equilibrium in a karst river. *Chemical Geology*, 283(1–2), pp.31–43.

De Paoli, M., 2021. Influence of reservoir properties on the dynamics of a migrating current of carbon dioxide. *Physics of Fluids*, 33(1), p.016602.

Dean, W.E. and Gorham, E., 1998. Magnitude and significance of carbon burial in lakes, reservoirs, and peatlands. *Geology*, 26(6), pp.535–538.

Deemer, B.R., Harrison, J.A., Li, S., Beaulieu, J.J., DelSontro, T., Barros, N., Bezerra-Neto, J.F., Powers, S.M., Dos Santos, M.A. and Vonk, J.A., 2016. Greenhouse gas emissions from reservoir water surfaces: a new global synthesis. *BioScience*, 66(11), pp.949–964.

Del Giorgio, P.A., Cole, J.J., Caraco, N.F. and Peters, R.H., 1999. Linking planktonic biomass and metabolism to net gas fluxes in northern temperate lakes. *Ecology*, 80(4), pp.1422–1431.

Donald, D.B., Parker, B.R., Davies, J.M. and Leavitt, P.R., 2015. Nutrient sequestration in the Lake Winnipeg watershed. *Journal of Great Lakes Research*, 41(2), pp.630–642.

Downing, J.A., Cole, J.J., Middelburg, J.J., Striegl, R.G., Duarte, C.M., Kortelainen, P., Prairie, Y.T. and Laube, K.A., 2008. Sediment organic carbon burial in agriculturally eutrophic impoundments over the last century. *Global Biogeochemical Cycles*, 22(1), pp.1–10.

Einsele, G., Yan, J. and Hinderer, M., 2001. Atmospheric carbon burial in modern lake basins and its significance for the global carbon budget. *Global and Planetary Change*, 30(3–4), pp.167–195.

Elser, J.J., Bracken, M.E., Cleland, E.E., Gruner, D.S., Harpole, W.S., Hillebrand, H., Ngai, J.T., Seabloom, E.W., Shurin, J.B. and Smith, J.E., 2007. Global analysis of nitrogen and phosphorus limitation of primary producers in freshwater, marine and terrestrial ecosystems. *Ecology Letters*, 10(12), pp.1135–1142.

Evans, C.D., Futter, M.N., Moldan, F., Valinia, S., Frogbrook, Z. and Kothawala, D.N., 2017. Variability in organic carbon reactivity across lake residence time and trophic gradients. *Nature Geoscience*, 10(11), pp.832–835.

Fearnside, P.M. and Pueyo, S., 2012. Greenhouse-gas emissions from tropical dams. *Nature Climate Change*, 2(6), pp.382–384.

Finlay, K., Leavitt, P.R., Wissel, B. and Prairie, Y.T., 2009. Regulation of spatial and temporal variability of carbon flux in six hard-water lakes of the northern Great Plains. *Limnology and Oceanography*, 54(6 part 2), pp.2553–2564.

Fuentes, N., Güde, H., Wessels, M. and Straile, D., 2013. Allochthonous contribution to seasonal and spatial variability of organic matter sedimentation in a deep oligotrophic lake (Lake Constance). *Limnologica*, 43(2), pp.122–130.

Geraldes, A.M. and Boavida, M.J., 2005. Seasonal water level fluctuations: implications for reservoir limnology and management. *Lakes & Reservoirs: Research & Management*, 10(1), pp.59–69.

Gougoulias, C., Clark, J.M. and Shaw, L.J., 2014. The role of soil microbes in the global carbon cycle: tracking the below-ground microbial processing of plant-derived carbon for manipulating carbon dynamics in agricultural systems. *Journal of the Science of Food and Agriculture*, 94(12), pp.2362–2371.

Grasset, C., Mendonça, R., Villamor Saucedo, G., Bastviken, D., Roland, F. and Sobek, S., 2018. Large but variable methane production in anoxic freshwater sediment upon addition of allochthonous and autochthonous organic matter. *Limnology and Oceanography*, 63(4), pp.1488–1501.

Groeger, A.W. and Kimmel, B.L., 1984. Organic matter supply and processing in lakes and reservoirs. *Lake and Reservoir Management*, 1(1), pp.282–285.

Gudasz, C., Bastviken, D., Steger, K., Premke, K., Sobek, S. and Tranvik, L.J., 2010. Temperature-controlled organic carbon mineralization in lake sediments. *Nature*, 466(7305), pp.478–481.

Gudasz, C., Sobek, S., Bastviken, D., Koehler, B. and Tranvik, L.J., 2015. Temperature sensitivity of organic carbon mineralization in contrasting lake sediments. *Journal of Geophysical Research: Biogeosciences*, 120(7), pp.1215–1225.

Hansen, A.M., Kraus, T.E., Pellerin, B.A., Fleck, J.A., Downing, B.D. and Bergamaschi, B.A., 2016. Optical properties of dissolved organic matter (DOM): effects of biological and photolytic degradation. *Limnology and Oceanography*, 61(3), pp.1015–1032.

Harrison, J.A., Deemer, B.R., Birchfield, M.K. and O'Malley, M.T., 2017. Reservoir water-level drawdowns accelerate and amplify methane emission. *Environmental Science & Technology*, 51(3), pp.1267–1277.

Hayes, N.M., Deemer, B.R., Corman, J.R., Razavi, N.R. and Strock, K.E., 2017. Key differences between lakes and reservoirs modify climate signals: a case for a new conceptual model. *Limnology and Oceanography Letters*, 2(2), pp.47–62.

He, D., Wang, K., Pang, Y., He, C., Li, P., Li, Y., Xiao, S., Shi, Q. and Sun, Y., 2020a. Hydrological management constraints on the chemistry of dissolved organic matter in the Three Gorges Reservoir. *Water Research*, 187, p.116413.

He, H., Liu, Z., Chen, C., Wei, Y., Bao, Q., Sun, H. and Yan, H., 2020b. The sensitivity of the carbon sink by coupled carbonate weathering to climate and land-use changes: sediment records of the biological carbon pump effect in Fuxian Lake, Yunnan, China, during the past century. *Science of the Total Environment*, 720, p.137539.

Holgerson, M.A. and Raymond, P.A., 2016. Large contribution to inland water CO_2 and CH_4 emissions from very small ponds. *Nature Geoscience*, 9(3), pp.222–226.

Hosen, J.D., Aho, K.S., Appling, A.P., Creech, E.C., Fair, J.H., Hall Jr, R.O., Kyzivat, E.D., Lowenthal, R.S., Matt, S., Morrison, J. and Saiers, J.E., 2019. Enhancement of primary production during drought in a temperate watershed is greater in larger rivers than headwater streams. *Limnology and Oceanography*, 64(4), pp.1458–1472.

Jacinthe, P.A., Filippelli, G.M., Tedesco, L.P. and Raftis, R., 2012. Carbon storage and greenhouse gases emission from a fluvial reservoir in an agricultural landscape. *Catena*, 94, pp.53–63.

Jin, H., Yoon, T.K., Lee, S.H., Kang, H., Im, J. and Park, J.H., 2016. Enhanced greenhouse gas emission from exposed sediments along a hydroelectric reservoir during an extreme drought event. *Environmental Research Letters*, 11(12), p.124003.

Johnson, E.R., Inamdar, S., Kan, J. and Vargas, R., 2018. Particulate organic matter composition in stream runoff following large storms: role of POM sources, particle size, and event characteristics. *Journal of Geophysical Research: Biogeosciences*, 123(2), pp.660–675.

Keaveney, E.M., Radbourne, A.D., McGowan, S., Ryves, D.B. and Reimer, P.J., 2020. Source and quantity of carbon influence its sequestration in Rostherne Mere (UK) sediment: a novel application of stepped combustion radiocarbon analysis. *Journal of Paleolimnology*, 64(4), pp.347–363.

Keller, P.S., Marcé, R., Obrador, B. and Koschorreck, M., 2021. Global carbon budget of reservoirs is overturned by the quantification of drawdown areas. *Nature Geoscience*, 14, pp.1–7.

Kellerman, A.M., Dittmar, T., Kothawala, D.N. and Tranvik, L.J., 2014. Chemodiversity of dissolved organic matter in lakes driven by climate and hydrology. *Nature Communications*, 5(1), pp.1–8.

Kellerman, A.M., Kothawala, D.N., Dittmar, T. and Tranvik, L.J., 2015. Persistence of dissolved organic matter in lakes related to its molecular characteristics. *Nature Geoscience*, 8(6), pp.454–457.

Kemenes, A., Forsberg, B.R. and Melack, J.M., 2007. Methane release below a tropical hydroelectric dam. *Geophysical Research Letters*, 34(12), pp.1–5.

Kim, Y., 2015. Effect of thaw depth on fluxes of CO_2 and CH_4 in manipulated Arctic coastal tundra of Barrow, Alaska. *Science of the Total Environment*, 505, pp.385–389.

Kim, B., Choi, K., Kim, C., Lee, U.H. and Kim, Y.H., 2000. Effects of the summer monsoon on the distribution and loading of organic carbon in a deep reservoir, Lake Soyang, Korea. *Water Research*, 34(14), pp.3495–3504.

Kim, J., Han, M.S. and Jheong, W., 2005. A comparative study on the phytoplankton dynamics between two large artificial reservoirs in Korea. *Internationale Vereinigung für theoretische und angewandte Limnologie: Verhandlungen*, 29(2), pp.754–759.

Knoll, L.B., Vanni, M.J., Renwick, W.H., Dittman, E.K. and Gephart, J.A., 2013. Temperate reservoirs are large carbon sinks and small CO_2 sources: results from high-resolution carbon budgets. *Global Biogeochemical Cycles*, 27(1), pp.52–64.

Knoll, L.B., Vanni, M.J., Renwick, W.H. and Kollie, S., 2014. Burial rates and stoichiometry of sedimentary carbon, nitrogen and phosphorus in Midwestern US reservoirs. *Freshwater Biology*, 59(11), pp.2342–2353.

Koehler, B., Von Wachenfeldt, E., Kothawala, D. and Tranvik, L.J., 2012. Reactivity continuum of dissolved organic carbon decomposition in lake water. *Journal of Geophysical Research: Biogeosciences*, 117(G1), pp.1–14.

Kunz, M.J., Anselmetti, F.S., Wüest, A., Wehrli, B., Vollenweider, A., Thüring, S. and Senn, D.B., 2011a. Sediment accumulation and carbon, nitrogen, and phosphorus deposition in the large tropical reservoir Lake Kariba (Zambia/Zimbabwe). *Journal of Geophysical Research: Biogeosciences*, 116(G3), pp.1–13.

Kunz, M.J., Wüest, A., Wehrli, B., Landert, J. and Senn, D.B., 2011b. Impact of a large tropical reservoir on riverine transport of sediment, carbon, and nutrients to downstream wetlands. *Water Resources Research*, 47(12), pp.1–16.

Kuo, Y.M., Chuang, I., Chu, H.J., Wu, J.T. and Jang, C.S., 2013. Identifying the factors affecting phytoplankton abundance dynamics in Shihmen Reservoir, Taiwan. In *EGU General Assembly Conference Abstracts* (pp. EGU2013-65).

Lee, Y., Ha, S.Y., Park, H.K., Han, M.S. and Shin, K.H., 2015. Identification of key factors influencing primary productivity in two river-type reservoirs by using principal component regression analysis. *Environmental Monitoring and Assessment*, 187(4), pp.1–12.

Li, S. and Zhang, Q., 2014. Carbon emission from global hydroelectric reservoirs revisited. *Environmental Science and Pollution Research*, 21(23), pp.13636–13641.

Li, S.L., Liu, C.Q., Li, J., Lang, Y.C., Ding, H. and Li, L., 2010. Geochemistry of dissolved inorganic carbon and carbonate weathering in a small typical karstic catchment of Southwest China: isotopic and chemical constraints. *Chemical Geology*, 277(3–4), pp.301–309.

Li, J., Crowe, S.A., Miklesh, D., Kistner, M., Canfield, D.E. and Katsev, S., 2012. Carbon mineralization and oxygen dynamics in sediments with deep oxygen penetration, Lake Superior. *Limnology and Oceanography*, 57(6), pp.1634–1650.

Li, Z., Zhao, Y., Xu, X., Han, R., Wang, M. and Wang, G., 2018. Migration and transformation of dissolved carbon during accumulated cyanobacteria decomposition in shallow eutrophic lakes: a simulated microcosm study. *PeerJ*, 6, p.e5922.

Li, Y., Meng, F., Wang, B., Yang, M., Liu, C.Q. and Xu, S., 2021. Regulation of particulate inorganic carbon by phytoplankton in hydropower reservoirs: evidence from stable carbon isotope analysis. *Chemical Geology*, 579, p.120366.

Liu, Z., Yan, H. and Zeng, S., 2021. Increasing autochthonous production in inland waters as a contributor to the missing carbon sink. *Frontiers in Earth Science*, 9, pp.1–11.

Luo, Z., Ma, J.M., Zheng, S.L., Nan, C.Z. and Nie, L.M., 2016. Different hydrodynamic conditions on the deposition of organic carbon in sediment of two reservoirs. *Hydrobiologia*, 765(1), pp.15–26.

Maavara, T., Lauerwald, R., Regnier, P. and Van Cappellen, P., 2017. Global perturbation of organic carbon cycling by river damming. *Nature Communications*, 8(1), pp.1–10.

Maavara, T., Chen, Q., Van Meter, K., Brown, L.E., Zhang, J., Ni, J. and Zarfl, C., 2020. River dam impacts on biogeochemical cycling. *Nature Reviews Earth & Environment*, 1(2), pp.103–116.

Maberly, S.C., Barker, P.A., Stott, A.W. and De Ville, M.M., 2013. Catchment productivity controls CO_2 emissions from lakes. *Nature Climate Change*, 3(4), pp.391–394.

Marcé, R., Obrador, B., Gómez-Gener, L., Catalán, N., Koschorreck, M., Arce, M.I., Singer, G. and von Schiller, D., 2019. Emissions from dry inland waters are a blind spot in the global carbon cycle. *Earth-Science Reviews*, 188, pp.240–248.

Mayorga, E., Aufdenkampe, A.K., Masiello, C.A., Krusche, A.V., Hedges, J.I., Quay, P.D., Richey, J.E. and Brown, T.A., 2005. Young organic matter as a source of carbon dioxide outgassing from Amazonian rivers. *Nature*, 436(7050), pp.538–541.

Melack, J.M. and Engle, D.L., 2009. An organic carbon budget for an Amazon floodplain lake. *Internationale Vereinigung für theoretische und angewandte Limnologie: Verhandlungen*, 30(8), pp.1179–1182.

Mendonça, R., Barros, N., Vidal, L.O., Pacheco, F., Kosten, S. and Roland, F., 2012a. Greenhouse gas emissions from hydroelectric reservoirs: what knowledge do we have and what is lacking. *Greenhouse Gases–Emission, Measurement and Management*, edited by: Liu, DG, pp.55–78. InTech Open.

Mendonca, R., Kosten, S., Sobek, S., Barros, N., Cole, J.J., Tranvik, L. and Roland, F., 2012b. Hydroelectric carbon sequestration. *Nature Geoscience*, 5(12), pp.838–840.

Mendonça, R., Kosten, S., Sobek, S., Cole, J.J., Bastos, A.C., Albuquerque, A.L., Cardoso, S.J. and Roland, F., 2014. Carbon sequestration in a large hydroelectric reservoir: an integrative seismic approach. *Ecosystems*, 17(3), pp.430–441.

Mendonça, R., Kosten, S., Sobek, S., Cardoso, S.J., Figueiredo-Barros, M.P., Estrada, C.H.D. and Roland, F., 2015. Organic carbon burial efficiency in a large tropical hydroelectric reservoir. *Biogeosciences Discussions*, 12(22), pp.18513–18540.

Mendonça, R., Kosten, S., Sobek, S., Cardoso, S.J., Figueiredo-Barros, M.P., Estrada, C.H.D. and Roland, F., 2016. Organic carbon burial efficiency in a subtropical hydroelectric reservoir. *Biogeosciences*, 13(11), pp.3331–3342.

Mendonça, R., Müller, R.A., Clow, D., Verpoorter, C., Raymond, P., Tranvik, L.J. and Sobek, S., 2017. Organic carbon burial in global lakes and reservoirs. *Nature Communications*, 8(1), pp.1–7.

Mendoza–Lera, C., Larrañaga, A., Pérez, J., Descals, E., Martínez, A., Moya, O., Arostegui, I. and Pozo, J., 2012. Headwater reservoirs weaken terrestrial-aquatic linkage by slowing leaf-litter processing in downstream regulated reaches. *River Research and Applications*, 28(1), pp.13–22.

Meybeck, M., 1993. Riverine transport of atmospheric carbon: sources, global typology and budget. *Water, Air, and Soil Pollution*, 70(1), pp.443–463.

Milner, A.M., Khamis, K., Battin, T.J., Brittain, J.E., Barrand, N.E., Füreder, L., Cauvy-Fraunié, S., Gíslason, G.M., Jacobsen, D., Hannah, D.M. and Hodson, A.J., 2017. Glacier shrinkage driving global changes in downstream systems. *Proceedings of the National Academy of Sciences*, 114(37), pp.9770–9778.

Mineeva, N.M. and Metelyeva, N.Y., 2019. Comparative characteristics of phytoplankton and epiphyton productivity in the upper volga reservoirs. *Inland Water Biology*, 12(1), pp.37–44.

Minor, E.C., Steinbring, C.J., Longnecker, K. and Kujawinski, E.B., 2012. Characterization of dissolved organic matter in Lake Superior and its watershed using ultrahigh resolution mass spectrometry. *Organic Geochemistry*, 43, pp.1–11.

North, R.L., Johansson, J., Vandergucht, D.M., Doig, L.E., Liber, K., Lindenschmidt, K.E., Baulch, H. and Hudson, J.J., 2015. Evidence for internal phosphorus loading in a large prairie reservoir (Lake Diefenbaker, Saskatchewan). *Journal of Great Lakes Research*, 41, pp.91–99.

Park, H.K., Byeon, M.S., Shin, Y.N. and Jung, D.I., 2009. Sources and spatial and temporal characteristics of organic carbon in two large reservoirs with contrasting hydrologic characteristics. *Water Resources Research*, 45(11), pp.1–12.

Passow, U. and Carlson, C.A., 2012. The biological pump in a high CO_2 world. *Marine Ecology Progress Series*, 470, pp.249–271.

Pekel, J.F., Cottam, A., Gorelick, N. and Belward, A.S., 2016. High-resolution mapping of global surface water and its long-term changes. *Nature*, 540(7633), pp.418–422.

Phyoe, W.W. and Wang, F., 2019. A review of carbon sink or source effect on artificial reservoirs. *International Journal of Environmental Science and Technology*, 16(4), pp.2161–2174.

Phyoe, W.W., Qin, Y., Yu, N. and Wang, F., 2020. Sediment distribution and organic carbon burial in a subtropical hydroelectric reservoir. *Aquatic Sciences*, 82(4), pp.1–13.

Prairie, Y.T., Alm, J., Beaulieu, J., Barros, N., Battin, T., Cole, J., Del Giorgio, P., DelSontro, T., Guérin, F., Harby, A. and Harrison, J., 2018. Greenhouse gas emissions from freshwater reservoirs: what does the atmosphere see?. *Ecosystems*, 21(5), pp.1058–1071.

Qin, Y., Hao, F., Zhang, D., Lang, Y. and Wang, F., 2020. Accumulation of organic carbon in a large canyon reservoir in Karstic area, Southwest China. *Environmental Science and Pollution Research*, 27(20), pp.25163–25172.

Quadra, G.R., Sobek, S., Paranaíba, J.R., Isidorova, A., Roland, F., do Vale, R. and Mendonça, R., 2020. High organic carbon burial but high potential for methane ebullition in the sediments of an Amazonian hydroelectric reservoir. *Biogeosciences*, 17(6), pp.1495–1505.

Rahman, M.M., Tsukamoto, J., Rahman, M.M., Yoneyama, A. and Mostafa, K.M., 2013. Lignin and its effects on litter decomposition in forest ecosystems. *Chemistry and Ecology*, 29(6), pp.540–553.

Ran, L., Lu, X.X. and Xin, Z., 2014. Erosion-induced massive organic carbon burial and carbon emission in the Yellow River basin, China. *Biogeosciences*, 11(4), pp.945–959.

Randerson, J.T., Chapin Iii, F.S., Harden, J.W., Neff, J.C. and Harmon, M.E., 2002. Net ecosystem production: a comprehensive measure of net carbon accumulation by ecosystems. *Ecological Applications*, 12(4), pp.937–947.

Romera-Castillo, C., Sarmento, H., Álvarez-Salgado, X.A., Gasol, J.M. and Marrasé, C., 2011. Net production and consumption of fluorescent colored dissolved organic matter by natural bacterial assemblages growing on marine phytoplankton exudates. *Applied and Environmental Microbiology*, 77(21), pp.7490–7498.

Saklani, U., Shrestha, P.P., Mukherji, A. and Scott, C.A., 2020. Hydro-energy cooperation in South Asia: prospects for transboundary energy and water security. *Environmental Science & Policy*, 114, pp.22–34.

Siriwardana, C., Cooray, A.T., Liyanage, S.S. and Koliyabandara, S.M.P.A., 2019. Seasonal and spatial variation of dissolved oxygen and nutrients in Padaviya Reservoir, Sri Lanka. *Journal of Chemistry*, 2019, 5405016.

Sobek, S., Söderbäck, B., Karlsson, S., Andersson, E. and Brunberg, A.K., 2006. A carbon budget of a small humic lake: an example of the importance of lakes for organic matter cycling in boreal catchments. *AMBIO: A Journal of the Human Environment*, 35, pp.469–475.

Sobek, S., Durisch-Kaiser, E., Zurbrügg, R., Wongfun, N., Wessels, M., Pasche, N. and Wehrli, B., 2009. Organic carbon burial efficiency in lake sediments controlled by oxygen exposure time and sediment source. *Limnology and Oceanography*, 54(6), pp.2243–2254.

Sobek, S., Zurbrügg, R. and Ostrovsky, I., 2011. The burial efficiency of organic carbon in the sediments of Lake Kinneret. *Aquatic Sciences*, 73(3), pp.355–364.

Sobek, S., DelSontro, T., Wongfun, N. and Wehrli, B., 2012. Extreme organic carbon burial fuels intense methane bubbling in a temperate reservoir. *Geophysical Research Letters*, 39(1).

St. Louis, V.L., Kelly, C.A., Duchemin, É., Rudd, J.W. and Rosenberg, D.M., 2000. Reservoir surfaces as sources of greenhouse gases to the atmosphere: a global estimate: reservoirs are sources of greenhouse gases to the atmosphere, and their surface areas have increased to the point where they should be included in global inventories of anthropogenic emissions of greenhouse gases. *BioScience*, 50(9), pp.766–775.

Stallard, R.F., 1998. Terrestrial sedimentation and the carbon cycle: coupling weathering and erosion to carbon burial. *Global Biogeochemical Cycles*, 12(2), pp.231–257.

Stefanidis, K. and Papastergiadou, E., 2010. Influence of hydrophyte abundance on the spatial distribution of zooplankton in selected lakes in Greece. *Hydrobiologia*, 656(1), pp.55–65.

Stets, E.G., Striegl, R.G., Aiken, G.R., Rosenberry, D.O. and Winter, T.C., 2009. Hydrologic support of carbon dioxide flux revealed by whole-lake carbon budgets. *Journal of Geophysical Research: Biogeosciences*, 114(G1), pp.1–14.

Stimson, A.G., Allott, T.E.H., Boult, S. and Evans, M.G., 2017. Reservoirs as hotspots of fluvial carbon cycling in peatland catchments. *Science of the Total Environment*, 580, pp.398–411.

Sun, H.G., Han, J., Lu, X.X., Zhang, S.R. and Li, D., 2010. An assessment of the riverine carbon flux of the Xijiang River during the past 50 years. *Quaternary International*, 226(1-2), pp.38–43.

Teodoru, C.R., Bastien, J., Bonneville, M.C., del Giorgio, P.A., Demarty, M., Garneau, M., Hélie, J.F., Pelletier, L., Prairie, Y.T., Roulet, N.T. and Strachan, I.B., 2012. The net carbon footprint of a newly created boreal hydroelectric reservoir. *Global Biogeochemical Cycles*, 26(2), pp.1–14.

Thomas, R.L., 1988. Lake sediments as indicators of changes in land erosion rates. In *Physical and Chemical Weathering in Geochemical Cycles* (pp. 143–164). Springer, Dordrecht.

Thornton, K.W., Kimmel, B.L. and Payne, F.E., 1990. *Reservoir Limnology: Ecological Perspectives.* John Wiley & Sons.

Thothong, W., Huon, S., Janeau, J.L., Boonsaner, A., De Rouw, A., Planchon, O., Bardoux, G. and Parkpian, P., 2011. Impact of land use change and rainfall on sediment and carbon accumulation in a water reservoir of North Thailand. *Agriculture, Ecosystems & Environment*, 140(3–4), pp.521–533.

Tranvik, L.J., Downing, J.A., Cotner, J.B., Loiselle, S.A., Striegl, R.G., Ballatore, T.J., Dillon, P., Finlay, K., Fortino, K., Knoll, L.B. and Kortelainen, P.L., 2009. Lakes and reservoirs as regulators of carbon cycling and climate. *Limnology and Oceanography*, 54(6part2), pp.2298–2314.

Urabe, J., Iwata, T., Yagami, Y., Kato, E., Suzuki, T., Hino, S. and Ban, S., 2011. Within-lake and watershed determinants of carbon dioxide in surface water: a comparative analysis of a variety of lakes in the Japanese Islands. *Limnology and Oceanography*, 56(1), pp.49–60.

Venkatesh, M. and Anshumali, 2020. Appraisal of the carbon to nitrogen (C/N) ratio in the bed sediment of the Betwa River, Peninsular India. *International Journal of Sediment Research*, 35(1), pp.69–78.

Wang, F., Maberly, S.C., Wang, B. and Liang, X., 2018. Effects of dams on riverine biogeochemical cycling and ecology. *Inland Waters*, 8(2), pp.130–140.

Wang, F., Lang, Y., Liu, C.Q., Qin, Y., Yu, N. and Wang, B., 2019a. Flux of organic carbon burial and carbon emission from a large reservoir: implications for the cleanliness assessment of hydropower. *Science Bulletin*, 64(9), pp.603–611.

Wang, W., Li, S.L., Zhong, J., Li, C., Yi, Y., Chen, S. and Ren, Y., 2019b. Understanding transport and transformation of dissolved inorganic carbon (DIC) in the reservoir system using δ^{13}CDIC and water chemistry. *Journal of Hydrology*, 574, pp.193–201.

Ward, N.D., Bianchi, T.S., Medeiros, P.M., Seidel, M., Richey, J.E., Keil, R.G. and Sawakuchi, H.O., 2017. Where carbon goes when water flows: carbon cycling across the aquatic continuum. *Frontiers in Marine Science*, 4, p.7.

Wetzel, R.G., 2001. *Limnology: Lake and River Ecosystems.* Gulf Professional Publishing.

Wetzel, R.G., Hatcher, P.G. and Bianchi, T.S., 1995. Natural photolysis by ultraviolet irradiance of recalcitrant dissolved organic matter to simple substrates for rapidbacterial metabolism. *Limnology and Oceanography*, 40(8), pp.1369–1380.

Wilkinson, B.H. and McElroy, B.J., 2007. The impact of humans on continental erosion and sedimentation. *Geological Society of America Bulletin*, 119(1–2), pp.140–156.

Williamson, C.E., Saros, J.E., Vincent, W.F. and Smol, J.P., 2009. Lakes and reservoirs as sentinels, integrators, and regulators of climate change. *Limnology and Oceanography*, 54(6part2), pp.2273–2282.

Williamson, T.J., Vanni, M.J. and Renwick, W.H., 2021. Spatial and temporal variability of nutrient dynamics and ecosystem metabolism in a hyper-eutrophic reservoir differ between a wet and dry year. *Ecosystems*, 24(1), pp.68–88.

Yang, L., 2019. Contrasting methane emissions from upstream and downstream rivers and their associated subtropical reservoir in eastern China. *Scientific Reports*, 9(1), pp.1–10.

Yang, Y.X., Xiang, P., Lu, W.Q. and Wang, S.L., 2017. The sedimentation rate and burial fluxes of carbon and nitrogen in Wujiangdu Reservoir, Guizhou, China. *Earth Environment*, 45, pp.66–73.

Yi, Y., Zhong, J., Bao, H., Mostofa, K.M., Xu, S., Xiao, H.Y. and Li, S.L., 2021. The impacts of reservoirs on the sources and transport of riverine organic carbon in the karst area: a multi-tracer study. *Water Research*, 194, p.116933.

Zarfl, C., Lumsdon, A.E., Berlekamp, J., Tydecks, L. and Tockner, K., 2015. A global boom in hydropower dam construction. *Aquatic Sciences*, 77(1), pp.161–170.

Zuijdgeest, A. and Wehrli, B., 2017. Carbon and nutrient fluxes from floodplains and reservoirs in the Zambezi basin. *Chemical Geology*, 467, pp.1–11.

12

Role of Microbes in Controlling the Geochemical Composition of Aquatic Ecosystems

Shilky (ORCID: 0000-0001-8292-0385)[1], Subhashree Patra (ORCID: 0000-0002-4624-6968)[1], Anamika Harshvardhan (ORCID: 0000-0002-0540-1220)[2], Amit Kumar (ORCID: 0000-0002-4582-5677)[3], and Purabi Saikia (ORCID: 0000-0001-5481-282X)[1]

[1] *Department of Environmental Sciences, Central University of Jharkhand, Ranchi, Jharkhand, India*
[2] *Silviculture and Forest Management Division, Institute of Wood Science and Technology, Bengaluru, Karnataka, India*
[3] *Department of Geoinformatics, Central University of Jharkhand, Ranchi, Jharkhand, India*

12.1 Introduction

The most indispensable concern for humankind is the quality of water as it is directly linked with human well-being and prosperity and therefore, designated as one of the most tremendous natural resources on Earth (Kumar 2004). The quality of aquatic bodies and composition of water depend on geological, hydrological, topographical, biological, physicochemical, and meteorological factors (Hamid et al. 2020) and are essential for agricultural, industrial, and domestic activities. Archaea and bacteria play a significant role in the circulation of C, N, P, and S and also act as an early indicator of pollution (Le et al. 2016). On the other hand, zooplanktons inhabiting lakes and ponds contribute to various ecological processes such as material transfer between habitats and play a vital role in the food chain, energy flow, and nutrient cycling (Park and Shin 2007). Small aquatic bodies such as springs, ponds, and small lakes are considered critical for freshwater biodiversity and also recognized for their role in ecosystem delivery in terms of ecosystem goods and services (Biggs et al. 2017). Fish culture being practiced at the commercial level helps in the improvement of the socioeconomic condition of fisher folks (Pal and Chakraborty 2014). Based on the differences in salt content, aquatic ecosystems are broadly grouped into freshwater (salt content $< 0.5\%$), marine or saline water (salt content $> 3.5\%$), and estuarine ecosystems with brackish water (a salt level halfway between fresh and saline water) that cover about three-quarters of the Earth's surface. Aquatic bodies are further divided into lentic and lotic ecosystems that contain wide varieties of life-form including fungi, bacteria, and protozoa along with its bottom-dwelling organisms such as snails, worms, insect larvae, planktons, and macrophytes such as grasses, reeds, and animals like fishes, birds,

Hydrogeochemistry of Aquatic Ecosystems, First Edition. Edited by Sughosh Madhav, Virendra Bahadur Singh, Manoj Kumar, and Sandeep Singh.

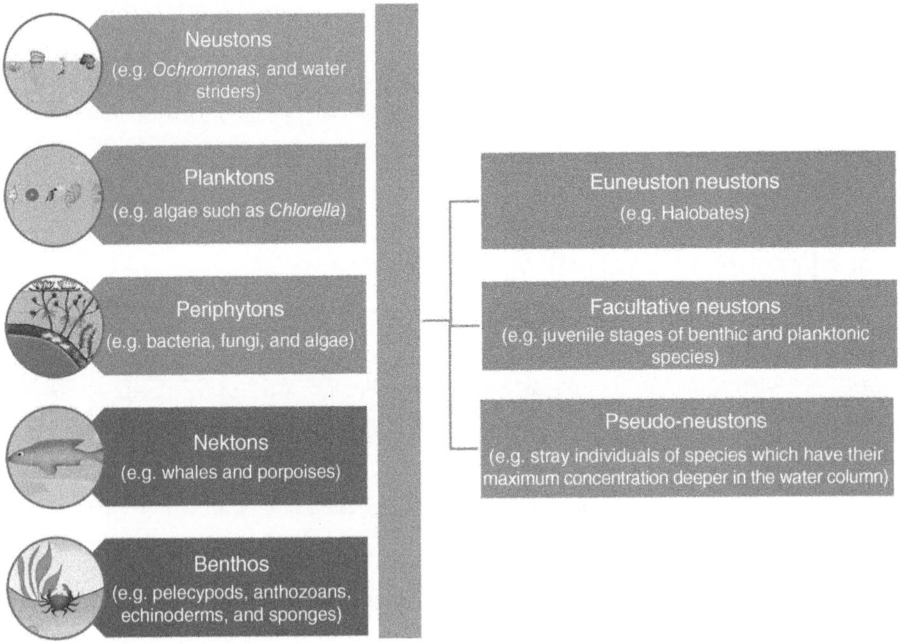

Figure 12.1 Various types of aquatic organisms with examples.

amphibians, and reptiles (Ramachandra and Solanki 2007). The whole complex of plant and animal communities inhabiting aquatic bodies influence the biological properties of the freshwater ecosystem that are closely interconnected with the environment. Based on their occurrence, aquatic organisms are classified into neustons, nektons, periphytons, benthos, and planktons (Figure 12.1). Neustons are free-floating plants and animals that live at the air–water interface. These are important food sources to zooplankton and fishes and are affected by a wide range of endogenous and exogenous processes, although they occupy a limited ecological niche (Albuquerque et al. 2021). Neuston is further divided into three ecological categories viz. euneuston neustons that have maximum prolificity in the vicinity of the surface during both day and night, facultative neustons that have abundance at the surface only during a few hours of the day, and pseudo-neustons that reach the surface for a few hours, although it has a maximum concentration in deeper water (Marshall and Burchardt 2005). On the other hand, nektons are free-floating organisms and a critical component of the estuarine ecosystem that are good sources of food for birds, mammals, and fishes, while some estuarine nekton species are commercially and recreationally important (Litvin et al. 2018). Periphytons are conglomerations of microscopic organisms like algae, filamentous algae, bacteria, and free-floating microorganisms such as rotifers, growing on the surface of engulfed objects in water to cover them with slimy coating that helps in improving water quality by producing oxygen, by taking up nitrate, ammonia, and also by trapping different types of substrates. It is a dominant contributor to nutrient cycling, carbon fixation, and an excellent indicator for changes occurring in aquatic ecosystems (Verdegem et al. 2005). Benthos refer to flora and fauna occurring at bottom substrata of aquatic bodies such as lakes, streams, rivers, and seas having maximum biomass and biodiversity of the Earth (Soto 2020) and are considered as good bioindicators of unfavorable conditions prevailing in marine ecosystems. Similarly, planktons are small organisms composed of viruses, bacteria, phytoplankton, zooplankton, and fishes that dwell in freshwater bodies, oceans, and sea that display an extensive range of capabilities in behaviors that assemble the transition from being a prime species in the ocean to a passive particle (McManus and Woodson 2012). Microorganisms and biogeochemical cycles are the two most important parts of any processes taking place in oceans, lands, and both open and confined environments and simultaneously produce and use greenhouse gases (GHGs) (Abatenh et al. 2018). Microorganisms are good indicators for monitoring drinking water quality because of their various adaptable characteristics and their usefulness in terms of local applications (Wen et al. 2020). Microbial activities have a higher impact on the process of wastewater remediation in wetlands through floating treatment methods (Shahid et al. 2020). Treatments of wastewater using microorganisms can be applied in both in situ and ex situ conditions by involving certain processes such as bioattenuation, biostimulation, and bioaugmentation (Bharagava et al. 2020). The use of molecular techniques to generate transgenic microorganisms for the remediation of environmental pollutants may help in further progress in bioremediation technology (Rabbani et al. 2021). Microorganisms are also used in advanced technologies like microbial fingerprinting for detection, monitoring, and management of drinking water quality (Favere et al. 2020).

Tectonics and atmospheric photochemical processes form geochemical cycles on Earth, which continuously supply substrates and remove products (Williams and Silva 1996). Tectonics, particularly volcanism and rock weathering, are responsible for replenishing C, S, and P on geological time spans (Falkowski et al. 2008). Consequently, global biogeochemical cycles have evolved into a nested system of abiotically driven acid-base and biologically driven redox processes, decreasing the amount of external energy requirements to keep running the cycles and profoundly altering the planet's surface redox state (Falkowski and Godfrey 2008). Photosynthesis is the major known energy transduction technique and does not rely on preformed bond energy to drive the Earth's biological oxidation. Although the quantity of cell and carbon turnover rates of microorganisms are frequently equal between hadal and abyssal locations in the outermost surface layer, they are substantially higher at the hadal trench bottom in marine ecosystems than in abyssal plain sediment below the surface layer (Glud et al. 2013).

12.2 Microbial Communities of Aquatic Bodies and Their Role in Geochemical Processes

Microorganisms play an essential part in biogeochemical cycles as a result of long evolutionary processes (Falkowski et al. 2008). The major elemental cycles and biogeochemical processes on Earth have evolved over the two billion years as the Earth was inhabited initially by prokaryotic bacteria exclusively (Jorgensen 2000). Remineralization, denitrification, nitrogen fixation, methanogenesis, and other essential components of biogeochemical cycling on land and in the water are frequently parameterized bacterial and archaeal processes (Coles and Hood 2016). Many anaerobic biogeochemical activities, including methanogenesis and sulfate reduction, are enhanced in eutrophic systems, and microbial population and their activities are typically higher in eutrophic waters than in oligotrophic waters (Cotner and Biddanda 2002). Cyanobacteria contribute significantly to the balance of nutrients in marine ecosystems, particularly C and N, and produce biologically active substances such as amino acids, enzymes, nucleotides, nucleosides, vitamins, and antibiotics, which they use to interact with other organisms (Andreeva et al. 2020). Planktonic cyanobacteria are important contributors to organic carbon cycling, despite their low diversity and small cell size. *Prochlorococcus* strains, in particular, are excellent in scavenging ammonium from seawater, regardless of their inability to metabolize nitrate (Martiny et al. 2009). They have a significant impact on life on Earth as O_2 generators, N_2 fixers, and primary producers of organic materials in ecosystems (Gaysina et al. 2019). Moreover, they are of particular importance in the aeration of the oceans (Andreeva et al. 2020) and in the formation of calcareous rocks. The actions of cyanobacteria and eukaryotic photosynthetic organisms have generated the current ocean-atmosphere system (Butterfield 2015). Many oxidation-reduction reactions required in geochemical cycling are catalyzed by microorganisms, in which oxidized species like O_2, No^{3-}, So_4^{2-}, and As(V) are reduced during respiration for energy conservation, while many reduced species including H_2, H_2S, Fe(II), and As(III) serve as electron donors (Amend and Shock 2001). Despite the fact that fungi are important decomposers of particulate detritus and play important roles in detrital food webs in freshwater environments, their contributions to biogeochemical cycles in most freshwater settings have been understudied (Gulis et al. 2006). On the other hand, viruses are important agents in the aquatic ecosystems, and they play a key role from microscale microbial processes to global biogeochemical cycling including water quality management (Suttle 2007). Viruses kill both bacteria and phytoplankton, suggesting that viral lysis may exert tremendous control over microbial populations in eutrophic lakes (Tijdens et al. 2008). Microbial communities of quarry lake sediment are more diverse than those found in the other water bodies (Kumar et al. 2019a) as sediments are more stable than surface interstitial waters, and operate as a substrate for strong biogeochemical processes (Savidge et al. 2016). *Geobacter,* a metal-reducing microbe found in soils and sediments, is believed to participate in both direct and interspecies electron transfer to methane-producing archaea (Smith et al. 2014). Spirochaetes bacteria like *Leptospira*, *Treponema*, and *Aminobacterium* may help with organic waste decomposition in sludge and sediments (Domaizon et al. 2013). Furthermore, *Euryarchaeota* is involved in methanogenesis, while two significant archaea species, *Thaumarchaeota* and *Crenarchaeota*, are involved in nitrification (Kerou et al. 2016). In a soft water lake, up to 55% of carbon input was regenerated as methane over the course of a year, with methane oxidation recycling 36% of this total carbon input (Rudd and Hamilton 1978).

Heterotrophic bacteria play a crucial role in the decomposition of organic substances, which aids in the release of CO_2, mineral nutrients (Cole et al. 1994), and the reinclusion of dissolved organic matter (DOM) into the food chain via a microbial loop. Organic matter decomposition, carbon fixation, and the generation of nutrients for heterotrophic organisms are all common roles for actinomycetes and cyanobacteria (Bai et al. 2017) (Table 12.1). Bacterial production and respiration are two important activities necessary for the aquatic carbon cycle, and these microbial activities may also be influenced by the availability of nutrients like N and P, quality of DOM, stoichiometry, and pressure of bacterial predation (Sarmento and Gasol 2012). Six primary elements, viz. C, O, H, N, P, and S, make up all biological macromolecules, with the first five being primarily driven by microbially catalyzed, thermodynamically constrained redox reactions (Borch et al. 2010). Primary

Table 12.1 Various microbial groups used in various geological processes in aquatic ecosystems.

Microbial group	Microorganisms	Utilized in	Sources
Bacteria	*Anabaena azollae*	N_2 fixation	Carrapico (2002)
	Arthrobacter spp.		Stirk and van Staden (2003)
	Pseudomonas sp.	Degradation of polymers	Tribedi and Sil (2013)
	Members of *Cytophaga-Flavobacteria-Bacteroidetes*	Allelopathic activity against algae and cyanobacteria	Hempel et al. (2008)
	Pseudomonas sp.	Denitrification	Ying-ru et al. (2013)
	Nitrosomonas sp.	Ammonification	Okabe et al. (2012)
	Rhodococcus sp.	Phenol degradation in biotreatment	Paisio et al. (2012)
	Alcaligenes odorans, Bacillus subtilis, Corynebacterium propinquum, Pseudomonas aeruginosa	Bioremediation of oil	Singh et al. (2013)
	Pseudomonas cepacia, Bacillus cereus, B. coagulans, Citrobacter koseri, Serratia ficaria	Bioremediation of diesel oil and crude oil	Maliji et al. (2013); Kehinde and Isaac (2016)
	Exiguobacterium indicum, E. aurantiacums, Bacillus cereus, Acinetobacter baumannii	Bioremediation of azo dyes effluents	Kumar et al. (2016)
	Azotobacter sp.	Bind Cr and Cd	Joshi and Juwarkar (2009)
Cyanobacteria	*Richelia intracellulans*	N_2 fixation	Hay et al. (2004)
Fungi	*Aspergillus niger, Candida glabrata, C. krusei, Saccharomyces cerevisiae*	Bioremediation of crude oil	Burghal et al. (2016)
	S. cerevisiae	Act as biosorbent and remove Zn and Cd	Talos et al. (2009)
	Cunninghamella elegans	Remediation of textile effluent with heavy metals	Tigini et al. (2010)
	Allescheriella sp., *Klebsiella oxytoca, Stachybotrys* sp., *Pleurotus pulmonarius, Phlebia* sp., *Botryosphaeria rhodina*	Bind heavy metals	D'Annibale et al. (2007)
	Aspergillus parasiticus, Cephalosporium aphidicola	Biosorption of Pb-contaminated soil	Akar et al. (2007)
	Hymenoscyphus ericae, Neocosmospora vasinfecta, Verticillium terrestre	Convert toxic Hg(II) to nontoxic form	Kelly et al. (2006)
	Phanerochaete chrysosporium	Lignin decomposition	Kirk et al. (1976)
Algae	*Chlorella vulgaris, Spirulina maxima*	Wastewater treatment	Chan et al. (2014)

producers and bacteria assimilate and fix the dissolved orthophosphate of water, which is then passed up to the aquatic food chain, helping in fish production (Maitra et al. 2015). Citric, gluconic, keto-gluconic, lactic, and other organic acids are produced by phosphate-solubilizing bacteria to dissolve mineral-bound insoluble phosphate (Lin et al. 2006). P is negatively charged and prefers to bind with cations, and heterotrophic bacteria play a great role in P cycling in both seawater and freshwater ecosystems, especially organic phosphorus-mineralizing bacteria play a prime part in the P cycle in lake sediment (Zhou et al. 2011). Microbial solubilization of Ca–P varies from the liberation of other mineral-bound P forms by redox or anaerobic microbes (Maitra et al. 2015). Seasonality is a deciding factor in the immobilization of total phosphorus and metals by the periphyton in the eutrophic ecosystem and wet period is the best time for it (Balle et al. 2021). The heterotrophic and autotrophic prokaryotic microorganisms produce an enormous amount of reactive N by N_2 fixation in the wetlands (Vitousek et al. 2002) and majority of inorganic N is found in the form of ammonium in sediments of wetlands (Zhu et al. 2010). The ability of cyanobacteria to fix nitrogen in wetlands is dependent on the amount of solar radiation available (Vitousek et al. 2002). *Azospirillum, Enterobacter, Klebsiella, Pseudomonas,* and *Vibrio* are some of the most common N-fixing bacteria found in wetlands (Lamers et al. 2012). Most aquatic macrophytes, including *Juncus effusus, J. balticus, Phragmites australis, Sagittaria trifolia,* and *Zostera marina*, used to have nitrogen-fixing bacteria such as *Azospirillum* sp. in their rhizospheres or roots (Knapp 2012) and stimulate the production of auxins, to affect pH and redox potential, as well as to speed up the nitrogen fixation process (Oyewole 2012). No extra N is required by the decomposer at a C:N ratio of around 20 : 1 (Daubenmire 1974), ensuring positive net mineralization through ammonium generation (Bowden 1987). Under aerobic conditions, microbes immobilize N, but under anaerobic conditions, they mineralize it in the litter (Bowden 1986). Eventually, the aerobic litter converted to anaerobic peat, which serves as a net source of available N for plant absorption (Bowden 1987). Fungi are involved in a variety of biogeochemical processes including nutrient cycling and mycogenic mineral production and accomplish a significant role in heavy metal remediation by augmenting their bioavailability and modification into less toxic forms (Aung et al. 2018). Fungal biochemical conversion of nutrients and toxins of household effluents have regional to global implications, as they frequently affect other species and also help in the mobility of various harmful substances (Wurzbacher et al. 2020). Heavy metal removal from polluted water has been demonstrated by various fungal genera viz. *Penicillium, Aspergillus,* and *Rhizopus* (Ahalya et al. 2003). Besides, *Hymenoscyphus ericae, Neocosmospora vasinfecta,* and *Verticillium terrestre* showed the ability to transform the toxic state of Hg(II) to a nontoxic form (Kelly et al. 2006). On the other hand, metal-binding bacteria include *Allescheriella* sp., *Botryosphaeria rhodina, Klebsiella oxytoca, Phlebia* sp., *Pleurotus pulmonarius,* and *Stachybotrys* sp. (D'Annibale et al. 2007). Similarly, by influencing the benthic microbial geochemical processes in numerous pathways, *Escherichia coli* causes niche structure alterations in the aquatic ecosystem and *Bacteroides, Candidatus, Dechloromonas, Desulfomicrobium, Methanofastidiosum,* and *Pseudomonas* are among the benthic nutrient cycling taxa that are being affected by it (Gu et al. 2021).

Primary producers including both phototrophic and chemotrophic microorganisms, as well as heterotrophic secondary producers, such as marine algae, cyanobacteria, bacteria, archaea, fungi, and viruses, recycle dissolved organic carbon and nutrients and aid in the processing of ~50% of the global biogeochemical flux of essential elements like C, N, P, S, and Fe (Suttle 2007; Falkowski et al. 2008). Photoautotrophic microbes are responsible for almost half of CO_2 fixation and O_2 synthesis on Earth, whereas heterotrophic microbes are responsible for the oxidation of organic compounds back into CO_2 (Heimann and Reichstein 2008). The biological sequestration of carbon, the lowering of atmospheric CO_2, and the maintenance of increased atmospheric and oceanic O_2 are driven by the temporal and geographic separation of photo-autotrophy and heterotrophy in the global environment (Oschlies et al. 2018). Chemoautotrophic bacteria fix CO_2 and in combination with anaerobic heterotrophic metabolisms, perform a variety of chemical transformations, including nitrogen fluxes from biologically available states to the production of strong GHGs including nitrous oxide (N_2O) (Devol 2008). Because marine sediments have larger CH_4 and S fluxes than freshwater sediments, therefore, denitrification is more efficient and P fluxes are higher in marine ecosystems (Harris 1999).

Estuaries are one of the most productive ecosystems and key sources of biogenic CH_4 emissions (Abril and Borges 2005) despite covering only 0.4% of the total global pelagic areas (Cloern et al. 2014). Because continental freshwater runoff combines with coastal water, estuaries are responsible for providing terrigenous Si, P, and N to the ocean, resulting in better primary productivity in marine environments (Harrison et al. 2008). With high salinity, redox potential, better organic matter content, and higher nutrient recycling, mangrove habitats are among the world's most prolific ecosystems and hotspots for microorganisms with high abundance and distinctive microbial resources (Holguin et al. 2001). Microorganisms and mangrove plants have a strong interaction and many microbes that reside in mangrove soils, especially nitrogen-fixing, phosphate-solubilizing, and sulfate-reducing bacteria, are beneficial and advantageous for the growth of mangrove trees (Kathiresan and Bingham 2001). In the mangrove ecosystem, S conversion is one of the most active chemical cycles, since microorganisms mediate S transformation in its various forms (Lyimo et al. 2002). Sulfate reduction by microbial clusters related to the *Deltaproteobacteria*, specifically the Desulfobacterales, is vital for anaerobic decomposition of biological materials in soils and sediments of the coast (Meyer and Kuever 2007).

12.3 Role of Microorganisms in Water Quality Management

Microorganisms play an essential role in pollutant removal from the soil, water, and sediments, because of their benefits over traditional remediation procedures (Abatenh et al. 2017). They provide sensitive, relevant, and measurable indicators of ecological change, both in terms of detection and consequences, and act as principal engineers in managing all ecological processes (Rani et al. 2019). Heavy metal toxicity is determined not only by the element itself but also simultaneously by its chemical composition and oxidation state (Aracic et al. 2015). Microorganisms can also reduce heavy metals and other hazardous contaminants to a less toxic form (Kot and Namiesnik 2000). Bioremediation is a cost-effective, environmentally friendly, and one of the best alternative techniques to traditional wastewater treatment and to safeguard our aquatic ecosystems (Maszenan et al. 2011) by using microorganisms to break down contaminants in contaminated soil and wastewater (Rani et al. 2019). Microbes make up the majority of aquatic biomass and are responsible for the majority of aquatic yield and nutrient cycling. Mostly the organic pollutants on the roots of the aquatic plants are degraded by the microorganisms by several biological processes (Shahid et al. 2020). They are engaged as biocatalysts in biochemical reactions to destroy targeted contaminants (Abatenh et al. 2017) and help preserve the natural environment by avoiding additional contamination. The luminescence emitted by *Vibrio fischeri*, a marine bacterium, is used to detect phenols in water bodies as the presence of phenols causes a quantitative reduction in the microorganism's luminescence effect (Stolper et al. 2008). Cyanobacteria, which live in both marine and freshwater habitats, have been genetically modified to identify and respond to a variety of toxins by producing a detectable signal (Aracic et al. 2015). Microbial biomass and their enzymatic activities are effective tools for promoting warming because they rapidly degrade carbon-based organic material and release hazardous substances into the environment (Abatenh et al. 2018). They play a significant role as a producer or consumer of GHGs by regenerating and changing the key elements like N and C (Joshi and Shekhawat 2014) by giving positive and negative responses to global warming and changing climate, thus serving as a universal stimulus for climatic change (Rani et al. 2019). Removal of contaminants from effluents using microbes accomplished by physical, chemical, and biological techniques helps in emphasizing lower cost-effective combinations of technologies or innovative alternatives (Crini and Lichtfouse 2019).

Biological wastewater treatment mainly depends on the association of microorganisms especially heterotrophic and autotrophic bacterial populations (Aragaw 2021) that mainly eliminate pollutants such as biochemical oxygen demand (BOD), excess nitrates and phosphates, and heavy metal from the wastewater (Vajda et al. 2011). The diversity of bacteria is influenced by the diversity of aquatic plant species and the amount of contaminants in wastewater, and simultaneously, the availability of nutrients in wastewater affects the bacterial metabolism and their ability to remove pollutants (Shahid et al. 2020). Microbial bioremediation technology is a type of environmental remediation technique that uses microorganisms and their enzymes to degrade and detoxify contaminants in a contaminated medium (Rabbani et al. 2021). Microorganisms and their enzymes are used to perform a major role in the treatment of toxic wastes and the decomposition of organic chemicals in wastewater on an industrial scale, as part of waste recycling and energy generation (Rani et al. 2019). The most often utilized enzyme for degrading biodegradable organic pollutants in wastewater treatment is oxidoreductase (Kumar et al. 2019b) and predominantly extremophilic bacteria have the highest potential as a wastewater degrader in the quality check and management processes (Aragaw 2021). Bioremediation also involves microbes in the degradation, augmentation, and absorption of contaminants in wastewater treatment (Ramadass et al. 2015). The crucial parameters for bioremediation processes are oxygenation, mixing rate and duration, nutrients, carbon sources, and pH (Aragaw 2021). Besides, the type, source, and condition of the contaminants all play a role in microbial wastewater remediation (Silva-Bedoya et al. 2016). The morphological and cellular properties of various extremophiles and their application have phenomenal adaptive responses due to excellent genetic material that can turn them into an interesting source of precursors for wastewater treatment and other industrial processes (Orellana et al. 2018). Algal, yeast, and bacterial biosensors have been used to monitor phenols, nonionic lubricants, pesticides, heavy metals, and industrial effluents in domestic wastewater and natural waters (Girotti et al. 2008). Mainly during the time of infiltration, breakdown of contaminants and stabilization of downstream treatment are done by the soil microbes and fecal bacteria (Aragaw 2021). The downstream treatment method improves the state of the collecting system by altering the physical and chemical features of the treatment plant, therefore improving the system's efficacy in protecting aquatic bodies (Del Nery et al. 2016). Heavy metal elimination by *Pseudomonas* sp. has been used particularly for arsenic-contaminated groundwater and industrial effluent having As(III) compound (Orellana et al. 2018). Mainly the involvement of rhizospheric bacteria, algae, and endophytes in the elimination of pollutants in wastewater treatment has a greater impact on the aquatic ecosystem (Shahid et al. 2020). Microbial cell concentration and community composition with their abundance and stability help to develop various parameters for water quality check and management processes (Favere et al. 2020).

12.4 Problems and Prospects of Using Microbes in Water Quality Management

With economic growth, the problem of water pollution is becoming increasingly significant, affecting the quality of groundwater especially for drinking purposes (Wen et al. 2020). Microbial growth and their activities in water quality management can cause changes in taste and odor of the aquatic body and simultaneously, can increase the risk of health hazards (Wingender and Flemming 2011). Microbes' association in pipe walls can cause pressure loss and can hasten pipe corrosion (Liu et al. 2016). Contamination of drinking water causes many disorders like digestive diseases, infectious diseases, skin diseases, cancer, and others (Wen et al. 2020). Using microbes in water quality management requires sensitive detection systems which can detect the low concentration of contaminants in the environment as exposure of any kind of viruses to the surrounding will be extremely hazardous (Harwood et al. 2013). Thus, the lack of a proper detection system is one of the major limitations of not using microorganisms in water management (Symonds and Breitbart 2015). Microbial enrichment in aquatic ecosystems causes eutrophication that leads to remarkable changes in the biogeochemical cycle, water quality, aquatic diversity, and coastal nutrients (Conley 1999). Excessive microbial proliferation can cause a range of issues during the distribution of drinking water (Favere et al. 2020). The ability to sustain and resist disturbances in the natural environment and fast regeneration ability (Saxena et al. 2015) are some of the other characteristics of microbes that create various problems in their utilization in water quality management.

12.5 Future Research Prospects and Recommendation

- The progress in research in the microbial world evidently demonstrates the occurrence of wide varieties of microorganisms in different aquatic ecosystems and their role in the management of water quality in vivid aspects.
- Industrial sectors of developing countries are experiencing a prime share of economic benefits from microbial resources which make it obligatory to conserve these microorganisms.

- The application of microorganisms to reform contaminated water can be an innovative approach in the coming decades. Usage of microorganisms can be a cost-effective and eco-friendly approach to resolve various water quality-related concerns.
- Identification of appropriate microbial strain is a mandatory prerequisite for the use of microorganisms toward reducing water pollution and water quality management as it is a more convenient, easy, and low-cost technology than several traditional methods.

12.6 Conclusions

The biogeochemical cycling in aquatic bodies is influenced by the interactions between microbial processes and the physiological characteristics of the microorganisms. Heterotrophic bacteria contribute to the carbon cycle by releasing CO_2 and other organic matter, as well as helping recover DOM by re-integrating it into their nutrition. Therefore, phytoplankton growth is stimulated by increased eutrophication of lakes and estuaries. Freshwater ecosystems are important components of the global carbon cycle and they help in minimizing the organic matter transported from various terrestrial ecosystems. Microorganisms have a substantial influence on the chemical and physical composition of subsurface environments and consequently on the fate and environmental impact of CO_2. Their growth and reproduction in aquatic bodies are due to the presence of nutrients from floating plants that enhance metabolism and pollutant removal efficiency of microbes. Microorganisms reduce the heavy metals and other hazardous contaminants in contaminated soil and wastewater to less toxic form by the process of bioremediation. It is one of the best alternatives to traditional wastewater treatment and to safeguard our aquatic ecosystems. They can easily break down complex organic compounds into simpler forms and simultaneously encourage plant growth by alleviating metal toxicity. Temperature, pH, and nutrient availability have a significant impact on the pollutant removal ability of a microorganism. Bioremediation is an eco-friendly, energy-efficient, profitable, and appealing technology for cleaning, controlling, and restoring polluted aquatic bodies through microbial activity. However, microbes can also cause changes in taste and odor of the aquatic body and simultaneously, can increase the risk of health hazards. Still, the use of microbes to reform tainted water could be a novel strategy in the coming decades which warrants further research to identify the appropriate microbial strain.

References

Abatenh, E., Gizaw, B., Tsegaye, Z. and Wassie, M., 2017. The role of microorganisms in bioremediation – a review. *Open Journal of Environmental Biology*, 2(1), pp.38–46.

Abatenh, E., Gizaw, B., Tsegaye, Z. and Tefera, G., 2018. Microbial function on climate change – a review. *Open Journal of Environmental Biology*, 3(1), pp.1–7.

Abril, G. and Borges, A.V., 2005. Carbon dioxide and methane emissions from estuaries. In: Tremblay, A., Varfalvy, L., Roehm, C., Garneau, M. (eds) *Greenhouse Gas Emissions-Fluxes and Processes* (pp. 187–207). Springer, Berlin, Heidelberg.

Ahalya, N., Ramachandra, T.V. and Kanamadi, R.D., 2003. Biosorption of heavy metals. *Research Journal of Chemistry and Environment*, 7(4), pp.71–79.

Akar, T., Tunali, S. and Cabuk, A., 2007. Study on the characterization of lead (II) biosorption by fungus *Aspergillus parasiticus*. *Applied Biochemistry and Biotechnology*, 136(3), pp.389–405.

Albuquerque, R., Bode, A., González-Gordillo, J.I., Duarte, C.M. and Queiroga, H., 2021. Trophic structure of Neuston across tropical and subtropical oceanic provinces assessed with stable isotopes. *Frontiers in Marine Science*, 7, p.1205.

Amend, J.P. and Shock, E.L., 2001. Energetics of overall metabolic reactions of thermophilic and hyperthermophilic Archaea and Bacteria. *FEMS Microbiology Reviews*, 25(2), pp.175–243.

Andreeva, A., Kulesha, E., Gough, J. and Murzin, A.G., 2020. The SCOP database in 2020: expanded classification of representative family and superfamily domains of known protein structures. *Nucleic Acids Research*, 48(D1), pp.D376–D382.

Aracic, S., Manna, S., Petrovski, S., Wiltshire, J.L., Mann, G. and Franks, A.E., 2015. Innovative biological approaches for monitoring and improving water quality. *Frontiers in Microbiology*, 6, p.826.

Aragaw, T.A., 2021. Functions of various bacteria for specific pollutants degradation and their application in wastewater treatment: a review. *International Journal of Environmental Science and Technology*, 18(7), pp.2063–2076.

Aung, K., Jiang, Y. and He, S.Y., 2018. The role of water in plant–microbe interactions. *The Plant Journal*, 93, pp.771–780.

Bai, L., Cao, C., Wang, C., Xu, H., Zhang, H., Slaveykova, V.I. and Jiang, H. 2017. Toward a quantitative understanding of the bioavailability of dissolved organic matter in freshwater lakes during Cyanobacteria blooming. *Environmental Science and Technology*, 51, pp.6018–6026.

Balle, M.G., Ferragut, C., Coelho, L.H.G. and Jesus, T.A.D., 2021. Phosphorus and metals immobilization by periphyton in a shallow eutrophic reservoir. *Acta Limnologica Brasiliensia*, 33, p. e11.

Bharagava, R.N., Saxena, G. and Mulla, S.I., 2020. Introduction to industrial wastes containing organic and inorganic pollutants and bioremediation approaches for environmental management. In *Bioremediation of Industrial Waste for Environmental Safety* (pp. 1–18). Springer, Singapore.

Biggs, J., von Fumetti, S. and Kelly-Quinn, M., 2017. The importance of small waterbodies for biodiversity and ecosystem services: implications for policy makers. *Hydrobiologia*, 793(1), pp.3–39.

Borch, T., Kretzschmar, R., Kappler, A., Cappellen, P.V., Ginder-Vogel, M., Voegelin, A. and Campbell, K., 2010. Biogeochemical redox processes and their impact on contaminant dynamics. *Environmental Science Technology*, 44(1), pp.15–23.

Bowden, W.B., 1986. Gaseous nitrogen emissions from undisturbed terrestrial ecosystems: An assessment of their impacts on local and global nitrogen budgets. *Biogeochemistry*, 2(3), pp.249–279.

Bowden, W.B., 1987. The biogeochemistry of nitrogen in freshwater wetlands. *Biogeochemistry*, 4(3), pp.313–348.

Burghal, A. A., Abu-Mejdad, N. M. J. A., & Al-Tamimi, W. H., 2016. Mycodegradation of crude oil by fungal species isolated from petroleum contaminated soil. *International Journal of Innovative Research in Science, Engineering and Technology*, 5(2), pp. 1517–1524.

Butterfield, N.J., 2015. Proterozoic photosynthesis – a critical review. *Palaeontology*, 58(6), pp.953–972.

Carrapico, F.J., 2002. Azolla-anabaena-bacteria system as a natural microcosm. In *Instruments, Methods, and Missions for Astrobiology IV* (Vol. 4495, pp. 261–265). International Society for Optics and Photonics.

Chan, A., Salsali, H. and McBean, E., 2014. Heavy metal removal (copper and zinc) in secondary effluent from wastewater treatment plants by microalgae. *ACS Sustainable Chemistry and Engineering*, 2(2), pp.130–137.

Cloern, J.E., Foster, S.Q. and Kleckner, A.E., 2014. Phytoplankton primary production in the world's estuarine-coastal ecosystems. *Biogeosciences*, 11(9), pp.2477–2501.

Cole, J.J., Caraco, N.F., Kling, G.W. and Kratz, T.K., 1994. Carbon dioxide supersaturation in the surface waters of lakes. *Science*, 265(5178), pp.1568–1570.

Coles, V.J. and Hood, R.R., 2016. Approaches and challenges for linking marine biogeochemical models with the "omics" revolution. In Glibert, P. & Kana, T. (eds.) *Aquatic Microbial Ecology and Biogeochemistry: A Dual Perspective* (pp. 45–63). Springer, Cham.

Conley, D.J., 1999. Biogeochemical nutrient cycles and nutrient management strategies. In Garnier, J. & Mouchel, JM. (eds) *Man and River Systems* (pp. 87–96). Springer, Dordrecht.

Cotner, J.B. and Biddanda, B.A., 2002. Small players, large role: microbial influence on biogeochemical processes in pelagic aquatic ecosystems. *Ecosystems*, 5(2), pp.105–121.

Crini, G. and Lichtfouse, E., 2019. Advantages and disadvantages of techniques used for wastewater treatment. *Environmental Chemistry Letters*, 17(1), pp.145–155.

D'Annibale, A., Leonardi, V., Federici, E., Baldi, F., Zecchini, F. and Petruccioli, M., 2007. Leaching and microbial treatment of soil contaminated by sulphide ore ashes and aromatic hydrocarbons. *Applied Microbiology and Biotechnology*, 74(5), pp.1135–1144.

Daubenmire, R.F., 1974. Plants and Environment: A Textbook of Plant Autecology. John-Wiley, New York.

Del Nery, V., Damianovic, M.H.Z., Moura, R.B., Pozzi, E., Pires, E.C. and Foresti, E., 2016. Poultry slaughterhouse wastewater treatment plant for high-quality effluent. *Water Science and Technology*, 73(2), pp.309–316.

Devol, A.H., 2008. Nitrogen uptake and assimilation. In Capone, D.G., Bronk, D.A., Mulholland, M.R. and Carpenter, E.J. (eds.) *Nitrogen in the Marine Environment* (pp.263–301). Academic Press.

Domaizon, I., Savichtcheva, O., Debroas, D., Arnaud, F., Villar, C., Pignol, C., Alric, B. and Perga, M.E., 2013. DNA from lake sediments reveals the long-term dynamics and diversity of Synechococcus assemblages. *Biogeosciences*, 10, pp.3817–3838.

Falkowski, P.G. and Godfrey, L.V., 2008. Electrons, life and the evolution of Earth's oxygen cycle. *Philosophical Transactions of the Royal Society B: Biological Sciences*, 363(1504), pp.2705–271.

Falkowski, P.G., Fenchel, T. and Delong, E.F., 2008. The microbial engines that drive Earth's biogeochemical cycles. *Science*, 320(5879), pp.1034–1039.

Favere, J., Buysschaert, B., Boon, N. and De Gusseme, B., 2020. Online microbial fingerprinting for quality management of drinking water: full-scale event detection. *Water Research*, 170, p.115353.

Gaysina, L.A., Saraf, A. and Singh, P., 2019. Cyanobacteria in diverse habitats. In A.K. Mishra, D.N. Tiwari and A.N. Rai. *Cyanobacteria* (pp. 1–28). Academic Press.

Girotti, S., Ferri, E.N., Fumo, M.G. and Maiolini, E., 2008. Monitoring of environmental pollutants by bioluminescent bacteria. *Analytica Chimica Acta*, 608(1), pp.2–29.

Glud, R.N., Wenzhöfer, F., Middelboe, M., Oguri, K., Turnewitsch, R., Canfield, D.E. and Kitazato, H., 2013. High rates of microbial carbon turnover in sediments in the deepest oceanic trench on Earth. *Nature Geoscience*, 6(4), pp.284–288.

Gu, L., Wu, J.Y. and Hua, Z.L., 2021. Benthic prokaryotic microbial community assembly and biogeochemical potentials in *E. coli*-Stressed aquatic ecosystems during plant decomposition. *Environmental Pollution*, 275, p.116643.

Gulis, V., Ferreira, V. and Graça, M.A.S., 2006. Stimulation of leaf litter decomposition and associated fungi and invertebrates by moderate eutrophication: implications for stream assessment. *Freshwater Biology*, 51(9), pp.1655–1669.

Hamid, A., Bhat, S.U. and Jehangir, A., 2020. Local determinants influencing stream water quality. *Applied Water Science*, 10(1), pp.1–16.

Harris, G.P., 1999. Comparison of the biogeochemistry of lakes and estuaries: ecosystem processes, functional groups, hysteresis effects and interactions between macro-and microbiology. *Marine and Freshwater Research*, 50(8), pp.791–811.

Harrison, P.J., Yin, K., Lee, J.H.W., Gan, J. and Liu, H., 2008. Physical–biological coupling in the Pearl River Estuary. *Continental Shelf Research*, 28(12), pp.1405–1415.

Harwood, V.J., Boehm, A.B., Sassoubre, L.M., Vijayavel, K., Stewart, J.R., Fong, T.T., Caprais, M.P., Converse, R.R., Diston, D., Ebdon, J. and Fuhrman, J.A., 2013. Performance of viruses and bacteriophages for fecal source determination in a multi-laboratory, comparative study. *Water Research*, 47(18), pp.6929–6943.

Hay, M.E., Parker, J.D., Burkepile, D.E., Caudill, C.C., Wilson, A.E., Hallinan, Z.P. and Chequer, A.D., 2004. Mutualisms and aquatic community structure: the enemy of my enemy is my friend. *Annual Review of Ecology, Evolution, and Systematics*, 35, pp.175–197.

Heimann, M. and Reichstein, M., 2008. Terrestrial ecosystem carbon dynamics and climate feedbacks. *Nature*, 451(7176), pp.289–292.

Hempel, M., Blume, M., Blindow, I. and Gross, E.M., 2008. Epiphytic bacterial community composition on two common submerged macrophytes in brackish water and fresh water. *BMC Microbiology*, 8(1), pp.1–10.

Holguin, G., Vazquez, P. and Bashan, Y., 2001. The role of sediment microorganisms in the productivity, conservation, and rehabilitation of mangrove ecosystems: an overview. *Biology and Fertility of Soils*, 33(4), pp.265–278.

Jorgensen, B.B., 2000. Bacteria and marine biogeochemistry. In Schulz, H.D., Zabel, M. (eds) *Marine Geochemistry* (pp. 173–207). Springer, Berlin, Heidelberg.

Joshi, P.M. and Juwarkar, A.A., 2009. in vivo studies to elucidate the role of extracellular polymeric substances from Azotobacter in immobilization of heavy metals. *Environmental Science and Technology*, 43(15), pp.5884–5889.

Joshi, P.A. and Shekhawat, D.B., 2014. Microbial contributions to global climate changes in soil environments: impact on Carbon cycle (Short Review). *Annals of Applied Bio-Sciences*, 1(1), pp.R7–R9.

Kathiresan, K. and Bingham, B.L., 2001. Biology of mangroves and mangrove ecosystems. *Advances in Marine Biology*, 40, pp.81–251.

Kehinde, F.O. and Isaac, S.A., 2016. Effectiveness of augmented consortia of *Bacillus coagulans*, *Citrobacter koseri* and *Serratia ficaria* in the degradation of diesel polluted soil supplemented with pig dung. *African Journal of Microbiology Research*, 10(39), pp.1637–1644.

Kelly, D.J., Budd, K. and Lefebvre, D.D., 2006. The biotransformation of mercury in pH-stat cultures of microfungi. *Botany*, 84(2), pp.254–260.

Kerou, M., Offre, P., Valledor, L., Abby, S.S., Melcher, M., Nagler, M., Weckwerth, W. and Schleper, C. 2016. Proteomics and comparative genomics of *Nitrososphaera viennensis* reveal the core genome and adaptations of archaeal ammonia oxidizers. *Proceedings of the National Academy of Sciences*, 113, pp.E7937–E7946.

Kirk, T.K., Connors, W.J. and Zeikus, J.G., 1976. Requirement for a growth substrate during lignin decomposition by two wood-rotting fungi. *Applied and Environmental Microbiology*, 32(1), pp.192–194.

Knapp, A., 2012. The sensitivity of marine N_2 fixation to dissolved inorganic nitrogen. *Frontiers in Microbiology*, 3, p.374.

Kot, A. and Namiesnik, J., 2000. The role of speciation in analytical chemistry. *Trends in Analytical Chemistry*, 19, pp. 69–79.

Kumar, A. (ed.), 2004. *Environment Management*. APH Publishing, New Delhi, India.

Kumar, S., Chaurasia, P. and Kumar, A., 2016. Isolation and characterization of microbial strains from textile industry effluents of Bhilwara, India: analysis with bioremediation. *Journal of Chemical and Pharmaceutical Research*, 8(4), pp.143–150.

Kumar, A., Ng, D.H., Wu, Y. and Cao, B., 2019a. Microbial community composition and putative biogeochemical functions in the sediment and water of tropical granite quarry lakes. *Microbial Ecology*, 77(1), pp.1–11.

Kumar, S.S., Ghosh, P., Malyan, S.K., Sharma, J. and Kumar, V., 2019b. A comprehensive review on enzymatic degradation of the organophosphate pesticide malathion in the environment. *Journal of Environmental Science and Health, Part C*, 37(4), pp.288–329.

Lamers, L.P., van Diggelen, J.M., Op Den Camp, H.J., Visser, E.J., Lucassen, E.C., Vile, M.A., Jetten, M.S., Smolders, A.J. and Roelofs, J.G., 2012. Microbial transformations of nitrogen, sulfur, and iron dictate vegetation composition in wetlands: a review. *Frontiers in Microbiology*, 3, p.156.

Le, H.T., Ho, C.T., Trinh, Q.H., Trinh, D.A., Luu, M.T., Tran, H.S., Orange, D., Janeau, J.L., Merroune, A., Rochelle-Newall, E. and Pommier, T., 2016. Responses of aquatic bacteria to terrestrial runoff: effects on community structure and key taxonomic groups. *Frontiers in Microbiology*, 7, p.889.

Lin, T.F., Huang, H.I., Shen, F.T. and Young, C.C., 2006. The protons of gluconic acid are the major factor responsible for the dissolution of Tricalcium phosphate by *Burkholderia cepacia* CC-Al74. *Bioresource Technology*, 97(7), pp.957–960.

Litvin, S.Y., Weinstein, M.P., Sheaves, M. and Nagelkerken, I., 2018. What makes nearshore habitats nurseries for nekton? An emerging view of the nursery role hypothesis. *Estuaries and Coasts*, 41(6), pp.1539–1550.

Liu, S., Gunawan, C., Barraud, N., Rice, S.A., Harry, E.J. and Amal, R., 2016. Understanding, monitoring, and controlling biofilm growth in drinking water distribution systems. *Environmental Science and Technology*, 50(17), pp.8954–8976.

Lyimo, T.J., Pol, A. and den Camp, H.J.O., 2002. Sulfate reduction and methanogenesis in sediments of Mtoni mangrove forest, Tanzania. *AMBIO: A Journal of the Human Environment*, 31(7), pp.614–616.

Maitra, N., Manna, S.K., Samanta, S., Sarkar, K., Debnath, D., Bandopadhyay, C., Sahu, S.K. and Sharma, A.P., 2015. Ecological significance and phosphorus release potential of phosphate solubilizing bacteria in freshwater ecosystems. *Hydrobiologia*, 745(1), pp.69–83.

Maliji, D., Olama, Z. and Holail, H., 2013. Environmental studies on the microbial degradation of oil hydrocarbons and its application in Lebanese oil-polluted coastal and marine ecosystems. *International Journal of Current Microbiology and Applied Sciences*, 2(6), pp.1–18.

Marshall, H.G. and Burchardt, L., 2005. Neuston: its definition with a historical review regarding its concept and community structure. *Archiv Fur Hydrobiologie*, 164(4), pp.429–448.

Martiny, A.C., Kathuria, S. and Berube, P.M., 2009. Widespread metabolic potential for nitrite and nitrate assimilation among Prochlorococcus ecotypes. *Proceedings of the National Academy of Sciences*, 106(26), pp.10787–10792.

Maszenan, A.M., Liu, Y. and Ng, W.J., 2011. Bioremediation of wastewater with recalcitrant organic compounds and metals by aerobic granules. *Biotechnology Advances*, 29, pp.111–123.

McManus, M.A. and Woodson, C.B., 2012. Plankton distribution and ocean dispersal. *Journal of Experimental Biology*, 215(6), pp.1008–1016.

Meyer, B. and Kuever, J., 2007. Molecular analysis of the diversity of sulfate-reducing and sulfur-oxidizing prokaryotes in the environment, using aprA as a functional marker gene. *Applied and Environmental Microbiology*, 73, pp.7664–7679.

Okabe, S., Nakamura, Y. and Satoh, H., 2012. Community structure and in situ activity of nitrifying bacteria in Phragmites root-associated biofilms. *Microbes and Environments*, 27(3), pp.242–249.

Orellana, R., Macaya, C., Bravo, G., Dorochesi, F., Cumsille, A., Valencia, R., Rojas, C. and Seeger, M., 2018. Living at the frontiers of life: extremophiles in Chile and their potential for bioremediation. *Frontiers in Microbiology*, 9, p.2309.

Oschlies, A., Brandt, P., Stramma, L. and Schmidtko, S., 2018. Drivers and mechanisms of ocean deoxygenation. *Nature Geoscience*, 11(7), pp.467–473.

Oyewole, O., 2012. Microbial communities and their activities in paddy fields: a review. *Journal of Veterinary Advances*, 2, pp.74–80.

Paisio, C.E., Talano, M.A., González, P.S., Busto, V.D., Talou, J.R. and Agostini, E., 2012. Isolation and characterization of a *Rhodococcus* strain with the phenol-degrading ability and its potential use for tannery effluent biotreatment. *Environmental Science and Pollution Research*, 19(8), pp.3430–3439.

Pal, P. and Chakraborty, K., 2014. Importance of some physical and chemical characteristics of water bodies in relation to the incidence of zooplanktons: a review. *Indian Journal of Social and Natural Sciences*, 3, pp.102–116.

Park, K.S. and Shin, H.W., 2007. Studies on Phyto-and-zooplankton composition and its relation to fish productivity in a west coast fish pond ecosystem. *Journal of Environmental Biology*, 28(2), p.415.

Rabbani, A., Zainith, S., Deb, V.K., Das, P., Bharti, P., Rawat, D.S., Kumar, N. and Saxena, G., 2021. Microbial technologies for environmental remediation: potential issues, challenges, and future prospects. In Ajay Kumar, Vipin Kumar Singh, Pardeep Singh, Virendra Kumar Mishra *Microbe Mediated Remediation of Environmental Contaminants* (pp. 271–286). Woodhead Publishing, UK & US.

Ramachandra, T.V. and Solanki, M., 2007. Ecological assessment of lentic water bodies of Bangalore. *The Ministry of Science and Technology*, 25, p.96.

Ramadass, K., Megharaj, M., Venkateswarlu, K. and Naidu, R., 2015. Ecological implications of motor oil pollution: earthworm survival and soil health. *Soil Biology and Biochemistry*, 85, pp.72–81.

Rani, N., Sangwan, P., Joshi, M., Sagar, A. and Bala, K., 2019. Microbes: a key player in industrial wastewater treatment. In Shah, M.P. and Couto, S.R. (eds.) *Microbial Wastewater Treatment* (pp. 83–102). Elsevier, Netherlands, UK, USA.

Rudd, J.W. and Hamilton, R.D., 1978. Methane cycling in a eutrophic shield lake and its effects on whole-lake metabolism 1. *Limnology and Oceanography*, 23(2), pp.337–348.

Sarmento, H. and Gasol, J.M., 2012. Use of phytoplankton derived dissolved organic carbon by different types of bacterioplankton. *Environmental Microbiology*, 14(9), pp.2348–2360.

Savidge, W.B., Wilson, A. and Woodward, G 2016. Using a thermal proxy to examine sediment-water exchange in mid-continental shelf sandy sediments, *Aquatic Geochemistry*, 22(5–6), pp.419–441.

Saxena, G., Bharagava, R.N., Kaithwas, G. and Raj, A., 2015. Microbial indicators, pathogens and methods for their monitoring in a water environment. *Journal of Water and Health*, 13(2), pp.319–339.

Shahid, M.J., AL-surhanee, A.A., Kouadri, F., Ali, S., Nawaz, N., Afzal, M., Rizwan, M., Ali, B. and Soliman, M.H., 2020. Role of microorganisms in the remediation of wastewater in floating treatment wetlands: a review. *Sustainability*, 12(14), p.5559.

Silva-Bedoya, L.M., Sánchez-Pinzón, M.S., Cadavid-Restrepo, G.E. and Moreno-Herrera, C.X., 2016. Bacterial community analysis of an industrial wastewater treatment plant in Colombia with screening for lipid-degrading microorganisms. *Microbiological Research*, 192, pp.313–325.

Singh, A., Kumar, V. and Srivastava, J.N., 2013. Assessment of bioremediation of oil and phenol contents in refinery wastewater via bacterial consortium. *Journal of Petroleum & Environmental Biotechnology*, 4(3), pp.1–4.

Smith, J.A., Tremblay, P.L., Shrestha, P.M., Snoeyenbos-West, O.L., Franks, A.E., Nevin, K.P. and Lovley, D.R., 2014. Going wireless: Fe (III) oxide reduction without pili by *Geobacter sulfurreducens* strain JS-1. *Applied and Environmental Microbiology*, 80, pp.4331–4340.

Soto, L. (ed.), 2020. *Advances in the Studies of the Benthic Zone*. BoD–Books on Demand.

Stirk, W.A. and van Staden, J., 2003. Occurrence of cytokinin-like compounds in two aquatic ferns and their exudates. *Environmental and Experimental Botany*, 49(1), pp.77–85.

Stolper, P., Fabel, S., Weller, M.G., Knopp, D. and Niessner, R., 2008. Whole-cell luminescence-based flow-through biodetector for toxicity testing. *Analytical and Bioanalytical Chemistry*, 390(4), pp.1181–1187.

Suttle, C.A., 2007. Marine viruses-major players in the global ecosystem. *Nature Reviews Microbiology*, 5(10), pp.801–812.

Symonds, E.M. and Breitbart, M., 2015. Affordable enteric virus detection techniques are needed to support changing paradigms in water quality management. *CLEAN–Soil, Air, Water*, 43(1), pp.8–12.

Talos, K., Pager, C., Tonk, S., Majdik, C., Kocsis, B., Kilar, F. and Pernyeszi, T., 2009. Cadmium biosorption on native *Saccharomyces cerevisiae* cells in aqueous suspension. *Agriculture and Environment*, 1, pp.20–30.

Tigini, V., Prigione, V., Giansanti, P., Mangiavillano, A., Pannocchia, A. and Varese, G.C., 2010. Fungal biosorption: an innovative treatment for the decolourisation and detoxification of textile effluents. *Water*, 2(3), pp.550–565.

Tijdens, M., van de Waal, D.B., Slovackova, H., Hoogveld, H.L. and Gons, H.J., 2008. Estimates of bacterial and phytoplankton mortality caused by viral lysis and microzooplankton grazing in a shallow eutrophic lake. *Freshwater Biology*, 53(6), pp.1126–1141.

Tribedi, P. and Sil, A.K., 2013. Low-density polyethylene degradation by *Pseudomonas* sp. AKS2 biofilm. *Environmental Science and Pollution Research*, 20(6), pp.4146–4153.

Vajda, A.M., Barber, L.B., Gray, J.L., Lopez, E.M., Bolden, A.M., Schoenfuss, H.L. and Norris, D.O., 2011. Demasculinization of male fish by wastewater treatment plant effluent. *Aquatic Toxicology*, 103(3-4), pp.213–221.

Verdegem, M.C.J., van Dam, A.A., Azim, M.E. and Beveridge, M.C.M., 2005. Periphyton ecology, exploitation and management: knowledge gaps and directions for future research. In *Periphyton: Ecology, Exploitation and Management* (pp.299–307). CABI Publishing, Wallingford, UK.

Vitousek, P. M., Cassman, K. E. N., Cleveland, C., Crews, T., Field, C. B., Grimm, N. B., . . . and Sprent, J. I. (2002). Towards an ecological understanding of biological nitrogen fixation. *Biogeochemistry*, 57(1), pp.1–45.

Wen, X., Chen, F., Lin, Y., Zhu, H., Yuan, F., Kuang, D., Jia, Z. and Yuan, Z., 2020. Microbial indicators and their use for monitoring drinking water quality – a review. *Sustainability*, 12(6), p.2249.

Williams, R.J.P. and Silva, F.D.J.J.R., 1996. *The Natural Selection of the Chemical Elements*. Clarendon Press, Oxford, UK, pp. 672.

Wingender, J. and Flemming, H.C., 2011. Biofilms in drinking water and their role as reservoirs for pathogens. *International Journal of Hygiene and Environmental Health*, 214(6), pp.417–423.

Wurzbacher, C., Kreiling, A.K., Svantesson, S., van den Wyngaert, S., Larsson, E., Heeger, F., Nilsson, H.R. and Pálsson, S., 2020. Fungal communities in groundwater springs along the volcanic zone of Iceland. *Inland Waters*, 10(3), pp.418–427.

Ying-Ru, Z., Yu-Fang, L., Hai-Lin, Z. and Wei-Ming, S., 2013. Aerobic denitrifying characteristics of duckweed rhizosphere bacterium RWX31. *African Journal of Microbiology Research*, 7(3), pp.211–219.

Zhou, C., Song, C., Huang, D., Liu, Y., Cao, X. and Zhou, Y., 2011. Isolation and characterization of organic phosphorus-mineralizing bacteria in the sediment of a Chinese large shallow eutrophic lake (Lake Taihu). *Geomicrobiology Journal*, 28(8), pp.660–666.

Zhu, G., Jetten, M.S., Kuschk, P., Ettwig, K.F. and Yin, C., 2010. Potential roles of anaerobic ammonium and methane oxidation in the nitrogen cycle of wetland ecosystems. *Applied Microbiology and Biotechnology*, 86(4), pp.1043–1055.

13

Impacts of Pollution on the Hydrogeochemical and Microbial Community of Aquatic Ecosystems in Bayelsa State, Southern Nigeria

Sylvester Chibueze Izah[1], Adams Ovie Iyiola[2], and Glory Richard[3]

[1] *Department of Microbiology, Faculty of Science, Bayelsa Medical University, Yenagoa, Bayelsa State, Nigeria*
[2] *Department of Fisheries and Aquatic Resources Management; Faculty of Renewable Natural Resources Management; College of Agriculture and Renewable Natural Resources, Osun State University, Osogbo, Osun State, Nigeria*
[3] *Department of Community Medicine, Faculty of Clinical Sciences, Niger Delta University, Wilberforce Island, Bayelsa State, Nigeria*

13.1 Introduction

Water is a component of the environment, and its responses are influenced by several biotic (plankton and nekton) and abiotic factors (temperature, dissolved oxygen, pH, etc.). The ocean and other surface water bodies are warmed by the sun and then water evaporates as vapor into the atmosphere which becomes moist. This moist air in the atmosphere helps to cool the Earth and condenses into a cloud which precipitates to the Earth. When this water gets to the soil it is absorbed and becomes groundwater. But when the water is not absorbed into the soil, it will return as storm water or runoff to surface water bodies (including rivers, creeks, creeklets, rivulets, streams, and lakes). This often repeats itself as the surface of bodies of water once again evaporates. Apart from these processes, moisture that vegetation captures or absorbs returns to the atmosphere through evapotranspiration. Therefore, the study of water (hydrology) is made up of the distribution of water in the environment as well as the impacts of anthropogenic activities on the quality of the water and the requisite management systems to the impacts.

Hydrology is an important field because it deals with one of the most important resources needed for human existence. Because of its uniqueness, experts in diverse fields including geologists, biologists, microbiologists, toxicologists, chemists, environmentalists, public health experts, ecologists, conservationists, and engineers have been involved in the study of water especially surface water and its numerous resources especially the biota. The study of how water interacts with the resources (abiotic and biotic) surrounding it is crucial to environmentalists. Experts in the different fields have shown interest in water management systems which is mainly because of the understanding that human activities could impact their sustainability in the supply of water for domestic and industrial applications.

Globally, water is variously for drinking, cooking, washing, cleaning, and bathing in the home (Ogamba et al. 2015a, 2015b, 2015c; Ben-Eledo et al. 2017a, 2017b). Water is also

Hydrogeochemistry of Aquatic Ecosystems, First Edition. Edited by Sughosh Madhav, Virendra Bahadur Singh, Manoj Kumar, and Sandeep Singh.

essential in agricultural activities such as the irrigation of vegetation during cultivation; it is used in the livestock industry and for the generation of power. In many coastal areas of the Niger Delta region in Nigeria, water is an essential mode of transportation for movement of goods and services from one community to another. Surface water ecosystem is a home to several species of living organisms such as plants, animals (mammals, reptiles, amphibians, pisces, zoo- and phytoplanktons, benthic micro- and macro-organisms, and even microbes including some species of protozoans, Protista, fungi, and bacteria.

In Bayelsa State, there are several communities that are only accessible via water transportation using boats (Seiyaboh and Izah 2017a). In addition, the oil exploration activities have a potential impact on the environment and have caused various conflicts between the natives and the drilling companies as a result of the negative impacts on the environment such as pollution of the air and water (Ejumudo and Amede 2019). To this end, this review is aimed at identifying the impacts of human activities on the microbial community, hydrogeochemical quality and functions of aquatic ecosystems in Bayelsa State, Nigeria. The microbial and hydrogeochemical characteristics of the wastes shall be compared with regulatory guidelines for drinking water by World Health Organization (WHO) and standard organization of Nigeria (SON). The study concluded by suggesting potential ways of mitigating the effects of anthropogenic activities on the microbial and hydrogeochemistry of aquatic ecosystems in Bayelsa State, Nigeria.

13.2 Characteristics of Aquatic Ecosystems in Bayelsa State

13.2.1 Hydrogeochemistry of Aquatic Ecosystems in Bayelsa State

The general physicochemical characteristics and nutrient composition of surface waters in Bayelsa State, Nigeria, are shown in Tables 13.1 and 13.2, respectively. Temperature plays an important role in the health of aquatic ecosystem and influences the speed of chemical reactions. The temperature change depends on prevailing weather conditions. The temperature recorded in the different surface waters in the area is within the ambient condition of the area which is in consonance with the SON limits. Also, it is less than WHO limits of 40 °C (Table 13.1).

The pH of water plays an essential role in the growth of microorganisms. It plays a critical role in the photosynthetic activity of phytoplanktons, which removes dissolved carbon dioxide from the water and causes a shift in the acidity toward alkaline level (Henneman and Petrucio 2011; Jayakumar et al. 2013; Nazneen et al. 2019). The pH of surface water in Bayelsa State including Epie creek (Izonfuo and Bariweni 2001), Nun River (Seiyaboh et al. 2013; Daka et al. 2014; Agedah et al. 2015; Ogamba et al. 2015a, Aghoghovwia et al. 2018), Taylor creek (Daka et al. 2014; Aigberua and Tarawou 2019), Kolo creek (Ogamba et al. 2015b; Ebuete and Bariweni 2019; Eremasi et al. 2015), Ikoli creek (Ogamba et al. 2015c), and Epie creek (Ben-Eledo et al. 2017a) were close to the allowable limits as specified by WHO and SON (Table 13.1).

Salinity is the sum of all non-carbonate salts (with Cl and sodium ions as the principal salt) dissolved in water. Also, marine waters may contain other ions such as potassium, magnesium, or sulfate ions. Salinity influences the types of organisms, including fishes, reptiles, mammals, amphibians, planktons, microbes, macrophytes, and plants that live in an aquatic ecosystem. The salinity values previously reported by authors (Agedah et al. 2015; Ogamba

Table 13.1 Some studies on the physicochemical characteristics of rivers and creeks in Bayelsa State, Southern Nigeria.

References	Water sources	Temperature (°C)	pH	Salinity (ppt)	Turbidity (NTU)	Conductivity (µS/cm)	DO (mg/L)	BOD (mg/L)	COD (mg/L)	TDS (mg/L)	TSS (mg/L)	TH (mg/L)	Alkalinity (mg/L)
Soleimani et al. (2020)	WHO limits	40	6.5–8.5	100–200	5	1000	6	50	–	500	–	–	–
SON (2007)	SON limits	Ambient condition	6.5–8.5	–	5	1000	–	–	–	500	–	150	–
Izonfuo and Bariweni (2001)	Epie creek	27.30–30.50	6.90–7.570	–	11.67–23.00	47.73–89.33	1.38–9.05	0.31–6.77	–	33.00–62.00	–	2.27	15.33–37.33*
Seiyaboh et al. (2013)	Nun River	26.00	7.40–7.60	–	5.00–64.00	89.00–95.00 umhos/cm	4.80–7.20	2.60–4.30	–	62.10–67.90	–	–	–
Daka et al. (2014)	Nun River and Taylor creek	–	7.00–7.50	–	24.18–130.42	54.00–63.00	5.20–6.80	2.50–4.80	–	27.00–32.00	–	–	–
Agedah et al. (2015)	Nun River	27.25–27.33	6.53–7.11	0.01–0.05‰	103.75–117.25	56.08–64.95	10.20–14.23	–	–	28.18–32.55	–	–	–
Eremasi et al. (2015)	Kolo creek	27.75–31.70	5.87–6.21	–	3.50–19.25	12.50–32.50	2.78–4.08	1.00–2.40	–	–	–	–	–
Ogamba et al. (2015a)	Nun River	25.00	6.91–7.35	0.00–0.017	25.70–40.53	33.17–68.00	–	–	–	10.33–34.33	3.43–5.47	0.90–3.33	–
Ogamba et al. (2015b)	Kolo creek	–	6.61–6.79	0.00–0.01‰	27.37–31.80	31.60–39.17	–	–	–	16.10–19.23	1.03–1.37	1.75–3.43	–
Ogamba et al. (2015c)	Ikoli creek	25.50–30.10	5.53–6.85	0.01–0.03	28.80–48.80	57.30–78.80	–	–	–	28.70–53.00	2.70–4.35	–	–
Ben-Eledo et al. (2017a)	Epie creek	–	5.80–7.01	0.12-2.58	31.29–105.04	265.00–1096.75	6.23–7.17	65.17–94.95	121.72–252.80	133.00–548.50	3.74–10.56	52.86–280.00	19.20–62.00
Aghoghovwia et al. (2018)	Nun River	–	6.27–6.45	0.03	18.52–18.99	59.70–71.65	5.96–6.71	148.80–157.13	–	29.83–35.83	1.96–2.13	22.20–23.30	13.40–15.50
Aigberua and Tarawou (2019)	Taylor creek	–	5.85–6.20	–	–	73.00–79.00	6.20–8.50	0.50–1.71		40.20–43.65		12.00–16.50	7.00–8.00
Ebuete and Bariweni (2019)	Kolo creek	28.30–32.00	6.50–7.70	–	0.00–3.60	–	4.00–10.81	0.01–1.20		45.80–168.00			

* Iroha et al. (2020).

Table 13.2 Some studies on the ions in rivers and creeks in Bayelsa State, Southern Nigeria.

References	Water sources	K (mg/L)	Na (mg/L)	Ca (mg/L)	Mg (mg/L)	HCO_3 (mg/L)	NO_3 (mg/L)	Cl (mg/L)	SO_4 (mg/L)	PO_4 (mg/L)
Soleimani et al. (2020)	WHO limits	12	200	75	75	500	50	200	250	**0.1*
SON (2007)	SON limits	–	–	–	–	–	50	–	250	–
Izonfuo and Bariweni (2001)	Epie creek	2.55–3.35	2.27–5.27	3.20–7.53	1.77–3.60	–	0.02–0.28	1.65–4.62	1.98–6.27	0.09–0.47
Seiyaboh et al. (2013)	Nun River	–	–	–	–	–	0.32–4.12	–	–	0.06–0.14
Daka et al. (2014)	Nun River and Taylor creek	–	–	–	–	–	0.01–0.61	5.24–8.64	–	<0.01–0.29
Agedah et al. (2015)	Nun River	–	–	–	–	–	0.33–0.81	–	–	–
Eremasi et al. (2015)	Kolo creek	–	–	–	–	–	1.62–3.60	–	–	0.08–0.21
Ogamba et al. (2015a)	Nun River	0.33–0.82	0.53–1.31	0.80–2.33	0.39–1.47	–	0.12–0.39	0.50–3.47	0.567–0.87	–
Ogamba et al. (2015b)	Kolo creek	0.31–0.36	0.58 – 0.68	1.11–1.18	0.37–0.50	–	0.13–0.15	1.26–1.47	0.412–0.57	–
Ogamba et al. (2015c)	Ikoli creek	–	–	–	–	–	0.12–0.26	1.60–3.40	1.40–3.50	–
Ben-Eledo et al. (2017a)	Epie creek	2.06–5.26	3.34–16.02	16.53–43.09	3.07–10.46	2.18–4.25	0.63–1.69	15.25–61.75	6.50–14.30	–
Aghoghovwia et al. (2018)	Nun River	1.17–1.38	3.55–4.84	–	–	1.79–2.53	0.18–0.35	11.10–14.33	1.79–2.53	–
Aigberua and Tarawou (2019)	Taylor creek	–	–	2.10–2.73	0.58–1.07	–	3.00–3.60	16.50–24.74	–	–
Ebuete and Bariweni (2019)	Kolo creek	–	–	–	–	–	86.00–100.00	–	–	–

* Iroha et al. (2020).

et al. 2015a; Ben-Eledo et al. 2017a; Aghoghovwia et al. 2018) in surface water in Bayelsa State are within allowable limits by WHO (Table 13.1). However, these studies were carried out in freshwater ecosystems.

Turbidity is often used to measure the rate at which water losses its transparency because of the presence of suspended materials. Turbidity level of surface water affects the rate of photosynthetic activity. The values recorded by authors in different surface water in Bayelsa State are within the SON and WHO limit of 5.00 NTU except for the value of 0.00–3.60 NTU recorded in Kolo creek by Ebuete and Bariweni (2019) which is within the acceptable level.

The water conductivity values reflect the freshness of the water and provide preliminary information about the salt content. The conductivity of surface water is mainly influenced by temperature as well as the amount of rainfall the surface water receives, an indication of seasonal influence. The conductivity values previously reported in surface water in Bayelsa State by authors (Izonfuo and Bariweni 2001; Seiyaboh et al. 2013; Daka et al. 2014; Agedah et al. 2015; Eremasi et al. 2015; Ogamba et al. 2015a, 2015b, 2015c; Aghoghovwia et al. 2018; Aigberua and Tarawou 2019; Ebuete and Bariweni 2019) are less than the permissible limits of WHO and SON, while some of the values recorded by Ben-Eledo et al. (2017a) exceed the allowable limits (Table 13.1).

Dissolved oxygen is a measure of the suitability of the water for aquatic organisms. The dissolved oxygen content of the water often has a concentration less than the WHO permissible limit of 6.00 mg/L. Instances of dissolved oxygen having a value below 6 mg/L in some of the sampling locations have been recorded in some of the surface water including Epie creek (Izonfuo and Bariweni 2001), Nun River (Seiyaboh et al. 2013; Aghoghovwia et al. 2018), and Kolo creek (Eremasi et al. 2015; Ebuete and Bariweni 2019). However, dissolved oxygen values exceeding the limits have been recorded in some of the water bodies including Taylor creek (Aigberua and Tarawou 2019) and Nun River (Agedah et al. 2015). The concentrations of dissolved oxygen recorded support life in the aquatic habitat.

Biological oxygen demand is a vital water quality parameter that provides an index to evaluate the effect of discharged wastewater on the aquatic ecosystem. The biological oxygen demand often recorded in the aquatic ecosystems in the study area (Bayelsa State) is within the WHO limit of 50 mg/L, though authors have recorded different values for biological oxygen demand in several surface water resources in Bayelsa State (Table 13.1).

Chemical oxygen demand is another important water quality parameter. Chemical oxygen demand usually shows the amount of oxygen equivalents consumed in the chemical oxidation of organic matter. Chemical oxygen demand is not a commonly studied water quality parameter in the study location. However, a concentration range of 121.72–252.8 mg/L has been recorded in Epie creek by Ben-Eledo et al. (2017a).

Water total dissolved solids provides information about the level of inorganic salts as well as other dissolved substances in the water. The total dissolved solids value recorded is far lower than the SON and WHO limits (500 mg/L) except for few locations in the study by Ben-Eledo et al. (2017a) in Epie creek where some of the samples exceeded the allowable limits. The total dissolved solid concentration below allowable limits in surface water in Bayelsa State have been reported in surface water in Bayelsa State by authors (Izonfuo and Bariweni 2001; Seiyaboh et al. 2013; Daka et al. 2014; Agedah et al. 2015; Eremasi et al. 2015; Ogamba et al. 2015a, 2015b, 2015c; Aghoghovwia et al. 2018; Aigberua and Tarawou 2019; Ebuete and Bariweni 2019)

Total suspended solids are undissolved solids in water that can be trapped by a filter. The total suspended solid in water emanate from diverse sources including decaying plant and animal matter, industrial wastes, and sewage. The concentrations of total suspended solids in water have been recorded and ranged from 1.03 to 10.56 mg/L (Table 13.1).

Total hardness is the amount of the mineral content in a water sample that is irremediable by boiling. The hardness has been recorded in diverse concentration in the surface waters of Bayelsa State with a concentration range of 0.00–16.50 mg/L (Izonfuo and Bariweni 2001; Ogamba et al. 2015a, 2015b; Aigberua and Tarawou 2019) and 20.00 – 280.00mg/L (Ben-Eledo et al. 2017a; Aghoghovwia et al. 2018). However, apart from the values recorded by Ben-Eledo et al. (2017a) in Epie creek, other values were within the permissible limits recommended by SON (Table 13.1).

Alkalinity is the ability of water to resist acidification (acid changes in water). The dissolution of carbonate, bicarbonate, or hydroxide compounds in water causes an increased alkalinity of the water. The concentration of alkalinity that have been recorded in surface water in Bayelsa State ranged from 7.00 to 62.00 mg/L (Izonfuo and Bariweni 2001; Ben-Eledo et al. 2017a; Aghoghovwia et al. 2018; Aigberua and Tarawou 2019).

Nitrate is one of the products of the decomposition of nitrogenous materials in water. The values recorded in the several surface water (including Epie creek, Nun River, Taylor creek, Kolo creek, Ikoli creek) in Bayelsa state are less than the WHO and SON limit of 50 mg/L except the study of Ebuete and Bariweni (2019) in Kolo creek that recorded a value range of 86.00–100.00 mg/L in nitrate (Table 13.2).

Chloride is among the vital inorganic anions commonly analyzed in water. It originates from the dissociation of salts (including sodium chloride or calcium chloride). The values recorded in the several surface water in Bayelsa State including Epie creek (Izonfuo and Bariweni 2001; Ben-Eledo et al. 2017a), Nun river and Taylor creek (Daka et al. 2014), Nun River (Ogamba et al. 2015a; Aghoghovwia et al. 2018), Kolo creek (Ogamba et al. 2015b; Ebuete and Bariweni 2019), Ikoli creek (Ogamba et al. 2015c), and Taylor creek (Aigberua and Tarawou 2019) are within the WHO limit (Table 13.2).

Sulfate is second to bicarbonate as the major anion in hard water reservoirs. Sulfate can occur naturally and through human activities such as municipal and industrial discharges. The sulfate concentration recorded in the several water bodies including Epie creek, Nun River, Kolo creek, and Ikoli creek in Bayelsa State are within with the SON and WHO limits (Table 13.2).

Phosphate is one of the most common pollutants that can lead to eutrophication in the aquatic ecosystem. Phosphate is not a commonly studied water quality parameter. However, a concentration range of <0.01–0.47 mg/L have been recorded in some surface water including Epie creek, Nun River, Nun River and Taylor creek, and Kolo creek in Bayelsa State (Table 13.2).

Bicarbonate is a natural component of mineral waters. Bicarbonate is also not a commonly studied water quality parameter in the study area; however, in few instances that were measured, their concentration ranged from 1.79 to 4.25 mg/L (Ben-Eledo et al. 2017a; Aghoghovwia et al. 2018). These values are less than the WHO permissible limit of 500 mg/L (Table 13.2).

The most abundant cations present in water are calcium, magnesium, sodium, and potassium. The permissible limit of these cations in water is 12, 200, 75, and 75 for calcium, magnesium, sodium, and potassium, respectively. However, the concentrations of these cations in the surface water in the study area are within the allowable limits. These cations have been studied in Epie creek (Izonfuo and Bariweni 2001; Ben-Eledo et al. 2017a), Nun River (Ogamba et al. 2015a), and Kolo creek (Ogamba et al. 2015b).

13.2.2 Microbial Community of Aquatic Ecosystems in Bayelsa State

Biological contaminants (microbes) can alter the wholesomeness of water apart from the physical and chemical contaminants. The surface water resources in the state have recorded several microbial genera including *Staphylococcus, Bacillus, Escherichia, Vibrio, Shigella, Klebsiella, Citrobacter, Serratia, Corynebacterium, Streptococcus, Micrococcus, Salmonella, Enterococcus, Clostridum, Bifobacterium bifidium, Providencia regretti, Shewanella putrefaciens, Alkaligens, Favobacterium, Chromobacterium, Enterobacter, Pseudomonas, Campylobacter, Aeromonas, Streptomyces, Lactobacillus, Cytophaga* and *Erminia* species (bacteria) and *Penicillium, Aspergillus, Saccharomyces, Geotrichum, Fusarium, Candida, Mucor, Trichoderma, Cunnighamella, Phymatotrichum, Pleurostromella, Endosporostible, Varicosporium, Rhizopus, Cladosporium*, and *Rhodotorula* species (fungi) (Izah et al. 2022). These microbial species are mainly opportunistic pathogens which imply that they found themselves in the surface water through natural processes (such as soil erosion) or human activities in the water resources. The microbial population and microbial species in aquatic ecosystem in Bayelsa State are presented in Tables 13.3 and 13.4, respectively.

13.3 Anthropogenic Activities Causing Pollution of Aquatic Ecosystem in Bayelsa State

Human activities are the principal sources of pollutants to the environment. A pollutant is a substance that can cause an alteration in the biological (microbial), chemical, and physical characteristics of receiving environment such as surface water have been reported in several environmental matrices (soil, water, and air). Water is the basic element of life and is used for many activities such as agriculture, transportation commerce, and recreation (Fleşeriu 2010). It is a principal source of energy which can be used for various activities. Water pollutants are mainly from human activities on the water and to a lesser extent through surface runoff after precipitation (Izah and Angaye 2016). The improper use of land and water has also reduced the health of watersheds drastically, leading to all forms of contamination in rural and urban areas (Bunnag et al. 2010). Water pollutants are major threats to the survival of many aquatic organisms as well as humans who depend on the aquatic resources for food (Madhav et al. 2020a). Water sanitation is therefore essential for sustenance of the resource on which life subsists. The major activities that cause pollution in surface water in Bayelsa State are discussed in this section.

13.3.1 Houseboat on the Water

Houseboats are mobile houses on water which can serve as dwelling and similar to houses on land; it is seen as a home kept on the water (MI News Network 2021). They are operated in minor water bodies and are of two major types:

Table 13.3 Microbial population in aquatic ecosystem of Bayelsa State southern Nigeria.

Aquatic system/ microbial species	Nun River[a]	Taylor creek[b]	Ikoli creek[c]	Epie creek[d]	Nun River[e]	Nun River[f]	WHO/ UNEPA standard
Total viable bacteria count (cfu/mL)	$2.45–2.72 \times 10^{6}$*	$0.34–6.58 \times 10^{5}$*	$0.74–8.43 \times 10^{6}$*	$0.24–5.50 \times 10^{6}$	$1.50–2.50 \times 10^{3}$	$1.78–9.30 \times 10^{6}$	1.0×10^{2}
Total coliform (MPN/100mL)	–	$0.20–4.20 \times 10^{4}$*	8.10–206.67	$1.02–6.66 \times 10^{2}$	$1.00–3.00 \times 10^{2}$	23.00–28.00	0
Fecal coliform (MPN/100mL)	–	–	5.03–170.00	$0.35–1.66 \times 10^{2}$	$1.00–3.009 \times 10^{2}$ CFU/100 mL	13.67–18.33	0
Salmonella (CFU/mL)	–	–	–	$0.53–3.47 \times 10^{2}$*	–	0.00–3.23 × 102	1.0×10^{2}

[a] Agedah et al. (2015).
[b] Seiyaboh et al. (2020).
[c] Seiyaboh and Izah (2017b).
[d] Ben-Eledo *et al.* (2017b).
[e] Mbah et al. (2017).
[f] Seiyaboh et al. (2017).
* Determined as total bacteria belonging to the Enterobacteriaceae counts and expressed as CFU/mL.

Table 13.4 Microbial species in aquatic ecosystems in Bayelsa State.

Microbial species	Taylor creek[a]	Epie creek[b]	Ikoli creek[c]	Nun River[d]	River Nun[e]
Staphylococcus aureus	P	P	P	P	A
Bacillus species	A	P	P	P	A
Escherichia coli	P	P	P	P	P
Shigella species	A	P	P	P	A
Proteus species	A	P	P	P	A
Pseudomonas species	P	P	P	P	P
Enterobacter species	P	P	P	P	A
Streptococcus species	P	A	A	**A**	A
Salmonella species	A	P	P	P	P
Citrobacter species	A	P	P	A	A
Klebsiella species	P	P	P	A	A
Micrococcus species	A	P	P	P	A
Corynebacterium species	A	P	A	P	A
Serratia species	A	P	A	A	A
Aeromonas species	P	A	A	A	A
Erminia species	A	P	A	A	A

P, present, A, absent.
[a] Seiyaboh et al. (2020).
[b] Ben-Eledo *et al.* (2017b).
[c] Seiyaboh and Izah (2017b).
[d] Seiyaboh et al. (2017).
[e] Ezekwe et al. (2013).

- Static houseboats which are constantly harbored and do not have propelling gadgets and are majorly residential
- Mobile houseboats which have propelling gadgets and are majorly for vacation

These two types of boat are often spotted on Bayelsa waters and there is dearth of documented information on the effects of houseboats in Bayelsa State but pollutants such as oils from boats were observed on the surface of the water where boats are situated (Personal communication, 2021). Since they are mostly situated on water, the sewage generated are mostly discharged into the water from the boat's holding tanks. Traditionally, wastes from toilet of boats are flushed on land from storage tanks in the boat but of late, discharge into water is rampant because the law states that wastes must be held on boats and discharged into a waste disposal system onshore (Holmes 2021).

Discharge into the waterways poses serious impact on the aquatic ecosystem. It can affect the water quality by increasing the growth of algae and upwell sediments which can affect light penetration and make the water turbid. Boats can leach out toxic compounds into the surface water. Depending on the constituents of the toxic compounds, the acidity of the water may be affected, thereby posing a serious environmental concern to the aquatic resources (Environmental Protection 2017).

Chemicals that are used to clean and run houseboats may also leach into the aquatic system thereby affecting aquatic resources. The requirements for biotic and abiotic factors are also affected by toxins from detergents, petroleum products, metals, and batteries. These may also cause disastrous effects such as mutations, cancers, and death of biota. It is suggested that wastes should be treated before discharge into the aquatic environment so as to avert environmental problems.

There are six major pollutants from houseboats as highlighted by Sullivan (2021):

- Oil and gas pollution
- Carbon dioxide emissions from engines
- Boat maintenance and repairs
- Cleaning substances
- Human wastes and garbage
- Noise pollution which is a threat to aquatic organisms

13.3.1.1 Oil and Gas Pollution

This results majorly from improper engine maintenance in which the oil and gas spill into the aquatic environment. Engine leaks may occur into the ocean either when the boat is steered or stationary. This form of pollution may be caused by both shipping and recreational houseboats and the amount of discharge is a function of the motor type, with older engines discharging more emissions than newer engines.

The following steps may be taken to reduce these pollutants in water so as to preserve the ecosystem:

- The faster oil spills are cleaned on board the cleaner the environment because they are less likely to flow into the water.
- Old oil filters for motorized engines must be properly disposed.
- Extra fuel tanks should be filled on land and not on board so as to avert spill and they should not be overfilled.
- The use of four-stroke engine is advised rather than two-stroke engine which causes more environmental damage.
- In cases of accidental spills, it must be reported to the appropriate agency for cleanup.

13.3.1.2 Emissions of Carbon Dioxide

Boats and large shipping vessels are one of the biggest contributors of carbon dioxide into the atmosphere and most of the emissions are released when on the waterways. Engine upgrade to the latest available technology is the major way to reduce this emission because older engines release tonnes of emissions and it is dangerous to marine wildlife and seabirds. Humans living close to the port are also affected by these emissions from older engines.

The following steps may be undertaken to reduce emissions:

- It is advised not to constantly move at top engine speed because the combustion power of engines will increase thereby generating more emissions and the possibility of oil spill is high
- Upgrade engine from two-stroke to four-stroke
- Proper maintenance of engine is required

13.3.1.3 Cleaning Solutions and Other Chemicals

These items are used to keep boats in good appearance and pristine conditions; therefore, these substances are applied on the surface and can be a source of pollution to the water ways. The major polluting cleaners are:

- Detergents, e.g. soaps, cleaning agents, and solvents, which are used to breakdown grease or oil on surface of boats
- Hull paints such as antifouling paint which is used on the boat's hull which contain toxic chemicals that can pollute the water thereby affecting aquatic life
- Antifreeze chemicals which are mostly applied during winter is nontoxic but its excessive use can pollute the waterbodies

Before cleaning substances are used, it is essential to check the label for the ingredients and best to avoid acidic contents or toxic chemicals. The following steps can be employed to reduce pollution from cleaning chemicals:

- It is important to wax the boat regularly which will reduce the need to wash harsh chemicals off the surface of boats
- Use soft sponge and avoid abrasive tools and detergents
- It is advisable to clean and wax boat when on land, so the excessive substances can percolate into the soil
- Use nontoxic and natural substances
- Avoid paints that are antifouling because they can affect the ecosystem; it is advised to use regular paints

13.3.1.4 Impacts of Boat Repairs and Maintenance

All forms of repairs should be done when boat is on land so that all forms of dirt, chemicals, and dust can be spilled on the land. This will greatly reduce pollution in water because wastes are collected and properly disposed on land. This can greatly prevent the ocean pH from becoming more acidic by preventing chemical and oil spills from the aquatic system. Acidification is harmful to most calcifying species of animals, e.g. oysters, clams, corals, and crabs, and can cause overgrowth of algae and seaweeds.

13.3.1.5 Human Wastes and Garbage

This is a major source of pollution especially in riverine areas and the major pollutant is plastics. They can suffocate animals while some mistake microplastics for food and ingest it. Birds and whales have been known to consume plastic trash thereby hindering life processes. Metal trash is also a problem as they do not degrade and remain in the environment for a long time. The following steps may be employed to reduce this form of pollution:

- Trash and wastes in the boat should be secured from flying off into the water
- Trash should be properly disposed on land so as to prevent dispersion into the water
- Do not throw wastes off the boats' deck into the water
- Use reusable and recyclable materials and items
- Waterways should be constantly cleaned

13.3.1.6 Noise Pollution

This is a major source of pollution to fish and other aquatic organisms. It can disrupt communication signals between aquatic animals when in search of food and other survival processes. It may also prevent aquatic animal's ability to sense the environment and may not be able to learn survival instincts which make them more susceptible to predators in water. Animals may have ruptured eardrums when noise is so intense thereby resulting to internal bleeding and death may occur when the noise is at extreme. Noise pollution does not require cleanup, once the noise goes away, it is gone. Therefore, rules and regulations should be geared toward reduction in noise pollution. The steps to reduce this form of pollution are:

- Engines should be in optimal condition because faulty engines make a lot of noise
- Signs of "no noise" may be inscribed in areas that have abundance of aquatic animals for the sake of conservation
- Noise reduced instruments can be used as it will reduce the use of noise creating equipment.

13.3.2 Agricultural Activities

The coastal and swampy lands along the ridges of aquatic ecosystem in Bayelsa State are important ecosystems and rich in organic materials and minerals capable for plant growth (Imoge et al. 2012). It is believed that the Niger Delta area contains one of the highest biodiversity in the world habiting abundant flora and fauna diversities and has a terrain which can sustain varieties of arable crops, agricultural trees, diverse fish species, palm oil plantation, banana and plantain, and sugarcanes (Uche et al. 2016). The region is rich in cash crop production which serves as source of economic returns to the region and foreign exchange. Some of the cash crops include oil palm, rubber, cashew, Raphia palm, groundnut, and coconut (Uche et al. 2016). The region is rich in diverse edible fruits which is a staple food for humans and are good sources of vitamins and minerals (Foraminifera Market Research Limited 2016). This region has conducive atmosphere for fruit production and one of the most conducive areas for food production in Nigeria. A list of fruits in this region include cashew, avocado, pineapple, banana, grapes, guava, lime, peach, tangerine, bread fruit, bush mango, soursop, pawpaw, mangoes, and oranges (Uche et al. 2016).

With the increasing attention to agriculture along the swamps of the state, the rate of pollution has also been on the increase. These activities release all forms of chemicals and heavy metals which pose a negative influence on the aquatic ecosystem. Agricultural activities such as deforestation, increased use of chemicals for crop and animal production, and slash-and-burn agriculture can threaten the environment (Uche et al. 2016). Olaniyi (2011) reported that evidences of food insecurity in this area is apparent despite having physical, social, and economic access to nutritious food to meet their daily requirements. The increase in demand for food is a major cause of the increased pressure on the environment and the contamination of water are a major issue of concern.

The pollutant sources to aquatic ecosystem in this region are

- Agricultural residues
- The use of fertilizers and animal manures: these substances are rich in phosphorus and nitrogen which are essential for plant growth as well as the major causes of aquatic

pollution. When there are heavy rains, excess manures are washed from the soil into the water thereby affecting the quality of water negatively
- The use of pesticides: excessive use of this can be washed into the aquatic environment

13.3.3 Lumbering Activities

Lumbering involves the felling of trees for production of timber and it is a major activity in the riverine areas that have forests which consists of huge, dense trees with wide canopies. However, this activity poses a negative impact on the environment because the environment is left bare and residues of these chemicals due to excessive use can be washed into the aquatic environment. Trees are felled for paper, pulp, furniture making, construction of houses, fuel, income generation, etc. The major tree species and hardwood found in this area include *Chlorophora excela* (iroko or African teak), *Entandrophragma cylindricum* (sapele wood), *Tripochitan scleroxylon* (obeche), *Khaya spp* (mahogany), *Mitragyna ciliata* (abura), *Lovoa trichilioides* (African walnut), and *Terminalia superba* (Afara) (Igben and Ohiembor 2015).

There have been various reports on the impacts of deforestation from lumbering on the environment: Adekunle et al. (2010) reported the effects in south western Nigeria; Ibimilua (2012) in Ekiti state; Enaruvbe and Atafo (2014) in Niger delta regions; and Igben and Ohiembor (2015) in Delta state. They reiterated a negative effect on the environment as well as leaving the land bare thereby exposing the aquatic environment to various surges from land.

Illicit lumbering in Bayelsa State is a major issue and very worrisome because loggers have depleted almost all the natural economical trees such mahogany, iroko, iron tree, cotton tree, and Afara as in the state. The most alarming is the felling of Ogbono tree (Bush mango – *Irvingla gabonensis*), which has been reported to cure high blood pressure, for wood; mahogany and iroko which is abundant in the area is also going extinct. The freshwater trees are almost depleted and loggers now started cutting down salt water trees like *Rhizophora apiculata* thereby exposing the environment to all forms of surges and pollution (Vanguard News 2021).

13.3.4 Transportation

The use of water ways as a medium of transportation in riverine areas is a major activity. The use of water transportation has existed from time immemorial. Goods, people, and services were transported using small dug-out canoes introduced by imperialists, and later steam engines were introduced by colonial masters as means for water transportation. In the seventeenth century, British vessels came around to the Niger Delta regions for the trade of palm oil and this stimulated the use of water transportation.

In Bayelsa State which is surrounded by water, most of the communities are accessed by the use of boats to convey goods and luggage. It has been observed that over the years, the government of the state is yet to develop the water transportation systems; emphasis has been on land transportation. Those living along the river banks devise facilities to convey goods from one place to another, most of which are crude and mostly do not take into consideration the effects on the ecosystem (Brown et al. 2015). This constant activity by community dwellers often causes disturbance to the ecosystem by the outboard engines with several forms of spills and waste disposal from passengers into the water.

In Bayelsa State, water transportation plays an important role in the following:

- It can contribute to the socioeconomic activities of the communities
- It facilitates marketing of products between communities and the nation at large
- It helps to create employment and entrepreneurship since boat making is the trending business in the state
- It makes communication between coastal communities very easy
- It encourages tourism and recreation in riverine areas thereby contributing to the economic power of the state

Based on all of these important roles, there is need to improve on this form of transportation so as to ensure sustainability of the aquatic environment. To avert pollution from transportation sources, the following may influence the success of water transportation:

- Weak institutional support
- Problem of goal formulation
- Insufficient public transport services
- Problems of planning response
- Transport problem of the urban poor

The ways of improving waterways for navigation are as follows:

- Effective waterways policing and ensuring security for users
- Good sewage plants may be provided by the government to reduce further water pollution by fecal materials
- Boats may be provided by the government for navigation

13.3.5 Oil and Gas Exploration

This activity is predominant in the Niger Delta regions of Nigeria, and such intense activities without proper environmental consciousness result in negative impact on the environment. This natural resource is very rich in Bayelsa with a lot of illegal activities associated with it as several makeshift structures were observed between 2011 and 2015 that processed crude into diesel and kerosene using crude technology (Gijo et al. 2017). These activities release various materials into the environment and may result in death of aquatic animals in extreme cases. Ipinmoroti et al. (2018) reported similar occurrence of oil exploration along the coastline of Ondo state and stated the negative effects of oil spills on the water quality in this area.

13.3.6 Dredging

Dredging is an activity that destroys the sea bed and alters the quality of water. There have been reports by Seiyaboh et al. (2013) that dredging has a negative effect on water quality. It also affects algal bloom (Ohimain et al. 2002) and microbenthic and zooplankton composition (Ohimain et al. 2005). This activity is very common in Bayelsa State and has been reported in most of the creeks in the state. It is carried out by artisans using manual means and a few motorized methods, and the materials used for this activity may leach into the water and alter the sediment composition (Seiyaboh et al. 2013).

13.3.7 Wastes

This is a principal source of pollutants by man to the environment and can be of various forms. Nigeria is lacking in adequate waste management procedures with heaps of dumps observed in the environment. In Bayelsa State, activities such as the use of pesticides, oil and gas exploration, and industrial and pharmaceutical processes can contaminate the aquatic resources (Izah and Angaye 2016).

13.3.8 Flooding

This is majorly a natural effect but its intensity is stimulated by human activities especially in areas with inadequate drainage network and it is therefore a major issue. Furthermore, social and economic interactions between man and the environment play a crucial role in the issue of flooding.

13.3.9 Marketing Activities

In Bayelsa State, markets are mostly built close to water areas and various forms of solid and liquid wastes are discharged into the water bodies. The Etegwe and Old Akenfa markets are located close to the Epie Creek as well as the abattoir wastes all of which are washed into the surface water thereby altering its characteristics (Ben-Eledo et al. 2017a, 2017b).

13.3.10 Sources of Pollution

There are two major sources of pollution in aquatic systems. It can be from point or non-point sources.

13.3.11 Point Source Pollutants

Pollutants such as toxic chemicals when intentionally dumped into the water body are called point source of pollution (Oroian and Viman 2010). These pollutants have various effects on the ecosystem such as:

- Destruction of the trees and grassland species surrounding the pollution points
- Pollutants can come from agricultural lands with excessive fertilizer application
- Pollutants can reduce dissolved oxygen levels which affect fish species and wildlife

13.3.12 Non-point Pollution Source

It occurs from runoffs over land washing pollutants and deposits into the water system. This form of polluting poses much problem to the environment as it washes all forms of materials from land into water bodies. In areas where water level is high, runoff may occur. In areas with low water table, chemicals and toxic materials may slip into the ground water polluting the water table. Agriculture also has a great influence on water quality, as the excessive use of chemicals can runoff into the water ways or percolate into the soil.

Non-point pollutants can also be referred to as natural pollution and have a great impact on the environment than point source pollutants which is a form of artificial pollutants.

Rivers may be polluted via three processes viz:

- Rain from the atmosphere
- Condensation in the atmosphere
- Residual or wastewater

Rainfall and condensation are two processes which take place majorly in the atmosphere while the latter results from the water falling on the ground. The level of point source pollution is a function of the rate of urbanization and industrialization which results in the release all forms of materials to the environment. The main sources of non-point pollutants are:

- Domestic sources: this form of waste originates from all forms of household services which generate waste and are discharged to the environment. Examples are the solid wastes (refuse) and sewage wastes (liquid).
- Industrial sources: these are generated from activities by industries in manufacturing and processing plants. This form of waste is majorly from activities which create a new product or service.
- Agricultural/farm activities: this form of waste is generated from agricultural activities such as crop production, animal husbandry, and all forms of processes within these confines. Manures from livestock production maybe washed into the aquatic system and may generate increased bacterial activities.

Non-point pollutants can be classified based on two criteria:

- The nature of the pollutants which may be physical, biological, or chemical forms and are discharged through different sources.
- The route of pollutant emission which may be categorized as accidental or permanent emissions.

13.3.13 Effects of Point and Non-point Pollutants

Both sources of pollutants have a negative effect on the environment causing several depletions in both aquatic and terrestrial biodiversity thereby affecting sustainability. It is therefore essential to maintain discharge limits to avert changes in the environment. It is easier to control pollutants from a point than non-point sources because most of them are usually from runoff sources from rainfall. They are difficult to identify and monitor and it requires special methods and best approaches to avert this. The following are some specific problems caused by point and non-point pollutants:

- The ecosystem of aquatic plants may be destroyed thereby affecting the processes of photosynthesis which is the important and principal source of food production for sustenance of the food chain.
- Pollutants may be absorbed by both aquatic and terrestrial plants thereby affecting their physiology and life processes.
- Vegetation may be destroyed by mud, clay, or chemicals which contain materials that are harmful to the plant.

13.4 Impacts of Pollutants on Microbial Ecology and Hydrogeochemistry in Bayelsa State

13.4.1 Hydrogeochemistry of Aquatic Ecosystems

The hydrogeochemical properties of water are very crucial in determining the suitability of the water for use and as well as the survival of aquatic life-forms. The hydrogeochemical characteristics of water are influenced by several factors including precipitation (rainfall pattern), soil erosion and weathering processes, level of industrialization, urbanization, population growth, and agricultural activities within a surface water body (Singh et al. 2015). The water pollutants that can alter the hydrogeochemistry of water resources include inorganic, organic, radioactive, metallic pollutants, and pathogens (Madhav et al. 2020a). The alteration in the water quality can in turn affect the survival of aquatic life-forms. Some of the life-forms also need nutrients such as free nitrogen, nitrate, phosphorus (phosphate), and other minerals and nutrients for survival (Dubey et al. 2012).

The water ecosystems are the major recipients of emissions, effluents, and solid wastes resulting from human activities, especially those emanating from inhabitants residing in the coastal areas of the Niger Delta region of Nigeria. For instance, flare gases that end up in the atmosphere where it combines with water or moisture to form acid rain, which could be detrimental to life-forms in both terrestrial and aquatic ecosystems. In addition, when acidified water reaches the soil, it easily ends up in surface water bodies (streams, creeks, creeklets, rivers, ponds, etc.) through surface runoff or storm water. Solid wastes and effluents resulting from household, agricultural and industrial activities, etc., are dumped at river banks and/or directly into river systems by inhabitants of coastal or riverine communities in the Niger Delta, especially in Bayelsa State (Ben-Eledo et al. 2017a, 2017b; Seiyaboh and Izah 2017a, 2017b). Human waste is also discharged directly into the surface water through pier toilet system in the area (Izah and Angaye 2016; Seiyaboh and Izah 2017a). In addition, surface water pollution could also occur due to weathering processes and decomposition of plant and animal remains.

Some of the parameters commonly used to assess hydrogeochemical characteristics of water systems include pH, total dissolved solid, electrical conductivity, total hardness, sodium, fluoride, chloride, potassium, sulphate, nitrate, carbonate, lead, manganese, chromium, cadmium, copper, iron, arsenic, mercury, zinc, cobalt, and nickel (Madhav et al. 201 8, 2020b, 2021a, 2021b; Ahamad et al. 2019; Ahamad et al. 2020; Ahamad et al. 2021). The focus of this chapter is on the physiochemical parameters excluding the heavy metals. Previous studies on the Niger Delta have comprehensively discussed the impact of anthropogenic activities on trace metals in river systems (Izah et al. 2022). Globally salinity in water is a source of concern especially in coastal regions in marine environment. Salinity in the environment is induced by precipitation of soluble salt in soils and water systems (Etikala et al. 2021). Salinity of water has a deleterious effect on the environment. For instance, it has the tendency to hinder the growth and productivity of certain species of plants. In drinking water resources of humans, it has negative impacts on human health. However, the concentration recorded in the freshwater resources is low to induce any major effect on human health. However, in the brackish and marine environment, the concentrations often recorded may be deleterious to man. As such caution should be exercised during exposure to high salt content in the marine environment. The impact of human activities including waste dumping have been reported to influence the physico-chemical characteristics of surface water systems (Ben-Eledo et al. 2017a).

13.4.2 Microbial Ecology of Aquatic Ecosystems

Because on environmental degradation (habitat destruction and pollution of the various environmental matrices – soil, water, and air) due to human activities, there has been a decline in biodiversity and ecosystem-based resources. Within the last five decades, there have been awareness creations about the deleterious effects of human activities on the environment. Furthermore, the awareness also focused on the conservation of biodiversity mainly plants and animals (wildlife). But little or no attention is shown toward the adverse effect of human activities on the natural microbial densities (Labbate et al. 2016). The authors further stated that this oversight is due to the fact that microorganisms are usually the first responders to ecological trepidation and can either enhance or cushion environmental alterations through frequently multifaceted, progressive, and deleterious response.

According to Labbate et al. (2016), microorganisms have the highest population in aquatic environment. These microbes play essential role in the productivity of the ecosystem as well as biogeochemical nutrient cycle (Labbate et al. 2016; Izah et al. 2018). Before now, little or nothing are reported about the impact of human activities on the ecology and function of aquatic microorganisms, but the advent of powerful biotechnological advancement using next-generation sequencing and novel modeling methods have shown the importance of studying the microbial community in nearly all environments.

The contamination of surface water by pollutant and nutrients can alter the microbial community (species diversity and populations) which could lead to reduction in species diversity. Some changes in the water quality parameters could also enhance the growth of potential toxic microorganisms which may have deleterious effects on the users of the water if it is not properly treated. Furthermore, when antibiotic-resistant microorganisms are introduced into the water bodies and its genes are deposited in the water, it may worsen the antimicrobial resistance. This concept may inextricably associate with aquatic animals. As such aquatic organisms may be a major reservoir for resistant microorganisms and their genes. Michael et al. (2014) and Labbate et al. (2016) reported that extreme use of antimicrobials and inadequate treatment of wastewater may make the recipient of the wastewater a host of antimicrobial resistant microbes and genes. Labbate et al. (2016) further reported that antibiotics in aquatic ecosystem enhance antibiotic resistance in bacteria and may lead to mutation in the genetic material of the organisms which may alter their role in the ecosystem.

13.5 Mitigation of Pollution on the Aquatic Ecosystem

With the ever-increasing human population, the demand for water which is a principal source of life is increasing. It is therefore essential to maintain the resource in its right quantity and quality and much more conservation which involves a continual supply to future generations as the human population has been predicted to increase (Bunnag et al. 2010). The following strategies may be used to prevent pollution:

- Holding of soil in place by planting trees and grasses which can prevent erosion into the aquatic systems
- Soil tests may be performed to ascertain the actual amount of fertilizer to be applied to farm lands thereby averting the indiscriminate use and environmental pollution
- Chemicals that are hazardous must be kept from rain most especially if outdoor and disposed properly
- The use of more permeable surfaces in landscape like bricks and wood should be encouraged rather than impermeable surfaces which may prevent the percolation of water into the soil and cause erosion
- Structures may be erected or planting of trees and shrubs along slopes which may reduce all forms of pollutant runoff into the aquatic environment especially when close to water ways
- Manure should be properly managed and livestock may be sited away from water ways.
- The activities of houseboat and dredging among others should be carefully carried out to minimize potential leaching of toxicant into the surface water.

13.6 Conclusion

Water is a resource which must be preserved because it can have numerous uses such as transportation, domestic, agricultural, commercial, and industrial purposes. It is responsible for all forms of productivity on Earth, and it is important to be smart about the use of water so as not to alter the hydrogeochemical processes as well as the microbial ecology. Waste discharge should be within the recommended limits by environmental agencies and activities that can pollute the environment must be avoided. To this end, Bayelsa State which is a rich oil producing area must be well managed with regular sensitization and necessary cleanups in areas prone to pollution thereby reducing the effects on aquatic system. The government should also live up to their responsibilities by provision of necessary incentives and resources needed for survival by the natives so as to avert cases of vandalism of oil and gas installations which pose danger to the environment.

References

Adekunle, V.A.J., Olagoke, A.O., Ogundare, L.F., 2010. The rate of timber production. In a tropical rainforest ecosystem of southwestern Nigeria and its implications on sustainable forest management. *Journal of Forestry Research*, 21(2), 225–230

Agedah, E.C., Ineyougha, E.R., Izah, S.C., Orutugu, L.A., 2015. Enumeration of total heterotrophic bacteria and some physico-chemical characteristics of surface water used for drinking sources in Wilberforce Island, Nigeria. *Journal of Environmental Treatment Techniques*, 3(1), 28–34.

Aghoghovwia, O.A., Umoru, O.D., Izah, S.C., 2018. Physicochemical characteristics of Nun river at Gbarantoru and Tombia Axis in Bayelsa State, Nigeria. *Bioscience Methods*, 9(1), 1–11.

Ahamad, A., Raju, N.J., Madhav, S., Gossel, W., Wycisk, P., 2019. Impact of non-engineered Bhalswa landfill on groundwater from Quaternary alluvium in Yamuna flood plain and potential human health risk, New Delhi, India. *Quaternary International*, 507, 352–369.

Ahamad, A., Raju, N.J., Madhav, S., and Khan, A.H., 2020. Trace elements contamination in groundwater and associated human health risk in the industrial region of southern Sonbhadra, Uttar Pradesh, India. *Environmental Geochemistry and Health*, 42(10), 3373–3391.

Ahamad, A., Raju, N.J., Madhav, S., Gossel, W., Ram, P., Wycisk, P., 2021. Potentially toxic elements in soil and road dust around Sonbhadra industrial region, Uttar Pradesh, India: Source apportionment and health risk assessment. *Environmental Research*, 202, 111685.

Aigberua, A., Tarawou, T., 2019. Water quality index (WQI) assessment along inland fresh waters of Taylor Creek in Bayelsa State, Nigeria. *Journal of Environmental Treatment Techniques*, 7(3), 260–269.

Ben-Eledo, V.N., Kigigha, L.T., Izah, S.C., Eledo, B.O., 2017a. Water quality assessment of Epie creek in Yenagoa metropolis, Bayelsa state, Nigeria. *Archives of Current Research International*, 8(2), 1 – 24.

Ben-Eledo, V.N., Kigigha, L.T., Izah, S.C., Eledo, B.O., 2017b. Bacteriological quality assessment of water from Epie Creek, Niger Delta Region of Nigeria. *International Journal of Ecotoxicology and Ecobiology*, 2(3),102–108.

Brown, I., Chikagbum, W and Chigozie, D.E., 2015. Development of waterways for effective transportation system and recreation enhancement in selected navigable creeks and waterways in Rivers state. *International Journal of Scientific and Engineering Research*, 6(9), 1016–1026

Bunnag, S., Pimda, W., Pongpera, S., 2010 Utilization of *Spirulina platensis* for wastewater treatment in fermented rice noodle factory. *ELBA Bioflux*, 2(2), 39–44.

Daka, E.R., Amakiri-Whyte, B., Inyang, I.R., 2014. Surface and groundwater in some oil field communities in the Niger Delta: implications for domestic use and building construction. *Research Journal of Environmental and Earth Sciences*, 6(2), 78–84.

Dubey, V.K., Srivastav, A.L., Singh, P.K., Sharma, Y.C., 2012. The nutrients level in middle Ganga Basin, India. *Journal of Applied Technology and Environmental Sanitation*, 2(2), 121–128.

Ebuete, A.W., Bariweni, P.A., 2019. Water quality index of Kolo Creek, Bayelsa State, Nigeria. *Journal of Applied Science and Environmental Management* 23 (11), 1923–1927

Ejumudo, K.B.O., Amede, H.O., 2019. Oil production and the problematic of water pollution in the Niger Delta: a study of selected communities in Bayelsa State. *Research in Social Change*, 11(3), 68–86

Enaruvbe, G.O., Atafo, O.P., 2014. Analysis of deforestation pattern in the Niger Delta Region of Nigeria. *Journal of Land use Science*, 11(1), 113–130

Environmental Protection 2017. The Environmental Impacts of Boating. https://eponline.com/Articles/2017/03/27/The-Environmental-Impacts-of-Boating.aspx?m=1

Eremasi, Y.B., Alagoa, K.J., Daworiye, P., 2015. Water quality evaluation and heavy metals concentration of Kolo creek, Imiringi, Bayelsa state. *International Journal of Current Research in Bioscience and Plant Biotechnology*, 2(2), 61–66.

Etikala, B., Adimalla, N., Madhav, S., Somagouni, S.G., Keshava Kiran Kumar, P. L., 2021. Salinity problems in groundwater and management strategies in arid and semi-arid regions. In Madhav, S., Singh, P. (eds.) *Groundwater Geochemistry: Pollution and Remediation Methods*, 42–56. Wiley.

Ezekwe, I.C., Ezekwe, A.S., Endoro, P.O., 2013. Biological contaminants in the River Nun and environmental ethics of riverside communities in the Niger Delta: the case of Amassoma, Bayelsa, Nigeria. *Estudos de Biologia*, 35(84), 67–75.

Fleşeriu, A., 2010 Endocrine disrupting pesticides and their impact on wildlife and human health. *HVM Bioflux*, 2(1), 1–4.

Foraminifera Market Research Limited 2016. Fruit Juice Production in Nigeria: The Opportunity. *Soutech Consults* 61-65 Egbe -isolo Road, Lagos, Nigeria.

Gijo, A.H., Hart, A.I., Seiyaboh, E.I., 2017. The impact of makeshift oil refineries on the physico- chemistry of the sediments of the nun river estuary, Niger Delta, Nigeria. *Sky Journal of Soil Science and Environmental Management*, 6 (1), 19–25.

Henneman, M.C., Petrucio, M.M., 2011. Spatial and temporal dynamic of trophic relevant parameters in a subtropical coastal lagoon in Brazil. *Environmental Monitoring Assessment* 181, 347–361.

Holmes, J., 2021. How do houseboat toilet work? https://crowsurvival.com/how-do-houseboat-toilets-work/

Ibimilua, A.F., 2012. Appraisal of the causes and consequences of human induced deforestation in Ekiti State, Nigeria. *Journal of Sustainable Development in Africa*, 14(3), 37–52.

Igben, J.L., Ohiembor, O.M., 2015. Lumbering as a factor of deforestation in the freshwater swamp forest in Delta State Nigeria. *International Journal of Research in Agriculture and Forestry*, 2(9), 1–9.

Imoge, A.L., Udosen, C.V., Arinde, C.I., Enomiya, D.O.M (2012. Utilization of Niger Delta basin coastal fresh water swamps for meaningful agricultural production. *Greener Journal of Agricultural Sciences*, 2(5), 180–187.

Ipinmoroti, M.O., Iyiola, A.O., Idowu, B., 2018), Economic analysis of artisanal fisheries in some selected fishing communities of Ilaje local government area, Ondo State, Nigeria. *International Journal of Development and Sustainability*, 7(2),716–723.

Iroha, I.R., Ude, I.U., Okoronkwo, C., Ovia, K., Okafor, C.O.O., Akuma, S.O., 2020. Comparative assessment of physicochemical characteristics, metal levels and anion contents of water from different aquatic environments in Ebonyi State. *Biomedical Journal of Science and Technology Research*, 29(5), 22834–22865.

Izah, S.C., Angaye, T.C., 2016. Heavy metal concentration in fishes from surface water in Nigeria: potential sources of pollutants and mitigation measures. *Sky Journal of Biochemistry Research*, 5(4), 31–47.

Izah, S.C., Bassey, S.E., Ohimain, E.I., 2018. Impacts of Cassava mill effleunts in Nigeria. *Journal of Plant Animal Ecology*, 1(1), 14–42.

Izah, S.C., Aigberua, A.O., Srivastav, A.L. 2022. Factors influencing the alteration of microbial and heavy metal characteristics of river systems in the Niger Delta region of Nigeria. In: Madhav, S., Kanhaiya, S., Srivastav, A.L., Singh, V.B., and Singh, P. (eds.) *Ecological Significance of Riparian Ecosystems*, 59–71. Elsevier. https://doi.org/10.1016/C2020-0-02111-0.

Izonfuo, L.W.A., Bariweni, A.P., 2001. The effect of urban runoff water and human activities on some physico-chemical parameters of the Epie creek in the Niger Delta. *Journal of Applied Science and Environmental Management*, 5(1), 47–55.

Jayakumar, R., Steger, K., Chandra, T.S., Seshadri, S., 2013. An assessment of temporal variations in physicochemical and microbiological properties of barmouths and lagoons in Chennai (southeast coast of India). *Marine Pollution Bulletin* 70(1–2), 44–53.

Labbate, M., Seymour, J.R., Lauro, F., Brown. M.V., 2016. Anthropogenic impacts on the microbial ecology and function of aquatic environments. *Frontier in Microbiology*, 7, 1044. https://doi.org/10.3389/fmicb.2016.01044.

Madhav, S., Ahamad, A., Kumar, A., Kushawaha, J., Singh, P., Mishra, P.K., 2018. Geochemical assessment of groundwater quality for its suitability for drinking and irrigation purpose in rural areas of Sant Ravidas Nagar (Bhadohi), Uttar Pradesh. *Geology, Ecology, and Landscapes*, 2(2), 127–136.

Madhav, S., Ahamad, A., Singh, A.K., Kushawaha, J., Chauhan, J.S., Sharma, S, Singh, P., 2020a. Water pollutants: sources and impact on the environment and human health. In *Sensors in Water Pollutants Monitoring: Role of Material* (pp. 43–62), Pooja D, Kumar P, Singh P, Patil S (Editors). Springer, Singapore.

Madhav, S., Kumar, A., Kushawaha, J., Ahamad, A., Singh, P., Dwivedi, S.B., 2020b. Geochemical assessment of groundwater quality in Keonjhar City, Odisha, India. *Sustainable Water Resources Management*, 6, 1–11.

Madhav, S., Raju, N.J., Ahamad, A., 2021a. A study of hydrogeochemical processes using integrated geochemical and multivariate statistical methods and health risk assessment of groundwater in Trans-Varuna region, Uttar Pradesh. *Environment, Development and Sustainability*, 23(5), 7480–7508.

Madhav, S., Raju, N.J., Ahamad, A., Singh, A.K., Ram, P., Gossel, W., 2021b. Hydrogeochemical assessment of groundwater quality and associated potential human health risk in Bhadohi environs, India. *Environmental Earth Sciences*, 80(17), 1–14.

Mbah, E.I., Ibe, S.N., Abu, G.O., 2017. Water quality assessment of the freshwater ecosystem at communities' points-of-use in Yenagoa Metropolis of Bayelsa State, Nigeria. *Research Journal of Pure Science and Technology*, 1(1), 15–22.

MI News Network, 2021. What are houseboats? *Marine Insight*. https://marineinsight.com/types-of-ships/what-are-houseboats/amp/

Michael, C.A., Dominey-Howes, D., Labbate, M., 2014. The antimicrobial resistance crisis: causes, consequences and management. *Front. Public Health*, 2, 145. https://doi.org/10.3389/fpubh.2014.00145.

Nazneen, S., Raju, N.J., Madhav, S., Ahamad, A., 2019. Spatial and temporal dynamics of dissolved nutrients and factors affecting water quality of Chilika lagoon. *Arabic Jorunal of Geosciences*, 12, 243. https://doi.org/10.1007/s12517-019-4417-x.

Ogamba, E.N., Izah, S.C., Oribu, T., 2015a. Water quality and proximate analysis of Eichhornia crassipes from River Nun, Amassoma Axis, Nigeria. *Research Journal of Phytomedicine*, 1(1), 43–48.

Ogamba, E.N., Seiyaboh, E.I., Izah, S.C., Ogbugo, I., Demedongha, F.K.,2015b. Water quality, phytochemistry and proximate constituents of Eichhornia crassipes from Kolo creek, Niger Delta, Nigeria. *International Journal of Applied Research and Technology*, 4(9), 77–84.

Ogamba, E.N., Izah, S.C., Toikumo, B.P., 2015c. Water quality and levels of lead and mercury in *Eichhornia crassipes* from a tidal creek receiving abattoir effluent, in the Niger Delta, Nigeria. *Continental Journal of Environmental Sciences*, 9(1), 13–25.

Ohimain, E.I., Imoobe, T.O.T., Benka-Coker, M.O., 2002. Impacts of dredging on zooplankton communities of Warri River, Niger Delta. *African Journal of Environmental Pollution and Health*, 1, 37–45.

Ohimain, E.I., Benka-Coker, M.O., Imoobe, T.O.T., 2005. The impacts of dredging on macrobenthic invertebrates in a tributary of the Warri River, Niger Delta. *African Journal of Aquatic Sciences*, 30, 49–53.

Olaniyi, E., 2011. Food insecurity: implications for Nigeria and the world at large. Department of Economics, University of Lagos, 2011.

Oroian, I., Viman, O., 2010, Damaging effects of overall water pollution. *Pro-Environment*, 3, 203–207.

Seiyaboh, E.I., Izah, S.C., 2017a. Review of impact of anthropogenic activities in surface water resources in the Niger Delta Region of Nigeria: a case of Bayelsa State. *International Journal of Ecotoxicology and Ecobiology*, 2(2), 61–73.

Seiyaboh, E.I., Izah, S.C., 2017b. Bacteriological assessment of a tidal creek receiving slaughterhouse wastes in Bayelsa state, Nigeria. *Journal of Advances in Biology & Biotechnology*, 14, 1–7.

Seiyaboh, E.I., Inyang, I.R., Gijo, A.H., 2013. Environmental impact of Tombia bridge construction across Nun River in Central Niger Delta, Nigeria. *The International Journal of Engineering and Science*, 2(11), 32–41.

Seiyaboh, E.I., Izah, S.C., Bokolo, J.E., 2017. Bacteriological quality of water from river nun at Amassoma Axises, Niger Delta, Nigeria. *ASIO Journal of Microbiology, Food Science & Biotechnological Innovations*, 3(1), 22–26.

Seiyaboh, E.I., Youkparigha, F.O., Izah, S.C., Mientei, K., 2020. Assessment of bacteriological characteristics of surface water of Taylor creek in Bayelsa state, Nigeria. *Noble International Journal of Scientific Research*, 4(4), 25–30.

Singh, P., Chaturvedi, R.K., Mishra, A., Kumari, L., Singh, R., Pal, D.B., Giri, D.D., Singh, N.L., Tiwary, D., Mishra, P.K., 2015. Assessment of ground and surface water quality along the river Varuna, Varanasi, India. *Environmental Monitoring Assessment*, 187(4), 170.

Soleimani, H., Nasri, O., Ghoochani, M., Azhdarpoor, A., Dehghani, M., Radfard, M., Darvishmotevalli, M., Oskoei, V., Heydari, M., 2020. Groundwater quality evaluation and risk assessment of Nitrate using monte carlo simulation and sensitivety analysis in rural areas of Divandarreh County, Kurdistan province, Iran. *International Journal of Environmental and Analytical Chemistry*. https://doi.org/10.1080/03067319.2020.1751147

Standard organization of Nigeria (SON), 2007. Nigerian Standard for Drinking Water Quality. Nigerian industrial standard, Nigeria.

Sullivan, S., 2021. How DO Boats Pollute? 6 Ways You Should Know Explained. https://www.godownsize.com/boats-pollute-oceans-environment/

Uche, C., Aje, E.N., Familusi, I.C. 2016. A survey of the agricultural prospects and potentials of the Niger Delta in Post oil era. *Global Journal of Science Frontier Research*, 16(8), 25–30

Vanguard News, 2021. Illicit loggers defy gov, threates Bayelsa communities, vegetation. https://www.vanguardngr.com/2021/05/illicit-loggers-defy-gov-threaten-bayelsa-communities-vegetation/.

14

Aquatic Eco-systems Under Influence of Climate Change and Anthropogenic Activities: Potential Threats and Its Mitigation Strategies

Suryadipta Mukherjee[1], Syed Shams Rizvi[2], George Biswas[1], Abhilash Kumar Paswan[3,4], Suantak Paolalsiam Vaiphei[4], Taufique Warsi[5], and Tarik Mitran[6]

[1] Department of Geology, Presidency University, Kolkata, West Bengal, India
[2] Department of Geology, Central University of Karnataka, Kalaburagi, Karnataka, India
[3] Department of Physical Science, Academy of Scientific and Innovative Research, Ghaziabad, Uttar Pradesh, India
[4] Department of Hydrogeochemistry, CSIR- National Geophysical Research Institute, Hyderabad, Telangana, India
[5] WOTR Centre for Resilience Studies (W-CReS), Watershed Organisation Trust (WOTR), Pune, Maharashtra, India
[6] Soil and Land Resources Assessment Division, National Remote Sensing Centre, Balanagar, Hyderabad, Telangana, India

14.1 Introduction

Aquatic ecosystems are valuable sources of food and recreation; however, they are experiencing unprecedented levels of changes due to human activities particularly those of pollution and climate variations. The rising population, industrialization, and anthropogenic pressures on the natural environment have escalated the ecosystem degradation and widespread extinction of plant and animal species. Because of these changes, a new geological time period emerged, which is known as Anthropocene, where humans have significantly impacted the Earth's geological and ecological systems (Waters et al. 2016). However, due to human interventions, the poor dispersal capacities between the plants and animals reduce the natural corridors and the opportunities for recolonization. As a result, fragmentation of the landscape and loss of rivers and lakes have been observed worldwide (Hughes 2007).

The aquatic ecosystems play a key role in maintaining the ecological balance of nature. In recent years, it has been observed that human activities along with changing climate have adversely influenced the functioning of such ecosystems. The severity and frequency of its impact are increasing and reported globally (Malhi et al. 2020; Gissi et al. 2021; Kappes et al. 2021; Sowman and Rebelo 2022). In the last few decades, changes in the aquatic ecosystem have been observed in almost all parts of the world and influenced by both anthropogenic and natural factors. Some of the observed indicators are increases in streamflow, stream temperature, lake temperature, and geographical shifting of some of the marine species (Pletterbauer et al. 2018). Most of the natural habitats have adapted to different environmental conditions because of variations and changes under natural processes. Aquatic ecosystems have been impacted by both climate change and human

Hydrogeochemistry of Aquatic Ecosystems, First Edition. Edited by Sughosh Madhav, Virendra Bahadur Singh, Manoj Kumar, and Sandeep Singh.

activities such as pollution, landscape variations, or changes in the hydrological systems worldwide (Srivastav et al. 2021; Wang et al. 2021). Hence, it is hard to envisage the effect of disturbances at a local scale due to their complexities and linkages within them. As a result, it is observed that the damage to one component of the ecosystem can alter the other components of the ecosystem. Hence, there is a need to reduce the impact of climate change and the adverse effects of anthropogenic activities by adopting various mitigation strategies. The possible mitigation strategies that can be adopted to lessen the impact are migration to a cleaner source of energy (Owusu and Asumadu-Sarkodie 2016), judicious treatment of wastewater from different industries (Muralikrishna and Manickam 2017), reducing overdependence on chemical fertilizers for increasing agricultural output, setting up and adaptation of better practices for tannery (Chen et al. 2014), etc. In this chapter, an attempt has been made to review and discuss the impacts of anthropogenic and climate change on aquatic systems with special emphasis on the Lotic ecosystem, Lentic ecosystem, Wetland and Swamps, Ocean and Marine ecosystem.

14.2 Aquatic Ecosystems

The ecosystems formed around a static or movable body of water are said to be aquatic ecosystems and can be of terrestrial (lake, river, stream, etc.), marine, and wetlands. In this chapter, we will focus on aquatic ecosystems (Figure 14.1).

Aquatic ecosystems are predominantly three types:

- Freshwater: Lotic (fast-moving) and Lentic ecosystem (slow-moving)
- Mixed water: Wetlands (inland and coastal)
- Saline water: Marine ecosystem (oceanic ecosystem)

14.2.1 Freshwater Ecosystem

Freshwater ecosystems are those which have a supply of fresh water from various sources such as rain, glacier, and groundwater. The major freshwater ecosystems are Lotic and Lentic ecosystems. These ecosystems are very important for biotic and abiotic interactions.

Lentic or standing water ecosystem: Refers to calm water ecosystem. Examples: lakes and ponds

Lotic or flowing water ecosystem: Refers to running water ecosystem. Examples: rivers, springs, and streams

Wetland ecosystem: Refers to ecosystems where water level fluctuates up and down frequently during different seasons and annually. Examples: marshes and swamps

Figure 14.1 Lentic, Lotic, and Wetland ecosystems.

14.2.1.1 Lotic Ecosystem

The Lotic ecosystem is one the most important environmental systems that provide life and habitat to all living beings. The term Lotic defines the state of movement (such as springs, rivers, and streams) where water is potentially active to create new landforms through chemical and physical reactions. The human advancements and tremendous population contribute to huge man-made anthropogenic activities which interrupted the natural environmental flow with new constructions, shifting cultivations, and diversions of rivers, produced chemical waste as well as hazardous gas, etc. Though the natural process of contaminants is considered as slow degradation, it still adds analytic values through geochemical processes, precipitation from atmospheric suspended particles, and rotting organic matter from soil (Vaiphei et al. 2020). Some of the common diseases which can result from the pollution of the Lotic ecosystem are cholera, teeth fluorosis, infantile paralysis, diarrhea, dracunculiasis, dysentery, lead poisoning, typhoid, and skin infection (WHO 2011). So, before it is too late, it is time that we need to protect and manage the Lotic bodies from further degradation. Because our mother Earth has access to only about 3% of surface water which is again not accessible to all due to geographical regions. Besides, India is listed in the world's hotspot for groundwater consumption with usages of about $250 \times 109\ m^3$/year (Shah et al. 2007). As per CGWB (2013) report, it is recorded that approximately $245 \times 109\ m^3$ groundwater is consumed in agriculture activities in India.

The Lotic ecosystem bodies include springs, streams, and rivers. Brief discussions are given in the following sections.

14.2.1.1.1 Springs It is defined as a point on the hillside or at the mountain foothills where water gets out naturally from the aquifer. Springs can be found in any rock bed but considering big springs, they formed under limestone and dolomite dominant in karst morphology. The amount of water flow depends on aquifer water pressure, the size of a particular spring basin, and the quantity of rainfall. Springwater may not always be clear as it depends on the minerals dissolved, e.g. high iron content gives to rusty/reddish-brown, while manganese and iron dissolved tends to black color.

14.2.1.1.2 Streams and Rivers Though both streams and rivers sound synonymous, in actuality they are quite different, despite their same source of origin, i.e. glacial melt and rainfall. They are considered as one of the versatile and most important primary sources of livelihood as well as easily accessible to all organisms. Streams are characterized by their narrow, speed, fall, and shallow flow which finally reach and join the bigger flow called rivers. Stream has also been termed a creek. Streams are classified into different orders (first order, second order, third order, fourth order, and so on) depending on the stages of flow while meeting the higher-order streams and finally joining main rivers.

14.2.1.1.3 River Rivers are comparatively wider, slower, and deeper. Sediments are brought down by streams and continue to transport through suspended particles by rivers to reach their final destination and deposited in low land basins, lakes, and oceans. River basins are considered to have the most fertile soil and dense agriculture area, for example, the Indo-Gangetic basin (Great plains) including major parts of northern and eastern India. It also forms a large sediment deposit called delta. Unfortunately, we human beings tend to forget the value of a naturally balanced ecosystem and start to disturb the natural cycle by releasing toxic gases, solid and liquid chemical wastes, deforestation, construction of dams, and settlements.

14.2.1.2 Lentic Ecosystem

Lentic ecosystems refer to all systems which are static bodies of water, like lakes, impoundments, and wetlands. The different types are as follows:

Lakes are most common in areas that are affected by geological changes through the different periods on the Earth's surface, but a few lakes are also ancient like lake Baikal in Russia. Lakes can be formed in several ways, like tectonic action, depressions caused due to craters, and glacial activity. The tectonic lakes are formed due to faulting or warping activities in a basin; most lakes of the Himalayas are of tectonic type. Lakes can also form in depressions caused by impact craters, e.g. Lonar Lake in Maharashtra, or extinct volcanic depression, e.g. Toba Lake in Indonesia; the world's largest volcanic crater lake. Glacial lakes are also called tarns, which are formed due to the filling up of glacial meltwater in the cirques; famous lakes include Gurudongmar Lake (India) and Tasman Lake (New Zealand).

14.2.1.3 Wetlands (Mixed Type)

Wetlands refer to areas that are permanently/periodically underwater, either in submerged condition or saturated either naturally or artificially with freshwater or brackish/saline water up to 6 m depth (IUCN 1971, Ramsar Convention). The wetlands can be divided into inland wetlands and coastal wetlands. Inland wetlands consist of freshwater, consisting of bogs and swamps. Coastal wetlands are saline/brackish which include mangrove forests, and mangrove swamps.

Swamps, marshes, and mangrove forests play an important role in coastal protection, providing habitat for different organisms and helping in carbon sequestration. Wetland habitats have a natural tendency to adapt to changing climatic conditions and thus protect the coast from various dangers (Salimi et al. 2021). These ecosystems should be saved from any type of human encroachment activities as their importance is paramount. A study showed that there is a linear relationship between loss of wetland habitat and increase in carbon emissions (Bindoff et al. 2019).

14.2.2 Marine Ecosystem

The Earth has formed around 4.6 Ga years ago and was completely covered with water, at the present state it is around 70% water and 30% land. Most of the available water occurs in the vast seas and oceans of the Earth which constitute the marine ecosystem and these waters are high in salt content so the water is typically saline in nature. Among all the ecosystems, this is the biggest. Marine ecosystems are segregated into different subsections based on oceanic depth and setting of the shoreline. They are

- Deep ocean
- Open ocean
- Benthic zone
- Inter-tidal zone
- Neritic zone

Marine ecosystems supply important nutrients needed by the organism of land and water. Different Physico-biogeochemical factors such as salinity trends, temperature changes, acidity, oceanic nutrient content, and dissolved gases can act as indicators of climate change and for anthropogenic activities. We will know more about these factors in the following sections.

14.3 Emerging Issues in the Aquatic Ecosystem

The continuously growing human populations and activities put pressure on multiple aquatic resources. Most of the species are disappearing due to human activities in which the aquatic species are more common than mammals and birds (Allan and Flecker 1993; Gangloff et al. 2016). The overgrowing rate of exploitation of aquatic organisms poses a serious issue to the marine environment and in this concern; it has become necessary to manage the sustainable exploitation by Environmental Defence Fund (EDF) to protect marine biodiversity. Mining, industries, forestry, water pollution, climate change, weathering, erosion, etc., are also responsible for diminishing natural habitats (Tewari and Bisht 2020). The genetic diversity of different species has been lost due to overexploitation. It has also affected the population size of the species because of age, structure, and sex composition (Tewari and Bisht 2020).

Injudicious groundwater use, environmental demands for water, and water quality issues have also affected the availability of water for agricultural production. In addition, aquaculture may also influence the biodiversity, water quantity/quality, and health of ecosystems. These can put pressure on fisheries and fish culture operations (Mark et al. 2006). Environmental strains such as drought, soil infertility, salinity, and groundwater pollution also influenced agricultural productivity (World Water Vision 1999; FAO 2000; USERC 2005; Mark et al. 2006; Srivastav et al. 2021). There are generally four types of potential risks linked with biotechnologies such as ethical issues, ecological influences, socioeconomic concern, and impact on human and animal health, and in this concern, the transmission rate of toxins from one life to another could be higher that can further lead to the allergic problems (Hossain et al. 2004; Mark et al. 2006). Possible sources of emerging contaminants and effective management strategies are shown in Figure 14.2.

Apart from that, physical alteration of habitat is also responsible for diminishing the species. The main causes are deforestation, water diversion, dam structure, alteration of marshy land, change in irrigation strategies, modification in the small water bodies, etc. Another major threat to biodiversity is pollution load such as toxic contaminants, thermal and organic contaminants, and sewage contaminants (Tewari and Bisht 2020).

14.3.1 Food Security

There is no doubt that the growing population also puts pressure on the food sector. Food production is generally dependent on irrigation but due to increasing demand from different sectors, the availability of water for food production is limited; thus it may be a serious concern for food security. Declining groundwater table, changes in water quality scenarios, and growing environmental water demands are further constraints. Few regions make

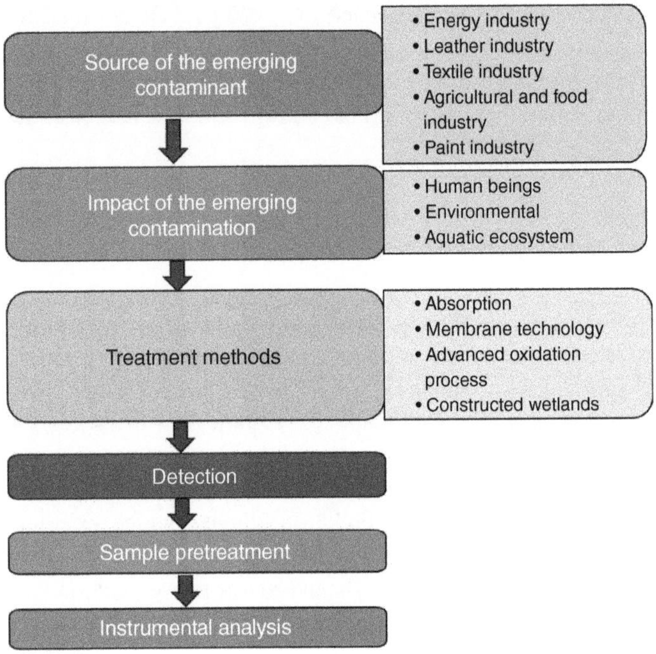

Figure 14.2 Emerging contaminants and management strategies.

improvements in childhood malnutrition but still millions of them suffer particularly in South Asian and Sub-Saharan African regions (World Water Vision 1999; FAO 2000; USERC 2005; Mark et al. 2006).

14.3.2 Fisheries

There is no doubt that human activities influenced habitat in physical, mechanical, and chemical ways. Physical alterations include artificial structures, aquaculture fixing, and oil rigs, whereas mechanical and chemical alterations include doses of nutrients, pesticides, heavy metals, and cultivating impact of dredges and trawls. In addition, fishing problems can reduce the productivity of resources and can distress the allied species. The other impact of fisheries could be the improper management of fishing practices, pollution from fish processing plants, macro- or micro-level plastic contamination that can confound the marine animals which can further lead to loss of fishing gear, etc. (Garcia et al. 2003).

14.3.3 Marine and Ocean Life

The atmosphere and ocean are getting warmer and the amounts of snow and ice are diminishing because of rising sea levels. The increasing concentration of CO_2 and greenhouse gases is responsible for the additional energy in the climate system. These changes in the CO_2 level are generally because of fossil fuel emissions and land-use changes which will enhance global warming and this can further influence the aquatic systems (fisheries and aquaculture). The water bodies' temperature across the globe is increasing which is further responsible for changes that occur in the freshwater system than in the ocean. Due to the increasing temperature, the dissolved oxygen (DO) level is decreasing (Allan et al. 2021). At coastal and offshore regions, the thinnest zones of O_2 across the ocean have extended, which can further influence the life cycle of different marine species. In addition, the ocean circulation pattern across the globe is also varying (Upwelling of Gulf stream and California current); intensification can be observed in the Benguela, Humboldt, and Canary current systems. The absorption level of anthropogenic CO_2 via the ocean is continuously growing. In this concern, the acidification of water could further affect the shell-forming aquatic life, particularly in low latitude to middle latitudes warm regions (Bahri et al. 2019).

14.3.4 Water Quality

Geogenic and anthropogenic activities are the chief factors for the weakening of groundwater quality, and they can pose both positive and negative impacts on the aquatic ecosystem and human life (Madhav et al. 2021). The intensification of nutrient concentrations concerning the warmer water temperature may hasten the biological productivity, plant biomass, and the intensity of algal blooms (IPCC 1995; Jacobs et al. 2001; Rizvi and Mohammed-Aslam 2019; Mohammed-Aslam and Rizvi 2020). In addition, the higher level of ultraviolet (UV) radiation may influence the growth and reproductivity of aquatic biota (Hader et al. 1995).

14.3.5 Climate Change

There is no doubt that climate change has affected freshwater biodiversity. Approximately 50% of the worldwide freshwater fish species are affected due to climate variation in which the reductions of body size structure, alteration in distribution pattern, changes in phenology, algal blooms, etc., are the most common. Temperature variation is also one of the most important factors in the alteration of different species (Parmesan 2006; Darwall and Freyhof 2015; Scheffers et al. 2016; Srivastav et al. 2021). Hydrogeochemistry of the freshwater regime is also affected due to climate variation which further threatens freshwater biodiversity.

14.3.6 Infectious Diseases

Though freshwater ecosystems are the hotspots for multiple species interactions, it is also responsible for the transmission of foci for human and wildlife pathogens. Many microparasites containing fungi, viruses, bacteria, and protozoans, and macroplastics consist of roundworms, flukes, tapeworms, arthropods, etc., are responsible for transmission This can further influence the freshwater biodiversity (Marcogliese 2008; Johnson and Paull 2011; Okamura and Feist 2011; Reid et al. 2018).

14.3.7 Destructive Algae

The newly developed freshwater algae are responsible for offering energy and nutrients to the associated food webs. The algal species that are selected via environmental ecological forces may accumulate biomass and these are known as harmful algal blooms which further threaten freshwater biodiversity. They can kill the fish by reducing the DO level or by toxicity. These algal blooms can also affect the development, existence, and reproduction of different organisms (D'Alelio et al. 2016; Reid et al. 2018).

14.3.8 Emerging pollutants

The key sources of surface water contaminants are mining, agriculture, paper production, oil and gas production, urban production, etc., which can directly or indirectly affect freshwater biodiversity by the impact on habitats or toxicity. Most of the antimicrobial compounds that originate under municipal wastewater or through the agricultural overflow are premeditated to eliminate the microorganisms in humans. This may further disturb the natural microbial communities (Nietch et al. 2013; Barra Caracciolo et al. 2015; Reid et al. 2018). These contaminants can disturb the algal diversity and periphyton.

14.3.9 Microplastic Contamination

Apart from biodegrading, plastics are broken down into smaller pieces through mechanical forces and UV rays. These are known as microplastics which include microbeads, nurdles, fragments, and microfibers (Barnes et al. 2009; Browne et al. 2011). Under the marine environment, these microplastics are swallowed by freshwater organisms (fish, birds, invertebrates, etc.), which can enhance the emerging risk on the organisms (Rios et al. 2007; Sigler 2014; Provencher et al. 2015; Holland et al. 2016; Campbell et al. 2017; Windsor et al. 2019).

14.3.10 Freshwater Salinization

Another problem with freshwater biodiversity is the occurrence of salinization. Most of the surveys were done and the result showed an unprecedented rate and scale. It is predicted that climate change could be responsible for strengthening salinization. In the lowland aquifer, discharge rises from salty water tables that may drive the dried-land salinization, whereas, in semi-arid regions, low rainfall may be responsible for increasing the rate of salts in surface waters during summer (Mills et al. 2013; Herbert et al. 2015; Reid et al. 2018). In addition, dumping or unintentional spilling of salty wastewater (by the production of coal seam gas and shale oil), excessive use of fertilizer in agriculture, etc., may also be responsible for the salinity (Vengosh et al. 2014; Etikala et al. 2021). Most freshwater species may lose their lives by absorbing the toxic level of salinization in their cells. This may reduce the species diversity of food resources for the patrons (Finlayson et al. 2013).

14.3.11 Calcium Deficiency

The main source of calcium to the freshwater is the low weathering rate of the parent rock which supplies the calcium pool inside the catchment soils that are further distributed to lakes and rivers. The acid rain may enhance the calcium leaching within the lakes that may further increase the lake-water calcium concentrations, whereas, in high calcium concentrations regions such as limestones, etc., calcium may be easily leached into waterways (Allen et al. 1997; Watmough et al. 2003; Reid et al. 2018). This may influence the freshwater biodiversity in the species by declining the percentage of calcium as calcium is a very important macronutrient for snail, salamander, and Arthropoda population present in the lakes and create an imbalance within the delicate ecosystem.

14.4 Impact of Climate Change on Aquatic Ecosystems

Climate change is a potential threat to global biodiversity and the functioning of the local ecosystem and the fact is well accepted by researchers across the globe and international bodies like Intergovernmental Panel on Climate Change (IPCC) and United Nations Framework Convention on Climate Change (UNFCCC). In the present time, climate-driven, physiochemical changes can be witnessed easily in every ecosystem, though the intensity may vary, none of them is unaffected. Aquatic ecosystems are more vulnerable to climate change and mainly occur through sea-level temperature rise, changes in frequency and amount of precipitation, extreme weather events, and water stress having both direct and indirect impacts on aquatic species (Figure 14.3). It directly affects the physiological behavior as well as the growth pattern of organisms, which further decreases the reproductive capacity, and causes mortality. Indirectly, it may change the productivity, structure, function, and composition of aquatic ecosystems (Ghosh et al. 2020).

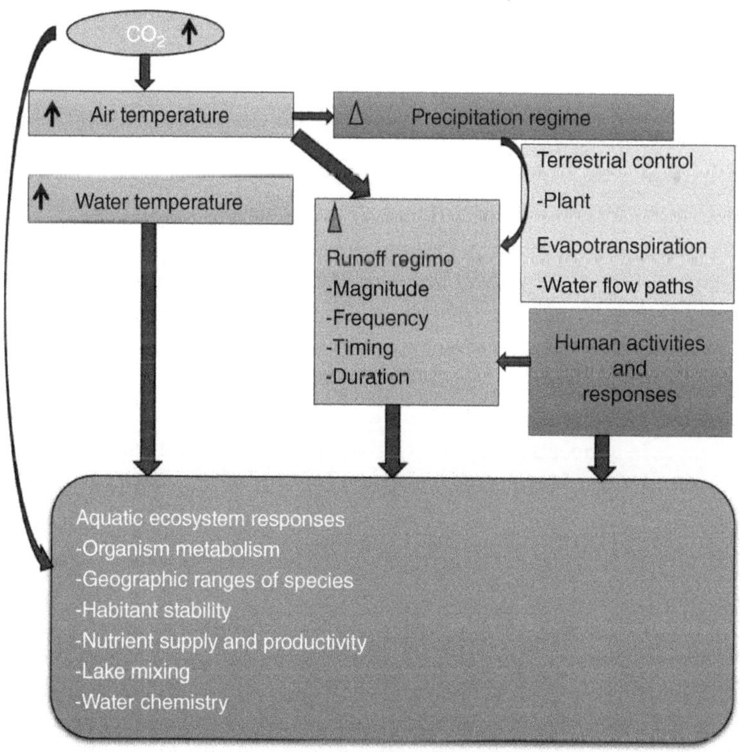

Figure 14.3 Inter-relationship between increased carbon dioxide (CO_2), climate and ecological processes. *Source:* Adapted from Poff et al. (2002).

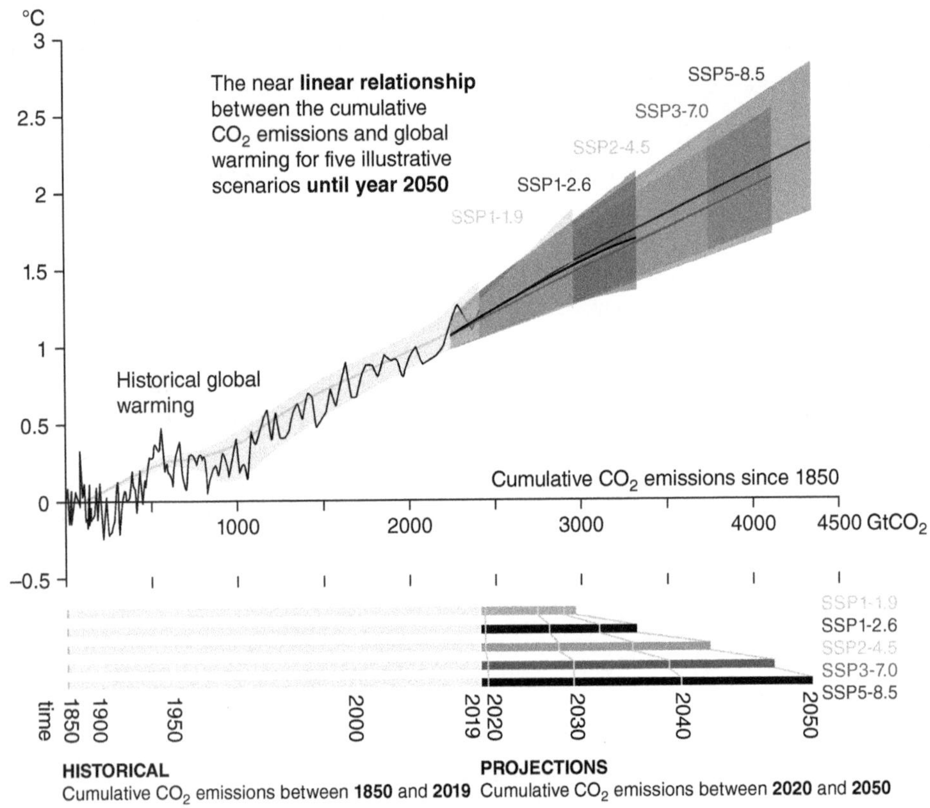

Figure 14.4 The global surface temperature increased since 1850–1900 (°C) as a function of cumulative CO_2 emission ($GtCO_2$). *Source:* Allan et al. (2021).

In the recent climate assessment report of Allan et al. (2021), a warmed climate is predicted by all climate scenarios (Figure 14.4); though it would not be uniform everywhere, impacts can be seen everywhere. The rise in global air temperature and changing precipitation patterns have a measurable impact on the physical characteristics of aquatic ecosystems, across the globe, and these changes are expected to continue in the coming decades too. These changes in temperature and precipitation are not uniformly distributed, and hence, their impact on aquatic ecosystems varies region by region, and changes in biogeochemical cycles are expected.

14.4.1 Climate Change and Freshwater Ecosystems

Freshwaters are such an aquatic ecosystem, which is relatively isolated and physically fragmented over the landmass, and are heavily exploited by humans for the provision of "goods and services" (e.g. drinking water and food) (Woodward et al. 2009). Though it covers only 0.8% of the Earth's surface, a home for an estimated 6% of all species, disproportionality is at risk on a global scale (Dudgeon et al. 2006). The changing climate shows its impact on the freshwater ecosystem by altering precipitation, temperature, and wind regimes. These

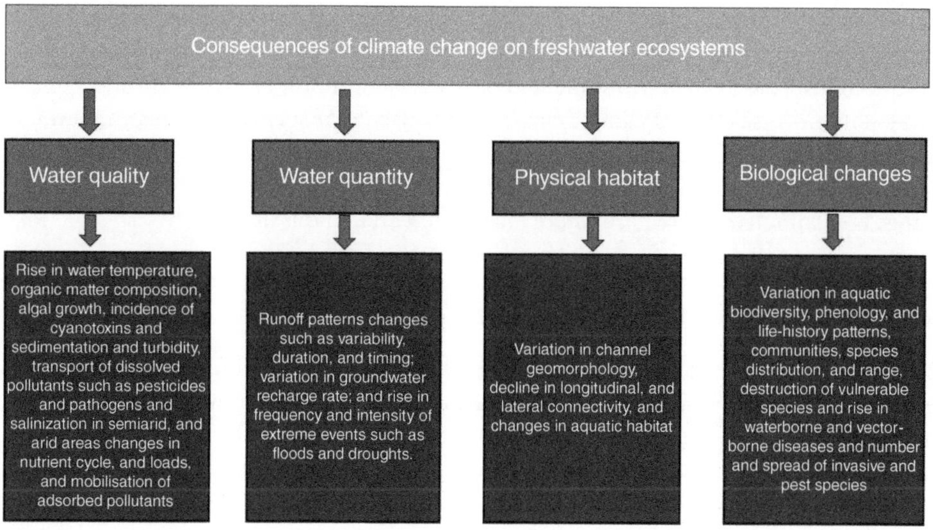

Figure 14.5 Consequences of climate change on freshwater ecosystems. *Source:* Adapted from Dallas and Rivers-Moore (2014).

changes mostly show in terms of alter in physical characteristics such as changes in ice cover, stratification, and mixing on water columns in lakes, regional, or catchment hydrology, which in turn alter the chemical and biological behaviors. The changes in chemical behavior can be observed in alteration in oxygen concentration, nutrient cycling, water-color, odor, etc. On the other hand, biological responses include the changes in the phenology and species distribution of most organism groups. The recent IPCC report advocates that the changes in the climate system will continue in the upcoming decades too and we can expect more changes in the freshwater ecosystem (Figure 14.5).

The changing climate is altering the energy budget of the freshwater ecosystem. The rise in global air temperature is changing the hypolimnetic temperature in lakes which resulted in a threat to deepwater anoxia. On the other hand, altered wind patterns are changing the input of mixing energy to lakes and overall effect of their heat balance and internal heat distribution (Kirillin et al. 2012; Schwefel et al. 2016; Mammarella et al. 2018). These changes together are reflected in changes in the physical behavior of lakes, which furthermore, modify the biological and chemical characteristics of surface water. Altering rainfall patterns, like fluctuation in intensity, amount seasonality is altering the hydrological cycles and thus biogeochemical cycles of the ecosystem (Kernan et al. 2011). The changes in the runoff pattern and thus the streamflow are also affecting the wetland and marine ecosystems.

Aquatic ecosystems are more sensitive and have a limited ability to adapt to climate change and hence are under stressed. Increasing water temperature as a consequence of climate change is altering the fundamental ecological processes and the geographic distribution of aquatic species. Such impacts may ameliorate the species to adapt to migration and increase the likelihood of species extinction and loss of biodiversity (Poff et al. 2002). The irreversible change in the climate system in the present time makes freshwater susceptible to climate change in an unprecedented way. This is because a) water temperature and availability are climate dependent; b) many systems are already exposed to numerous anthropogenic stressors; c) many species within these fragmented habitats have limited abilities to disperse as the environment changes (Woodward et al. 2010). In the coming decades, the adverse effect of climate change can be seen over such ecosystems and several freshwater species may face threats to its extinction. Though several studies have been done so far, our understanding of interactions between freshwater ecosystems and climate change is still limited and requires more focused research in this domain.

14.4.2 Climate Change and Marine Ecosystem

The ocean covers approximately 70% of the Earth's surface and is crucial for climate and weather patterns globally. The marine ecosystem, like others, has also been impacted by climate change and the changes are associated with concurrent shifts in ocean stratification, circulation, temperature deoxygenation, nutrient input, and ocean acidification, with potentially wide-ranging biological effects (Doney et al. 2012, IUCN 2016, IPCC 2014, Abram et al. 2019; Allan et al. 2021). With the rise in global atmospheric CO_2, the ocean, which is considered being the CO_2 sink, has been disproportionately impacted in recent decades. The increased CO_2 uptake is the predominant factor driving ocean acidification (Dore et al. 2009) and reflects a series of chemical changes such as reduced pH, carbonate ion, and $CaCO_3$ saturation states and elevated aqueous CO_2 and total inorganic carbon (Doney et al. 2009). The estimated pH of sea surface has dropped by 0.1 pH units since the preindustrial era (Feely et al. 2009), and over the past 2–3 decades, rate of pH change is unprecedented; and this decline in the ocean interior has been observed in all ocean basins (Allan et al. 2021).

14.4.2.1 Salinity

Changes in ocean salinity can be used to measure changes in Earth's water cycle as it reflects the net exchange of freshwater between the ocean and the atmosphere. The ocean accounts for ~80% of the global surface freshwater flux (Durack 2015) and changes in these fluxes have a great impact on ocean salinity and thus on the flora and fauna of marine ecosystems. The observed changes in salinity in the world's oceans (Figure 14.6a, b) indicate shifts and acceleration in the global rainfall and the evaporation cycle and are tied directly to climate change. In the coming decade, all the climate scenarios have been projected as an increase in rainfall and thus changes in salinity can be expected.

In both figures (Figure 14.6a, b), the black contours depicting the 0–200 m mean salinity climatology whereas the dotted areas indicate that the linear trends are statically significant at the 95% confidence level based on a modified Mann–Kendall test. The mean contour value of 35 or more indicates the saltier regions whereas the mean contour value of less than 35 depicts the fresher regions. The trend in Figure 14.6a is EN4 data whereas in Figure 14.6b is the ensemble mean of all observational datasets such as SCRIPPS, JAMSTEC, EN4, BOA, and NCEI (units: psu/year).

14.4.2.2 Ocean Heat and Sea Surface Temperature

It is one of the crucial factors in a marine ecosystem which helps to detect the changing climatic conditions and their adverse effect on the marine ecosystem. The rise in oceanic

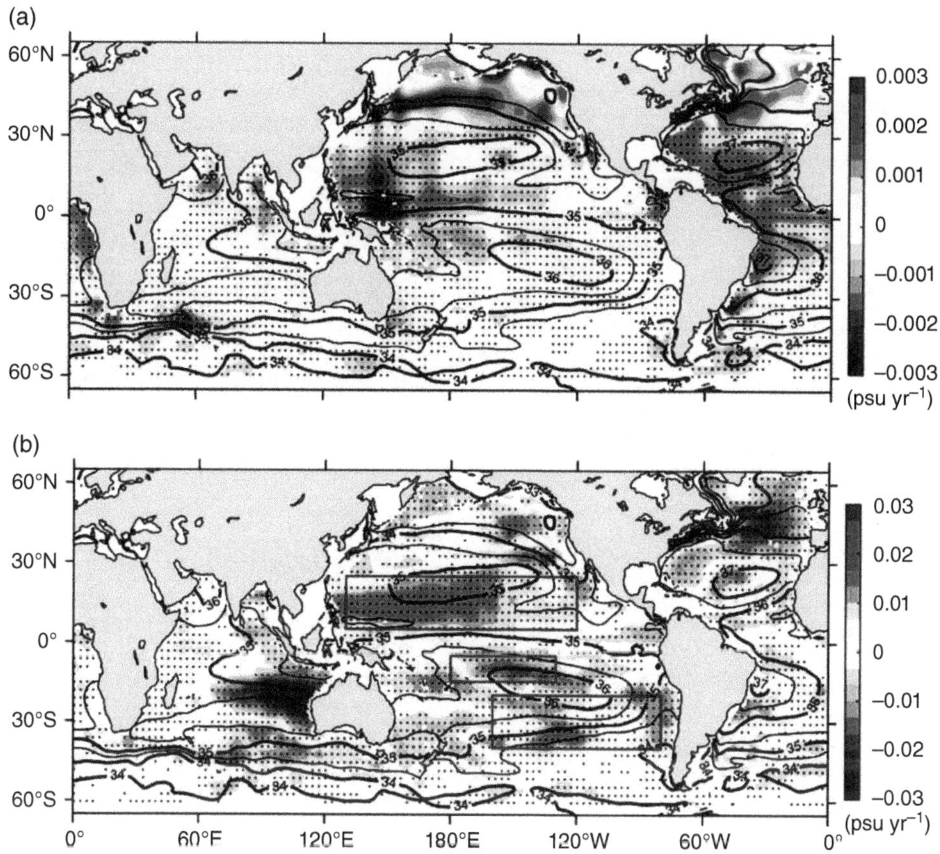

Figure 14.6 (a) The linear trends of (0–200 m) mean salinity during 1950–2004. (b) The linear trends of (0–200 m) mean salinity during 2005–2015. *Source:* Li et al. (2019a)/Springer Nature/ CC BY 4.0.

temperature (Figure 14.7) can be attributed directly to various anthropogenic activities (Bindoff et al. 2019). From 1901 through 2020, sea surface temperature rose at an average rate of 0.14°F per decade.

On the other hand, the long-term trend shows that the top 700 m of the oceans have become warmer since 1955 which indicates that the heat absorbed by surface waters extends to much lower depths over time (Levitus et al. 2012, EPA). It is important to note that the rate of change in ocean heat content can vary from year to year (Figure 14.8a, b) and these changes are influenced by events such as volcanic eruptions and recurring ocean-atmosphere patterns such as El Niño, Maden Jullian Oscillation.

14.4.2.3 pH

pH is considered to be directly related to oceanic carbon uptake as it is one of the biggest sinks of carbon dioxide. The recent trends indicate an increase of oceanic pH as observed CO_2 concentration in the atmosphere is rising unprecedentedly since the last couple of decades and is expected to continue in projected changes (Figure 14.9).

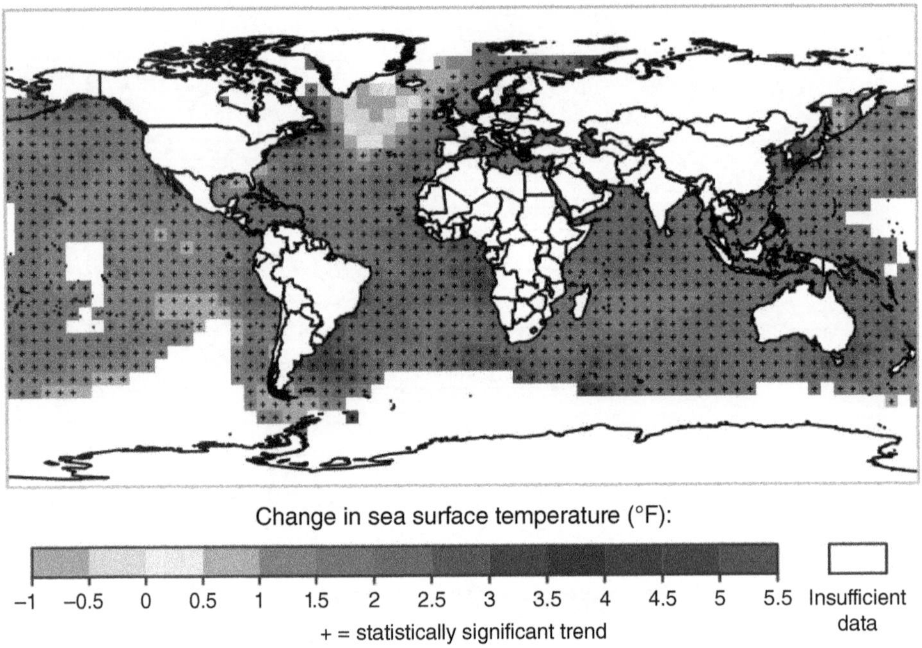

Figure 14.7 Changes in average SST between 1901 and 2020. *Source:* NOAA.

14.4.2.4 Oceanic Nutrients

Temperature and wind circulation play a crucial role in nutrient mixing and distribution, and changes have been observed and reported by several researchers. It has been observed that due to oceanic stratification, the nutrient cycles are disturbed and have altered the marine food chain.

14.4.2.5 Dissolved Gases

The oceans act as a huge repository of oxygen, carbon dioxide, nitrogen, etc., and careful monitoring of these gases can help us identifying the root causes of changes taking place in the oceanic ecosystems and would help to plan the mitigation strategies. With an increase of carbon dioxide emission, though carbon uptake has increased the oxygen level has decreased 0.5–3.3% between 1970 and 2010 (Bindoff et al. 2019) which eventually is an alarming situation for marine species.

14.4.3 Climate Change and Wetland Ecosystem

Wetlands are mostly found in every country with diverse climates across the globe except in Antarctica. It provides critical ecosystem services like cleaning water, absorbing floodwater, and providing habitat for wildlife, and is considered to be a natural solution to fight against climate change (Mitra et al. 2005; Mitsch et al. 2013). Wetlands are areas of high groundwater environments with permanent (shallow water bodies) or temporary inundation and vital for both humankind and nature (Dawson et al. 2003) and consist of 5–8% of the terrestrial land surface (Mitsch and Gosselink 2015), out of which, 92.8% of continental

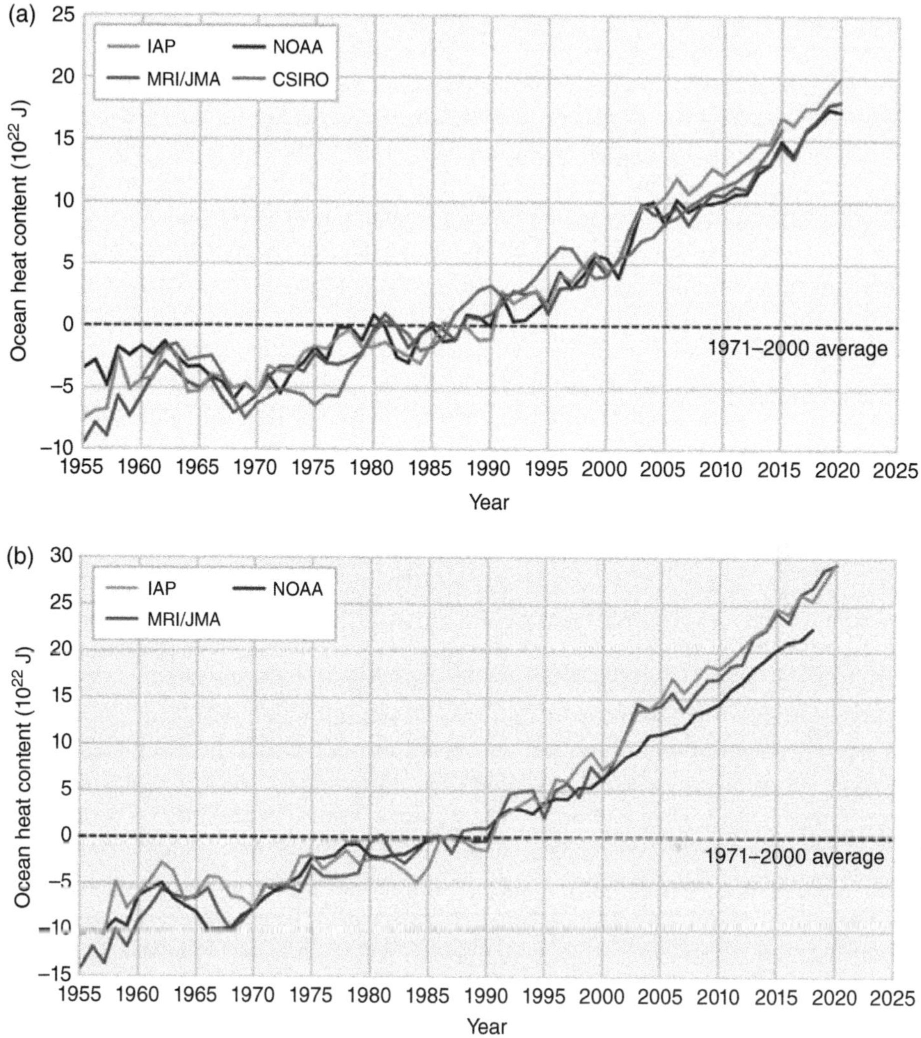

Figure 14.8 Heat content in the world ocean between 1955 and 2020 a) top 700 m, b) top 2000 m. *Source:* EPA.

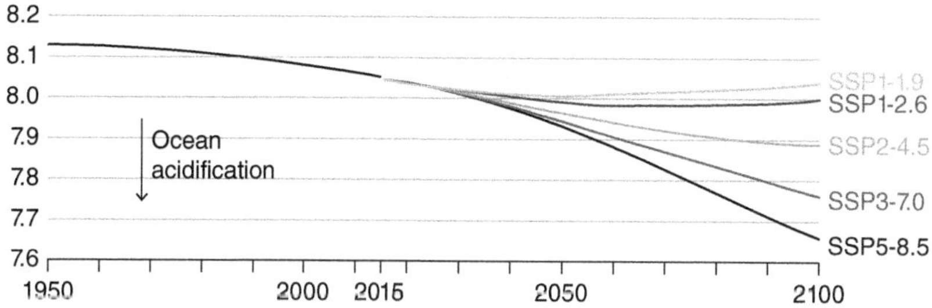

Figure 14.9 Global ocean surface pH (a measure of acidity) under observed and different climate scenarios. *Source:* Allan et al. (2021).

wetland area is inland and only 7.2% is coastal (Davidson et al. 2018). In an anoxic wet condition, the wetlands are optimum natural environments for confiscating and storing carbon from the atmosphere and are considered an important global carbon sink (Mitsch et al. 2013). In all terrestrial ecosystems, wetlands consist of the highest carbon density that led to global biogeochemical and carbon cycles, and therefore climate change (Kayranli et al. 2010). In the present era, climate change has been recognized as a chief threat to the wetlands system. Wetlands have witnessed rapid changes in water temperature, flow, and runoff rate in the last couple of decades and projected those changes in the climate system are likely to affect wetlands significantly, in their spatial extent, distribution, and function (Allan et al. 2021).

Wetland ecosystems are vulnerable to climate change and can be affected by both direct and indirect stressors. Climate change has a pronounced effect on wetlands through alterations in hydrological regimes as well as nutrient budgets with great global variability. Extreme climate events, increasing temperature, variations in rainfall and intensity and frequency, and anthropogenic activities make wetlands more susceptible to climate change. Such changes lead to decomposing and release nutrients to the surface water causing problems such as acidification, eutrophication, and brownification in the water bodies. Thus, this change is expected to make future efforts to restore and manage wetlands more complex (Roulet and Moore 2006; Erwin 2009; Corman et al. 2018). Wetlands are considered the main interface for moving nutrients, pollutants, and sediments from land to water. A small alteration in hydrologic fluctuations has a profound effect on the wetland ecosystem. The rising temperature provides warmer conditions which can accelerate microbial activity and eventually may result in more nitrous oxide emission. The changing climate might accelerate the rate of decomposition than production which can further lead to a shift from carbon sink to a source. Wetlands are affected by climate change in several ways such as rising water temperature and sea level, changes in rainfall pattern and distribution, land-use changes, favoring more invasive species, alteration in hydroperiod and hydrology, extreme weather events, possible trophic changes, and human water consumption patterns. Though wetlands are key components of global greenhouse budgets, its role in the climate feedback cycle is still poorly understood.

14.5 Protection, Adaptation, and Mitigation Strategies

Human interventions are comprehensively impacting the aquatic ecosystem and causing the disappearance of species on a large scale (Konrad et al. 2011; Castello and Macedo 2016). Aquatic habitats are more prone to extinction than birds and mammals, which result impacts the whole ecosystem; hence, close monitoring is required to prevent the redo loss which has already been made. The driving forces for freshwater challenges are agricultural and urban runoff, the introduction of exotic species, construction of dams, and diversion of water. In addition, other factors also play a vital role in accelerating the disturbance through urbanization, upscaling of industries, mining, and deforestation. The growing water demand has put immense pressure on water managers to rethink protecting source water in the form of rivers, lakes, and other aquatic systems. Aquatic ecosystems cover a range of physical configurations, ranging from an open system like the ocean to isolated water bodies in a terrestrial environment (Postel and Carpenter 1997). To protect the aquatic ecosystem and lead it to a sustainable stage, strategic support is needed, for which, conservation of biological resources could make the difference to preserve the ecosystem. A target-based restoration of aquatic species can be taken which have been declined or suffered habitat loss due to human intervention.

- Watershed management is proven to be a holistic approach to conserve the ecosystem and make it sustainable for the near future. Watershed practices are being done on a local scale which in turn is proving to be an effective practice to make the watershed more resilient and sustainable (Tezer et al. 2020).
- Afforestation is an effective exercise to prevent soil loss and holding it to avoid the siltation in the catchment water bodies; in addition, the enhancement of recharge can also take place and can enhance the soil moisture-holding capacity.
- The industrial waste disposal settlement must be taken care of before selecting a suitable site. However, surface and sub-surface water bodies should be disconnected from industrial waste disposal to restore the water quality and aquatic biodiversity (Gopal 2005; Pompeu et al. 2005; Trivedi and Verma 2016).
- Public awareness programs should be run to educate and aware the local inhabitant and stakeholders through government schemes and NGOs to protect the aquatic ecosystem.
- Public authorities must accelerate the water restorations and management program to conserve and protect the water bodies at local levels (villages, etc.)

It has become necessary to monitor and mitigate the emerging threats through newly developed tools. The conservation of biodiversity protection must be the primary concern with different strategies. The aquatic regions which have been harmed or suffered from habitat loss/degradation, species populations that have been declined, may be restored. Aquatic bio-reserve could be installed to protect the fisheries, plants, animals, habitats, etc. (Tewari and Bisht 2020). Bioregional management is important because it normalizes factors that affect aquatic biodiversity by maintaining the conservation of a given region (Irfan and Alatawi 2019). This may be further useful in the small- and large-scale biosphere reserves. The practices for the management of watersheds must be in the proper way to conserve aquatic diversity. Wastewater discharge must be monitored continuously for healthy water quality and to protect biodiversity.

14.6 Worst Case Scenario: Evidence of Anthropogenic Impact on Bengaluru Lakes: From "City of Lakes" to the "City of Froth Lakes"

Bengaluru is situated on a ridge which is above 900 m Above Mean Sea Level (AMSL) and occupied by three main valleys, namely, Vrishabhavathi Valley, Koramangala-Challaghatta Valley (KC Valley), and Hebbal Valley. During precipitation, water flows in the upper catchment area and feeds it to these valleys and each valley accommodates several lakes and water from high relief regions flows across the lakes during stormwater. The generation of sewage in Bengaluru is around 1400 MLD, but the worry is with a capacity to treat sewage,

Figure 14.10 Bellandur lake outlet foaming which flows into Varthur lake, 2019. *Source:* Courtesy of Sumita Bhattacharyya, ATREE.

60% of sewage goes untreated into the streams and lakes through the city drain system (Sanjana Alex 2020). In addition to domestic sewage, the by-product and effluent of industries are also contributing a major part to the damage, which is of real concern to think about (Figure 14.10).

There is a history of lakes catching fire several times in the recent past, the highlighted example is Bellandur Lake, which caught fire for four consecutive years and was first witnessed in 2015; the longest duration of fire flame reported in 2018, which lasted for 30 mins and created chaos in neighboring habitat. Bellandur and Varthur are situated at the lowest elevation in the lower catchment; hence, it receives the maximum effluent from upstream, causing more threat to aquatic biodiversity.

14.6.1 Cause of Fire in an Aquatic Environment

Domestic sewage and industrial untreated effluents pass through municipal drains, which finally open into streams and lakes. Inadequacy of effluent treatment plants (ETPs) and common ETPs. Moreover, these treatment plants are designed to treat specific types of contaminants but not all. Small- to medium-scale industries do not own their ETPs and transport their effluents far away from the source, which is not safe at all and the cost of treatment at ETPs and Common Effluent Treatment Plants (CETPs) are way higher and small-scale industries can feel for it. The dumping of solid waste into the lake from the demolition site and household gives rise to a thick layer on the lake surface, which in turn prevents the oxygen diffusion into the lake from the environment. As a result, it may generate an anaerobic condition and produce methane accumulation and toxic froth (Li et al. 2019b).

14.6.2 Protection and Mitigation to Conserve Aquatic Ecosystem

Chemical surfactants, biological surfactants, and filamentous bacteria are responsible to minimize the surface tension and maximize the anaerobic condition. Hence, proper treatment of sewage and effluents are needed before discharging into the lakes.

- Domestic and industrial waste could be centralized separately to treat them accordingly as per the nature of waste.
- Deployment of constructed wetland at the mouth of lakes to undertake secondary treatment before it finally goes to the lakes.
- Track the source of contaminated water and make it mandatory to monitor water quality regularly.
- Proper removal and treatment of anaerobic sediment from the lake bed to restore and conserve aquatic ecosystems.

14.7 Conclusion

With a rising population, our dependence on aquatic ecosystems has increased manifold to fulfil the demand and is under stress at several places. Unethical and nonscientific practices from individual to industrial level are intensifying the changes in the aquatic ecosystem, and many adverse impacts are emerging as a threat to food security, water security, and thus livelihood. Climate change is likely to further stress every ecosystem and is expected to create adverse impacts. In the context of ongoing climate change, improving our understanding of the physical conditions in aquatic systems, both in the present time as well as in the future, is important for maintaining the ecological balance. Developing predictions in the cause-and-effect pathways and how specific populations and communities are expected to respond would benefit the management of such biota. Changes in the terrestrial hydrological cycle may intensify the functioning of the freshwater and wetlands ecosystem and could adversely affect the habitats and biodiversity. It has been projected that mean ocean temperature, marine heatwaves, and ocean acidity will continue to increase; however, ocean salinity and DO is expected to decrease in the twenty-first century. As a result, more disruption in the marine ecosystem can be expected in near future. In the current context, it is paramount to adopt sustainable management of such ecosystems.

References

Abram, N., Carolina, A., Bindoff, N. L., & Cheng, L. (2019). IPCC Report, 2019: Special Report on the Ocean and Cryosphere in a Changing Climate. *Intergov. Panel Clim. Chang*, 1, 1–36.

Allan, J. D. and Flecker, A. S. (1993). Biodiversity conservation in running waters: identifying the major factors that threaten destruction of riverine species and ecosystems. *BioScience*, 43 (1), 32–43, https://doi.org/10.2307/1312104

Allen, R. B., Clinton, P. W. and Davis, M. R. (1997). Cation storage and availability along a Nothofagus forest development sequence in New Zealand. *Canadian Journal of Forest Research*, 27, 323–330.

Allan, R.P., Hawkins, E., Bellouin, N. and Collins, B. (2021). IPCC, 2021: Summary for Policymakers.

Bahri, T., Barange, M. and Moustahfid, H. (2019). Climate change and aquatic systems. In: *Impacts of Climate Change on Fisheries and Aquaculture, Synthesis of Current Knowledge, Adaptation and Mitigation Options* (eds M. Barange, T. Bahri, M. C. M. Beveridge, K.L. Cochrane, S. Funge-Smith, and F. Poulaine), pp. 1–17. FAO, Rome.

Barnes, D. K. A., Galgani, F., Thompson, R. C. and Barlaz, M. (2009). Accumulation and fragmentation of plastic debris in global environments. *Philosophical Transactions of the Royal Society B: Biological Sciences*, 364, 1985–1998.

Barra Caracciolo, A., Topp, E. and Grenni, P. (2015). Pharmaceuticals in the environment: biodegradation and effects on natural microbial communities. A review. *Journal of Pharmaceutical and Biomedical Analysis*, 106, 25–36.

Bindoff, N. L., Cheung, W. W., Kairo, J. G., Arístegui, J., Guinder, V. A., Hallberg, R., Hilmi, N. J. M., Jiao, N., Karim, M. S., Levin, L. and O'Donoghue, S. (2019). Changing ocean, marine ecosystems, and dependent communities. *IPCC Special Report on the Ocean and Cryosphere in a Changing Climate*, pp.477–587.

Browne, M. A., Crump, P., Niven, S. J., Teuten, E., Tonkin, A., Galloway, T. and Thompson, R. (2011). Accumulation of microplastic on shorelines worldwide: sources and sinks. *Environmental Science & Technology*, 45, 9175–9179.

Campbell, S. H., Williamson, P. R. and Hall, B. D. (2017). Microplastics in the gastrointestinal tracts of fish and the water from an urban prairie creek. *FACETS*, 2, 395–409.

Castello, L. and Macedo, M. N. (2016). Large-scale degradation of Amazonian freshwater ecosystems. *Global Change Biology*, 22(3), 990–1007.

Central Ground Water Board (CGWB), (2013). Groundwater brochure Medak District, Andhra Pradesh. http://cgwb.gov.in/District_Profile/Telangana/Medak.pdf

Chen, K. W., Lin, L. C. and Lee, W. S. (2014). Analyzing the carbon footprint of the finished bovine leather: a case study of aniline leather. *Energy Procedia*, 61, 1063–1066.

Corman, J. R., Bertolet, B. L., Casson, N. J., Sebestyen, S. D., Kolka, R. K. and Stanley, E. H. (2018). Nitrogen and phosphorus loads to temperate seepage lakes associated with allochthonous dissolved organic carbon loads. *Geophysical Research Letters*, 45(11), 5481–5490.

D'Alelio, D., Montresor, M., Mazzocchi, M. G., Margiotta, F., Sarno, D. and Ribeira d'Alcalà, M. (2016). Plankton food-webs: to what extent can they be simplified? *Advances in Oceanography and Limnology*, 7, 67–92.

Dallas, H. F. and Rivers-Moore, N. (2014). Ecological consequences of global climate change for freshwater ecosystems in South Africa. *South African Journal of Science*, 110(5–6), 1–11.

Darwall, W. R. and Freyhof, J. (2015). Lost fishes, who is counting? The extent of the threat to freshwater fish biodiversity. In: *Conservation of Freshwater Fishes* (eds G. P.Closs, M. Krkosek, and J. D. Olden), pp. 1–35. Cambridge University Press, Cambridge.

Davidson, N. C., Fluet-Chouinard, E. and Finlayson, C. M. (2018). Global extent and distribution of wetlands: trends and issues. *Marine and Freshwater Research*, 69(4), pp.620–627.

Dawson, T. P., Berry, P. M. and Kampa, E. (2003). Climate change impacts on freshwater wetland habitats. *Journal for Nature Conservation*, 11(1), 25–30.

Doney, S. C., Fabry, V. J., Feely, R. A. and Kleypas, J. A. (2009). Ocean acidification: the other CO_2 problem. *Annual Review of Marine Science*, 1, 169–192.

Doney, S. C., Ruckelshaus, M., Emmett Duffy, J., Barry, J. P., Chan, F., English, C. A., Galindo, H.M., Grebmeier, J.M., Hollowed, A.B., Knowlton, N., Polovina, J., Rabalais, N.N., Sydeman, W.J. and Talley, L. D. (2012). Climate change impacts on marine ecosystems. *Annual Review of Marine Science*, 4(1), 11–37. https://doi.org/10.1146/annurev-marine-041911-111611

Dore, J. E., Lukas, R., Sadler, D. W., Church, M. J. and Karl, D. M. (2009). Physical and biogeochemical modulation of ocean acidification in the central North Pacific. *Proceedings of the National Academy of Sciences*, 106(30), 12235–12240.

Dudgeon, D., Arthington, A. H., Gessner, M. O., Kawabata, Z. I., Knowler, D. J., Lévêque, C., Naiman, R. J., Prieur-Richard, A. H., Soto, D., Stiassny, M. L. and Sullivan, C. A. (2006). Freshwater biodiversity: importance, threats, status and conservation challenges. *Biological Reviews*, 81(2), 163–182.

Durack, P. J. (2015). Ocean salinity and the global water cycle. *Oceanography*, 28(1), 20–31.

Erwin, K. L. (2009). Wetlands and global climate change: the role of wetland restoration in a changing world. *Wetlands Ecology and Management*, 17(1), 71–84.

Etikala, B., Adimalla, N., Madhav, S., Somagouni, S. G., & Keshava Kiran Kumar, P. L. (2021). Salinity problems in groundwater and management strategies in arid and semi-arid regions. In: *Groundwater Geochemistry: Pollution and Remediation Methods* (eds S. Madhav, and P. Singh), pp. 42–56. John Wiley & Sons Ltd. https://doi.org/10.1002/9781119709732.ch3

FAO (Food and Agriculture Organization of the United Nations) (2000). FAO's statement on biotechnology. Press release during the Codex Alimentarius Ad Hoc Intergovernmental Task Force on Foods Derived from Biotechnology, March 2000, Japan. http://www.fao.org/biotech/stat.asp (accessed April 2005).

Feely, R. A., Doney, S. C. and Cooley, S. R. (2009). Ocean acidification: present conditions and future changes in a high-CO_2 world. *Oceanography*, 22(4), 36–47.

Finlayson, C. M., Davis, J. A., Gell, P. A., Kingsford, R. T. and Parton, K. A. (2013). The status of wetlands and the predicted effects of global climate change: the situation in Australia. *Aquatic Sciences*, 75, 73–93.

Gangloff, M., Edgar, G. and Wilson, B. (2016). Imperilled species in aquatic ecosystems: emerging threats, management and future prognoses: imperilled species in aquatic ecosystems. *Aquatic Conservation: Marine and Freshwater Ecosystems*, 26, 858–871. https://doi.org/10.1002/aqc.2707.

Garcia, S., Zerbi, A., Aliaume, C. and Chi, T. (2003). The ecosystem approach to fisheries: issues, terminology, principles, institution foundations, implementation and outlook. *FAO Fish Tech Paper*, 443, 1–71. https://www.fao.org/publications/card/en/c/701c6108-e79d-57bb-87d8-eeb85effb2cd/

Ghosh, S., Chatterjee, S., Prasad, G. S. and Pal, P. (2020). Effect of climate change on aquatic ecosystem and production of fisheries. In: *Inland Waters-Dynamics and Ecology* (eds A. Devlin, J. Pan, and M.M. Shah, (Eds.). IntechOpen. 10.5772/intechopen.93784

Gissi, E., Manea, E., Mazaris, A. D., Fraschetti, S., Almpanidou, V., Bevilacqua, S., Coll, M., Guarnieri, G., Lloret-Lloret, E., Pascual, M. and Petza, D. (2021). A review of the combined effects of climate change and other local human stressors on the marine environment. *Science of the Total Environment*, 755, 142564.

Gopal, B. (2005). Does inland aquatic biodiversity have a future in Asian developing countries? In: *Aquatic Biodiversity II* (eds Segers, H., Martens, K.), vol. 180, pp. 69–75. Springer, Dordrecht. https://doi.org/10.1007/1-4020-4111-X_10

Hader, D.-P., Worrest, R. C., Kumar, H. D., and Smith, R. C. (1995). Effects of increased solar ultraviolet radiation on aquatic ecosystems. *Ambio*, 24, 174–180.

Herbert, E. R., Boon, P., Burgin, A. J., Neubauer, S. C., Franklin, R. B., Ardón, M., Hopfensperger, K. N., Lamers, L. P. M. and Gell, P. (2015). A global perspective on wetland salinization: ecological consequences of a growing threat to freshwater wetlands. *Ecosphere*, 6, art206.

Holland, E. R., Mallory, M. L. and Shutler, D. (2016). Plastics and other anthropogenic debris in freshwater birds from Canada. *Science of the Total Environment*, 571, 251–258.

Hughes, J. M. (2007). Constraints on recovery: using molecular methods to study connectivity of aquatic biota in rivers and streams. *Freshwater Biology*, 52, 616–631.

IPCC (Intergovernmental Panel on Climate Change). (1995). IPCC Second Assessment: Climate Change 1995.

IPCC (2014). Summary for policymakers. In: *Climate Change 2014: Impacts, Adaptation, and Vulnerability. Part A: Global and Sectoral Aspects. Contribution of Working Group II to the Fifth Assessment Report of the Intergovernmental Panel on Climate Change* (ed Field C.B. et al.). Cambridge University Press, Cambridge, United Kingdom and New York, NY, USA, pp. 1–32

Irfan, S. and Alatawi, A. M. M. (2019). Aquatic ecosystem and biodiversity: a review. *Open Journal of Ecology*, 9(1), 1.

IUCN (1971). Education and the environment. Papers of the Nevada Conference of 1970 and the Zurich Conference of December 1971. IUCN Publication New Series, Morges.

IUCN (2016). International Union for Conservation of Nature annual report 2016. https://portals.iucn.org/library/node/46619

Jacobs, K., Adams, D. B. and Gleick, P. (2001). Potential consequences of climate variability and change for the water resources of the United States. In: *National Assessment Synthesis Team, Climate Change Impacts on the United States: The Potential Consequences of Climate Variability and Change. Report for the US Global Change Research Program.* Cambridge University Press, Cambridge, UK. https://irma.nps.gov/DataStore/Reference/Profile/2225255

Johnson, P. T. J. and Paull, S. H. (2011). The ecology and emergence of diseases in freshwaters. *Freshwater Biology*, 56, 638–657.

Kappes, P. J., Benkwitt, C. E., Spatz, D. R., Wolf, C. A., Will, D. J. and Holmes, N. D. (2021). Do invasive mammal eradications from islands support climate change adaptation and mitigation? *Climate*, 9(12), 172.

Kayranli, B., Scholz, M., Mustafa, A. and Hedmark, Å. (2010). Carbon storage and fluxes within freshwater wetlands: a critical review. *Wetlands*, 30(1), 111–124.

Kernan, M., Battarbee, R. W. and Moss, B. R. eds. (2011). *Climate Change Impacts on Freshwater Ecosystems*. John Wiley & Sons.

Kirillin, G., Leppäranta, M., Terzhevik, A., Granin, N., Bernhardt, J., Engelhardt, C., Efremova, T., Golosov, S., Palshin, N., Sherstyankin, P. and Zdorovennova, G. (2012). Physics of seasonally ice-covered lakes: a review. *Aquatic Sciences*, 74(4), 659–682.

Konrad, C. P., Olden, J. D., Lytle, D. A., Melis, T. S., Schmidt, J. C., Bray, E. N., Freeman, M. C., Gido, K. B., Hemphill, N. P., Kennard, M. J. and McMullen, L. E. (2011). Large-scale flow experiments for managing river systems. *BioScience*, 61(12), 948–959.

Levitus, S., Antonov, J., Boyer, T., Baranova, O., Garcia, H., Locarnini, R., Mishonov, A., Reagan, J., Seidov, D., Yarosh, E., Zweng, M. (2012). World ocean heat content and thermosteric sea level change (0–2000 m), 1955–2010. *Geophysical Research Letters*, 39. 1–5. https://doi.org/10.1029/2012GL051106.

Li, G., Zhang, Y., Xiao, J., Song, X., Abraham, J., Cheng, L. and Zhu, J. (2019a). Examining the salinity change in the upper Pacific Ocean during the Argo period. *Climate Dynamics*, 53(9), 6055–6074.

Li, Y., Chen, Y. and Wu, J. (2019b). Enhancement of methane production in anaerobic digestion process: a review. *Applied energy*, 240, 120–137.

Madhav, S., Raju, N. J., Ahamad, A., Singh, A. K., Ram, P. and Gossel, W. (2021). Hydrogeochemical assessment of groundwater quality and associated potential human health risk in Bhadohi environs, India. *Environmental Earth Sciences*, 80(17), 1–14.

Malhi, Y., Franklin, J., Seddon, N., Solan, M., Turner, M. G., Field, C. B. and Knowlton, N. (2020). Climate change and ecosystems: threats, opportunities and solutions. *Philosophical Transactions of the Royal Society B*, 375 (1794), 20190104.

Mammarella, I., Gavrylenko, G., Zdorovennova, G., Ojala, A., Erkkilä, K.M., Zdorovennov, R., Stepanyuk, O., Palshin, N., Terzhevik, A., Vesala, T. and Heiskanen, J., 2018. Effects of similar weather patterns on the thermal stratification, mixing regimes and hypolimnetic oxygen depletion in two boreal lakes with different water transparency. *Boreal Environment Research*, 23, 237–247.

Marcogliese, D. J. (2008). The impact of climate change on the parasites and infectious diseases of aquatic animals. *Revue Scientifique et Technique (International Office of Epizootics)*, 27, 467–484.

Mark, R., Sarah, C., and Andrea, S. R. (2006). Global Water and Food Security: Emerging Issues, https://www.researchgate.net/publication/254385599.

Mills, K., Gell, P., Gergis, J., Baker, P. J., Finlayson, C. M., Hesse, P. P., Jones, R., Kershaw, P., Pearson, S., Treble, P. C., Barr, C., Brookhouse, M., Drysdale, R., McDonald, J., Haberle, S., Reid, M., Thoms, M. and Tippy, J. (2013). Paleoclimate studies and natural-resource management in the Murray-Darling basin II: unravelling human impacts and climate variability. *Australian Journal of Earth Sciences*, 60, 561–571.

Mitra, S., Wassmann, R. and Vlek, P. L. (2005). An appraisal of global wetland area and its organic carbon stock. *Current Science*, 88(1), 25–35.

Mitsch, W. and Gosselink, J. (2015). *Wetlands*, 5th edition. John Wiley & Sons, Inc., Hoboken, New Jersey.

Mitsch, W. J., Bernal, B., Nahlik, A. M., Mander, Ü., Zhang, L., Anderson, C. J., Jørgensen, S.E. and Brix, H. (2013). Wetlands, carbon, and climate change. *Landscape Ecology*, 28(4), 583–597.

Mohammed-Aslam, M. A. and Rizvi, S. S (2020). Hydrogeochemical characterisation and appraisal of groundwater suitability for domestic and irrigational purposes in a semi-arid region, Karnataka state, India. *Applied Water Science*, 10, 237. https://doi.org/10.1007/s13201-020-01320-1.

Muralikrishna, I. V. and Manickam, V. (2017). Chapter – 13 Industrial wastewater treatment technologies, recycling, and reuse. In: Iyyanki V. Muralikrishna, Valli Manickam (eds.), Environmental Management, Butterworth-Heinemann 295–336. https://doi.org/10.1016/B978-0-12-811989-1.00013-0.

Nietch, C. T., Quinlan, E. L., Lazorchak, J. M., Impellitteri, C. A., Raikow, D. and Walters, D. (2013). Effects of a chronic lower range of triclosan exposure on a stream mesocosm community. *Environmental Toxicology and Chemistry*, 32, 2874–2887.

Okamura, B. and Feist, S. W. (2011). Emerging diseases in freshwater systems. *Freshwater Biology*, 56, 627–637.

Owusu, P. A. and Asumadu-Sarkodie, S. (2016). A review of renewable energy sources, sustainability issues and climate change mitigation. *Cogent Engineering*, 3(1), 1167990.

Parmesan, C. (2006). Ecological and evolutionary responses to recent climate change. *Annual Review of Ecology, Evolution, and Systematics*, 37, 637–669.

Pletterbauer, F., Melcher, A. and Graf, W. (2018). Climate change impacts in riverine ecosystems. *Riverine Ecosystem Management. Aquatic Ecology Series*, 8, pp.203–223.

Poff, N. L., Brinson, M. M. and Day, J. W. (2002). Aquatic ecosystems and global climate change. *Pew Center on Global Climate Change, Arlington, VA*, 44, 1–36.

Pompeu, P. S., Alves, C. B. M. and Allisto, M. C., 2005. The effects of urbanization on biodiversity and water quality in the Rio das Velhas basin, Brazil. *American Fisheries Society Symposium*, 47, 11–22).

Postel, S. and Carpenter, S. (1997). Freshwater ecosystem services. In: *Nature's Services* (ed G. Daily), 195–214. Island Press, Washington, DC.

Provencher, J. F., Bond, A. L. and Mallory, M. L. (2015). Marine birds and plastic debris in Canada: a national synthesis and a way forward. *Environmental Reviews*, 23, 1–13.

Reid, A., Carlson, A., nd Creed, I., Eliason, E., Gell, P., Johnson, P., Kidd, K., Maccormack, T., Olden, J., Ormerod, S., Smol, J., Taylor, W., Vermaire, J., Dudgeon, D. and Cooke, S. (2018). Emerging threats and persistent conservation challenges for freshwater biodiversity. *Biological Reviews*. https://doi.org/10.1111/brv.12480.

Rios, L. M., Moore, C. and Jones, P. R. (2007). Persistent organic pollutants carried by synthetic polymers in the ocean environment. *Marine Pollution Bulletin*, 54, 1230–1237.

Rizvi, S. S. and Mohammed-Aslam, M. A (2019). Microbiological quality of drinking water in Amarja reservoir catchment, Aland taluk, Karnataka, India. *Current Science*, 117(1), 114–121. https://doi.org/10.18520/cs/v117/i1/114-121.

Roulet, N. and Moore, T. R., 2006. Browning the waters. *Nature*, 444(7117), 283–284.

Salimi, S., Almuktar, S. A. and Scholz, M. (2021). Impact of climate change on wetland ecosystems: a critical review of experimental wetlands. *Journal of Environmental Management*, 286, 112160.

Sanjana Alex (2020). https://bengaluru.citizenmatters.in/author/sanjanaalex.

Scheffers, B. R., De Meester, L., Bridge, T. C. L., Hoffmann, A. A., Pandolfi, J. M., Corlett, R. T., Butchart, S. H. M., Pearce-Kelly, P., Kovacs, K. M., Dudgeon, D., Pacifici, M., Rondinini, C., Foden, W. B., Martin, T. G., Mora, C., Bickford, D. and Watson, J. E. M. (2016). The broad footprint of climate change from genes to biomes to people. *Science*, 354, aaf7671.

Schwefel, R., Gaudard, A., Wüest, A. and Bouffard, D. (2016). Effects of climate change on deepwater oxygen and winter mixing in a deep lake (Lake Geneva): Comparing observational findings and modeling. *Water Resources Research*, 52(11), 8811–8826.

Shah, T., Burke, J. and Villholth, K. (2007). Groundwater: a global assessment of scale and significance. In: *Water for Food, Water for Life* (ed. D. Molden). Earthscan, London, UK and IWMI, Colombo, Sri Lanka, pp. 395–423.

Sigler, M. (2014). The effects of plastic pollution on aquatic wildlife: current situations and future solutions. *Water, Air, & Soil Pollution*, 225, 2184.

Sowman, M. and Rebelo, X. (2022). Sustainability, disaster risk reduction and climate change adaptation: building from the bottom up – a South African perspective from the small-scale fisheries sector. In: *Creating Resilient Futures* (eds. S. Flood, Y. Jerez Columbié, M. Le Tissier, and B. O'Dwyer), pp. 151–181. Palgrave Macmillan, Cham.

Srivastav, A. L., Dhyani, R., Ranjan, M., Madhav, S. and Sillanpää, M., 2021. Climate-resilient strategies for sustainable management of water resources and agriculture. *Environmental Science and Pollution Research*, 28(31), 41576–41595.

Tewari, G. and Bisht, A. (2020), Aquatic Biodiversity: Threats and Conservation http://aquafind.com/articles/aquatic_biodiversity.php

Tezer, A., Turkay, Z., Uzun, O., Terzi, F., Koylu, P., Karacor, E., Okay, N. and Kaya, M. (2020). Ecosystem services-based multi-criteria assessment for ecologically sensitive watershed management. *Environment, Development and Sustainability*, 22(3), pp.2431–2450.

Trivedi, R. and Verma, A. (2016). Threat to aquatic biodiversity of arid regions: a review on dyeing units in Rajasthan. In: *Freshwater Environmental Impact on Biodiversity* (eds B. R. Bamniya and B. R. Gadi), 1st Edition, pp. 21–40.

USERC (United States Environmental Resource Center) (2005). Biotechnology and Water Security Opportunities. http://www.us-erc.org/greenchannel/gc7/biotechnology-watersecurity.php (accessed April 2005)

Vaiphei, S. P., Kurakalva, R. M. and Sahadevan, D. K. (2020). Water quality index and GIS-based technique for assessment of groundwater quality in Wanaparthy watershed, Telangana, India. *Environmental Science and Pollution Research*, 27(36), 45041–45062. https://doi.org/10.1007/s11356-020-103457.

Vengosh, A., Jackson, R. B., Warner, N., Darrah, T. H. and Kondash, A. (2014). A critical review of the risks to water resources from unconventional shale gas development and hydraulic fracturing in the United States. *Environmental Science and Technology*, 48, 8334–8348.

Wang, J., Soininen, J. and Heino, J. (2021). Ecological indicators for aquatic biodiversity, ecosystem functions, human activities and climate change. *Ecological Indicators*, 132. https://doi.org/10.1016/j.ecolind.2021.108250.

Waters, C. N., Zalasiewicz, J., Summerhayes, C., Barnosky, A. D., Poirier, C., Gałuszka, A., Cearreta, A., Edgeworth, M., Ellis, E.C., Ellis, M. and Jeandel, C. (2016). The Anthropocene is functionally and stratigraphically distinct from the holocene. *Science*, 351(6269), aad2622. https://doi.org/10.1126/science. aad2622.

Watmough, S. A., Aherne, J. and Dillon, P. J. (2003). Potential impact of forest harvesting on lake chemistry in south-central Ontario at current levels of acid deposition. *Canadian Journal of Fisheries and Aquatic Sciences*, 60, 1095–1103.

WHO. (2011). Guidelines for drinking-water quality. *WHO Chronicle*, 38(4), 104–108.

Windsor, F. M., Tilley, R. M., Tyler, C. R. and Ormerod, S. J. (2019). Evidence of microplastic entry routes into freshwater food webs. *Science of the Total Environment*, 646, 69–74.

Woodward, G., Friberg, N. and AG, H. (2009). The need for scientific rigour in biomonitoring and conservation of freshwaters.In: *Freshwater Ecosystems and Aquaculture Research* (eds. F. de Carlo and A. Bassano), pp. 277–288. Nova Science Publishing. ISBN: 978-1-60741-707-1

Woodward, G., Perkins, D. M. and Brown, L. E. (2010). Climate change and freshwater ecosystems: impacts across multiple levels of organization. *Philosophical Transactions of the Royal Society B: Biological Sciences*, 365(1549), 2093–2106.

World Water Vision. (1999). Biotechnology and Water Security in the 21st Century. Report of the Panel on Biotechnology of the World Commission on Water for the 21st Century, 3–5 February 1999. M.S. Swaminanthan Research Foundation, Chennai, India.

15

Role of Stable Isotopes in Groundwater Resource Management

Syed Shams Rizvi[1], *M. A. Mohammed-Aslam*[1], *Zameer Ahmad Shah*[2], *Taufique Warsi*[3], *Mohammad Muqtada Ali Khan*[4], *Tarik Mitran*[5], *and Shyam Kanhaiya*[6]

[1] *Department of Geology, Central University of Karnataka, Kalaburagi, Karnataka, India*
[2] *Geological Survey of India, U.T. Jammu and Kashmir, NR-Srinagar, India*
[3] *WOTR Centre for Resilience Studies (W-CReS), Watershed Organisation Trust (WOTR), Pune, Maharashtra, India*
[4] *Department of Geoscience, Faculty of Earth Science, Universiti Malaysia Kelantan, Jeli, Kelantan, Malaysia*
[5] *Soil and Land Resources Assessment Division, National Remote Sensing Centre, Balanagar, Hyderabad, Telangana, India*
[6] *Department of Earth and Planetary Sciences, VBS Purvanchal University, Jaunpur, Uttar Pradesh, India*

15.1 Introduction

Isotopes are the naturally occurring substances found on the Earth. Hydrogen, carbon, nitrogen, oxygen, and sulfur are the most important elements of the hydrological and geological systems. Elements having the same number of protons with a different number of neutrons are called isotopes. The element can have two, three, or more stable naturally occurring isotopes (Sinha et al. 2019). The movement of water can be traced by the abundance of hydrogen and oxygen isotopes (Sinha et al. 2019; Gupta and Goel 2021). An element can have both stable as well as radioactive isotopes. Stable isotope does not undergo any radioactive decay because their nuclei are always stable (Guay et al. 2006; Sinha et al. 2019; Gupta and Goel 2021).

Precipitation forms an important component in the hydrologic cycle and plays a vital role to replenish the groundwater sustainably (Wan et al. 2012; Khairul Nizar et al. 2018). Precipitation is classified as rain, freezing rain, snow, snow pellets, hail, and ice pellets (Patel and Shah 2008). The study of precipitation is of much significance from a hydrological perspective (Warrier et al. 2010), and the process which controls the temporal and geographic disparity of isotopic composition of precipitation always play a significant role in isotope hydrology (Gupta and Deshpande 2003). "Isotopic fractionation" leads to the geographical and temporal disparities in the isotopic composition of precipitation. Therefore, for a particular environment of precipitation, the unique isotopic signature provides information on the provenance of groundwater. The ^{2}H or D and oxygen ^{18}O help track the movement of water in the hydrologic cycle. These two isotopes of water are also called "water isotopes"

Hydrogeochemistry of Aquatic Ecosystems, First Edition. Edited by Sughosh Madhav, Virendra Bahadur Singh, Manoj Kumar, and Sandeep Singh.

(Clark and Fritz 1997; Gupta 2011; Cayuela Linares et al. 2018; Valdivielso et al. 2020). Oxygen and hydrogen isotopes of precipitations are the important tracers of hydrological processes for understanding the source of vapor, locally recycled vapor, rainout history, condensation temperature, relative humidity, etc. (Gupta and Deshpande 2003; Srivastava et al. 2015). The difference between the compositions of water isotopes can be explained using Global Meteoric Water Line (GMWL). Global Network of Isotopes in Precipitation (GNIP) has been jointly established by World Meteorological Organization (WMO) and International Atomic Energy Agency (IAEA). This organization conducts isotopic studies of $\delta^{18}O$ and $\delta^{2}H$ for precipitation at various monitoring stations worldwide including stations in India. The database maintained by the GNIP network is available at the website http://isohis.iaea.org (IAEA/WHO 2004; Deshpande 2006; Deshpande and Gupta 2008).

The Indian monsoon is considered a major factor in the atmosphere-ocean circulation system. The oceanic moistures sources generated from the Bay of Bengal (BOB) and the Arabian Sea chiefly contribute to the rainfall in the Indian sub-continent. The southwest monsoon from June to September is influenced by the Arabian Sea and the Indian Ocean in the southern Indian peninsula. BOB, Indian Ocean, and continental vapor play an important role in the contribution of moisture for the precipitation during the northeast monsoon (Berkelhammer et al. 2012; Roxy et al. 2015; Saravanan et al. 2019). However, there is a major gap in the availability of stable isotopic data of precipitation in India as a result of limited stations.

15.2 Background

15.2.1 Isotope Systematics

The isotope partitioning, which is also known as isotope fractionation, results from the differences in mass numbers of the molecules. The disparities that occur in the number of neutrons of an element are responsible for the different mass number of the same element. Heavy water ($^{2}H_2\,^{16}O$) with a mass number of 20 and normal water ($^{1}H_2\,^{16}O$) having a mass number of 18 represent the example of this situation. Water molecules travel through different domains of the hydrological cycle having different isotopic combinations that may get partitioned (fractionated) among the three phases of matter (Nisi et al. 2014).

The ratios of the two isotopes are considered for the stable isotopic studies of water using ^{1}H and ^{2}H, and ^{16}O and ^{18}O (Clark and Fritz 1997). The relative difference in the ratio of heavy isotope to the more abundant, light isotope of a sample to a reference is measured, as the absolute abundance ratio is not ordinarily determined for natural water. This difference (δ) is expressed as Eq. 15.1 (Gonfiantini 1981):

$$\left\{\delta = \frac{R_{sample} - R_{reference}}{R_{reference}}\right\} \tag{15.1}$$

where R represents the ratio of isotopes. In the case of water, it can be $^{2}H/^{1}H$ or $^{18}O/^{16}O$.

The isotopic composition of a water sample is analyzed by considering Vienna Standard Mean Ocean Water (V-SMOW) as a reference (Clark and Fritz 1997). Since small differences occur between the samples and reference value, the δ value may be expressed as part per thousand or permil (‰) differences (Gonfiantini 1981) as given in Eq. 15.2. Equation 15.2 has been further modified as Eqs. 15.3 and 15.4 when the relative difference of ratio is considered in permil (‰).

$$\delta(\text{‰}) = \delta \times 1000 \tag{15.2}$$

$$\delta(\text{‰}) = \frac{R_{sample} - R_{reference}}{R_{reference}} \times 1000 \tag{15.3}$$

$$\delta(\text{‰}) = \frac{R_{Sample} - R_{VSMOW}}{R_{VSMOW}} \times 1000 \tag{15.4}$$

where the δ (‰) is the isotope ratio relative to a standard, R represents the ratio of isotopes. R_{VSMOW} is the ratio of the isotopes used in international standards (Gonfiantini 1978).

A positive [(+ve) sign] with δ values gain by the above equation elucidates that there is an increase of massive (heavier) isotope (i.e. ^{18}O) over the second (^{16}O) to a standard. The negative [(-ve) sign] of δ values represents the other situation (Clark and Fritz 1997).

15.2.2 Isotope Fractionations

The disparities between the velocity and bond strength between isotopes lead to isotopic partitioning or fractionation. It is also described as differential partitioning of isotopes between the two phases of liquid or vapor or solid (Kumar 2013). Sometimes isotopic partitioning is described as "isotope discrimination." The fractionation factor, "α," is characterized as the isotope proportions, which is expressed as Eq. 15.5 (Clark and Fritz 1997).

$$\left\{ \alpha = \frac{R_{reactant}}{R_{product}} \right\} \tag{15.5}$$

where $R_{reactant}$ and $R_{product}$ represent the isotopic ratios of reactant and product, respectively.

For instance, in the case of a water–water vapor system, Eq. 15.5 can be written as follows (Eq. 15.6).

$$\left\{ \alpha = \frac{\left(O^{18} / O^{16}{}_{water} \right)}{\left(O^{18} / O^{16}{}_{vapour} \right)} \right\} \tag{15.6}$$

Reactant and product interact for a longer time during the isotopic equilibrium establishment process. In a nonequilibrium state, rapid temperature change or addition or removal of product or reactant takes place. In the case of diffusive fractionation, the product under isotopic equilibrium disseminates slowly from the product reservoir. Equilibrium and kinetic fractionations describe these three types of fractionation processes (Araguas-Araguas et al. 2000; Chacko et al. 2001).

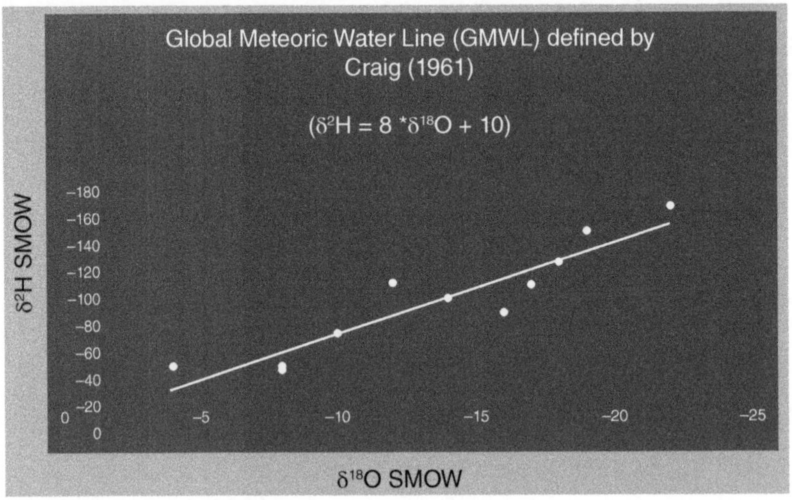

Figure 15.1 Global meteoric water line. *Source:* Adapted from Craig (1961).

15.2.2.1 ^{18}O and D in Precipitation

The stable isotopic composition of precipitation is controlled by parent vapor and temperature. It may also show higher variations due to evapotranspirational reprocess (Warrier et al. 2010). Based on the empirical observations, Craig (1961) correlated $\delta^{18}O$ and $\delta^{2}H$ in fresh water on a worldwide scale. The $\delta^{2}H$ and $\delta^{18}O$ are the potential tool to decipher the recharge processes and identify the diverse source of hydrological processes (Scholl and Murphy 2014; Oiro et al. 2018; Yeh and Lee 2018; Yang et al. 2019). It is known as the "Global Meteoric Water Line (GMWL)" that describes the relationship between $\delta^{18}O$ and δD in global fresh surface waters (Figure 15.1). The GMWL is expressed as Eq. 15.7.

$$\delta D = 8\delta^{18}O + 10 \tag{15.7}$$

The slope '8' in Eq. 15.7 is the ratio of isotopic equilibrium fractionation factors for D and ^{18}O (at 25°C) during the equilibrium condensation. The "D" enrichment in water is approximately "8" times greater than that of ^{18}O. The "intercept" 10 is a function of meteorological conditions in nonequilibrium evaporation of the ocean surface from where the precipitation begins (Clark and Fritz 1997).

The intercept of Eq. 15.7 has been further confirmed from several studies (Yurtsever and Gat 1981; Rozanski et al. 1993). They used all the isotopic results acquired by the "GNIP" program jointly initiated by IAEA, Vienna, and WMO. Those observations were given in Eq. 15.8.

$$\delta D = \left(8.17\delta^{18}O \pm 0.06\right) + \left(10.35 \pm 0.65\right) \tag{15.8}$$

Equation 15.8 was obtained by using the mean arithmetic values. When the weighted mean values were used, Eq. 15.9 was generated (Yurtsever and Gat 1981; Rozanski et al. 1993).

$$\delta D = \left(8.20\delta^{18}O \pm 0.07\right) + \left(11.27 \pm 0.65\right) \tag{15.9}$$

Deshpande et al. (2003) and Warrier et al. (2010) proposed the "Local Meteoric Water Line" for various parts of Indian regions. Kumar et al. (2010) provided an Indian Meteoric Water Line (IMWL) through the isotope ratios of oxygen ($^{18}O/^{16}O$) and hydrogen ($^{2}H/^{1}H$) in natural waters based on data from several stations in India. In addition, the "Regional Meteoric Water Line" (RMWL) for the different zones of the country like the southern part, western Himalayas, and the northern part of the country were also given by them. The IMWL made by them can be expressed as Eq. 15.10.

$$\delta D = \left(7.93 \pm 0.06\right)\delta^{18}O \pm \left(9.94 \pm 0.51\right) \tag{15.10}$$

The IMWL discussed in Eq. 15.10 is having a slight difference in slope and intercept compared to GMWL. IMWL has a great significance and it gives a baseline for groundwater investigation of a particular region. It differs from GMWL based on the disparities in climatic conditions and geographic constraints (Kumar et al. 2010).

15.2.2.2 Deuterium Excess

Deuterium excess is defined as the excess deuterium that cannot be accounted for kinetic fractionation between "water" and "vapor." ^{18}O and ^{2}H concentrate in a liquid phase, whereas ^{16}O and ^{1}H enter into the vapor phase based on the difference in mass number. The water vapor gets enriched in ^{16}O and ^{1}H and the remaining liquid water with ^{18}O and ^{2}H during the process of evaporation. Humidity, wind speed, salinity, and temperature are the important parameters that control the process of kinetic fractionation (Urey 1947; Clark and Fritz 1997). Among them, humidity is the most important parameter.

In the GMWL equation proposed by Craig (1961), the deuterium excess has a value of ~10 which represents the evaporation with ~ 85% of relative humidity. In regional precipitation, the value of deuterium excess is greater than 10 when the humidity is low (Gat and Carmi 1970). When the humidity (h = 50%) is low, the vapor is strongly depleted and the Local Meteoric Water Line (LMWL) plot would be above the GMWL, whereas when humidity is 85% (h = 85%), the precipitation plot is close to the GMWL (Clark and Fritz 1997).

Dansgaard (1964) studied the characteristics of the d-excess at global precipitation, expressed by Eq. 15.11.

$$d = \delta D - 8\delta^{18}O \tag{15.11}$$

Deuterium excess always provides valuable information about the moisture sources in the precipitation and the evaporation (Clark and Fritz 1997).

15.2.2.3 Continental Effect

Topographic condition in landmasses controls the rainout situation. Extreme temperature variation affects the isotopic composition in precipitation. The coastal precipitations are isotopically enriched with a minor range of $\delta^{18}O$ (−0‰ to −5‰), whereas the colder inner continental province receives isotopically depleted precipitation with strong seasonal differentiation. There is a strong seasonal variation in the temperature at continental regions which refers to a reflection of distance from moderating marine influences and latitude (Wolf et al. 2020; Saranya et al. 2021).

15.2.2.4 Seasonal Effect

Strong seasonal variation controls the water isotopes of precipitation due to seasonal variation of temperature. The values are found depleted in the colder months than the summer months suggested that there is a latitude variation from 0.5 ‰ per degree celsius in high latitude stations to 0 ‰ per degree celsius in tropical ocean islands (IAEA 1983; Ala-aho et al. 2018; Dublyansky et al. 2018; Vasil'chuk et al. 2022). Rozanski et al. (1993) also discussed the variations of $\delta^{18}O$ in a highly continental station and the sub-tropical region as a result of strong seasonal variation in temperature.

15.2.2.5 Latitude Effect

Precipitation at higher latitudes tends to have more negative $\delta^{18}O$ values as the air masses come to the inland, the vapor gets constantly depleted in the heavier isotopes due to the rainout effect process from the cloud during moisture transport from the tropics to the poles. The global study of $\delta^{18}O$ in precipitation demonstrated a relatively flat gradient over the tropic's regions. This gradient increases poleward with some distortion because of the continental effect and ocean currents (Yurtsever and Gat 1981; Clark and Fritz 1997; Kern et al. 2020; Leketa and Abiye 2020; Laonamsai et al. 2021; He et al. 2021). Rozanski et al. (1993) found that there is a (-0.6‰) latitude effect of $\delta^{18}O$ for continental stations of North America and Europe and (-2‰) $\delta^{18}O$ for the colder Antarctic regions.

15.2.2.6 Altitude Effect

Temperature change with altitude is called the "wet adiabatic lapse rate." At higher altitude having low average temperature, precipitation is isotopically depleted that varies from –0.15 ‰ to 0.5 ‰ per 100 m rise in altitude for ^{18}O and from −1 to −4‰ for ^{2}H. The altitude effect is also known as the "alpine effect" or "elevation effect." The altitude effect has great significance in groundwater recharge studies (IAEA 1983; Clark and Fritz 1997; Kern et al. 2020).

15.2.3 Methodology and Precautions in the Collection of Rainwater Samples

Appropriate sample collection techniques are important in any environmental study. In the case of isotope hydrological studies, enough precautions are needed during the collection of precipitation samples. To acquire dependable isotope data, it is fundamental that the standard testing system ought to be pursued. When the isotope technique is applied to a region to examine the various hydrological processes, all the accessible hydrological and meteorological data can also be obtained and this helps in fixing the sampling points and their intervals. Monitoring programs for precipitation are combined monthly like 12 samples per year. Daily samples of rain gauges are poured into a monthly sample bottle and are sealed and the accumulative amount can be noted for the weighting calculations (Clark and Fritz 1997; Krajcar Bronić et al. 2020; Wenninger 2020).

15.2.3.1 Rainwater Collection Techniques

Rainwater samples can be collected on a daily or weekly or monthly basis for isotopic studies using a specially designed rainwater collector. Rainwater collections on monthly basis are generally followed. In a monthly composite precipitation methodology, the accumulated precipitation for one full month period is collected. A thin layer of paraffin or silicon oil of about 2 mm thickness should be poured into the rainwater collector for preventing the loss of water through evaporation, which floats over the sample (Clark and Fritz 1997). Samples were collected from the specially designed collector instrument suggested by IAEA. The rainwater collector instrument contains 20 L polyvinyl chloride (PVC) carboy with a specially designed nylon cap consisting of two holes. These two holes have been further fixed with a plastic funnel (6 m long and 20 cm in diameter) and a 10-mm-thick PVC tube. A 6-mm plastic hosepipe inserted through a polymer cork and tightly fitted inside the funnel into the container up to the base. Paraffin oil of about 2 mm thickness should be poured into the rainwater collector apparatus before sampling so that it can form a layer of paraffin floating over water to protect it from evaporation. The instrument should be washed properly after collecting the sample and must be poured with paraffin oil into the collector before starting next month's sampling. A "separating funnel" should be used for separating water from the paraffin layer at the end of each month. Precipitation samples could have been taken in clean polythene (Torson) bottle of 60 mL capacity by filling up to the maximum level leaving only a little space for thermal expansion. The bottle must be closed tightly to avoid isotope exchange with air moisture.

15.2.3.2 Principle of Isotope Ratio Mass Spectrometer

Isotope Ratio Mass Spectrometer (IRMS) is specially designed to measure the mass differences of a molecular compound. The basic concept of IRMS is that the beam of the charged molecules can be produced by the thermal ionization of a solid source (solid sample) or by ionizing a gaseous sample. Source design generally depends on the isotopes of interest. IRMS can measure even a very small fraction of isotopes of $^{2}H/^{1}H$ and $^{18}O/^{16}O$. Oxygen-18 and deuterium (D) analysis of water samples can be done by equilibration methods. The analysis of oxygen isotopes ($^{18}O/^{16}O$ ratio) can be performed by CO_2–H_2O equilibration method, whereas deuterium analysis can be performed by H_2–H_2O equilibration method using platinum as a catalyst (Clark and Fritz 1997; Kaklamanos et al. 2020). Table 15.1 shows the different standards for the del (δ) value suggested by IAEA for V-SMOW.

15.2.4 Application of Isotopes in Water Resource Management

There are various applications of the environmental isotope in groundwater resource management as follows:

- Interconnections between surface and groundwater
- Tracking sources of groundwater recharge
- Groundwater age or groundwater dating
- Source and mechanism of groundwater pollution
- Sanitary landfills
- Agriculture watersheds
- Groundwater salinization

Table 15.1 Norms suggested for δ value by IAEA for V-SMOW (Vienna-SMOW, SLAP, and GISP agencies).

		Delta (δ) value for V-SMOW	
S. No.	**Norms**	**δ^2H (‰)**	**$\delta^{18}O$ (‰)**
1	V-SMOW	0	0
2	Standard Light Antarctic Precipitation (SLAP)	δ^2H_{SLAP} = -428.0 ‰ VSMOW	$\delta^{18}O_{SLAP}$ = -55.50 ‰ VSMOW
3	Greenland Ice Sheet Precipitation (GISP)	δ^2H_{GISP} = -189.5 ‰ VSMOW	$\delta^{18}O_{GISP}$ = -24.78 ‰ VSMOW

Source: Adapted from Gonfiantini (1978).

15.2.4.1 Interconnections Between Surface and Groundwater

The demand for freshwater resources has been increased due to the rapid growth of population, industrial development, and agricultural activities. The relationship between surface water (such as lakes, rivers, and streams) and groundwater can be well established using stable isotopes tracers. Isotopic compositions of groundwater are generally controlled by meteorological processes. Both environmental isotopes, stable isotopes and radioactive isotopes, are used as a tracer in hydrological investigations and provide essential information for the better management of water resources (Clark and Fritz 1997, Archana et al. 2014; Yang et al. 2019; Jung et al. 2020). Stable isotopic data can be well used to assess the relative proportions of surface water and precipitation in recharge. In arid and semiarid regions, due to the evaporation in surface water bodies, the stable isotope compositions of groundwater are useful in the detection of different recharge components. River water can also show a seasonal variation in stable isotope composition). In arid regions, river water may be enriched in δD and $\delta^{18}O$ compared to the groundwater, and then the fractionations of river water in groundwater can be evaluated based on differences in the isotopic composition of the mixing components (IAEA 2011). Therefore, the mechanism of stable isotope exchange in their composition can be well understood by surface and groundwater interactions.

15.2.4.2 Tracking Sources of Groundwater Recharge

It has been challenging for the hydrologist to track the sources of recharge to aquifers and the relative contributions from multiple sources. The rate of pumping has been increased in the past years which affects the declination of head and recharge changes. Therefore, it is indeed to understand the qualitative and quantitative characterization of groundwater recharge to ensure the sustainable development and management of groundwater resources (Zongyu et al. 2011; IAEA 2011). In the past years, stable and radioactive isotopes have been applied successfully to understand the problems related to groundwater recharge and delineation of flow systems. In addition, the advancement of nanotechnology imparts the novel tracers being used for groundwater dynamics (Cook and Herczeg 2000; Geyh et al. 2008; Kinzelbach et al. 2002; Warsi et al. 2020). The isotopic composition of groundwater can be evaluated based on the isotopic composition of recharge, and if the recharge is obtained through the direct infiltration of precipitation, then the groundwater will reflect the isotopic composition of that precipitation; and if it is through surface water instead of precipitation, it will reflect the mean isotopic composition of the contributing river or lake. Tritium is produced naturally by cosmic radiation and over the past three decades, the presence of thermonuclear tritium (^3H) was used as clear evidence for active recharge. In the arid and semiarid regions, isotope techniques constitute virtually the only approach for the identification and quantification of groundwater recharge (Clark and Fritz 1997; IAEA 2011; Aggarwal et al. 2021).

15.2.4.3 Groundwater Age or Groundwater Dating

Groundwater age means groundwater residence time, which means the length of time from which the water has been isolated from the atmosphere or the average travel time between the point of recharge and the point of discharge. Groundwater dating means estimating the groundwater age by using different methods. Radioactive isotopes play an important role in the estimation of groundwater age. Tritium (^3H) and carbon-14 (^{14}C) are the most well-known radioisotopes that are used in groundwater dating. In the confined aquifers, recharge is usually occurring in a horizontal or lateral gradient age which means the age increases with distance from the area of recharge. In addition to the unconfined aquifers, recharge is usually occurring in a vertical gradient of groundwater ages which means increasing the age with depth (Clark and Fritz 1997; IAEA 2011; Gupta 2011; Chacha et al. 2018; Ju et al. 2020).

15.2.4.4 Source and Mechanism of Groundwater Pollution

Stable isotope can also be helpful in the estimation of the source and mechanism responsible for groundwater pollution. Groundwater isotopes such as $\delta^{18}O$ and δ^2H, when used in combination with hydrochemical data, are very useful in characterizing and delineating various sources of groundwater contamination, which is rather difficult to ascertain from hydrochemical data alone (Datta et al. 1996). Pollution that occurs based on anthropogenic resources has a great influence on groundwater resources management. Stable isotopes of groundwater ($\delta^{18}O$ and δ^2H) are important in tracing the pathways of pollutants in aquifers and also their spatial and temporal variations. In addition, the sources of nitrate and sulfate pollution can be identified with the measurement of concentration and stable isotope composition of sulfate and nitrate. Chemical and isotopes tracers are the best tools to provide a better understating of groundwater origin (IAEA 2011; Vitvar et al. 2005; Rizvi and Mohammed-Aslam 2019; Mohammed-Aslam and Rizvi 2020)

15.2.4.5 Sanitary Landfills

Municipal garbage dumps were unlined and sited with little regard to the local hydrogeology and the consequences are that leachate and gas migration is now threatening groundwater resources. Environmental isotopes in conjunction with geochemistry play an important role in the identification of the migration of leachate from the older, non-engineered sites or modern landfills, identification of leachate contamination zone, and siting of new facilities to demonstrate the security of groundwater from contamination (Sinha et al. 2012; Aljaradin and Persson 2014; Ahamad et al. 2019; Madon et al. 2020).

15.2.4.6 Agriculture Watersheds

The use of excess fertilizers and pesticides can deteriorate the quality of groundwater and surface water in agricultural watersheds, and the predominance of the septic systems and municipal waste dumps adds a further complication to understanding the sources and pathways of the contaminant zones (Madhav et al. 2021). This can be well understood with the help of environmental isotopes with the recharge origin and mixing in groundwater, to determine the sources of nitrate and ammonium from fertilizer, to determine the fate of chlorocarbons pesticides, to determine the sources of salinity and oxidation processes, etc. (Clark and Fritz 1997; Bohlke 2002; IAEA 2013; Rizvi and Mohammed-Aslam 2018; Rizvi and Mohammed-Aslam 2019).

15.2.4.7 Groundwater Salinization

Saline groundwater is mainly dominated by chloride salts with sulfates and other anions in minor constituents. Sodium chloride salinity is dominated in the coastal areas where the groundwater is affected by seawater intrusion. Isotope techniques can be used in the measurement of the sources where salinization occurs in groundwater. Isotope techniques can be used to distinguish the leaching of salts by percolating water, the intrusion of saltwater bodies such as seawater, brackish surface water, or brines, and the concentration of dissolved salts through evaporation (Clark and Fritz 1997; IAEA 2011; Han et al. 2011; Han et al. 2014; Etikala et al. 2021).

15.3 Utilization of Water Isotopes in Hydrology: A Case Study

15.3.1 Study Site

An investigation has been carried out in the north-western part of the Kalburgi district to get a better understanding of water isotopes ($\delta^{18}O$ and $\delta^{2}H$) of precipitation of different monsoon seasons within the region. The entire region falls under the semiarid provinces of Karnataka State of India (Rizvi and Mohammed-Aslam 2018, 2019). The region is well known for water scarcity (Mohammed-Aslam and Rizvi 2020). This area receives rainfall generally from June till September. The average annual rainfall received by the area is 777 mm, whereas the annual minimum rainfall is 342 mm and the annual maximum rainfall is 1270 mm (CGWB 2012). Therefore, monthly composite precipitations samples (except pre-monsoon (March to May) and post-monsoon (October to February) periods) from 2016 to 2017 were collected. The rainfall collector was fixed at the PWD office, Aland located at latitude 17^{0}34'15" N and longitude 76^{0}34'7" E, having an elevation of 497 m from the mean sea level (MSL) as shown in Figure 15.2.

15.3.2 Findings and Discussions

Rainfall, wind speed, and relative humidity data were collected from the online portal (https://www.ksndmc.org) of Karnataka State Natural Disaster Monitoring Centre (KSNMDC) for 2016 and 2017. The average annual rainfall of 856.53 mm in 2016 and

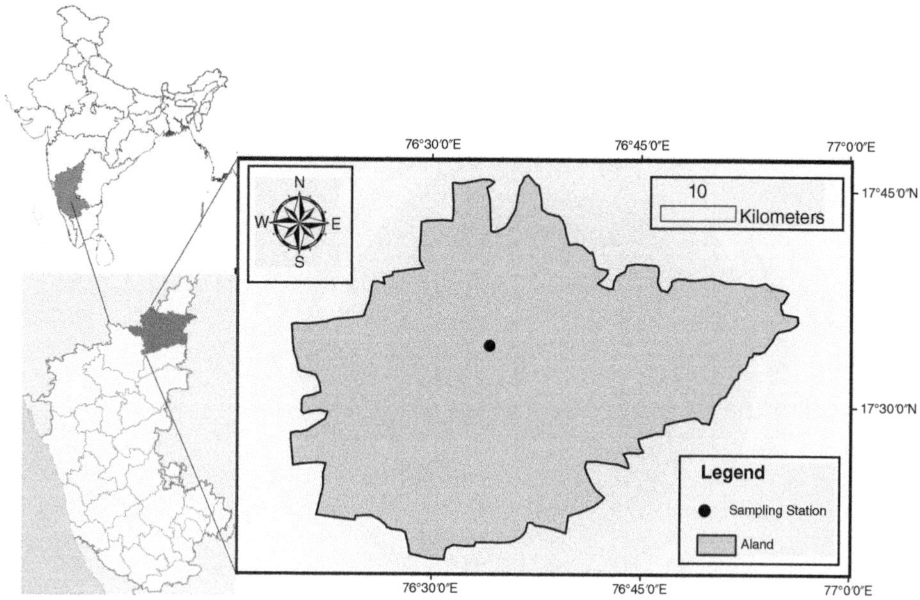

Figure 15.2 Location map of the Aland Taluk.

853.71 mm in 2017 were recorded. The annual maximum rainfalls recorded by the area were 585.50 mm in 2016 and 631.0 mm in 2017. A total number of eight precipitation samples were analyzed for water isotopes ($\delta^{18}O$ and δD) from 2016 to 2017. Southwest monsoon (June to September) is the major contributor (approximately 78%) in rainfall in the study area compared to post-monsoon (October to February) and pre-monsoon (March to May). The mean (monthly) temperature of the area varied from 33.8 °C in April to 23.3 °C in December. The maximum mean (monthly) value of humidity was 77% in September, whereas the minimum was 32% in April. At low humidity, the evaporation from the continental surface body may generate vapor masses with the highest deuterium excess (d-excess) values (Clark and Fritz 1997). The monthly variation of rainfall in the years 2016 and 2017 was shown in Figure 15.3a. Monthly average rainfall data with average relative humidity, wind speed, dew point, and average vapor pressure of the study area are given in Table 15.2.

The mean monthly highest saturated vapor pressure was 52.73 hPa in April, whereas the minimum was in December (28.68 hPa). The actual mean monthly vapor pressure showed from 14.74 in January to 28.29 in June during 2016–2017. The mean monthly "Dew Point Temperature" varied from 12.73 °C in January to 23.09 °C which was the maximum dew temperature in June. The results of the stable isotope analysis with rainfall data of the sampling location are shown in Table 15.3.

Water vapor over the continents and the average rainout history of air masses control the isotopic composition of local precipitation. The LMWL of Aland has been plotted in Figure 15.3b and was compared with GMWL and IMWL proposed by Kumar et al. (2010). The best-fitted regression line, passing through all the points of δ^2H and $\delta^{18}O$ in precipitation, was fairly correlated (R-square = 0.977 and n = 8 with standard error ± 0.46 and ± 1.79, respectively).

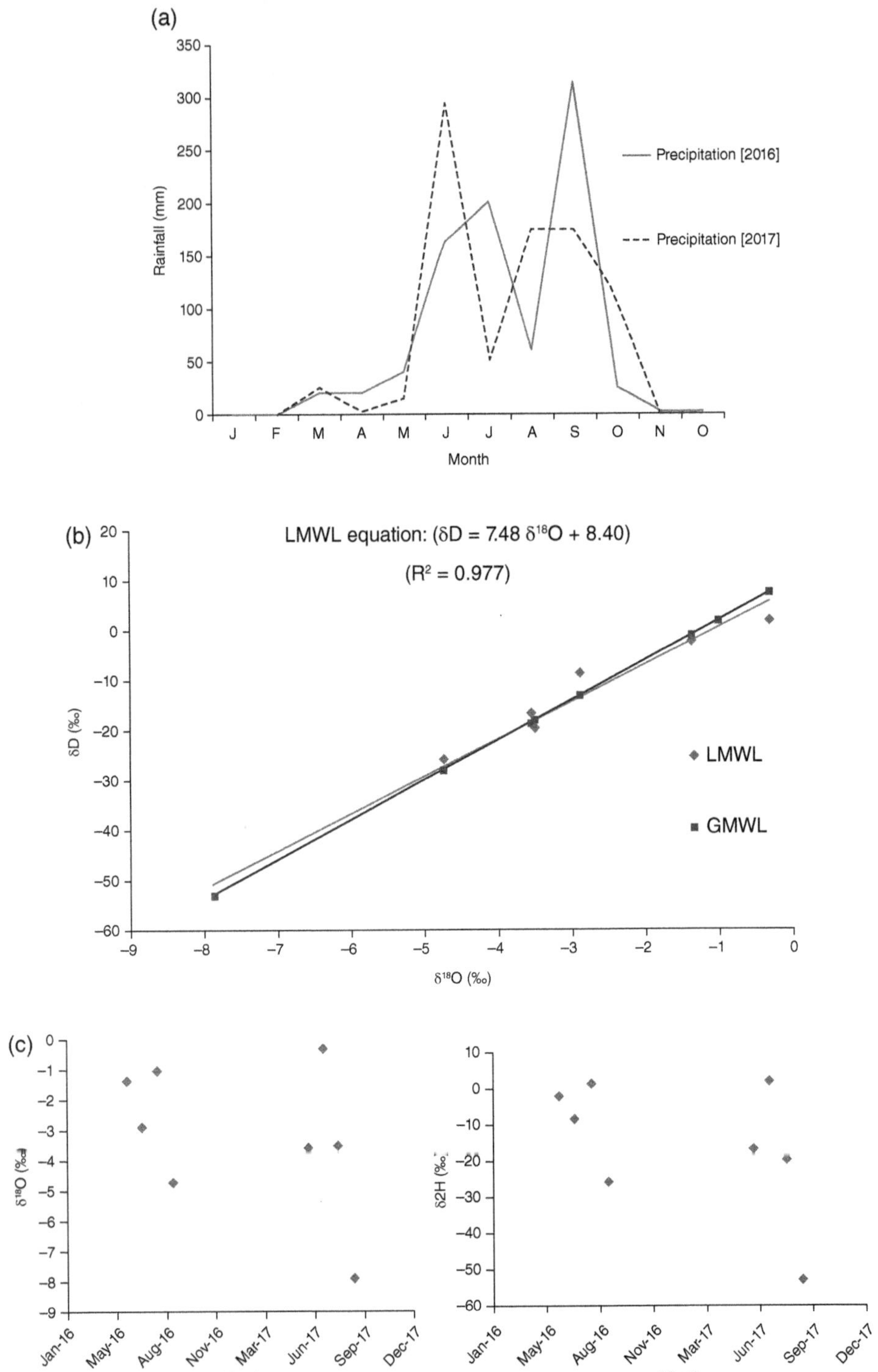

Figure 15.3 (a) Monthly variation of the rainfall in the study region. (b) Local Meteoric Water Line of Aland. (c) Monthly variations of $\delta^{18}O$ and $\delta^{2}H$ in precipitation in Aland.

Table 15.2 Average of meteorological parameters of the region (2016–2017).

Month	Rainfall (mm)	Temperature (°C)	Wind Speed (m/s)	Dew Point (°C)	Relative Humidity (%)	Vapor Pressure (h/Pa)	
						Saturated Vapor Pressure	Actual Vapor Pressure
January	0.39	23.9	0.73	12.7	49.58	29.80	14.74
February	0.99	27.4	0.91	13.2	41.73	36.48	15.18
March	22.6	30.2	0.94	14.7	39.18	42.95	16.76
April	12.38	33.8	1.18	15.02	32.41	52.73	17.10
May	27.8	33.8	1.1	20.03	44.57	52.66	23.44
June	228.96	29.3	0.90	23.09	69.51	40.68	28.29
July	126.59	27.7	0.77	22.5	73.27	37.28	27.25
August	118.16	27.4	0.66	22.4	74.47	36.52	27.20
September	245.94	26.8	0.54	22.4	77.21	35.27	27.28
October	67.8	26.2	0.55	20.4	70.45	34.14	24.14
November	1.73	24.4	0.40	15.7	58.74	30.57	18.02
December	1.76	23.3	0.60	14.8	58.68	28.68	16.83

Table 15.3 Stable isotopes and monthly rainfall of the study area from 2016 to 2017.

Year	Month of Collection	Rainfall (mm)	δD (‰)	$\delta^{18}O$ (‰)	d-excess
	June	163.28	−2.1	−1.38	8.94
	July	201.93	−8.54	−2.9	14.66
	August	60.07	1.5	−1.01	9.58
	September	315.30	−25.68	−4.74	12.24
	October	25.25	NA	NA	NA
	June	294.64	−16.79	−3.56	11.69
	July	51.26	1.97	−0.32	4.53
2017	August	176.25	−19.52	−3.51	8.49
	September	176.59	−52.92	−7.87	10.04

The LMWL of Aland gives the relationship which is expressed in Eq. 15.12.

$$\delta D = \left(7.48\ ^{18}O \pm 0.46\right) + \left(8.40 \pm 1.76\right) \quad \left(R^2 = 0.977, n = 8\right) \tag{15.12}$$

The equation of IMWL with which the results were compared is given as Eq. 15.13.

$$\delta D = \left(7.93\delta^{18}O \pm 0.06\right) + \left(9.94 \pm 0.51\right) \quad \left(R^2 = 0.98, n = 272\right) \tag{15.13}$$

RMWL for the northern, southern, and Himalayan regions were also developed by Kumar et al. (2010) is given as Eq. 15.14.

$$\delta D = \left(7.82\delta^{18}O \pm 0.17\right) + \left(10.23 \pm 0.85\right) \quad \left(R^2 = 0.97,\ n = 62\right) \tag{15.14}$$

The slope of the generated LMWL of Aland (Eq. 15.12) is close to the slope of IMWL (Eq. 15.13), RMWL (Eq. 15.14) and GMWL (Eq. 15.7) respectively. But, the intercept of the LMWL of Aland (Eq. 15.12) is much lower than the intercept of GMWL, IMWL, and RMWL. This condition confirmed that the modifications of water isotopic signatures are due to the "secondary evaporation" from the precipitation.

Monthly variations between $\delta^{18}O$ and δ^2H of precipitation are given in Figure 15.3c, and it indicated that there was a significant variation during the samples collected between southwest monsoons (June to September) for both years. There was no precipitation in the area of study for March to May (pre-monsoon) and October to February (post-monsoon) months. As a result, a correlation for these months was not possible. There was a considerable enrichment in deuterium and oxygen-18 values during southwest monsoon periods between 2016 and 2017. The deuterium (δ^2H) and oxygen-18 ($\delta^{18}O$) signatures of southwest monsoon periods of both years showed a uniform trend indicating the constant nature of their vapor sources. Secondary evaporation always decreases when relative humidity increases (Zongyu et al. 2011).

15.3.2.1 Estimation of d-excess

Deuterium excess can be used in finding out the sources of vapor, information on moisture sources in precipitation, and evaporation of water bodies. It may also be related to kinetic fractionation (Warrier et al. 2010). Deuterium excess is controlled by the relative humidity of air mass from its origin. Monthly variations of deuterium excess values are shown in Figure 15.4a and Table 15.3. The values of d-excess' were found varying from 8.94‰ in June 2016 to 14.66‰ in July 2016, whereas 11.69‰ in June 2017 to 4.53‰ in July 2017 during southwest monsoon periods. The higher variation in stable isotopes and "d-excess" values were observed in samples of 2017, indicating that the precipitation of this year has contributed to the addition of re-evaporated vapor and moistures from continental sources through inland evaporation as noticed by Warrier et al. (2010).

However, samples taken during southwest monsoon periods showed a uniform trend for both years indicative of their nature of the constant source. Kong and Pang (2012) explained that the deuterium excess study may be useful in the investigation of "moisture recycling processes" in arid regions. The relationship between $\delta^{18}O$ and temperature and $\delta^{18}O$ and humidity has been shown in Figure 15.4b. A poor correlation among these explains the "complex isotopic fractionation" during the rainout process. Such a condition is similar to the explanation given by Kumar et al. (2010) in respect of sources of condensing moisture and their effects.

15.3.2.2 Estimation of the Amount Effect

Dansgaard (1964) explained the term "amount effect" for stable isotopes studied for monthly rainfall. He found that there is an increment in the negative values of oxygen-18 with heavy rainfall for several worldwide stations. The variation between the "amount of rainfall – oxygen-18 ($\delta^{18}O$)" values and "amount of rainfall – deuterium (δ^2H)" values is

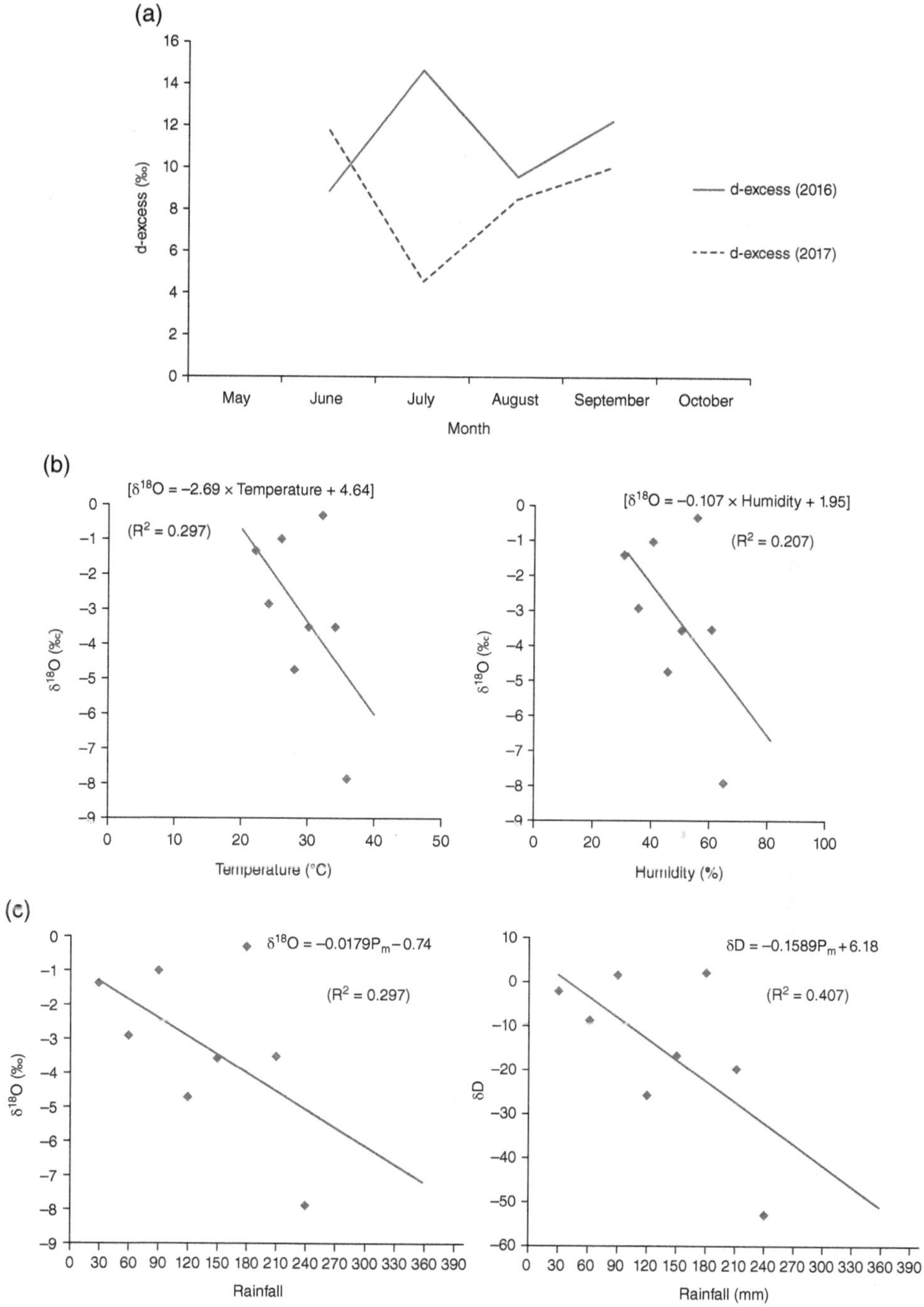

Figure 15.4 (a) Monthly variations of deuterium excess (d-excess) values of precipitation in Aland, (b) plot of $\delta^{18}O$ versus air temperature and humidity of Aland, and (c) variations between the amount of rainfall and $\delta^{18}O$, δD values in precipitation where $\delta^{18}O$ and δD are functions of monthly amount of precipitation in Aland

given in Figure 15.4c considering the monthly precipitation. A poor correlation exists between the monthly amount of precipitation and $\delta^{18}O$, and δ^2H of rainfall in the study region. It was observed that the precipitation versus δ^2H ($R^2 = 0.407$) is slightly better than that of $\delta^{18}O$ ($R^2 = 0.297$). The relationship between the monthly precipitation and $\delta^{18}O$, δ^2H of rainfall of the station at Aland is expressed by the Eqs. 15.15 and 15.16.

$$\delta^{18}O = -0.0179P_m - 0.74 \qquad \left(R^2 = 0.297\right) \tag{15.15}$$

$$\delta D = -0.1589P_m + 6.18 \qquad \left(R^2 = 0.407\right) \tag{15.16}$$

The above equation indicates that the average rate of depletion in $\delta^{18}O$ was 1.79‰ per 100 mm of precipitation, whereas 15.89‰ in δD per 100 mm of precipitation during southwest monsoon (June to September). The rate of depletion of δD in precipitation is higher than the depleting rate of $\delta^{18}O$ per 100 mm of precipitation. The poor correlation obtained by the Eqs. 15.15 and 15.16 indicates the existence of secondary evaporation in the region.

Relative humidity, precipitation, and temperature are the key factors influencing evaporation below the cloud base. Probably, secondary evaporation can be maximum during northwest and pre-monsoon season due to poor rainfall and minimum during southwest monsoon seasons, when a higher amount of rainfall is received (Zongyu et al. 2011).

15.3.2.3 Estimation of Seasonal Variations

The average and weighted mean values of $\delta^{18}O$, δ^2H, and d-excess of precipitation along with the average amount of rainfall during sampling periods (2016 and 2017) are shown in Table 15.4. The depth of rainfall in the container (rainwater collector) was insufficient to precede the study during March to May and October to February for the analysis of these stable isotopes. Therefore, the only available precipitation samples during southwest monsoon periods (June to September) were used for the study.

Table 15.4 Arithmetic mean and the weighted mean of stable isotopes during 2016 and 2017.

		Arithmetic mean			Weighted mean		
Season/period	Average rainfall (mm)	$\delta^{18}O$ (‰)	δ^2H (‰)	d-excess (‰)	$\delta^{18}O$ (‰)	δ^2H (‰)	d-excess
Northeast monsoon	71.51 (mm)	NA	NA	NA	NA	NA	NA
Southwest monsoon	719.66 (mm)	−3.16	−15.26	10.02	−3.67	−18.52	10.79
Pre-monsoon	62.78 (mm)	NA	NA	NA	NA	NA	NA

Note: NA, not available

The stable isotope values of southwest monsoon samples are given in Table 15.4. The $\delta^{18}O$ values were found varying from -0.32‰ in July 2017 to −7.87‰ in September 2017. The δD showed the range from −2.1‰ in June 2016 to 1.9‰ in July 2017. The values of d-excess were varying from 4.53‰ in July 2017 to 14.66 in July 2016 with an average value of 10.02‰ (Table 15.4). Timely variation of isotopic occurrence between the samples collected in June and September to the samples collected in July and August were noticed during the southwest monsoon. The values of $\delta^{18}O$ during July and August samples showed enriched isotopic values with an average value of -1.93‰ (average $\delta^{18}O_{J,A}$ = -1.93‰), whereas June and September samples were showing relatively depleted isotopic values with an average value of -4.38‰ (average $\delta^{18}O_{J,S}$ = -4.38‰). Precipitation taking place in July and August has almost maintained its marine identity with an average value of $\delta^{18}O$ (-1.93‰) and the low average value of deuterium excess (9.31‰). This situation may be taken as the best representation of southwest monsoon precipitation, whereas the $\delta^{18}O$ values of June and September of precipitation samples have shown slightly depleted isotopic signature with an average value of $\delta^{18}O$ (-4.38‰) and relatively high deuterium excess value (10.72‰).

The northern Indian Ocean and the Arabian Sea are the major sources of moisture during southwest monsoon precipitation in the southern part of India and the isotopic values of $\delta^{18}O$ in most of the southwest precipitation samples are relatively enriched [($\delta^{18}O_{SW}$ > (-3‰)], as suggested by Warrier et al. (2010), confirming that the rainfall occurring in the region during southwest monsoon periods. The higher value of deuterium excess was noticed in July 2016 (14.66‰) and September 2016 (12.24‰).

Northeast monsoon wind originating from central Asia draws moisture from depleted continental sources and oceanic sources such as the South China Sea and BOB (Deshpande et al. 2003) and is influenced by successive rainout effects. Therefore, it may be a strong reason for very low to no rainfall during the northeast monsoon season in the region as it has to travel over large landmasses (affected by a higher continental effect) before reaching the Aland. Pre-monsoon rain is generally formed by the storm over the BOB, but it is very complicated to predict their arrival and this may be a reason for low to almost no rainfall during this study. However, a detailed study can provide more about the "complex isotopic fractionation" during the rainout process.

15.4 Conclusion

The water isotope characterization of a semiarid region located in a north-western part of Kalburgi District, Karnataka State, during the study period in 2016 and 2017 has been discussed. The LMWL of Aland has been generated and was compared with the GMWL, IMWL, and RMWL. The slope (7.48) of LMWL was almost close (or less than) to the slope of GMWL (evaporation effect), whereas the intercept (8.40) was much lower than the intercept of GMWL, suggesting that the modifications of water isotopic signatures because of "secondary evaporation" from the falling raindrops from the cloud to surface. The deuterium (δ^2H), oxygen-18 ($\delta^{18}O$) signatures, and deuterium excess of southwest monsoon periods of both years show a uniform trend which indicates the constant nature of their vapor sources. The monthly amount of precipitation and $\delta^{18}O$, and δ^2H of rainfall were poorly correlated in the study region. The relationship indicates that the average rate of depletion in $\delta^{18}O$ was 1.7‰ per 100 mm of precipitation whereas 15‰ in δD per 100 mm of precipitation during southwest monsoon periods confirms that the rate of depletion of δD in precipitation was higher than the depleting rate of $\delta^{18}O$ per 100 mm of precipitation. The pre-monsoon and northeast monsoon periods in the region were lacking any distinctive isotopic signature because of very low to almost no rainfall. The nearest LMWL available to the study area is of the Belgaum region given by Kumar et al. (2010) which can be expressed as:

$$\delta D = \left(7.78\delta^{18}O \pm 0.62\right) + \left(11.32 \pm 2.28\right), \qquad \left(R^2 = 0.94,\ n = 12\right)$$

This study would also help open the knowledge gaps in the understanding the characterization of water isotopes of monsoon precipitation in the Indian sub-continent.

15.5 Summary and Conclusions

Environmental isotopes play a vital role in the management of groundwater resources. It has great importance in tracking groundwater resources using stable isotopes of hydrogen and oxygen; groundwater recharge using tritium; groundwater mixing; groundwater dating using tritium, carbon-14, etc. Nowadays, isotope techniques are widely used in groundwater management around the globe. To understand the isotopic variation of any region, continental effect, seasonal effect, latitude effect, and altitude effect should be summarized adequately because these factors always influence the isotopic fractionations. Deuterium excess (d-excess) is useful in finding the vapor sources, formation of moisture sources in precipitation, and evaporation of water bodies. Isotopic variations that occur in the natural waters are important in solving many critical hydrological problems. Among the various environmental isotopes, stable isotopes such as deuterium, oxygen-18, carbon-13, and radioisotopes such as tritium and carbon-14 are important in hydrological studies. Deuterium and oxygen-18 are known as the ideal geochemical tracers of water because their composition does not change with the interaction of aquifer substances. Environmental isotopes are also important in dating younger and older groundwater.

Acknowledgements

This work is a contribution to "Central University of Karnataka" and has been accomplished during the PhD research work entitled "Hydro-geological investigations and stable isotope analysis of Aland, Karnataka State, India." I would like to especially thank (late) Dr C. Unnikrishnan Warrier for his unconditional support, encouragement, and guidance for my research work. I further expressed my gratitude to Dr T. R. Resmi, Head i/c, Isotope Hydrology Division and Mr M. Sudheesh, Technical Officer, Isotope Hydrology Division, for providing the sample analysis details from the CWRDM, Kozhikode. At last but not least, I would also like to thank Dr R. D. Deshpande, Scientist-SG and Chairman, Geosciences Division, Physical Research Laboratory (PRL), for supporting and providing sample analysis details.

References

Aggarwal, P. K., Froehlich, K, & Kshitij, M. K. (2021). Environmental Isotopes in Groundwater Studies, Groundwater, Vol. II, EOLSS, https://www.eolss.net/sample-chapters/c07/E2-09-03-06.pdf.

Ahamad, A., Raju, N. J., Madhav, S., Gossel, W., & Wycisk, P. (2019). Impact of non-engineered Bhalswa landfill on groundwater from Quaternary alluvium in Yamuna flood plain and potential human health risk, New Delhi, India. *Quaternary International*, 507, 352–369.

Ala-aho, P., Soulsby, C., Pokrovsky, O. S., Kirpotin, S. N., Karlsson, J., Serikova, S., Vorobyev, S. N., Manasypov, R. M., Loiko, S., & Tetzlaff, D. (2018). Using stable isotopes to assess surface water source dynamics and hydrological connectivity in a high-latitude wetland and permafrost influenced landscape, *Journal of Hydrology*, 556, 279–293, https://doi.org/10.1016/j.jhydrol.2017.11.024.

Aljaradin, M., & Persson, K. (2014). Municipal landfilling practice and its impact on the water resources – Jordan. *World Environment*, 4. https://doi.org/10.5923/j.env.20140405.03.

Araguas-Araguas, L., Froehlich, K., & Rozanski, K. (2000). Deuterium and Oxygen-18 isotope composition of precipitation and the atmospheric moisture. *Hydrological Processes*, 14, 1341–1355.

Archana, S. D., Ansari, M. A., Sharma, S., Jacob, N., Kumar, U. S., & Singh, G. (2014). Isotope techniques for water resource management, *BARC Newsletter*, 337, 29–35.

Berkelhammer, M., Sinha, A., Stott, L., Cheng, H., Pausata, F. S., & Yoshimura, K. (2012). An abrupt shift in the Indian monsoon 4000 years ago. *Geophysical Monograph Series*, 198(7), 75–87.

Bohlke, J. (2002). Groundwater recharge and agricultural contamination. *Hydrogeology Journal*, 10, 153–179. https://doi.org/10.1007/s10040-001-0183-3.

Cayuela Linares, C., Latron, J., Geris, J., & Llorens, P. (2018). Spatio-temporal variability of the isotopic input signal in a partly forested catchment: Implications for hydrograph separation. *Hydrological Processes*, 33. https://doi.org/10.1002/hyp.13309.

CGWB (2012). Groundwater Information Booklet, Gulbarga District, Karnataka. South-Western Region, Bangalore, 2012. http://cgwb.gov.in/District_Profile/karnataka/2012/Gulbarga-2012.pdf.

Chacha, N., Njau, K. N., Lugomela, G. V. & Muzuka, A. N. N. (2018). Groundwater age dating and recharge mechanism of Arusha aquifer, northern Tanzania: application of radioisotope and stable isotope techniques. *Hydrogeology Journal*, 26, 2693–2706. https://doi.org/10.1007/s10040-018-1832-0.

Chacko, T., Cole, D. R., & Horita, J. (2001). Equilibrium oxygen, hydrogen and carbon isotope fractionation factors applicable to geologic systems. *Reviews in Mineralogy and Geochemistry*, 43(1), 1–81.

Clark, I. D., & Fritz, P. (1997). Environmental Isotopes in Hydrogeology, 352 pp. CRC Press, Boca Raton, Fla.

Cook, P. G., & Herczeg, A. L. (2000). Environmental Tracers in Subsurface Hydrology. Kluwer Academic Publishers, Boston, pp. 397–424.

Craig, M. (1961). Isotopic variation in meteoric waters. *Science*, 133, 1702–1703. https://doi.org/10.1126/science.133.3465.1702.

Dansgaard, W. (1964). Stable isotopes in precipitation. *Tellus*, 16, 436–468. https://doi.org/10.1111/j.2153-3490.1964.tb00181.x.

Datta, P. S., Deb, D. L., Tyagi, S. K. (1996). Stable isotope (^{18}O) investigations on the processes controlling fluoride contamination of groundwater. *Journal of Contaminant Hydrology*, 24(1), 85–96.

Deshpande, R. D. (2006). Groundwater in and around Cambay Basin Gujarat: Some Geochemical and Isotopic Investigations, MS University of Baroda, Vadodara, India p.158.

Deshpande, R. D., & Gupta, S. K. (2008). National programme on isotope fingerprinting of waters of India (IWIN) Glimpses of Geosciences Research in India, the Indian Report to IUGS. Indian National Science Academy (eds Singhvi, A. K., Bhattacharya, A., & Guha, S.). INSA, New Delhi, pp. 10–16.

Deshpande, R. D., Bhattacharya, S. K., Jani, R. A., & Gupta, S. K. (2003). Distribution of oxygen and hydrogen isotopes in shallow groundwaters from Southern India: influence of a dual monsoon system. *Journal of Hydrology*, 271 (1–4), 226–239.

Dublyansky, Y. V., Klimchouk, A. B., Tokarev, S. V., Amelichev, G. N., Langhamer, L., & Spötl, C. (2018). Stable isotopic composition of atmospheric precipitation on the Crimean Peninsula and its controlling factors, *Journal of Hydrology*, 565, 61–73, https://doi.org/10.1016/j.jhydrol.2018.08.006.

Etikala, B., Adimalla, N., Madhav, S., Somagouni, S. G., & Keshava Kiran Kumar, P. L. (2021). Salinity Problems in Groundwater and Management Strategies in Arid and Semi-arid Regions. Sughosh Madhav, Pardeep Singh Groundwater Geochemistry: Pollution and Remediation Methods, 42–56. John Wiley & Sons Ltd., USA.

Gat, J. R., & Carmi, I. (1970). Evolution of the isotopic composition of atmospheric waters in the Mediterranean Sea area. *Journal of Geophysical Research*, 75, 3039–3048.

Geyh, M., Amore, F. D., Darling, G., Paces, T., Pang, Z. and Sila, J. (2008). Groundwater Saturated and Unsaturated Zone. International Atomic Energy Agency, Vienna, Austria, pp. 1–117.

Gonfiantini, R. (1978) Standards for stable isotope measurements in natural compounds. *Nature*, 271, 534–536.

Gonfiantini, R. (1981). The δ-notation and the mass-spectrometric measurement techniques. In Stable Isotope Hydrology: Deuterium and Oxygen-18 in the Water Cycle. Technical Report Series No. 210 (eds Gat, J. R. and Gonfiantini, R.), IAEA, Vienna, pp. 103–142.

Guay, B. E., Eastoe, C. J., Bassett, R. L., & Long, A. (2006). Identifying sources of groundwater in the lower Colorado River valley, USA, with δ18O, δD, and 3H: Implications for river water accounting. *Hydrogeology Journal*, 14(1), 146–158.

Gupta, S. K. (2011). Modern Hydrology and Sustainable Water Development, Wiley-Blackwell Publishing. 978-1-4051-7124-3.

Gupta, S. K., & Deshpande, R. D. (2003). Synoptic hydrology of India from the data of isotopes in precipitation. *Current Science*, 85, 1591–1595.

Gupta, P., & Goel, M. (2021). Environmental tracers and isotopic techniques: tools for sustainable water management. In: Adhikary, P.P., Shit, P.K., Santra, P., Bhunia, G.S., Tiwari, A.K., Chaudhary, B.S. (eds) Geostatistics and Geospatial Technologies for Groundwater Resources in India. Springer Hydrogeology. Springer, Cham. https://doi.org/10.1007/978-3-030-62397-5_26.

Han, D., Kohfahl, C., Song, X., Xiao, G., & Yang, J. (2011). Geochemical and isotopic evidence for palaeo-seawater intrusion into the south coast aquifer of Laizhou Bay, China. *Applied Geochemistry*, 26. https://doi.org/10.1016/j.apgeochem.2011.02.007.

Han, D., Song, X., Currell, M., Yang, J. L., & Xiao, G. Q. (2014). Chemical and isotopic constraints on evolution of groundwater salinization in the coastal plain aquifer of Laizhou Bay, China. *Journal of Hydrology*, 508. 12–27. https://doi.org/10.1016/j.jhydrol.2013.10.040.

He, S., Jackisch, D., Samanta, D., Yi, P. K. Y., Liu, G., Wang, X., & Goodkin, N. F. (2021). Understanding tropical convection through triple oxygen isotopes of precipitation from the maritime continent. *Journal of Geophysical Research: Atmospheres*, 126, e2020JD033418. https://doi.org/10.1029/2020JD033418

IAEA (2011). Using Isotope effectively to support comprehensive groundwater management. https://www.iaea.org/sites/default/files/gc/gc55inf-5-att3_en.pdf.

IAEA (2013). Application of Isotope Techniques for assessing nutrient dynamics in river basins. https://www-pub.iaea.org/MTCD/Publications/PDF/TE-1695_web.pdf.

IAEA/WHO (2004). Global Network of Isotopes in Precipitation. The GNIP Database. http://isohis.iaea.org.

International Atomic Energy Agency (IAEA) (1983). Guidebook on Nuclear Techniques in Hydrology, IAEA Technical Reports Series No. 91. IAEA, Vienna, Austria, p. 439.

Ju, Y., Massoudieh, A., Kaown, D., Yoon, Y, & Lee, K.-K. (2020). Characterization of flow dynamics around highly-utilized agricultural wells in a fractured-rock aquifer: assessment of uncertainties lying on groundwater age-dating. *Journal of Hydrology*, 586, 124885. https://doi.org/10.1016/j.jhydrol.2020.124885.

Jung, H., Koh, D.-C., Kim, Y., Jeen, S.-W., & Lee, J. (2020). Stable isotopes of water and nitrate for the identification of groundwater flowpaths: a review. *Water*, 12. 138. https://doi.org/10.3390/w12010138.

Kaklamanos, G., Aprea, E., & Theodoridis, G. (2020). Mass spectrometry: principles and instrumentation. Encyclopedia of Food and Health. Academic Press, pp. 661 668. https://doi.org/10.1016/B978-0-12-813266-1.00011-5.

Kern, Z., Hatvani, I., Czuppon, G., Fórizs, I., Erdélyi, D., Kanduč, T., Palcsu, L., & Vreča, P. (2020). Isotopic 'Altitude' and 'Continental' effects in modern precipitation across the Adriatic-Pannonian Region. *Water*, 12, 1797. https://doi.org/10.3390/w12061797.

Khairul Nizar, M. S., Wan, N. A. S., Mohammad, F. R., Faradiella, M. K., & Kamarudin, S. (2018). Assessments of seasonal groundwater recharge and discharge using environmental stable isotopes at Lower Muda River Basin, Malaysia. *Applied Water Science*, 8, 120.

Kinzelbach, W., Aeschbach, W., Alberich, C., Goni, I.B., Beyerle, U., Brunner, P., Chiang, W.-H., Rueedi, J., & Zoellmann, K. (2002). A Survey of Methods for Groundwater Recharge in Arid and Semi-arid Regions. Early Warning and Assessment Report Series, UNEP/DEWA/RS.02-2. United Nations Environment Programme, Nairobi, Kenya, 1–20.

Kong, Y., & Pang, Z. (2012). Evaluating the sensitivity of Glacier Rivers to climate change based on hydrograph separation of discharge. *Journal of Hydrology*, 434–435, 121–129.

Krajcar Bronić, I., Barešić, J., Borkovic, D., Sironić, A., Mikelić, I. L., & Vreča, P. (2020). Long-term isotope records of precipitation in Zagreb, Croatia. *Water*, 12, 226. https://doi.org/10.3390/w12010226.

Kumar, C. P. (2013). Hydrogeological studies using isotopes. *International Journal of Innovative Research and Development*, 2(13), 8–13.

Kumar, B., Rai, S. P., Saravana, K. U., Verma, S. K., Garg, P., Vijaya, K. S. V., Jaiswal, R., Purendra, B. K., Kumar, S. R., & Pande, N. G. (2010). Isotopic characteristics of Indian precipitation. *Water Resources Research*, 46, W12548. https://doi.org/10.1029/2009WR008532.

Laonamsai, J., Ichiyanagi, K., Patsinghasanee, S., & Kamdee, K. (2021). Controls on stable isotopic characteristics of water vapor over Thailand. *Hydrological Processes*, 35. https://doi.org/10.1002/hyp.14202.

Leketa, K., & Abiye, T. (2020). Investigating stable isotope effects and moisture trajectories for rainfall events in Johannesburg, South Africa. *Water SA*, 46(3), 429–437. https://dx.doi.org/10.17159/wsa/2020.v46.i3.8653.

Madhav, S., Raju, N. J., Ahamad, A., Singh, A. K., Ram, P., & Gossel, W. (2021). Hydrogeochemical assessment of groundwater quality and associated potential human health risk in Bhadohi environs, India. *Environmental Earth Sciences*, 80(17), 1–14.

Madon, I., Drev, D., & Likar, J. (2020). Long-term groundwater protection efficiency of different types of sanitary landfills: model description, *MethodsX*, 7, 100810, https://doi.org/10.1016/j.mex.2020.100810.

Mohammed-Aslam, M. A., & Rizvi, S. S. (2020). Hydrogeochemical characterisation and appraisal of groundwater suitability for domestic and irrigational purposes in a semi-arid region, Karnataka state, India. *Applied Water Science*, 10, 237, https://doi.org/10.1007/s13201-020-01320-1.

Nisi, B., Raco, B., & Dotsika, E. (2014). Groundwater contamination studies by environmental isotopes: a review. In: Scozzari, A., Dotsika, E. (eds) Threats to the Quality of Groundwater Resources. The Handbook of Environmental Chemistry, vol. 40, 115–150. Springer, Berlin, Heidelberg. https://doi.org/10.1007/698_2014_281.

Oiro, S., Jean-Christophe, C., Chris, S., & Kristine, W. (2018). Using stable water isotopes to identify spatiotemporal controls on groundwater recharge in two contrasting East African aquifer systems. *Hydrological Science Journal*, 63(6), 862–877.

Patel, A. S., & Shah, D. L. (2008). Water Management: Conservation, Harvesting and Artificial Recharge. New Age International (P) Limited, Publishers, India.

Rizvi, S. S., & Mohammed-Aslam, M. A (2018). Estimation of hypsometric integral and groundwater potential zones of Amarja Reservoir Catchment, Karnataka, India using SRTM data and geospatial tools.*International Journal of Economic and Environmental Geology*, 9(2), 75–85.

Rizvi, S. S., & Mohammed-Aslam, M. A (2019). Microbiological quality of drinking water in Amarja reservoir catchment, Aland taluk, Karnataka, India. *Current Science*, 117(1), 114–121, https://doi.org/10.18520/cs/v117/i1/114-121

Roxy, M. K., Ritika, K., Terray, P., Murtugudde, R., Ashok, K., & Goswami, B. N. (2015). Drying of Indian subcontinent by rapid Indian Ocean warming and a weakening land-sea thermal gradient. *Nature Communications*, 6(1), 1–10.

Rozanski, K., Araguas-Araguas, L., & Gonfiantini, R. (1993). Isotopic patterns in modern global precipitation. In: Swart PK, Lohmann KL, McKenzie J, Savin S (eds). Climate Change in Continental Isotope Records, vol 78. American Geophysical Union, Washington, pp. 1–37. https://doi.org/10.1029/GM078p0001.

Saranya, P., Krishnakumar, A., Sinha, N., Sudhir Kumar, K., & Krishnan, A. (2021). Isotopic signatures of moisture recycling and evaporation processes along the Western Ghats orography, *Atmospheric Research*, 264,105863, https://doi.org/10.1016/j.atmosres.105863.

Saravanan, P., Gupta, A. K., Zheng, H., Panigrahi, M. K., & Prakasam, M. (2019). Late Holocene long arid phase in the Indian subcontinent as seen in shallow sediments of the eastern Arabian Sea. *Journal of Asian Earth Sciences*, 181, 103915.

Scholl, M. A., & Murphy, S. F. (2014) Precipitation isotopes link regional climate patterns to water supply in a tropical mountain forest, eastern Puerto Rico. *Water Resources Research*, 50, 4305–4322. https://doi.org/10.1002/2013WR014413.

Sinha, U. K., Mohokar, H. V., Tomar, B.S. Reddy, A. V. R., & Venu Babu, P. (Eds.). (2012). Application of Environmental Isotopes to Identify Landfill Leachate Contamination: Gazipur landfill, New Delhi, India. India: Bhabha Atomic Research Centre.

Sinha, U.K., Tirumalesh, K., Mohokar, H.V. (2019). Isotope hydrology. In: Ray, S. (eds) Ground Water Development – Issues and Sustainable Solutions. Springer, Singapore. https://doi.org/10.1007/978-981-13-1771-2_14.

Srivastava, R., Ramesh, R., Gandhi, N., Jani, R. A., & Singh, A. K. (2015). Monsoon onset signal in the stable oxygen and hydrogen isotope ratios of monsoon vapor. *Atmospheric Environment*, 108, 117–124.

Urey, H. C. (1947). The thermodynamics properties of isotopic substances. *Journal of Chemical Society (Resumed)*, 562–581. https://doi.org/10.1039/jr9470000562

Valdivielso, S., Vázquez-Suñé, E., Custodio, E. (2020). Origin and variability of oxygen and hydrogen isotopic composition of precipitation in the Central Andes: a review. *Journal of Hydrology*, 587, 124899, ISSN 0022-1694. https://doi.org/10.1016/j.jhydrol.2020.124899.

Vasil'chuk, Y., Chizhova, J., Budantseva, N. Vystavna, Y., & Eremina, I. (2022). Stable isotope composition of precipitation events revealed modern climate variability. *Theoritical and Appl Climatology*, 147(3), 1649–1661. https://doi.org/10.1007/s00704-021-03900-w.

Vitvar, T., Aggarwal, P. K., & McDonnell, J. J. (2005). A review of isotope applications in catchment hydrology. In: Aggarwal, P. K., Gat, J. and Froehlich, K. F. Eds., Isotopes in the Water Cycle, pp. 151–170. Springer, Dordrecht, Netherlands. https://doi.org/10.1007/1-4020-3023-1_12

Wan, Z. W. M. T., Johari, A. L., Kamaruzaman, M., Rohaimah, D., Seyed, R. S., Ismail, Y., & Zainudin, O. (2012). Preliminary assessment of groundwater recharge using environmental isotopes investigation (^{18}O, ^{2}H, ^{13}C) to support groundwater management in the north Kelantan River Basin. Conference Proceeding, on 2nd International Conference on Water Resources (ICWR 2012), Langkawi, Malaysia.

Warrier, C. U., Babu, M. P., Manjula, P., Velayudhan, K. T., Hameed, A. S., & Vasu, K. (2010). Isotopic characterisation of dual monsoon precipitation-evidence from Kerala, India. *Current Science*, 98, 1487–1495.

Warsi, T., Bhattacharjee, L., Thangamani, S., Jat, S. K., Mohanta, K., Bhattacharjee, R. R., Ramaswamy, R., Manikyamba, C. & Rao, T. V. (2020). Emergence of robust carbon quantum dots as nano-tracer for groundwater studies. *Diamond and Related Materials*, 103, 107701.

Wenninger, J. (2020). Handbook for Stable Isotope Data Interpretation in India, IHE Delft Institute for Water Education, Delft, The Netherlands. http://cgwb.gov.in/Handbook%20-%20Isotope.pdf.

Wolf, A., Roberts, W. H. G., Ersek, V. Johnson, K., & Griffiths, M. (2020). Rainwater isotopes in central Vietnam controlled by two oceanic moisture sources and rainout effects. *Scientific Reports*, 10, 16482. https://doi.org/10.1038/s41598-020-73508-z.

Yang, K., Guilin, H., Chao, S, & Peng, Z. (2019). Stable H-O isotopic composition and water quality assessment of surface water and groundwater: a case study in the Dabie Mountains, Central China. *International Journal of Environmental Research and Public Health*, 2019(16), 4076. https://doi.org/10.3390/ijerph16214076.

Yeh, H. F., & Lee, J. H. (2018). Stable hydrogen and oxygen isotopes for groundwater sources of Penghu Islands, Taiwan. *Geosciences*, 8, 84. https://doi.org/10.3390/geosciences8030084.

Yurtsever, Y., & Gat, J. R. (1981). Stable isotope hydrology, Tech. Rep. Ser. 210, pp. 103-142, Int. At. Energy Agency, Vienna, Guidebook on Nuclear Techniques in Hydrology, Technical Report Series Vienna.

Zongyu, C., Jinxiang, M, Jun, L., Wen, W., Ying, W., & Jian, C. (2011). Tracing sources of recharge to groundwater in the Northern Henan plain using environmental isotopes. *International Symposium on Water Resource and Environmental Protection*. https://doi.org/10.1109/iswrep.2011.5893000.

16

Assessment of Causes and Impacts of Sand Mining on River Ecosystem

Soumya Pandey[1], Gaurav Kumar[1], Neeta Kumari[1], and Richa Pandey[2]

[1] *Civil and Environmental Engineering, Birla Institute of Technology, Ranchi, Jharkhand, India*
[2] *Department of Mechanical Engineering, Birla Institute of Technology, Ranchi, Jharkhand, India*

16.1 Introduction

The growing urban sprawl and infrastructure development have resulted in the extraction of natural materials like river sand (Zare et al. 2017; Yatoo et al. 2020). It has been estimated that the sand extraction rate has increased by threefold in the past two decades (Stabnikov et al. 2013; Elmayel et al. 2019; Haghnazar et al. 2020). This could be seen globally and mostly in coastal areas. According to United Nations Environment Program (UNEP-GEAS 2014) assessment, in unindustrialized nations, a massive extraction of rock and sand has been taking place. This rate is estimated in the range of 32 to 50 billion globally (Aliu et al. 2022). Surface water resources that flow, like rivers, streams, and runoff, add a significant amount of sand on the sediment bed (Haghnazar et al. 2020; Wang et al. 2021). These riverine sources have sufficient energy to rack and crush the rock surfaces and carry them downstream (Walling and Collins 2008; Gao et al. 2020). In terms of construction, the infrastructure development becomes cost-effective, as most urban development occurs at the watershed level with cheap availability of soil and water resources (Chuenchum et al. 2020). The construction price reduces as the tedious task of mining, crushing, and managing basic construction materials like sand and aggregate (Giri et al. 2013) are taken care of by the rivers themselves. These sand and its aggregate have the advantage that they have high compression strength, better coarser shape (rakish), and better mineral composition (Manoj et al. 2012). In this chapter, the use of the term "sand mining" is done to collectively address the river soil products, irrespective of their physical sizes (Filho et al. 2021).

Sand mining activities are one of the most disturbing issues affecting the riverine biomes (Singh et al. 2020). As per world wildlife fund, many microbial species have been depleted in the riverine biomes (WWF 2018). Furthermore, the adverse ecological and social effects of sand mining have been reported by the media and research scientists. As per the Geological Survey of India (GSI), sand mining has been known to bring serious modifications in the physical attributes of both river bodies and sediment beds (Chowdary et al. 2013).

Hydrogeochemistry of Aquatic Ecosystems, First Edition. Edited by Sughosh Madhav, Virendra Bahadur Singh, Manoj Kumar, and Sandeep Singh.

Along with sand, the minerals like rutile, ilmenite, and zircon which are the basic source of titanium and zirconium, are extracted (Corre et al. 2002). This causes loss to industries that use them as raw materials like electroplating, aircraft, and ceramics.

For any traditional construction, every ton of cement would require approximately eight times more of sand and aggregate for a mortar mix, depending on the design mix adopted during construction (Prusty et al. 2016). This increases the demand in sand mining as sand resources require a fair share of construction budget finances. A country like China has seen an explosion in the construction sector with the addition of seven million homes and numerous infrastructures since 2000 (de Leeuw et al. 2010). Unfortunately, the sand mining process results in variation in the riverbed (Padmalal et al. 2008; Manoj et al. 2012) and change in its courses, and leads to bank erosion (Ustaoğlu et al. 2021) and flooding (Bhattacharyya et al. 2015; Ye et al. 2019). Demand for better sand mining strategies and strict laws have been demanded for a long time now (Filho et al. 2021). Another major issue occurs in the form of uncontrolled illegal sand mining (Ashraf et al. 2011). Most places in the world suffer from illegal sand mining activities, like in China and India, even causing the death of natives (Padmalal et al. 2008; Sinha and Singh 2016). Other than construction, sand is actively used by other industries like water filtration (Filho et al. 2021), plastics, glasses, dams, oil extraction, and electronics. It is one of the most consumed natural resources at present and accounts for 85% of minerals extracted by weight annually around the globe (UNEP-GEAS 2014).

Effects of sand mining on riverine resources and watersheds may or may not be immediate (Figure 16.1). They mostly depend on the morphology of the river, mining techniques, hydraulic characteristics, rate of extraction (Peduzzi 2014), rate of rejuvenation, sediment transport capacity (STC), etc. (Haghnazar et al. 2020). Direct effects occur when sand mining directly impacts the ecological balances and physical features of the riverine systems. For example, the explosion of floodplains boundaries resulted in floods with the rise in water levels, increased turbidity, plume dispersal (Ma et al. 2015), ship and machinery

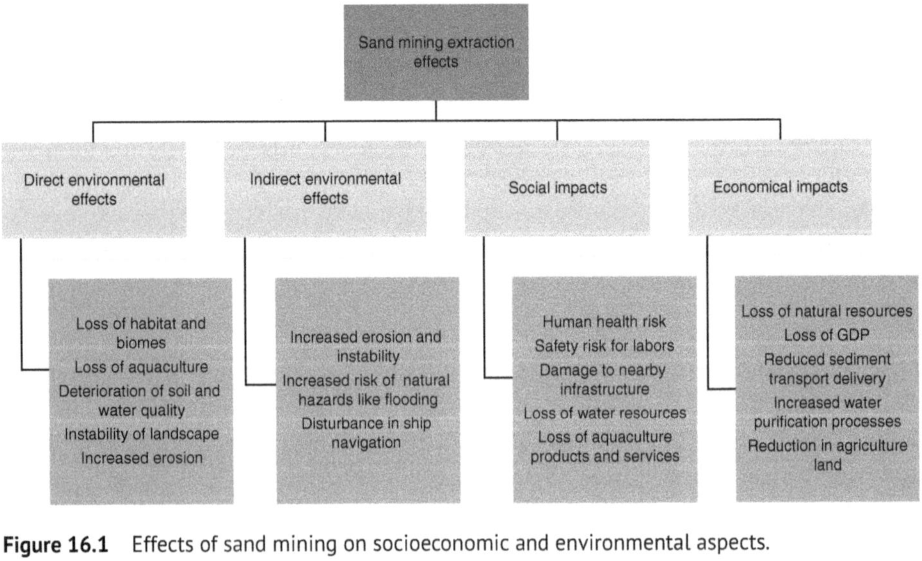

Figure 16.1 Effects of sand mining on socioeconomic and environmental aspects.

noise pollution, bathymetric changes (Rahmani et al. 2018), and change in tides and currents. The indirect or circuitous effects are mostly found in the form of environmental changes caused by sand mining. For example, destruction of animals and microorganisms in the riverbed, increase in pollution, change in pH (Selvam et al. 2020), and groundwater recharge are affected.

Lack of data on sand mining (including sand and aggregate both) especially from developed countries leads to a lack of estimation and planning for resource management (UNEP-GEAS 2014). Furthermore, the adverse effect of mining cause both socioeconomic and ecological losses. Some common ecological losses occurred are loss of biodiversity, loss of inland, coastal erosion, and carbon emission through transportation (Koondhar et al. 2020). It also causes damage to dams, intake structures, embankments, and bridges. The loss of coastal international boundaries (like in Indonesia) and natural hazards have occurred due to sand mining (Koondhar et al. 2020). Hence it is crucial to plan, monitor, and control the sand mining process for sustainable, cheap, and eco-friendly development.

For monitoring and assessment of sediment deposition, sand mining processes, and its environmental impact on riverine ecosystems, use of geographical information systems (GIS) (Arsyad et al. 2020; Bhattacharya et al., 2019a), remote sensing and satellite data, neural networks (ANN, CNN, etc.,), deep learning, etc., have become common recently (Kayet et al. 2016). Through satellite observation, it is easy to monitor the mining activities. The amount of sedimentation, slope drainage, runoff, etc., can also be easily calculated using GIS (Pal and Chakrabortty 2019). Based on these data, models can also be predicted using neural networking and deep learning for future prediction and analysis (Goralski and Tan 2020). The purpose of this study is to understand the impacts of sand mining on various aspects of riverine ecosystem components; also, to analyze the recent tools those are being used for assessment and monitoring purposes.

The objective of this study is to i) to analyze the sand mining activities and their processes; ii) to find the drivers of sand mining processes; iii) to find the current practices of monitoring these activities; iv) to find the impacts of sand mining on aquatic ecosystems; v) to find the alternative of sand as a material. Through these aspects one can understand the whole process and adverse impacts of sand mining on ecology and economy.

16.2 Why River Sand?

The sand mining done from the rivers is based on sediment characteristics. Most sand mining are done for construction purposes, for which it should have some ideal engineering characteristics (IS 383:2016). For construction, concrete mixes are produced using coarse aggregate (aggregate) and fine aggregate (sand) with cement in proper ratios (Reddy et al. 2020). Fine aggregate or sand (size between 4.75 and 0.075 mm as per Indian Standard codes) (IS 2386 (Part I) 1963) should have cohesion, rakish and coarser features, and sharp grains and should be chemically inert, strong, and durable (IS 2386 (Part I) 1963). Also, they should be free from organic matter, hygroscopic salt, silt, and clay. Whereas coarse aggregate (size > 4.75 mm as per IS 2386 (Part 1) 1963)) should also be chemically inert, durable (resistant to weathering), free of impurities, organic matter, salts, silts, and clays (IS 2386 (Part I) 1963). Furthermore, they should have mechanical strength (IS 2386 (Part 4) 1963), determined by aggregate impact value, aggregate abrasion value, and aggregate crushing value, and should have adequate soundness (IS 2386 (Part 5) 1963). The shape of aggregates tends to affect their density, particle arrangement, and other engineering properties. Abrasive and coarser aggregates help create better cohesion and friction in concrete mixes thus parting more strength to the finished concrete products (IS 2386 (Part IV) 2016; IS 2386 (Part III) 1963).

The sediment (aggregate and sand) found in rivers reach these ideal requirements in the process of formation (Koehnken et al. 2020). Sand is made from different weathered rocks (inland), carried, and transported with river, storm, wind, or runoff. The rough path of the river channel causes a lot of abrasion and weathering of sediments it carries throughout its course (Kim and Yoo 2020; Santhosh et al. 2021). In the full course of a river, the heavy and larger particles are weathered from their parent rocks, carried by the high flow intensity of the river, and transported to a newer location. Throughout the course of the river, the sediment particle keeps undergoing weathering, sorting, and deposition or sedimentation with the deceleration of flow intensity (Wahab et al. 2019). Furthermore, erosion also keeps taking place throughout the course of the river which adds more sediment to the riverbed. Even during storms or with surface runoff, during monsoon, the particles keep getting settled on the sand bed (Okur and Erturaç 2021).

Sand is abundantly present in deserts, yet they are not ideally used for construction, as their engineering (geotechnical) properties are not suitable for construction (Gong et al. 2015). The advantage with river sand is that they are clean, sorted, free from impurities, coarser (due to abrasion and weathering), durable, with high mechanical strength, index, and geotechnical properties (grain size, bulk density, compaction characteristics, permeability, shear strength, etc.). Desert sand has a mixed chemical composition, rounder and finer shape, open structure, and low bearing capacity (Pastore et al. 2021). They are unable to form links or create cohesion in concrete mixes due to their open structure. Hence, the concrete mixes become slurry and could result in bleeding (Pastore et al. 2021). Also, their texture and chemical composition need to be analyzed and may need pretreatment to pass as adequate construction material. Similar is the case with sea sand. Sea sands are like desert sand but have higher chemical impurities, like carbonate, chlorine, and sulfates (Park et al. 2021). The presence of these compounds can cause cracking of concrete and corrosion of reinforcement. This will massively reduce the compressive and tensile strength of concrete structures causing failure. Sea sand also has a very low bearing capacity and the impurities present in them, especially salts, can absorb moisture causing dampness (Kim and Yoo 2020). All these issues make river sand more ideal for any industry, especially construction.

16.3 Processes of Sand Mining

UNEP portrays sand mining as "the impermanent or extremely durable bringing down of the useful limit of land." "Mining is a damaging improvement movement where nature endures at the raised area of the economy" (Chauhan 2010). Mining pits that are formed postextraction create steep slopes or gradients which increase the rate of scouring. The scour formed migrate several kilometers, both upstream and downstream (Reddy et al. 2020)

In the process of extraction, the soil mass up to a depth of 2–3 m from the top is removed, which is rich in silts, fines clay or mud, and minerals, and is stacked separately in the form of piles (Sengani and Zvarivadza 2019). Then the next 10–12 m of the sand layer is extracted and stacked in separate piles. Thereafter, stacked fines and sands are blended homogenously for better strength and quality of sand resources. Furthermore, the mined areas are then sold to developers for construction (Tamura et al. 2020). The homogenous mixing of soil layers is formed and then uniformly spread over the sandpit created during mining. The leveling of the sandpit with blended topsoil evened and compacted makes the field ready for construction or any other development (Shrestha et al. 2017). One of the common processes of removal of sand from sediment bed is by dredging. It focuses on maintaining the depth of navigation channels for ships and contaminant exposure control to human and aquatic lives (Duan et al. 2019).

Stream residues or dregs involve a range of shapes and sizes like stones, cobbles, rocks, granules, sand, residue, and mud. As a rule, coarser particles being heavy are loosely packed and usually found on the upper stream side (Lenhart et al. 2018; Himanshu et al. 2019; Aydin et al. 2021); the finer particles like small stones and coarser sand are found on the lower streams, whereas fine sand, silt, and clays are found in the plains (Sreebha and Padmalal 2011; Maharana and Tripathi 2018). This distribution occurs due to the force or energy of the water is higher at the beginning of the river. Hence, rivers carry heavy masses as they flow. As the intensity of flow decreases, the sediment carrying capacity is decreased and only finer particles are carried downstream. The shear strength (Mamo and Bubenzer 2001; Seleiman et al. 2019) of the soil particles also plays a major role in the process of erosion and sedimentation (Kondolf 1994). The better the shear strength of the particle is likely to not get detached or eroded from parent sources or even while its downward journey (Rezaei et al. 2009). The process and functioning of the river soil erosion are presented in Figure 16.2.

On the global front, government authorities of almost every country, especially developing countries are trying to monitor the illegal activity of sand mining. Many laws have been introduced in countries like China and India which has made illegal sand mining punishable (Mahadevan 2019). But the real issue occurs as the sand mafias are hard to identify and deal with. This group of people may not necessarily be from any organization but could be the local native. For form population, it is an unlimited source of material that could be utilized as per choice. But lack of knowledge about the adverse impact of this activity has

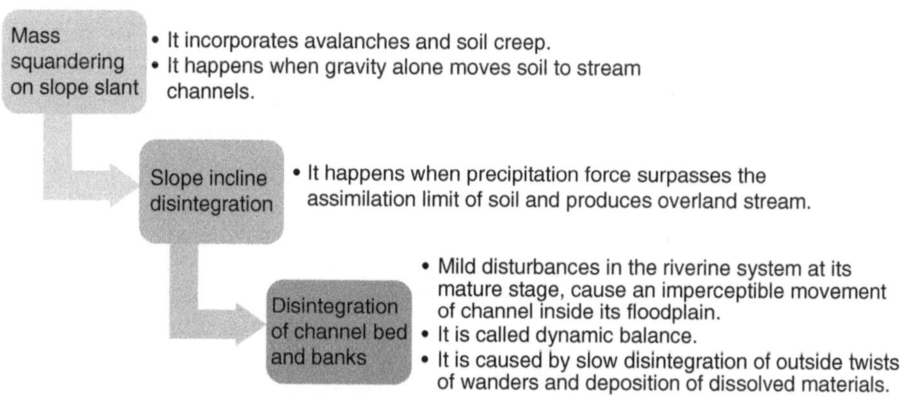

Figure 16.2 Processes of river soil erosion. *Source:* Adapted from Kondolf (1994).

magnified the ecological issues. In China, sand mining activities were permitted to 98 boats or ships, for extraction from the Yangtze River (midstream) in 2003. But in the annual assessment, it has been found that around 500 boats have been extracting sand from the river (de Leeuw et al. 2010; Xu et al. 2021). This caused the death of aquaculture in the region. Midstream extraction also results in deepening of the riverbed in the middle pulling the water on the borders inside. This may result in coastal erosion causing further damage to the river morphology (Arsyad et al. 2020). Also, the sinks created in sand and sandpits due to mining have been causing a drowning effect. A similar case was found in the study of Poyang Lake which derives its water from the major rivers in China like Gangjiang, Xiushui, and Xinjiang and drains into the Yangtze River. The purpose of the study was to assess the magnitude and impacts of sand mining using GIS. The Aster sensor was used to identify the moving vessels in the river basin in both monsoon and summer seasons. It was found that the river bodies are subjected to mining irrespective of season and water level. Also, the amount of extraction is way less than the sedimentation rate of the lake. Widening of the channel, immense turbidity, and deposition of sediments in Yangtze River from Poyang Lake were also noted (de Leeuw et al. 2010).

In a study on environmental impact assessment (EIA) of sand mining from river basins in the southwest coast of India, it was found that approximately 11.527 million ton per year of sand is being extracted. This amounts include both in-stream and floodplain sand. It was also found that manual mining is frequent in the midlands, and bar skimming was practiced in the upper land frequently while pit excavations was common in lowlands (Sreebha and Padmalal 2011). In the Neyyar River basin of Kerala, India, a socioeconomic impact assessment showed that the local conflicts, addiction to alcohol, and deaths due to drowning in the basin area have been prominent. Ecologically, deterioration in water quality and quantity has been observed. Severe river bank erosion, valley slumping, and channel widening in the area have put the nearby structures under the threat of collapse (Shaji and Anilkuar 2014). The Aghanashini River watershed suffers from both soil erosion of the catchment and mining from the river sediments. This has been causing the change in land use and land cover, and the fish breeding has been destroyed (Ramachandra et al. 2018). In India, in an analysis study of water quality along the South Bengal River, it was revealed that the river water quality was poor in areas with mining and pit sites (Bhattacharya et al., 2019b). Another study on effect of in-stream sand mining on hydraulic variable of bedload transport and channel platform was done on Kangsabati River. It was found that in-stream mining was majorly responsible for change in river ecosystem (Bhattacharya et al., 2019a). Ramya and Kumari (2020) found that the sand mining activities roughly take place in a radius of 50 km alongside water banks near most construction site. Hence, they proposed a sand mining detection technique using short message service (SMS) based on internet of things (IoT). The sand mining detector developed was made with Arduino Uno board and sensor.

16.4 Monitoring of Sand Mining Activities

The problem of sand mining has been blowing out of proportions due to a lack of adequate information, planning, policy, and support from locals. But in recent times, monitoring of erosion and sedimentation, sand deposition in water and sediment bed, and illegal activities like mining, have become much easier with the help of satellite and remote sensing data. Use of satellite data, obtained with help of sensors (active and passive), which can be categorized on many bases like electromagnetic spectrum, object, and spectrum, and use of satellite image help in identifying the movement of object on the land and water surface. A study on Hongze Lake of China used the satellite data from Moderate Resolution Imaging Spectroradiometer (MODIS)/Aqua, LANDSAT, and VIIRS day/night band nighttime light. The study revealed that the mining activities started around February 2012, and a massive increase was seen in terms of the monthly average in 2016. Also, the change in the suspended particulate matter was found to be consistent with the movement of the vessel (Duan et al. 2019). In a study by Reddy et al. (2020) on the Swarnamukhi watershed, in the Chittoor district of Andhra Pradesh, the sand mining activities were mapped by identifying mining sites in the region. The sites were identified using GIS tools like ArcGIS and Google Earth Engine (GEE). The study found that river mining activities have led to the deterioration of groundwater and the riverine environment (Reddy et al. 2020). Similarly, the river morphology is studied using GIS tools and software, with help of satellite and GEE data. A study on Sa'dang River in Indonesia studied the change in river morphology by varying years impacted by sand mining activities (2006, 2014, 2017, and 2019). The study found that channel width is inversely proportional to the river water body in case the sediment deposits decrease (Arsyad et al. 2020). The most authenticated study of the impacts of sand mining on river morphology and ecosystem should generally include field study along with experimental and software-based analysis. The authenticity of the satellite data is obtained by observing the ground truth from field data. River morphology can also be studied using the various spectral indices, like Normalized Difference Vegetation Index (NDVI) and Normalized Difference Water Index (NDWI) (Agapiou 2020; Arsyad et al. 2020; Kafy et al. 2021). These spectral indices help in identifying the quality of water, surrounding vegetation, erosion of banks, sedimentation in downstream areas, pollution, reduction in farmlands, etc. (Okur and Erturaç 2021). Through a spatial-temporal analysis, identifying the change patterns in river morphology, water flow pattern, sand mining zones, ecological deterioration, air, land, and water pollution levels can all be monitored (Arsyad et al. 2020). A study on the Nzhelele River in Limpopo province, South Africa, has been done to study the sand mining impacts on the riverine environment identified the adverse factors (Sengani and Zvarivadza 2019). The study was done based on data obtained from field observation, analysis of water and soil quality, particle size analysis, survey (questionnaire), and GIS data obtained from LANDSAT (ETM+), and NDVI. Research demonstrated that sand mining has been contributing to erosion, worsening of water quality, and reduction in grassland and agricultural land (Sengani and Zvarivadza 2019)

Excessive extraction of sand especially using tools like backhoes, power shovels, draglines, dredging boats, and suction pumps can cause more damage as their rate of extraction is more than manual tools. The use of this modern equipment and tools causes instability of shoreline, riverbanks, etc. In India, around 193 people have been reported to have died due to sand mining activities during 2019–2020, as per the report of Reuters Graphics, 2021 (Marco et al. 2021). The major issue that occurs with mining is that the location pits are never marked, their depth and width are unknown, and due to increased turbidity, they become invisible. Unfortunately, locals who enter the river even for daily activities like bathing and cleaning, are unaware of the mining pits and tend to fall or drown seldom. Yet on a global scale, the number of deaths occurring due to sand mining is massive.

Another aspect of change in river morphology is slope stability. Another piece of equipment used for monitoring and analysis is the unmanned aerial vehicles (UAVs) or drones (Xue and Su 2017). UAVs offer a better alternative to other aerial photography techniques. These are cheap in terms of operational cost, provide better land coverage and enhanced ground resolution, and are user friendly. A study by Darmajati et al. (2021) on slope stability of river morphology was done using UAVs multitemporal image. In the study, aerial photography was carried out with UAVs. digital elevation model (DEM) was developed by processing the multitemporal aerial images for slope geometry analysis. While the field sampling was done to obtain soil samples for experiments like shear parameters and bulk density, the data was analyzed using the Bishop slope stability method and the slopes were found unstable.

16.5 Sand Mining: A Threat to Aquatic Ecosystem

Sand mines or pits can affect the quality of groundwater and surface water (Singh et al. 2018). As the mining occurs, it disturbs the physical structure and chemical composition of the sediment beds. The sediment material is broken down and mineral leaching occurs. The pH and electric conductivity (Bhat et al. 2018) increase with the increase in turbidity, contamination, etc., in the river body surface. In surface water, the increased turbidity due to dregs is the real issue (Singhal et al. 2021). Dregs in the form of suspended solids can cover the sediment bed, affecting the fishes and other benthic creatures (Irandoost et al. 2021), decrease visibility, and prohibit the transmission of sunlight to the sediment bed. Sand and sediments mostly consist of silica and alumina in the form of quartz. Other minerals that are found in the sand are feldspar, clay (Kaolinite, gibbsite), carbonate, organic carbon, calcite, zircon, aragonite, magnetite, and kyanite minerals. These minerals have been identified commonly using modern equipment like X-Ray diffraction (XRD) (Deb and Sahu 2010), scanning electron microscope (SEM), flame atomic absorption spectrometry (FAAS), inductively coupled plasma mass spectroscopy (ICP-MS) (Nartey et al. 2012; Banerjee et al. 2016), and inductively coupled plasma optical emission spectrometry (ICP-OES).

During monsoon seasons, the surface runoff gets deposited in the rivers (Pandey et al. 2021). The runoff contains a heavy amount of suspended and eroded particles along with pollutants from point and non-point sources (Liu et al. 2014; Rodrigo-Comino et al. 2019; Zamora-Ledezma et al. 2021). Among these pollutants, mostly are metals, metalloids, and heavy metals that create major issues. Also, the leaching of oxides of sulfur (Seppälä et al. 2006; Doostmohammadi et al. 2017) can make the river body more acidic, be poisonous to aquatic life, and can affect the general climate of the water. Sediments are found to be sink and source of both heavy metals. Heavy metals easily get transported to the river body by getting attached to the suspended soil particles and get deposited on the sediment bed (Loredana et al. 2020; Salvo et al. 2018; Zhang et al., 2018a). They tend to bio-accumulate and bio-magnify throughout the food chain. Innumerous cases have been reported of poisoning of aquatic plants (Kale 2016; Ahn et al. 2018) and products with heavy metals especially mercury, lead, cadmium, etc. These products, if consumed by humans, can also affect human health and may be fatal. Heavy mining equipment tends to cause massive disturbances on the shoreline protection, infrastructure built on coastal areas, water intake structures, etc. Also, sometimes dewatering is done by miners for the extraction of sand, which affects the water table in the area. This leads to drying up of well and water stress in the region.

16.6 Impacts of Sand Mining on Components of Riverine Ecosystem

Sand mining is one of the major causes of the destruction of water resources. They are equally responsible for the loss of ecosystem products and services like aquaculture and agriculture but is one of the most underrated regions of study. Sand mining is a major problem especially in Asian countries, like China, India, Cambodia, South Korea, Malaysia, Indonesia, and Japan (Abidin et al. 2018; Arsyad et al. 2020). Other than this, developing countries in Africa and Asia are also facing this problem. For example, analysis of sand mining activities in the Mekong River and their environmental effect have been studied in almost all the countries, like China, Myanmar, Laos, Cambodia, and Vietnam, it flows through (Hackney et al. 2020). A study on dams of Cambodia based on the Mekong River revealed that the volume of the lake has decreased in a four-decade period (1980–2018) and the water storage has also decreased. The shrinkage of the lake has been attributed to the intensification of sand mining which is promptly done for construction, landfilling, and export of sand to other areas (Ng and Park 2021). Other countries lying in the course of the Mekong River have also presented high sand extraction rate from the riverbed, 50 metric tonnes per year which are way less than the river sand deposition rate (6.17 metric tonnes per year) (Chuenchum et al. 2020; Kim et al. 2020; Tamura et al. 2020; Ng and Park 2021; Park et al. 2021). Another study on water quality and planktons in Pantai Labu, Indonesia, found that there has been a tremendous decrease in aquaculture productivity, with a difference of approximately 239 tonnes per year from 2016 to 2017. The major reason has been attributed to intensive sand mining and other ecological issues that occur due to it (Doloksaribu et al. 2020)

The impact of sand mining is more severe in basins with ancient morphology. For example, the Deccan Plateau of southern India or the Deccan Trap has quite limited sediment reaches for the river flowing through it. The Deccan Plateau was formed at the end of the Cretaceous period. Hence, the land surface has become smoother with continuous erosion and runoff from flowing rivers as compared to Northern plains. Northern plains are formed from the deposition of major Himalayan rivers like the Ganges, Brahmaputra, and Indus. A study of the environmental impact of sand mining in the city of Kochi illustrated that an amount of 11.527 million tonnes per year, including both in-stream and floodplain sand, were mined for infrastructure development (Sreebha and Padmalal 2011). Another study in Vembanad Lake, Kochi city, revealed that in the lake that acts as a sink for sediments from seven rivers (length varying between 78 and 244 km), the quantity of sand extracted from the in-stream is around 40 times the rate of sediment (Padmalal et al. 2008). The processes of sand mining have been found to disturb the balance of removal and deposition of sand in the riverbed. The process of deposition and sand accumulation takes years to quantify. Hence, they are known to have "long recollection" which implies that the damage to rivers caused due to the sand mining processes may have a prolonged effect, even when the mining is stopped. Furthermore, mining damages the formation of dregs and triggers which help in balancing the export and import of soil on sediment beds. The major problem faced in solving the issues of sand mining is the lack of proper information about the quantity of extraction, deposition, characteristics, mineral composition, land use land cover (LULC) of the watershed, hydrology, siltation in the area, slope, flora and fauna, agriculture, aquaculture practices, etc. These data are crucial in understanding the status of the ecological and socioeconomic situation. Based on these data, future utilization and conservation plans can be developed (Lamchin et al. 2016).

16.6.1 Impacts on River Morphology and Sediment

A prolonged sand mining process impacts the features of river morphology and the sediment characteristics of the riverbed (Chandra et al. 2014). Among other issues like channel widening, channel narrowing, variation in sediment transport, channel hydraulics, flow intensity, and flood control, channel incision takes major impact from sand mining comparatively (Arsyad et al. 2020). Channel incision varying from 0.5 to over 30 m has been recorded in different studies. The over steepening and instability caused by continuous extraction of sand at a rate faster than sediment deposition have turned out to create havoc on the sediment bed by causing channel incision. In the case of the Bachang River in Taiwan, the increase in channel incision of 30 m led to channel narrowing by 1/6th of its original width (Yu and Shih 2015). Other such cases have been observed from rivers in Italy where the channel width has narrowed to 50% of its original width. Sand mining tends to change the topography and bathymetry of the rivers. Excessive in-stream mining increases the depth of rivers and the water from the coast or banks are pulled inward. This movement causes the erosion of banks and a decrease in the water width of the river resulting in the narrowing of rivers (Rahmani et al. 2018). Additionally, excessive erosion of banks causes a change in the river flow pattern and its floodplains. It also affects the wave transformation, flow intensity, spatial distribution of suspended sediments, natural sedimentation processes, and their rate. Furthermore, the increased turbidity creates havoc for aquatic plants and animals (Bhattacharya et al., 2019b).

Much research has illustrated the impact of sand mining on riverbed morphology and sediment characteristics. For example, Yao et al. (2019) quantified the hydrodynamic impacts of sand mining on Poyang Lake, China, which has a river-connected floodplain. The study demonstrated that continuous mining led to a decrease in the elevation of the outlet channel from 4 to 10 m along with a decrease in water head. Another major problem occurs with mine pits that are created which are invisible from the top and even their depth cannot be determined. These blind pits have caused many drowning incidents for humans especially in India, many daily (bathing, cleaning, etc.) and religious activities are done in rivers bodies, but sudden slopes created by mine pits cause people to slip and fall. Haghnazar et al. (2020) found that the depth of the mine pit should be kept restricted to 70% of channel flow depth for the pits to refill and the distance between pits should have an l/b ratio of 1.28. Channel incision also led to the relocation of Thalweg, deterioration in connectivity of floodplains to rivers, and decrease in groundwater recharge and groundwater levels. The groundwater levels, in general, tend to decrease as the water channel narrows and are unable to recharge the groundwater zones. The erosion of banks, increase in river depth, and narrowing of the channel can also impact the intake structures, bridges, piers, etc., constructed on the rivers. Sand mining also affects the grain distribution and sediment loading in the river. Continuous turbidity caused by mining leads to the removal of fine soil from the area and their accumulation in a more quiescent location. This leads to bed coarsening or removal of particle size (fines usually) from sediment bed (Haghnazar et al. 2020).

16.6.2 Impacts on Water Quality and Aquatic Life

The water quality of the river is majorly impacted by sand mining activities. The continuous movement of boats and ships throughout the day and night, usually in all seasons, leads to the creation of excessive waves in the river causing turbidity and bank erosion. Furthermore, water is contaminated from fuel exhaustion from ships. With the increase in anthropogenic activities, even the runoff reaching the rivers in the monsoon season tends to bring excessive contaminants along with it which specifically include heavy metals. These metals usually settle down on the sediment beds as they have a genuine attachment to fine sediment particles. This movement in water causes turbidity, the metals keep floating along with fines. These metals also spread wherever they go with the flow polluting more areas. Heavy metals are toxic to all forms of life. Hence, it is crucial to manage and control the turbulence in the water. Doloksaribu et al. (2020) studied the impact of sand mining on the seawater quality of Pantai Labu, Indonesia. The study demonstrated that the area has been prone to sediment pollution, with excessive turbidity (169–186 mg/L) which impacts the benthic community in the seawater. This affected the aquaculture in the area, with abrasion of the coast and loss of spawning ground. The analysis based on the STORET method showed that water can now be classified as lightly polluted. Through Shannon-Weaner analysis, it was found that the spread of each plankton species was of type medium.

The riverbed acts as a breeding and growth surface for aquatic animals, microbes, and plants. Loss of sediment in quantity and quality and change in their physical structure can cause severe loss for these aquatic lives. In a bigger frame, this affects the regenerating capacity of river and aquaculture services offered by the rivers and riverine sources (Ren et al. 2020). The rate of sand mining or extraction should be kept in proportion to the replenishment rate of the riverine system and the bed width. For example, if the extraction is done on a wider bed or floodplain (Okeke et al. 2021), in proportion to the replenishment rate, then the adverse effect of mining will be less as compared to narrow floodplains (Zhang et al., 2018b). Also, the use of heavy tools and equipment for extraction results in the compaction of riverbeds and the destruction of the breeding of aquatic animals. Loss in the aquaculture sector has been massive. Even the physical changes to habitats of fishes have been reported. The loss of spawning grounds has been resulting in a decrease in fish species. The increased turbidity suffocates the aquatic life below and decreases the sunlight transmission to the riverbed, and the loss of fines from soil disturbs the growth of aquatic plants and invertebrates.

16.6.3 Sand Bank Filtration

Sandbanks filtration is a type of filtration in which the water gets purified by passing through the banks of a river or lake. Then the water is extracted or drawn via extraction well at some distance from the river body. This is a good source of rechargeable groundwater (Piyadasa 2011). The water is automatically filtered and does not usually require any pretreatment. This process is quite effective as the water undergoes physical (straining through soil particles), biological (through soil microbes), and chemical (through ion exchange with soil fines) changes (McCarroll et al. 2020). But the process is dependent on sediment characteristics that are its hydraulic, chemical, and physical properties. Hydraulic properties of soil also vary with soil particles and excessive suspended fines can limit the aquifer recharge capacity (Wang et al. 2019; McCarroll et al. 2020). Also, if there is some contaminant leaching in the soil from any source, then the purification process may not be fully effective. Gutiérrez et al. (2018) studied natural recovery of infiltration capacity in simulated bank filtration in highly turbid water and found that the recovery was possible as long as the mixture of suspended sediments is present and the streambed sand is not coarse. Cui et al. (2021) studied the effect of riverbed sediment flushing and clogging on river water infiltration rate. It was found that the mechanism of change associated with flushing and clogging processes is revealed based on field monitoring, laboratory experiment, and field test. Sand mining activities have disturbed the texture of sediments and filtration of water in the aquifers. Also, the river bodies have shrunken, resulting in lowering of groundwater table.

16.7 Remediation Alternative

Scientists from all around the world have been working on finding new and modern materials and techniques to reduce the extraction and application of sand in various industries (Elmayel et al. 2019). It is given that the major application of sand occurs in the construction industry. China is both the largest producer and consumer of sand (Jiang et al. 2021). But in recent decades, other materials like M-sand, quarry rock dust, recycled sand and aggregate, and coconut shell have been all used as a replacement (partial or sometimes fully) of sand in concrete. Most of these experiments have found positive results (Mane et al. 2020; Moorthy et al. 2020; Reddy et al. 2020). These materials are some sort of waste causing pollution to other resources, like water, air, and soil, and can be properly disposed of in a cheap and eco-friendly manner in the construction sector. This will also decrease the load on landfills and prohibit other ecological issues such as leaching and contamination of water resources (Hamid 2015; Kankam et al. 2017). In a study conducted on the use of quarry dust as a replacement for sand, it has been found that quarry stone dust increases the durability of concrete by 10%. Quarry dust formed in abundance as a by-product of aggregate production. Also, it releases a lot of carbon and creates major air pollution. M-sand or manufactured sand is formed by crushing hard granite rock to a fineness of 4.75 mm as per Indian Code provision. M-sand is adequate for the replacement of fine aggregate in concrete, especially in places where river sand is not available. Mane et al. (2020) studied the properties of concrete produced with M-sand and electronic waste sand (E-sand) with varying the percentage of replacement from 0 to 40%, with an increment of 10%. The study showed that the replacement of 20% M-sand by E-sand gave maximum strength in concrete. Many such combinations of alternative materials have been used for the replacement of aggregate in concrete mixes (Table 16.1)

Table 16.1 Recent studies on alternative materials to be used as replacement of sand and aggregates.

Methodology	Findings	References
Replacement of river sand in varying proportions of 0, 5, 10, 15, 20, 25, 30, 35, and 40% with M-sand.	This study concluded that an increase in the percentage of manufacturing sand in the range of 35–40 will provide adequate engineering strength	(Senthil and Suresh 2020)
Performance of concrete by substituting manufactured sand by E-waste in different percentages (0, 10, 20, 30, and 40%), with 0.45 water–cement (WC) ratio	The study reveals that partly replacing 20% factory-made sand (M-sand) with E-waste (E- sand) yields maximum strength, better chloride permeability with improved microstructure than conventional concrete	(Mane et al. 2020)
Mechanical and durability properties of recycled aggregate concrete (RAC) was studied	The study revealed that replacement of recycled aggregate more than 10% results in lowering of compressive strength and depreciation in acid resistance and sulfate resistance.	(Surendar et al. 2021)
Strength analysis using pozzolanic materials such as silica fume, metakaolin, and Ground Granulated Blast-furnace Slag (GGBS) as partial replacement to the weight of cement by varying percentages (5, 10, and 15) and M–sand as a complete replacement to river sand.	At replacement of 10% to the weight of cement with silica fume, metakaolin, and GGBS using M–sand instead of river sand gives higher compressive and split tensile strengths	(Naga Venkat et al. 2020)
Investigates the properties of fly ash (FA)-based geopolymer mortars prepared with expanded vermiculite for thermal insulation applications in buildings	The study suggests that incorporating expanded vermiculite in FA-based geopolymer mortar can provide eco-friendly and lightweight building composites with improved sound and thermal insulation properties, contributing toward the reduction of the environmental effects of waste materials and conservation of natural sand	(Gencel et al. 2021)
126 geopolymer concrete samples were made for investigating the mechanical properties of geopolymer concrete with the sand replacement ratio of 0–15% by weight in increments of 5%	Observations revealed that the use of waste lime and waste glass aggregates reduced the compressive and tensile strength of geopolymer concrete	(Taher et al. 2021)
Specimens of the same size were prepared with different percentages of nanoalumina (0, 0.25, 0.5, 0.75, 1, 1.25, and 1.5%)	The optimum amount of alpha phase nanoalumina in M-sand cement mortar should be 1%	(Alex et al. 2021)

Many nations, like China and the United States, have tapped into the potential of "crushed rock" to be used as an alternative to aggregate in concrete (Santhosh et al. 2021). But the major problem is that the procedure of blasting, mining, and grinding of rocks to adequate construction size aggregates would require massive manpower and funding. This may not be ideal for small or large construction projects (Senthil and Suresh 2020). Also, not all countries can have sources of rock bed with high compressive strength and proper chemical composition. Furthermore, the extraction of rock also causes environmental problems like noise, air, water pollution, leaching, and loss of fertile topsoil. The types of machinery used in mining and extraction would require an equivalent amount of electricity which is costly and still inadequately supplied (Duncan and De Vries 2018; Moorthy et al. 2020).

16.8 Conclusion

Through this study, it was found that sand mining process has created havoc on the river aquatic biomes and is responsible for the death of many river bodies. The sand mining activities are mostly influenced by construction industries, mineral extraction (rutile, ilmenite, and zircon), glass, electroplating, and manufacturing industries. The impacts on river body and aquatic biomes are also significant. The river body morphology suffers from channel incision, narrowing, deepening of river mouth, instability of banks, and clogging of aquifers. Aquatic animals lose their spawning zones, and with increased turbidity, their survival becomes impossible. This has resulted in the death of aquaculture in many rivers globally, impacting the socioeconomic status of the locals. These activities are practiced illegally in major part of the developing countries like China, India, and Indonesia. Illegal activities have led to increase the number of death due to drowning, shrinking of the river bodies, and coastal erosion as reported by many research. The illegal sand mining activities are hard to monitor and assess as most locals are involved in the process. The application of GIS and neural networking is practiced actively for tracking, monitoring, and controlling the sand mining process. River sand has its superiority in terms of its shape and texture compared to desert sand or sea sand, hence it is required by the construction industries as its largest consumer. Although many alternatives have been found for sand like M-sand and quarry dust, they are more costly and cause environmental concerns like pollution. Hence their applicability is subjected to feasibility and budget of the construction project. Furthermore, lack of data on sand mining is a huge problem. There is an immediate need to monitor and draw up plans to conserve the riverine ecosystem from the adverse effects of sand mining.

References

Abidin, M. Z., Kutty, A. A., Lihan, T., & Zakaria, N. A. (2018). Hydrological change effects on Sungai Langat water quality. *Sains Malaysiana*, 47(7), 1401–1411. https://doi.org/10.17576/jsm-2018-4707-07

Agapiou, A. (2020). Estimating proportion of vegetation cover at the vicinity of archaeological sites using sentinel-1 and-2 data, supplemented by crowdsourced openstreetmap geodata. *Applied Sciences (Switzerland)*, 10(14). https://doi.org/10.3390/app10144764

Ahn, J., Kang, K. W., Kang, W., Lim, H., Cho, S., Moon, J., & Park, W. (2018). Mercury poisoning in a fisherman working on a pelagic fishing vessel due to excessive tuna consumption. Journal of Occupational Health, 60, 89–93. https://doi.org/10.1539/joh.16-0274-CS

Alex, A. G., Gebrehiwet, T., & Kemal, Z. (2021). M-Sand cement mortar with partial replacement of alpha phase nano alumina. *Journal of Building Pathology and Rehabilitation*, 6(1). https://doi.org/10.1007/s41024-021-00111-8

Aliu, I. R., Akoteyon, I. S., & Soladoye, O. (2022). Sustaining urbanization while undermining sustainability: the socio-environmental characterization of coastal sand mining in Lagos Nigeria. *GeoJournal*, 7. https://doi.org/10.1007/s10708-021-10563-7

Arsyad, Rukmana, D., Salman, D., & Alimuddin, I. (2020). Impact of sand mining on the changes of morphological and physical dynamics in Sa'dang River, Pinrang District, Indonesia. *J. Degrade. Min. Land Manage*, 8(1), 2451–2460. https://doi.org/10.15243/jdmlm

Ashraf, M. A., Maah, M. J., Yusoff, I., Wajid, A., & Mahmood, K. (2011). Sand mining effects, causes and concerns: a case study from Bestari Jaya, Selangor, Peninsular Malaysia. *Scientific Research and Essays*, 6(6), 1216–1231. https://doi.org/10.5897/SRE10.690

Aydin, H., Ustaoğlu, F., Tepe, Y., & Soylu, E. N. (2021). Assessment of water quality of streams in northeast Turkey by water quality index and multiple statistical methods. *Environmental Forensics*, 22(1–2), 270–287. https://doi.org/10.1080/15275922.2020.1836074

Banerjee, S., Kumar, A., Maiti, S. K., & Chowdhury, A. (2016). Seasonal variation in heavy metal contaminations in water and sediments of Jamshedpur stretch of Subarnarekha river, *India. Environmental Earth Sciences*, 75(3), 1–12. https://doi.org/10.1007/s12665-015-4990-6

Bhat, B., Parveen, S., & Hassan, T. (2018). Seasonal assessment of physicochemical parameters and evaluation of water quality of river Yamuna, India. *Advances in Environmental Technology*, 4(1), 41–49. https://doi.org/10.22104/aet.2018.2415.1121

Bhattacharya, R., Dolui, G., & Das Chatterjee, N. (2019a). Effect of instream sand mining on hydraulic variables of bedload transport and channel planform: an alluvial stream in South Bengal basin, India. *Environmental Earth Sciences*, 78(10), 1–24. https://doi.org/10.1007/s12665-019-8267-3

Bhattacharya, R. K., Das Chatterjee, N., & Dolui, G. (2019b). Consequences of sand mining on water quality and instream biota in alluvial stream: a case-specific study in South Bengal River, India. *Sustainable Water Resources Management*, 5(4), 1815–1832. https://doi.org/10.1007/s40899-019-00345-y

Bhattacharyya, R., Ghosh, B. N., Mishra, P. K., Mandal, B., Rao, C. S., Sarkar, D., Das, K., Anil, K. S., Lalitha, M., Hati, K. M., & Franzluebbers, A. J. (2015). Soil degradation in india: challenges and potential solutions. *Sustainability (Switzerland)*, 7(4), 3528–3570. https://doi.org/10.3390/su7043528

Chandra, P., Patel, P. L., Porey, P. D., & Gupta, I. D. (2014). Estimation of sediment yield using SWAT model for Upper Tapi basin. *ISH Journal of Hydraulic Engineering*, 20(3), 291–300. https://doi.org/10.1080/09715010.2014.902170

Chauhan, S. S. (2010). Mining, development and environment: a case study of Bijolia mining area in Rajasthan, India. *Journal of Human Ecology*, 31(1), 65–72. https://doi.org/10.1080/09709274.2010.11906299

Chowdary, V. M., Chakraborthy, D., Jeyaram, A., Murthy, Y. V. N. K., Sharma, J. R., & Dadhwal, V. K. (2013). Multi-criteria decision making approach for watershed prioritization using analytic hierarchy process technique and GIS. *Water Resources Management*, 27(10), 3555–3571. https://doi.org/10.1007/s11269-013-0364-6

Chuenchum, P., Xu, M., & Tang, W. (2020). Estimation of soil erosion and sediment yield in the lancang-mekong river using the modified revised universal soil loss equation and GIS techniques. *Water (Switzerland)*, 12(1). https://doi.org/10.3390/w12010135

Corre, M. D., Schnabel, R. R., & Stout, W. L. (2002). Effect of Hardness and Tensile properties of Al6061 based composites for Piston alloy using Stir casting method. *Journal of Chemical and Pharmaceutical Sciences*, 34(3), 445–457.

Cui, G., Su, X., Liu, Y., & Zheng, S. (2021). Effect of riverbed sediment flushing and clogging on river-water infiltration rate: a case study in the Second Songhua River, Northeast China. *Hydrogeology Journal*, 29(2), 551–565. https://doi.org/10.1007/s10040-020-02218-7

Darmajati, Y., Samodra, G., & Purwanto, T. (2021). Slope Stability Analysis Based on Change in Slope Geometry Using Slope Stability Analysis Based on Change in Slope Geometry Using Multitemporal UAV Image. November.

Deb, S., & Sahu, S. S. (2010). Study of Mineralogy and Weathering of Two Soil Profiles of Undulating Plateau Area of Jumar Subwatershed, Jharkhand, India. *International Journal of Agriculture, Environment and Biotechnology* 3(3), 295–301.

Doloksaribu, D. C. N., Barus, T. A., & Sebayang, K. (2020). The impact of marine sand mining on sea water quality in Pantai Labu, Deli Serdang Regency, Indonesia. *IOP Conference Series: Earth and Environmental Science*, 454(1). https://doi.org/10.1088/1755-1315/454/1/012086

Doostmohammadi, R., Olfati, M., & Roodsari, F. G. (2017). Mining pollution control using biogrouting. *Journal of Mining Science*, 53(2), 367–376. https://doi.org/10.1134/S1062739117022248

Duan, H., Cao, Z., Shen, M., Liu, D., & Xiao, Q. (2019). Detection of illicit sand mining and the associated environmental effects in China's fourth largest freshwater lake using daytime and nighttime satellite images. *Science of the Total Environment*, 647, 606–618. https://doi.org/10.1016/j.scitotenv.2018.07.359

Duncan, A. E., & Vries, N. De. (2018). Assessment of heavy metal pollution in the sediments of the River Pra and its tributaries. *Water, Air, & Soil Pollution*, 229(8), 1–10.

Elmayel, I., Higueras, P. L., Bouzid, J., Noguero, E. M. G., & Elouaer, Z. (2019). Recent Advances in Geo-Environmental Engineering, Geomechanics and Geotechnics, and Geohazards (Vol. 49, Issue January). Springer International Publishing. https://doi.org/10.1007/978-3-030-01665-4

Filho, W. L., Hunt, J., Lingos, A., Platje, J., Vieira, L. W., Will, M., & Gavriletea, M. D. (2021). The unsustainable use of sand: Reporting on a global problem. *Sustainability (Switzerland)*, 13(6), 1–16. https://doi.org/10.3390/su13063356

Gao, X., Ouyang, W., Lin, C., Wang, K., Hao, F., Hao, X., & Lian, Z. (2020). Considering atmospheric N_2O dynamic in SWAT model avoids the overestimation of N_2O emissions in river networks. *Water Research*, 174, 115624. https://doi.org/10.1016/j.watres.2020.115624

Gencel, O., Gholampour, A., Tokay, H., & Ozbakkaloglu, T. (2021). Replacement of natural sand with expanded vermiculite in fly ash-based geopolymer mortars. *Applied Sciences (Switzerland)*, 11(4), 1–18. https://doi.org/10.3390/app11041917

Giri, S., Singh, A. K., & Tewary, B. K. (2013). Source and distribution of metals in bed sediments of Subarnarekha River, India. *Environmental Earth Sciences*, 70(7), 3381–3392. https://doi.org/10.1007/s12665-013-2404-1

Gong, L., Ran, Q., He, G., & Tiyip, T. (2015). A soil quality assessment under different land use types in Keriya river basin, Southern Xinjiang, China. *Soil and Tillage Research*, 146(PB), 223–229. https://doi.org/10.1016/j.still.2014.11.001

Goralski, M. A., & Tan, T. K. (2020). Artificial intelligence and sustainable development. *International Journal of Management Education*, 18(1). https://doi.org/10.1016/j.ijme.2019.100330

Gutiérrez, J. P., van Halem, D., Uijttewaal, W. S. J., del Risco, E., & Rietveld, L. C. (2018). Natural recovery of infiltration capacity in simulated bank filtration of highly turbid waters. *Water Research*, 147, 299–310. https://doi.org/10.1016/j.watres.2018.10.009

Hackney, C. R., Darby, S. E., Parsons, D. R., Leyland, J., Best, J. L., Aalto, R., Nicholas, A. P., & Houseago, R. C. (2020). River bank instability from unsustainable sand mining in the lower Mekong River. *Nature Sustainability*, 3(3), 217–225. https://doi.org/10.1038/s41893-019-0455-3

Haghnazar, H., Sangsefidi, Y., Mehraein, M., & Tavakol-Davani, H. (2020). Evaluation of infilling and replenishment of river sand mining pits. *Environmental Earth Sciences*, 79(14), 1–18. https://doi.org/10.1007/s12665-020-09106-z

Hamid, M. A. (2015). Improved concrete properties using quarry dust as replacement for natural sand. *International Journal of Engineering Research and Development*, 11(3), 46–52.

Himanshu, S. K., Pandey, A., Yadav, B., & Gupta, A. (2019). Evaluation of best management practices for sediment and nutrient loss control using SWAT model. *Soil and Tillage Research*, 192(August 2018), 42–58. https://doi.org/10.1016/j.still.2019.04.016

Irandoost, F., Agah, H., Rossi, L., Calizza, E., Careddu, G., & Costantini, M. L. (2021). Stable isotope ratios (δ13C and δ15N) and heavy metal levels in macroalgae, sediment, and benthos from the northern parts of Persian Gulf and the Gulf of Oman. *Marine Pollution Bulletin*, 163(July 2020), 111909. https://doi.org/10.1016/j.marpolbul.2020.111909

IS:2386 (Part IV). (2016). Methods of test for Aggregates for Concrete, part 4 : Mechanical properties. Bureau of Indian Standards, New Delhi, 1–37.

IS:2386- Part I. (1963). Method of test for aggregate for concrete. Part I – Particle size and shape. Indian Standards, (Reaffirmed 2002).

IS:2386- Part III. (1963). Method of Test for aggregate for concrete. Part III- Specific gravity, density, voids, absorption and bulking. Bureau of Indian Standards, New Delhi, (Reaffirmed 2002).

Jiang, L., Liu, Y., Wu, S., & Yang, C. (2021). Analyzing ecological environment change and associated driving factors in China based on NDVI time series data. *Ecological Indicators*, 129. https://doi.org/10.1016/j.ecolind.2021.107933

Kafy, A.A., Dey, N. N., Al Rakib, A., Rahaman, Z. A., Nasher, N. M. R., & Bhatt, A. (2021). Modeling the relationship between land use/land cover and land surface temperature in Dhaka, Bangladesh using CA-ANN algorithm. *Environmental Challenges*, 4(May), 100190. https://doi.org/10.1016/j.envc.2021.100190

Kale, V. S. (2016). Consequence of temperature, pH, turbidity and dissolved oxygen water quality parameters. http://www.env.gov.bc.ca/wat/wq/BCguidelines/temptech/temperature.html#rec. *International Advanced Research Journal in Science, Engineering and Technology ISO*, 3(8), 186–190. https://doi.org/10.17148/IARJSET.2016.3834

Kankam, C. K., Meisuh, B. K., Sossou, G., & Buabin, T. K. (2017). Stress-strain characteristics of concrete containing quarry rock dust as partial replacement of sand. *Case Studies in Construction Materials*, 7, 66–72. https://doi.org/10.1016/J.CSCM.2017.06.004

Kayet, N., Pathak, K., Chakrabarty, A., & Sahoo, S. (2016). Spatial impact of land use/land cover change on surface temperature distribution in Saranda Forest, Jharkhand. *Modeling Earth Systems and Environment*, 2(3), 1–10. https://doi.org/10.1007/s40808-016-0159-x

Kim, J. H., & Yoo, S. H. (2020). Public perspective on the environmental impacts of sea sand mining: Evidence from a choice experiment in South Korea. *Resources Policy*, 69(August), 101811. https://doi.org/10.1016/j.resourpol.2020.101811

Kim, T. T., Huong, N. T. M., Huy, N. D. Q., Tai, P. A., Hong, S., Quan, T. M., Bay, N. T., Jeong, W. K., & Phung, N. K. (2020). Assessment of the impact of sand mining on bottom morphology in the Mekong river in an Giang province, Vietnam, using a hydro-morphological model with gpu computing. *Water (Switzerland)*, 12(10), 1–24. https://doi.org/10.3390/w12102912

Koehnken, L., Rintoul, M. S., Goichot, M., Tickner, D., Loftus, A. C., & Acreman, M. C. (2020). Impacts of riverine sand mining on freshwater ecosystems: A review of the scientific evidence and guidance for future research. *River Research and Applications*, 36(3), 362–370. https://doi.org/10.1002/rra.3586

Kondolf, M. G. (1994). Geomorphic and environmental effects of instream gravel mining. *Landscape and Urban Planning*, 28(2–3), 225–243. https://doi.org/10.1016/0169-2046(94)90010-8

Koondhar, M. A., Li, H., Wang, H., Bold, S., & Kong, R. (2020). Looking back over the past two decades on the nexus between air pollution, energy consumption, and agricultural productivity in China: a qualitative analysis based on the ARDL bounds testing model. *Environmental Science and Pollution Research*, 27(12), 13575–13589. https://doi.org/10.1007/s11356-019-07501-z

Lamchin, M., Lee, J. Y., Lee, W. K., Lee, E. J., Kim, M., Lim, C. H., Choi, H. A., & Kim, S. R. (2016). Assessment of land cover change and desertification using remote sensing technology in a local region of Mongolia. *Advances in Space Research*, 57(1), 64–77. https://doi.org/10.1016/j.asr.2015.10.006

de Leeuw, J., Shankman, D., Wu, G., de Boer, W. F., Burnham, J., He, Q., Yesou, H., & Xiao, J. (2010). Strategic assessment of the magnitude and impacts of sand mining in Poyang Lake, China. *Regional Environmental Change*, 10(2), 95–102. https://doi.org/10.1007/s10113-009-0096-6

Lenhart, C. F., Smith, D. J., Lewandowski, A., Belmont, P., Gunderson, L., & Nieber, J. L. (2018). Assessment of stream restoration for reduction of sediment in a large agricultural watershed. *Journal of Water Resources Planning and Management*, 144(7), 04018032. https://doi.org/10.1061/(asce)wr.1943-5452.0000908

Liu, R., Wang, J., Shi, J., Chen, Y., Sun, C., Zhang, P., & Shen, Z. (2014). Runoff characteristics and nutrient loss mechanism from plain farmland under simulated rainfall conditions. *Science of the Total Environment*, 468–469, 1069–1077. https://doi.org/10.1016/j.scitotenv.2013.09.035

Loredana, G., Torre, L., Cicero, N., Bartolomeo, G., Rando, R., Vadal, R., Santini, A., Durazzo, A., Lucarini, M., Dugo, G., & Salvo, A. (2020). Assessment and monitoring of fish quality from a coastal ecosystem under high anthropic pressure: A case study in Southern Italy. *International Journal of Environmental Research and Public Health*, 17(9), 3285

Ma, J., Nossa, C. W., & Alvarez, P. J. J. (2015). Groundwater ecosystem resilience to organic contaminations: Microbial and geochemical dynamics throughout the 5-year life cycle of a surrogate ethanol blend fuel plume. *Water Research*, 80, 119–129. https://doi.org/10.1016/j.watres.2015.05.003

Mahadevan, P. (2019). Sand mafias in India. Disorganized Crime in a Growing Economy.

Maharana, C., & Tripathi, J. K. (2018). The Son, A Vindhyan River. In: The Indian Rivers, pp. 191–197. Springer, Singapore. https://doi.org/10.1007/978-981-10-2984-4_15.

Mamo, M., & Bubenzer, G. D. (2001). Detachment rate, soil erodibility, and soil strength as influenced by living plant roots part II: Field study. *Transactions of the ASAE*, 44(5), 1167–1174.

Mane, K. M., Nadgouda, P. A., & Joshi, A. M. (2020). An experimental study on properties of concrete produced with M-sand and E-sand. *Materials Today: Proceedings*, 38(5), 2590–2595. https://doi.org/10.1016/j.matpr.2020.08.086

Manoj, K., Padhy, P. K., & Chaudhury, S. (2012). Study of heavy metal contamination of the river water through index analysis approach and environmetrics. *Bulletin of Environment, Pharmacology and Life Sciences*, 1(10), 7–15. www.bepls.com

Marco, H., Scarr, S., & Daigle, K. (2021). The messy business of sand mining explained. *Reuters Graphics*. https://graphics.reuters.com/GLOBAL-ENVIRONMENT/SAND/ygdpzekyavw/

McCarroll, R. J., Masselink, G., Valiente, N. G., Wiggins, M., Scott, T., Conley, D. C., & King, E. V. (2020). Impact of a headland-associated sandbank on shoreline dynamics. *Geomorphology*, 355, 107065. https://doi.org/10.1016/J.GEOMORPH.2020.107065

Moorthy, G. V., Thangaraj, P., & Shanmugam, T. (2020). Strength aspects of high-performance concrete by using waste material as replacement of fine aggregate in various percentages. *IOP Conference Series: Materials Science and Engineering*, 955(1). https://doi.org/10.1088/1757-899X/955/1/012047

Naga Venkat, G., Chandramouli, K., Ahmed, E., & Nagendrababu, V. (2020). Comparative study on mechanical properties and quality of concrete by part replacement of cement with silica fume, metakaolin and GGBS by using M-Sand as fine aggregate. *Materials Today: Proceedings*, 43(2), 1874–1878. https://doi.org/10.1016/j.matpr.2020.10.819

Nartey, V. K., Hayford, E. K., & Ametsi, S. K. (2012). Assessment of the impact of solid waste dumpsites on some surface water systems in the Accra Metropolitan Area, Ghana. *Journal of Water Resource and Protection*, 4(8), 605–615. https://doi.org/10.4236/jwarp.2012.48070

Ng, W. X., & Park, E. (2021). Shrinking Tonlé Sap and the recent intensification of sand mining in the Cambodian Mekong River. *Science of the Total Environment*, 777. https://doi.org/10.1016/j.scitotenv.2021.146180

Okeke, C., Abbey, S., Oti, J., Eyo, E., Johnson, A., Ngambi, S., Abam, T., & Ujile, M. (2021). Appropriate use of lime in the study of the physicochemical behaviour of stabilised lateritic soil under continuous water ingress. *Sustainability (Switzerland)*, 13(1), 1–26. https://doi.org/10.3390/su13010257

Okur, H., & Erturaç, M. K. (2021). Temporal monitoring of vast sand mining in NW Turkey: Implications on environmental/social impacts. https://doi.org/10.31223/X5HK5T

Padmalal, D., Maya, K., Sreebha, S., & Sreeja, R. (2008). Environmental effects of river sand mining: A case from the river catchments of Vembanad lake, Southwest coast of India. *Environmental Geology*, 54(4), 879–889. https://doi.org/10.1007/s00254-007-0870-z

Pal, S. C., & Chakrabortty, R. (2019). Simulating the impact of climate change on soil erosion in sub-tropical monsoon dominated watershed based on RUSLE, SCS runoff and MIROC5 climatic model. *Advances in Space Research*, 64(2), 352–377. https://doi.org/10.1016/j.asr.2019.04.033

Pandey, S., Kumari, N., & Priya, S. (2021). Soil quality and pollution assessment around Jumar watershed of Jharkhand, India. *Arabian Journal of Geosciences*. https://doi.org/10.1007/s12517-021-09091-y

Park, E., Loc, H. H., Van Binh, D., & Kantoush, S. (2021). The worst 2020 saline water intrusion disaster of the past century in the Mekong Delta: Impacts, causes, and management implications. *Ambio*. https://doi.org/10.1007/s13280-021-01577-z

Pastore, G., Baird, T., Vermeesch, P., Resentini, A., & Garzanti, E. (2021). Provenance and recycling of Sahara Desert sand. *Earth-Science Reviews*, 216(February), 103606. https://doi.org/10.1016/j.earscirev.2021.103606

Peduzzi, P. (2014). Sand, rarer and one thinks. *UNEP Global Environmental Alert Service*, 2012(March), 1–15.

Piyadasa, R. U. K. (2011). River sand mining and associated environmental problems in Sri Lanka. *IAHS-AISH Publication*, 349 (December 2004), 148–153.

Prusty, J. K., Patro, S. K., & Basarkar, S. S. (2016). Concrete using agro-waste as fine aggregate for sustainable built environment – A review. *International Journal of Sustainable Built Environment*, 5(2), 312–333. https://doi.org/10.1016/j.ijsbe.2016.06.003

Rahmani, V., Kastens, J. H., de Noyelles, F., Jakubauskas, M. E., Martinko, E. A., Huggins, D. H., Gnau, C., Liechti, P. M., Campbell, S. W., Callihan, R. A., & Blackwood, A. J. (2018). Examining storage capacity loss and sedimentation rate of large reservoirs in the Central U.S. great plains. *Water (Switzerland)*, 10(2), 1–17. https://doi.org/10.3390/w10020190

Ramachandra, T. V., Vinay, S., & Subash Chandran, M. D. (2018). Quantification of annual sediment deposits for sustainable sand management in Aghanashini river estuary. *Journal of Environmental Management*, 206, 1263–1273. https://doi.org/10.1016/j.jenvman.2017.07.060

Ramya, R., & Kumari, M. N. (2020). Sand mining detection – Using IOT. 2020 7th International Conference on Smart Structures and Systems, ICSSS 2020, 29–32. https://doi.org/10.1109/ICSSS49621.2020.9202088

Reddy, B. M., Guddeti, S. S. A. L., & Nagamani, K. (2020). Mapping of sand mining areas using remote sensing and Gis: A case study in parts of Swarnamukhi river bed, Chittoor District, Andhra Pradesh. *The Bioscan*, 15(1), 1–4.

Ren, Q., Wang, X., Li, W., Wei, Y., & An, D. (2020). Research of dissolved oxygen prediction in recirculating aquaculture systems based on deep belief network. *Aquacultural Engineering*, 90(September 2019), 102085. https://doi.org/10.1016/j.aquaeng.2020.102085

Rezaei, K., Guest, B., Friedrich, A., Fayazi, F., Nakhaei, M., Aghda, S. M. F., & Beitollahi, A. (2009). Soil and sediment quality and composition as factors in the distribution of damage at the December 26, 2003, Bam area earthquake in SE Iran ($M_s = 6.6$). *Journal of Soils and Sediments*, 9(1), 23–32. https://doi.org/10.1007/s11368-008-0046-9

Rodrigo-Comino, J., Keshavarzi, A., Bagherzadeh, A., & Brevik, E. C. (2019). The use of multivariate statistical analysis and soil quality indices as tools to be included in regional management plans. A case study from the Mashhad Plain, *Iran. Geographical Research Letters*, 45(2), 687–708. https://doi.org/10.18172/cig.3640

Salvo, A., Loredana, G., Torre, L., Mangano, V., Erminia, K., Bartolomeo, G., Santini, A., Granata, T., & Dugo, G. (2018). Ecotoxicology and Environmental Safety Toxic inorganic pollutants in foods from agricultural producing areas of Southern Italy : Level and risk assessment. *Ecotoxicology and Environmental Safety*, 148(May 2017), 114–124. https://doi.org/10.1016/j.ecoenv.2017.10.015

Santhosh, K. G., Subhani, S. M., & Bahurudeen, A. (2021). Cleaner production of concrete by using industrial by-products as fine aggregate: A sustainable solution to excessive river sand mining. *Journal of Building Engineering*, 42(January), 1–17. https://doi.org/10.1016/j.jobe.2021.102415

Seleiman, M. F., Kheir, A. M. S., Al-Dhumri, S., Alghamdi, A. G., Omar, E. S. H., Aboelsoud, H. M., Abdella, K. A., & Abou El Hassan, W. H. (2019). Exploring optimal tillage improved soil characteristics and productivity of wheat irrigated with different water qualities. *Agronomy*, 9(5), 1–11. https://doi.org/10.3390/agronomy9050233

Selvam, S., Jesuraja, K., Venkatramanan, S., Chung, S. Y., Roy, P. D., Muthukumar, P., & Kumar, M. (2020). Imprints of pandemic lockdown on subsurface water quality in the coastal industrial city of Tuticorin, South India: A revival perspective. *Science of the Total Environment*, 738, 139848. https://doi.org/10.1016/j.scitotenv.2020.139848

Sengani, F., & Zvarivadza, T. (2019). Proceedings of the 18th Symposium on Environmental Issues and Waste Management in Energy and Mineral Production. In Proceedings of the 18th Symposium on Environmental Issues and Waste Management in Energy and Mineral Production. Springer International Publishing. https://doi.org/10.1007/978-3-319-99903-6

Senthil, M. S., & Suresh, E. S. M. (2020). Sustainable construction material management using GIS (A case study of partial replacement of M-Sand with River Sand in Trichirappali District). *Journal of Critical Reviews*, 7(4), 1173–1179. https://doi.org/10.31838/jcr.07.04.211

Seppälä, J., Posch, M., Johansson, M., & Hettelingh, J. P. (2006). Country-dependent characterisation factors for acidification and terrestrial eutrophication based on accumulated exceedance as an impact category indicator. *International Journal of Life Cycle Assessment*, 11(6), 403–416. https://doi.org/10.1065/lca2005.06.215

Shaji, J., & Anilkuar, D. R. (2014). Socio-Environmental Impact of River Sand Mining: An Example from Neyyar River, Thiruvananthapuram District of Kerala, India. *IOSR Journal of Humanities and Social Science*, 19(1), 1–7. https://doi.org/10.9790/0837-19150107

Shrestha, V. M., Anandh, S., & Sindhu Nachiar, S. (2017). Experimental study on the strength parameter of Quarry Dust mixed Coconut Shell Concrete adding Coconut Fibre. *IOP Conference Series: Earth and Environmental Science*, 80(1). https://doi.org/10.1088/1755-1315/80/1/012021

Singh, R. K., Singha, M., Singh, S. K., Pal, D., Tripathi, N., & Singh, R. S. (2018). Land use/land cover change detection analysis using remote sensing and GIS of Dhanbad District, India. *Eurasian Journal of Forest Science*, 6(2), 1–12. https://doi.org/10.31195/ejejfs.428381

Singh, S., Bhardwaj, A., & Verma, V. K. (2020). Remote sensing and GIS based analysis of temporal land use/land cover and water quality changes in Harike wetland ecosystem,

Punjab, India. *Journal of Environmental Management*, 262(March), 110355. https://doi.org/10.1016/j.jenvman.2020.110355

Singhal, A., Pandey, S., Kumari, N., Chauhan, D. K., & Jha, P. K. (2021). Impact of climate change on soil microbes involved in biogeochemical cycling. 63–94. https://doi.org/10.1007/978-3-030-76863-8_5

Sinha, R., & Singh, P. (2016). Evaluation of infiltration behaviour and soil characteristics in Dhanbad – Jharia Township Area, Jharkhand, India. *Current World Environment*, 11(2), 584–591. https://doi.org/10.12944/cwe.11.2.29

Sreebha, S., & Padmalal, D. (2011). Environmental impact assessment of sand mining from the small catchment rivers in the Southwestern Coast of India: A case study. *Environmental Management*, 47(1), 130–140. https://doi.org/10.1007/s00267-010-9571-6

Stabnikov, V., Chu, J., Myo, A. N., & Ivanov, V. (2013). Immobilization of sand dust and associated pollutants using bioaggregation. *Water, Air, and Soil Pollution*, 224(9). https://doi.org/10.1007/s11270-013-1631-0

Surendar, M., Ananthi, G.B.G., Sharaniya, M., Deepak, M. S., & Soundarya, T. V. (2021). Mechanical properties of concrete with recycled aggregate and M-sand. *Materials Today: Proceedings*, 44, 1723–1730. https://doi.org/10.1016/j.matpr.2020.11.896

Taher, S. M. S., Saadullah, S. T., Haido, J. H., & Tayeh, B. A. (2021). Behavior of geopolymer concrete deep beams containing waste aggregate of glass and limestone as a partial replacement of natural sand. *Case Studies in Construction Materials*, 15(August), e00744. https://doi.org/10.1016/j.cscm.2021.e00744

Tamura, T., Nguyen, V. L., Ta, T. K. O., Bateman, M. D., Gugliotta, M., Anthony, E. J., Nakashima, R., & Saito, Y. (2020). Long-term sediment decline causes ongoing shrinkage of the Mekong megadelta, Vietnam. *Scientific Reports*, 10(1), 4–10. https://doi.org/10.1038/s41598-020-64630-z

UNEP-GEAS. (2014). Sand, rarer than one thinks. United Nations Environment Program (UNEP). *Global Environmental Alert Service (GEAS)*, 2012(March), 1–15. https://wedocs.unep.org/bitstream/handle/20.500.11822/8665/GEAS_Mar2014_Sand_Mining.pdf?sequence=3&isAllowed=y

Ustaoğlu, F., Taş, B., Tepe, Y., & Topaldemir, H. (2021). Comprehensive Assessment of Water Quality and Associated Health Risk by Using Physicochemical Quality Indices and Multivariate Analysis in Terme River, Turkey. Environmental Science and Pollution Research. https://doi.org/10.1007/s11356-021-15135-3

Wahab, N. A., Amri Kamarudin, M. K., Toriman, M. E., Juahir, H., Md Saad, M. H., Ata, F. M., Ghazali, A., Hassan, A. R., Abdullah, H., Maulud, K. N., Hanafiah, M. M., & Harith, H. (2019). Sedimentation and water quality deterioration problems at Terengganu Riverbasin, Terengganu, Malaysia. *Desalination and Water Treatment*, 149, 228–241. https://doi.org/10.5004/dwt.2019.23836

Walling, D. E., & Collins, A. L. (2008). The catchment sediment budget as a management tool. *Environmental Science and Policy*, 11(2), 136–143. https://doi.org/10.1016/j.envsci.2007.10.004

Wang, Y., Liu, Y., Jin, S., Sun, C., & Wei, X. (2019). Evolution of the topography of tidal flats and sandbanks along the Jiangsu coast from 1973 to 2016 observed from satellites. *ISPRS Journal of Photogrammetry and Remote Sensing*, 150, 27–43. https://doi.org/10.1016/J.ISPRSJPRS.2019.02.001

Wang, S. W., Munkhnasan, L., & Lee, W. K. (2021). Land use and land cover change detection and prediction in Bhutan's high altitude city of Thimphu, using cellular automata and Markov chain. *Environmental Challenges*, 2(November 2020). https://doi.org/10.1016/j.envc.2020.100017

WWF. (2018). Living Planet Report – 2018: Aiming higher. In Environmental Conservation (Vol. 26, Issue 4). https://www.wwf.eu/campaigns/living_planet_report_2018/

Xu, F., Jia, Y., Wang, Y., Zhang, F., Li, L., Li, Y., Ren, L., Wang, D., & Zhang, T. (2021). Does sand mining affect the remobilization of copper and zinc in sediments? – A case study of the Jialing River (China). *Environmental Research*, 200(April), 111416. https://doi.org/10.1016/j.envres.2021.111416

Xue, J., & Su, B. (2017). Significant remote sensing vegetation indices: A review of developments and applications. *Journal of Sensors*, 2017. https://doi.org/10.1155/2017/1353691

Yao, J., Zhang, D., Li, Y., Zhang, Q., & Gao, J. (2019). Quantifying the hydrodynamic impacts of cumulative sand mining on a large river-connected floodplain lake: Poyang Lake. *Journal of Hydrology*, 579(June), 124156. https://doi.org/10.1016/j.jhydrol.2019.124156

Yatoo, S. A., Sahu, P., Kalubarme, M. H., & Kansara, B. B. (2020). Monitoring land use changes and its future prospects using cellular automata simulation and artificial neural network for Ahmedabad city, India. *GeoJournal*, 5. https://doi.org/10.1007/s10708-020-10274-5

Ye, C., Butler, O. M., Chen, C., Liu, W., Du, M., & Zhang, Q. (2019). Geoderma Shifts in characteristics of the plant-soil system associated with flooding and revegetation in the riparian zone of Three Gorges Reservoir, China. *Geoderma*, 361, 114015. https://doi.org/10.1016/j.geoderma.2019.114015

Yu, P. H., & Shih, P. T. Y. (2015). A geomorphological study of Bachang Shih with airborne LiDAR. ACRS 2015 – 36th Asian Conference on Remote Sensing: Fostering Resilient Growth in Asia, Proceedings.

Zamora-Ledezma, C., Negrete-Bolagay, D., Figueroa, F., Zamora-Ledezma, E., Ni, M., Alexis, F., & Guerrero, V. H. (2021). Heavy metal water pollution: A fresh look about hazards, novel and conventional remediation methods. *Environmental Technology and Innovation*, 22, 101504. https://doi.org/10.1016/j.eti.2021.101504

Zare, M., Panagopoulos, T., & Loures, L. (2017). Simulating the impacts of future land use change on soil erosion in the Kasilian watershed, Iran. *Land Use Policy*, 67(June 2016), 558–572. https://doi.org/10.1016/j.landusepol.2017.06.028

Zhang, C., Shan, B., Zhao, Y., Song, Z., & Tang, W. (2018a). Spatial distribution, fractionation, toxicity and risk assessment of surface sediments from the Baiyangdian Lake in northern China. *Ecological Indicators*, 90(December 2017), 633–642. https://doi.org/10.1016/j.ecolind.2018.03.078

Zhang, Y., Yan, K., Ji, F., & Zhang, L. (2018b). Enhanced removal of toxic heavy metals using swarming biohybrid adsorbents. *Advanced Functional Materials*, 28(52), 1–9. https://doi.org/10.1002/adfm.201806340

17

Nutrient Dynamics in the Polar Ice Sheets and Mountain Glaciers: A Review

Monica Sharma Shamurailatpam, AL. Ramanathan, and N. Janardhana Raju

School of Environmental Sciences, Jawaharlal Nehru University, New Delhi, India

17.1 Introduction

Glaciers were considered as nutrient-poor environments with little or no life on them; however, with the advances in geoscience, they are widely recognized as unique biomes (Anesio et al. 2017) and play an important role in the biogeochemical cycling of nutrients (Jones 1999). Life on the glacier surface has been known ever since the late nineteenth century (Nordenskiöld 1872), although its role in biogeochemical cycling and in influencing the physical modifications of glaciers has been given importance only in recent times (Marshall 1996; Takeuchi et al. 2006; Anesio et al. 2009; Stibal et al. 2012; Lutz et al. 2016; Di Mauro et al. 2020; Williamson et al. 2020). Glacier surfaces, snowpack, and glacier beds are considered important areas and provide an active environment for microbial activity (Stibal et al. 2008b, 2009; Anesio et al. 2017). This microbial activity plays an important role in nutrient cycling in the glacier environment (Stibal et al. 2012). Cryoconite and bare ice in the supraglacial environment serve as a hotspot and harbor different types of microbes during the melt season (Cameron et al. 2012b). In comparison to the rest of the glacial environment, the cryoconite holes potentially promote biogeochemical cycles and facilitate storage for nutrients (Hodson et al. 2002; Webster-Brown et al. 2010; Cameron et al. 2012a). Numerous studies in the polar glaciers and ice sheets have confirmed that nutrient composition and production on the glacier surface are substantially linked with microbial productivity (Telling et al. 2012a; Lawson et al. 2014b; Musilova et al. 2017). The autotrophic microbes residing on the glacier surface and within the cryoconite holes have the ability to fix atmospheric carbon through in situ microbial gross photosynthesis activity and convert it into bioavailable dissolved organic carbon (DOC) (Tranter et al. 2004; Bhatia et al. 2013; Lawson et al. 2014b) once the liquid water is available during the melt-season (Stibal et al. 2008a). The nutrients produced in the supraglacial habitats have the potential to get further transformed by the heterotrophs (Stibal et al. 2008b; Cameron et al. 2012a; Telling et al. 2012b; Bagshaw et al. 2013; Bhatia et al. 2013; Lawson et al. 2014b; Wadham et al. 2016). As the melting advances, these nutrients may wash away along with the glacier meltwater flow and

Hydrogeochemistry of Aquatic Ecosystems, First Edition. Edited by Sughosh Madhav, Virendra Bahadur Singh, Manoj Kumar, and Sandeep Singh.

are exported to the downstream ecosystem (Hawkings et al. 2014, 2016; Lawson et al. 2014a; Musilova et al. 2017). Several studies in the Himalayan cryosphere also reported an increase in solute concentration in the meltwater with increased discharge (Sharma et al. 2013; Singh et al. 2016, 2020; Singh and Ramanathan 2018), which is potentially due to the advancement in climate, leading to the exposure of rocks to weathering and further hydrogeochemical processes (Singh et al. 2018). Thus, glaciers and ice sheets can be important contributors of solute and nutrients to subglacial and downstream marine ecosystems. Also, the high physical erosion nature of glaciers during the melt season induces biogeochemical weathering of rocks from glacier surface and the subglacial bed, exporting rock-originated nutrients like phosphorous to the downstream (Hawkings et al. 2016).

Rapid shrinkage of glaciers around the globe due to changes in climatic factors, such as a change in temperature and precipitation, has resulted in increased runoff discharge along with a shift in nutrients and solutes exports, resulting in global changes to the downstream systems (Bliss et al. 2014; Zhang et al. 2018; Boix Canadell et al. 2019). An outline of the glaciers and ice sheets from the polar and mountain regions around the world from the Randolph Glacier Inventory (RGI) (Pfeffer et al. 2014; RGI Consortium 2017) is given in Figure 17.1, and the details of the glaciers can be found in the Pfeffer et al. (2014).

Due to the rapid climate change of the twenty-first century, most studies have been focusing on the storage and export of nutrients from ice sheets and mountain glaciers (Bhatia et al. 2013; Lawson et al. 2014b; Hood et al. 2015; Hawkings et al. 2016; Krishna et al. 2016; Liu et al. 2016; Dubnick et al. 2017; Musilova et al. 2017; Li et al. 2018; Zhang et al. 2018; Gao et al. 2020), as biogeochemical changes due to the global warming can have a significant impact on the nutrient dynamics (Hawkings et al. 2015; Hood et al. 2015; Liu et al. 2016). However, the glacial response under the current climate change may have significant impact on both the polar ice sheets and mountain glaciers, and may respond differently to this climatic variation. This chapter has reviewed the various published data from the polar ice sheets and mountain glaciers

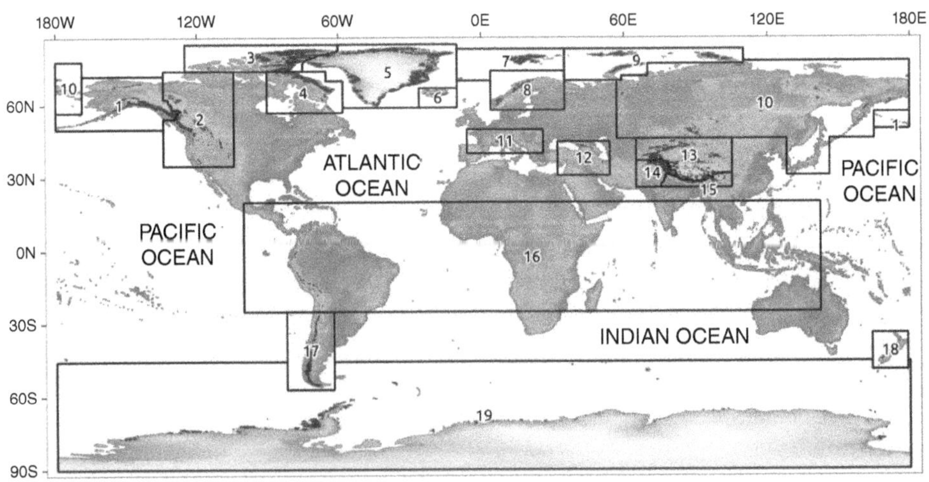

Figure 17.1 The Randolph Glacier Inventory (RGI): a digital outline of all the glaciers around the world. *Source:* Adapted from Pfeffer et al. (2014).

around the world, to understand the nutrient dynamics concerning its storage, transformation, and export in these regions.

17.2 Nutrient Dynamics: Transformation, Storage, and Export in the Glacial Environment

17.2.1 Dissolved Organic Carbon

A basic overview of nutrient cycling in the supraglacial environment is shown in Figure 17.2. The primary source of organic carbon (OC) in the supraglacial environment is potentially either from wind-deposited materials or microbial colonies surviving in the snowpack, cryoconite holes, and meltwater. Autotrophs residing on the glacier surface (in both cryoconite and bare ice) can directly fix the carbon dioxide from the atmosphere through photosynthesis and convert it into carbohydrates and part of it is deposited as biomass on the supraglacial environment (Cook et al. 2012; Cook 2016). This carbohydrates synthesis by the autotrophs provides the energy source for the other heterotroph microbes living in the supraglacial environment. The heterotrophs metabolize carbon and return it to the atmosphere in the form of carbon dioxide or remineralized to form dissolved inorganic carbon (DIC) (Figure 17.2).

The cryoconite holes on glacier surface can retain the sediment for several months or years depending on the availability of nutrient supplies for carrying out photosynthesis and the stability of cryoconite holes from redistribution of the sediment biomass, allowing prolonged periods of net carbon fixation on the glacial ice surface (Hodson et al. 2007; Anesio et al. 2009). Export of the cryoconite sediment biomass is crucial for microbial dynamics in the supraglacial and subglacial environments (Hodson et al. 2008). The OC synthesized from the supraglacial environment is distributed to the glacial surroundings and may enter the subglacial system through meltwater channels, thus providing substrate for

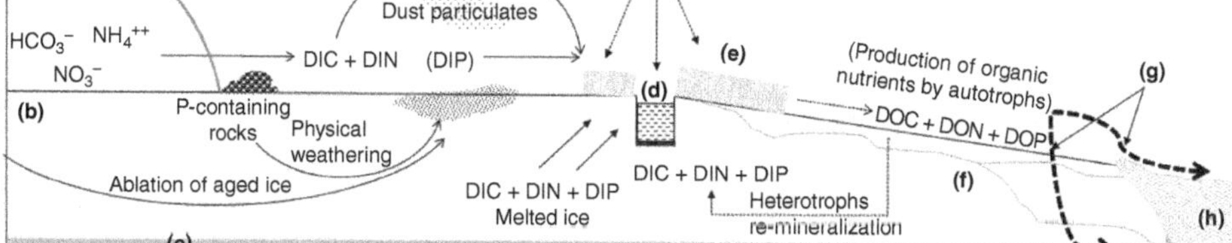

Figure 17.2 The supraglacial environment of a typical glacier showing the potential nutrient (C, N, and P) sources and its transformation on the glacier surface. *Source:* Adapted from Holland et al. (2019).

metabolism of microbial communities residing in this area, a potential source of CO_2 and acidity (or protons) for driving subglacial chemical weathering (Montross et al. 2013; Singh and Ramanathan 2017). This OC from the glacial surface may be subjected to changes in the subglacial microbial processes (either with consumption or synthesis) that may have significant implications for nutrients and potential pollutants absorption properties, thereby changing its bioavailability in the downstream aquatic environment (Hamilton et al. 2013; Anesio et al. 2017; Ren et al. 2019). Study of fluorescence characteristics of DOC in three glacial systems with varying hydrological and thermal regimes in the Mc Murdo Dry Valleys of Antarctica by Barker et al. (2006), attempted to study the characteristics of DOC in the supraglacial and subglacial meltwater system. The study highlights the different fluorescence characteristics of DOC originating in different glaciers and within the same environment to identify its dynamics and abundance. For example, in the study, Jon Evans had microbial originated DOC that dominate the supraglacial and the subglacial environments; however, with melt season advanced, the fluorescence characteristics shifted toward a more recalcitrant organic matter. Different fluorescence spectrum was observed for Outre glaciers, due to differences in the distribution and channelizing system in the glacier bed, while in Victoria Glaciers, the DOC contents and characteristics were not much different for the supraglacial and subglacial environments, indicating the lack of subglacial originated OC (Barker et al. 2006). Therefore, the transformation of DOC is duly dependent on the glacier meltwater routing and channelization during the melt season, and residence time of meltwater in the subglacial environment (Tranter et al. 1992; Barker et al. 2006; Dubnick et al. 2017).

An overview of the DOC concentration, carbon storage, flux, and yield from some of the published data from glaciers all around the world is given in Table 17.1. Hood et al. (2015) estimated that the global glacier ice storage of OC is roughly 6 Pg with most of the DOC repository in Antarctica (4.19 ± 2.78 Pg, 93%) followed by the Greenland ice sheet (0.22 ± 0.06 Pg, 5%) and mountain glaciers (0.07 ± 0.01 Pg, 2%). In the same study, Hood et al. (2015) reported that the majority of DOC concentrations were potentially on the glacial surface rather than in the englacial environment for ice sheet and mountain glaciers, with the exception of Antarctica ice sheet where the estimate of DOC storage including englacial ice data would be ~8.4 Pg, nearly doubling the global estimation. Since Hood et al. (2015) considered only one sampling site from Asia's mountain glaciers, Liu et al. (2016) re-estimated the storage of DOC from Chinese glaciers considering four glaciers from the different regions in western China. The study found that DOC storage in the Chinese mountain glaciers (3960 Gg C) account for ~6% of the total DOC in mountain glaciers (7000 Gg C).

DOC from mountain glaciers is an important source of bioavailable carbon in the downstream aquatic ecosystem and its impact is likely to increase due to the higher melting rate of glaciers (Feng et al. 2018). Ice sheets and glaciers transfer enormous amounts of labile DOC to downstream aquatic ecosystems; however, recent researches suggest that the release of DOC from mountain glaciers may be underestimated. For instance, the study by Li et al. (2018) proposed that the annual release of DOC from mountain glaciers is ~0.78 Tg C/year, which is 1.4 times higher than previously estimated by Hood et al. (2015). With the warming of Earth's surface due to global change, 15 Tg of glacial DOC is anticipated to be exported to the coastal seas by 2050 (Hood et al. 2015). Increased glacier melting will result in more nutrients being exported to the downstream ecosystem thereby increasing production in a nutrient-deficient environment.

Table 17.1 The area and volume, the concentration of DOC, and the yield estimation from the mountain glaciers and ice sheets around the globe.

Glacier names	[a]Area (km^2)	[b]Volume of ice (km^3)	Runoff (Gt/a)	Sampling period	Snow	Snow pit	Meltwater	Storage	Export	Yield ($kg\ km^2/a$)	Study by
					DOC (μgC/L)			DOC (Gg C/a)			
Arctic											
Icelandic Glacier	11,060	3,434	51	Jul 2016			525	1,532	8	700	(Chifflard et al. 2019)
Werenskioldbreen Glacier	27			2006			3400				(Stibal et al. 2008a)
Austre Brøggerbreen Glacier	6			Aug 2012			880				(Zhu et al. 2016)
Salvard and Jan Mayen	33,922	6,700	73	May–Sept 2013			2140	12,187	156	4599	(Li et al. 2018)
Arctic Canada North	145,767	24,351	212	May–Sept 2013			520	10,763	110	755	(Li et al. 2018)
John Evans Glacier	162.8	41		Jun–Jul 2001			230				(Barker et al. 2006)
John Evans Glacier	162.8	41		Jun–Aug 2002			800				(Bhatia et al. 2006)
Alaska											
Barrow				Apr–May 2017	170						(Zhang et al. 2020)
Across				Apr 2017	300						(Zhang et al. 2020)
Juneau Icefield	3800			May 2013	200						(Fellman et al. 2015)
Herbert Glaciers	24 –152			May–Oct 2012			230				(Fellman et al. 2014)
Mendenhall				May 2010		190	380				(Stubbins et al. 2012)
Southeast Alaska drainage basin	188,829	20,400	430.1	Model (1981–2010)	–	–	–		1170		(Edwards et al. 2021)
Gulf of Alaska	75,300		410	Jul 2008			1227		130	1726	(Hood et al. 2009)
European Alps											
French Alps	550		9	Sept 2012	110	300			0.003		(Legrand et al. 2013)

(*Continued*)

Table 17.1 (Continued)

Glacier names	[a]Area (km^2)	[b]Volume of ice (km^3)	Runoff (Gt/a)	Sampling period	Snow DOC (μgC/L)	Snow pit DOC (μgC/L)	Meltwater DOC (μgC/L)	Storage DOC (Gg C/a)	Export DOC (Gg C/a)	Yield (kg km^2/a)	Study by
Swiss Alps	724	74	1.0	Oct 2016–Sep 2017			145		0.06	80	(Boix Canadell et al. 2019)
Italian Alps											
(RGI 2017) European Alps	2092	118	9	Jul–Aug 2010		140	270	14	0.34	163	(Singer et al. 2012)
European Alps	2063	118	9	May–Sept 2013			310		2.8	1357	(Li et al. 2018)
Asian Mountain Glaciers											
Glacier No. 1				Aug 2014; May–Oct 2013	350	520					(Liu et al. 2016; Li et al. 2018)
Altai Mountains	7367.52	719		Nov 2016–Apr 2017	1010	710		434			(Zhang et al. 2019)
Musidao Altai				Aug 2014		760					(Liu et al. 2016)
LHG Glacier No. 12	20.4			Aug 2014		660					(Liu et al. 2016)
XDKMD				May–Sept 2013	1170						(Li et al. 2018)
XDKMD				Aug 2014		640					(Liu et al. 2016)
Chinese Glaciers	51,766	4,494		Aug 2014		1006		3,960	15.4	1439	(Liu et al. 2016)
Northern Tibetan Plateau				Aug 2006	380	220					(Hu et al. 2018)
Northern Tibetan Plateau	20.4			Jul–Aug 2015	229	332	238		0.192	9412	(Yan et al. 2016)
Southeast Tibetan Plateau	10,699		109	Jun 2015	160	160			1.96	183	(Zhang et al. 2018)

Tibetan Plateau	49,873	56		2013–2014			180		10.15		(Li et al. 2018)
Tibetan Plateau	49,873	67		Mean	-	332	193				(Xie et al. 2006; Yan et al. 2016)
Mountain Glacier Asia	121,694	14,721	359	May–Sept 2013			540	8,800–13,800	194	1594	(Li et al. 2018)
Mountain Glaciers	726,792		1167						786		(Li et al. 2018)
Mountain Glaciers	726,792		1430	Mean	490	–	370	70,000	580	798	(Hood et al. 2015)
Greenland Ice Sheet											
Greenland Ice Sheet (Periphery)	89,721			Summer 2006		485					(Hagler et al. 2007)
Greenland Ice Sheet (Periphery)	89,721			May–Aug 2012	60						(Musilova et al. 2017)
Greenland Ice Sheet	1,711,279	17,865	963	Mean	200		430	220,000	216	120	(Hood et al. 2015)
Antarctica											
Taylor Valley	84.4										(Bagshaw et al. 2007)
Joyce Glacier	90			Jan 2010; Dec 2008	240	–	654				(Bagshaw et al. 2013; Barker et al. 2013)
Dry Valleys	4800			Mean	460						(Hood et al. 2015)
Antarctica	12,295,000	59,158	1,621	Mean	983	270	510	4,190,000	243	20	(Hood et al. 2015; Nowak et al. 2018)

[a] (Farinotti et al. 2009; Pfeffer et al. 2014)
[b] (Radić and Hock 2010; Grinsted 2013)

Thousands of streams and rivers drain in southeast Alaska's coastal zone, which is one of the wettest, carbon-rich, and topographically diverse areas in North America. Identifying the timing, position, and source of water and DOC may allow understanding the impact of glacial runoff on the nearby marine environment. Previous studies by Hood et al. (2009) estimated that out of 130 Gg C exported annually, 77% (100 Gg C) of the DOC were highly bioavailable. The export DOC is usually unaltered and very labile to the marine heterotrophic organisms due to the glacier's proximity to the coastal environment and huge sediment volumes discharged (Hood et al. 2009). A recent study by Edwards et al. (2021) in the vast areas of the southeast Alaska drainage basin, also estimated that most of the DOC delivered to coastal water were in labile form (23–66%), which might alter the metabolism and productivity of the marine environment. The complex physiological structure of southeast Alaska, along with the variation and distribution of water and DOC input, large complex coastline, bounded estuarine volume, and nearshore shoreline tides, can have a significant influence on the DOC metabolism in this region (Edwards et al. 2021). Nevertheless, the export of DOC from mountain glaciers is widely variable due to a lack of data and regional variability in the magnitude of DOC (Li et al. 2018).

According to Hood et al. (2015), the annual export of DOC from global glacier runoff is 1.04 0.18 Tg C/year, with mountain glaciers contributing to more than half (0.58 0.07 Tg C/year, 56%), followed by Antarctica (0.24 0.16 Tg C/year, 23%), and Greenland ice sheet (0.22 0.04 Tg C/year, 21%). Over the past century, the global melting rate of mountain glaciers has accelerated dramatically, leading to an increase in the amount of melting water flowing into lakes and rivers. The DOC release from the Muz Taw glacier is estimated to be 3.47–18.5 t C/year. The mass of DOC storage in the Asian mountain glacier is estimated to be 8.8–13.8 $\times$ 10^6 t C (Gao et al. 2020) accounting for 11–20% of the total DOC storage in mountain glaciers 70 $\times$ 10^6 t C (Hood et al. 2015). The deposition of DOC on the glacier surface of Central Asia is influenced by the sources from anthropogenic inputs, wild biomass burning emissions, and dust inputs from local and long-distance transportation (Gao et al. 2020). Low C/N ratio of surface snow is widely reported in the Himalaya and Tibetan plateau (Feng et al. 2016; Gao et al. 2020), indicating its highly bioavailable form of DOC for microbial metabolism to the downstream ecosystem (Antony et al. 2014; Gao et al. 2019).

17.2.2 Nitrogen

In the nitrogen (N) cycle, ice and snow play as the interface between underlying Earth or ocean surface and the atmosphere (Figure 17.2). Inorganic nitrogen compounds (nitrate and ammonium) in the glacier are either directly deposited as dry/aerosols on the surface of snow or dissolved in the rainfall. Hodson et al. (2005) in their nutrient budget in the Arctic glacier found that snow and rain as sources of NO_3^- and NH_4^+ in the catchment area and runoff as sink and exports of N-compounds. It is also noticed in the study that NO_3^- and NH_4^+ are constantly lost from the snow, which could be because of its consumption or transformation or leaching from the snowpack as melting advances (Brooks et al. 1999). NH_4^+ are the least conservative species of N and are constantly lost during the early ablation season; however, microorganisms in the cryoconite are able to uptake the NH_4^+ from the snowmelt (Hodson et al. 2005). Transformation of N species to NO_3^- through the utilization of NH_4^+ could be one of the explanations for enriched NO_3^- in the glacial environment (Hodson et al. 2005). Various studies have reported rapid transformation of NH_4^+ through microbial activity or being oxidized and converted into NO_3^- in the snowpack (Hodson et al. 2005; Wynn et al., 2007; Telling et al., 2011, 2012b).

In the review by Wolff (2013), NO_3^- and NH_4^+ dominate in polar ice and their typical concentration is in the range of 0.01–0.93 and 0.0003–0.18 mg/L, respectively. Based on this concentration, Wolff (2013) established a total inventory of N-compounds in the polar ice sheet; he estimated the concentration of NO_3^- and NH_4^+ to be 30 and 10 Tg N, respectively, in the Greenland ice sheet and to be 200 and 20 Tg N in Antarctica ice sheet, with nitrate being the dominant species in both regions. The study also estimated total deposition of N-compounds from the atmosphere as NO_3^- and NH_4^+ to be 0.015 and 0.003 Tg N/a, respectively, in Greenland and 0.02 and 0.002 Tg N/a in Antarctica ice sheet with a total deposition of 0.02 Tg N/a.

Some of the published data on nitrogen concentration in various glacier environments are presented in Table 17.2. In the polar ice sheets, snow potentially plays a significant source of N, with the majority of nitrogen in the form of nitrate (Hodson 2006). Greater concentrations of N are found in the cryoconite holes compared to snow samples in the same sampling location (Table 17.2). Many types of research have proven that microbes in the cryoconite holes have the ability to utilize aqueous N and are also able to fix atmospheric N (Telling et al. 2011; MacDonell and Fitzsimons 2012; Segawa et al. 2014) which can potentially lead to nutrient enrichment in the cryoconite holes (Macdonell and Fitzsimons 2012).

From the research carried out so far, Asian mountain glaciers have higher inorganic N concentrations than Antarctica and Greenland ice sheets and to some extent from the Arctic glaciers. Inorganic nitrogen in the Asian mountain regions is largely originated from long distance transportation of anthropogenic pollutants, agricultural and biomass burning emissions (Shrestha et al. 1997; Xu et al. 2013; Niu et al. 2014; Zhang et al. 2019; Gao et al. 2020). Data on dissolved organic nitrogen (DON) are still limited; however, a recent study by Gao et al. (2020) from Central Asia reported DON concentration in snow surface and snow pit of 0.31 and 0.18 mg/L, respectively, which is comparable to those reported from the Tibetan Plateau but significantly higher than the Antarctic snow (Table 17.2). In the high mountain regions of Asia, there is large uncertainty in estimating the multiple air pollutants being deposited through the wind from arid and semiarid regions (Xu et al. 2013); therefore, it is quite challenging to study the exact source of these N species in this region.

From the experiment performed by Telling et al. (2012b), it was found that microorganisms on the glacier surface can fix nitrogen when alternative N sources become scarce or unavailable. The study also reported the extent of nitrogen exchange in Arctic glaciers has been affected since the preindustrial time indicating the anthropogenic influence on this fragile system.

Hodson et al. (2005) and Telling et al. (2011, 2012b) have provided in-depth analysis of nitrogen budgets and cycling in the Arctic glacier environment. The study by Wolff (2013) estimated the total storage of N in the polar ice sheets to be 260 Tg N with a larger fraction of N stored in the Antarctica ice sheet (220 Tg N, >80%) owing to its vast land area. The TDN flux from Greenland ice sheet is estimated to be ~142 t/a with an estimated yield of 236 kg/km^2, which is a much higher yield than that reported from the Arctic rivers (36–81 kg/km^2). In a recent study by Gao et al. (2020), the total storage is estimated to

Table 17.2 Total dissolved nitrogen (TDN), dissolved inorganic nitrogen (DIN), and dissolved organic nitrogen (DON) from different glaciers by various researchers.

Study area	Sample types	TDN (mg/L)	DIN (mg/L)	DON (mg/L)	References
Muz Taw Glacier, Central Asia	Surface snow	0.62	0.32	0.31	(Gao et al. 2020)
	Snowpit	0.31	0.13	0.18	
Snow cover, Central Asia	Snow cover	0.36			(Zhang et al. 2019)
Santopath, Himalayan glacier	ice		0.04		(Panicker et al. 2019)
Dokriana glacier	Snowpit (pre-monsoon 2013)		3.02		(Sundriyal et al. 2018)
	Snowpit (pre-monsoon 2015)		2.12		
Chhota Shigri, Himalaya	Snow		0.83		(Nijampurkar et al. 1993)
	Surface ice		0.43		
	Meltwater		0.69		
Gangotri glacier	Proglacial meltwater		0.44		(Kumar et al. 2009)
	Ice		0.02		
	Snow		0.05		
Northwest China	Meltwater (2006)		1.12		(Feng et al. 2012)
	Meltwater (2007)		0.98		
Tienshan	Snowpit		0.25		(Zhang et al. 2016)
Tienshan	Cryoconite		8.69		(Segawa et al. 2014)
Northern TP	Snowpit		0.21		(Zhang et al. 2016)
Central TP	Surface snow	0.32	0.18	0.13	(Hu et al. 2020)
Southeast TP	Surface snow		0.21		(Niu et al. 2016)
Arctic regions	Cryoconite debris	2.2–3.8	0.65–18.2	1–3.8	(Telling et al. 2011)
John Evans, Arctic	Spring snow		0.10		(Sharp et al. 2002)
	Winter snow		0.08		
Greenland	Surface snow	0.045	0.02	0.026	(Wadham et al. 2016)
Canada glacier, Antarctica	Glacier ice		0.02		(Bagshaw et al. 2007)
	Cryoconite		0.09		
Antarctica	Snow		0.07		(Dubnick et al. 2017)

be 2.8–4.4 Tg N in Asian mountain glaciers, with an annual export of 0.11 Tg N to the downstream aquatic environment which is of much magnitude compared to the deposition rate in polar ice sheets (0.02 Tg N/a).

Various factors affecting the export of N and other solutes could be the precipitation rate, for instance, rocky glaciers of Alps which flushes the microbial active debris following permafrost breakdown (Barnes et al. 2014) increased flush of microbial active sediment can increase N export from this glacier during the discharge season (Colombo et al. 2018). The increase of N export to downstream ecosystems has been reported by various research, for instance, glacial-fed lakes are found to have nitrate concentrations several times higher as compared to the lakes that are fed by only snow melting (Saros et al. 2010). The Asian mountains have been rapidly shrinking since the past few decades (Wang et al. 2011, 2014; Azam et al. 2012, 2014; Bolch et al. 2012; Yao et al. 2012; Wagnon et al. 2013; Mandal et al. 2016; Milner et al. 2017; Li et al. 2019); therefore N release from these regions can potentially influence the downstream environment (Gao et al. 2020). Consequently, many such nutrient-poor downstream habitats are more vulnerable to the effects of changing nutrient dynamics, climate change, and loss of biodiversity compared to the terrestrial and marine ecosystems.

17.2.3 Phosphorus

Phosphorus (P) is an important macronutrient with a very low concentration in the supraglacial environment and is often a limiting nutrient in the supraglacial microbial communities (Stibal and Tranter 2007; Stibal et al. 2008b). The source of P in the supraglacial environment is primarily derived from the crustal rock of the glacier surrounding (inorganic rock-derivedP) and aeolian inputs (both biologically derived P and rock-derived P) (Stibal et al. 2008a) (Figure 17.2). The concentration of soluble reactive P (SRP) in the supraglacial environment is typically <0.1 μM (Stibal et al. 2009). Therefore, the study of P cycling in the glacial environment is very limited as it is difficult to quantify P at low concentrations (Stibal et al. 2008a). In a nutrient-poor environment, like a glacial system, dissolved organic phosphate (DOP) often dominates the P pool (Karl and Tien 1997), therefore it is a critical step for the microbes to convert organic P to inorganic P in order to regulate the P availability which otherwise can ultimately lead to the nutrient limitation (Ridal and Moore 1992). The organic P fraction constitutes the majority of total P in the glacier surface of polar regions (Stibal et al. 2008a, 2008b; Bagshaw et al. 2013; Holland et al. 2019), with as high as ~90% of the total P in Svalbard glacier (Stibal et al. 2008a). The transformation of P begins with the microbes taking up and breaking down the complex P organic compounds into organic P and incorporation in the intercellular P-molecules in the living cell; the organic P thus produced is mostly in the bioavailable form depending on the molecular structure (Karl 2014). The presence of phosphatase activity in the cryoconite holes from the Arctic (Stibal et al. 2009) and Antarctic (Foreman et al. 2007) confirmed the microbial transformation of P from complex organic P molecules to inorganic P, thus facilitating the recycling of P. When compared to supraglacial snow and glacier ice, inorganic P content in the cryoconite holes water is generally higher (Bagshaw et al. 2013; Dubnick et al. 2017) (Table 17.3), enrichment of inorganic P in the cryoconite holes may be due to the assimilation of inorganic P through the microbial production of organic P (Hodson et al. 2005). Thus it represents an important

Table 17.3 Phosphorus [SRP, DOP, Particulate P (PP), and Total Phosphorus (TP)] concentration from various glaciers.

Glacier names	Sample type	SRP (µg/L)	DOP (µg/L)	PO_4^{3-} (µg/L)	P (mg/g)	Source
Werenskioldbreen	Supraglacial stream	0.91 ± 0.80	5.9			(Stibal et al. 2008a)
	Cryoconite holes	0.75 ± 0.47	6.1			
	Cryoconite debris				2.2 ± 0.53	(Stibal et al. 2008b)
	Local rock debris				0.7–1.1	
	Runoff	0.8 ± 0.83	5.7			(Stibal et al. 2008a)
Austre Broggerbreen	Proglacial sediment				0.4	(Hodson et al. 2004)
	Cryoconite hole water	3.80				(Anesio et al. 2007)
Commonwealth	Glacier ice			BDL		(Bagshaw et al. 2013)
	Cryoconite hole water	6.01		6.01		
Canada glacier	Glacier ice			BDL		
	Cryoconite hole water	3.2		BDL		(Bagshaw et al. 2013), (Telling et al. 2014)
	Suspended sediment				0.46	(Bagshaw et al. 2013)
Taylor glacier	Supraglacial stream	5.7				Mueller et al. (2001)
	Suspended sediment				0.47	(Bagshaw et al. 2013)
Greenland glacier	Suspended sediment				0.85 -1.21	(Hawkings et al. 2016)
	Runoff	7.28				
Swiss Alps	Suspended sediment				0.37 -0.62	(Föllmi et al. 2009)
French Alps	Suspended sediment				0.23	(Hodson et al. 2004)

source and storage for the bioavailable inorganic P to the glacial environment (Bagshaw et al. 2013), although the nutrient concentration in cryoconite is highly variable (Dubnick et al. 2017). The variability of P content in the cryoconite holes could be due to the flushing of meltwater and hydrological connectivity within the glacier surface during melting season (Bagshaw et al. 2011; Dubnick et al. 2017), that might affect the stability of cryoconite holes. Excess organic P was observed in the cryoconite debris compared to the local rock debris in the Arctic glacier, which Stibal et al. (2008b) explained it as inputs of organic P due to the additional sources like allochthones or from other biomass sources. A recent study by McCutcheon et al. (2021) also observed that heterotrophs like glacier ice algae biomass on the glacial surface can potentially convert (through bio-weathering or algal biomass accumulation) the rock minerals containing P to its organic form.

The study by Darcy et al. (2018) has shed more light on phosphate limitation in the cold arid environment, concluding that the addition of P can accelerate the succession of biota in these cold and fragile ecosystems. Further research by McCutcheon et al. (2021) also found a relationship between the mineral P in glacial surfaces and biomass of ice algae in the Greenland ice sheet, thereby increasing the algal growth and thus playing a secondary control on ice sheet melting. Thus, studying the transformation, storage, and export of P is important to understand its bioavailability and potential impact on the downstream system. From the literature, no research has so far been done on the total storage of P in the ice sheets and glaciers; however, Hawkings et al. (2016) have estimated P export from the Greenland ice sheet. The study by Hawkings et al. (2016) found that the ice sheet is a hotspot for phosphorus export, with SRP concentrations up to 0.35 M, which are equivalent to those seen in Arctic rivers. The ice sheet is also expected to contribute 15% of total bioavailable phosphorus input to Arctic water (11 Gg/year) and to dominate overall phosphorus input (408 Gg/year) according to the study. Over a 3000 kg P/km^2 of phosphorous yield is reported from the Leverette glacier in the same study, which is in higher concentration than those reported from some of the major riverine catchments of the world. With the increase in melt rate due to the projected climate change, like other nutrients (including DOC and N), the export of phosphorus is likely to increase with the quantity of ice melts (Hawkings et al. 2015, 2016; Dubnick et al. 2017). Studies on P biogeochemical cycles are still very limited and most of the studies have been carried out in the polar region (Downes et al. 1986; Hodson et al. 2004, 2005; Stibal et al. 2008b, 2009); however, despite its importance, studies in the mountain glaciers are yet to be explored.

17.3 Factors Influencing Nutrients Concentration and Flux

Several factors have been identified as being responsible for the export/production load of nutrients from the polar ice sheet and mountain glaciers in various studies (Mindl et al. 2007; Liu et al. 2008; Brown et al. 2011; Corriveau et al. 2013; Hotaling et al. 2021). The rate of glaciers and ice sheet mass loss has been accelerated since the last few decades (Paul et al. 2007; Kang et al. 2015; Mandal et al. 2016; Jin et al. 2017; Straneo et al. 2019), raising concern for the glacier's exports and storage of nutrients. The factors responsible for this variation can be broadly divided into the following, and these factors will be discussed in this section.

17.3.1 Climatically Driven Export of Nutrients

Climatic parameters like temperature and precipitation have a significant effect in controlling the annual glacier mass loss (Bliss et al. 2014), thereby affecting the nutrient exports through meltwater discharge to the downstream (Hawkings et al. 2015; Krishna et al. 2016). An increase in glacier surface temperature in addition to the bioavailable nutrients

(Buckeridge and Grogan 2008; Bergstrom et al. 2020) can accelerate the growth of microbes, thereby absorbing more solar radiation and substantial contributions to the melting of glacial surface (Feng et al. 2016; Hu et al. 2018; Zhang et al. 2019, 2020; Gao et al. 2020). The presence of DOC in snow can affect the surface albedo of glaciers especially in the high Himalayan mountains, where concentration is often higher compared to the rest of glaciers around the globe (Liu et al. 2016; Yan et al. 2016; Li et al. 2018; Zhang et al. 2019). Despite the growing interest in the influence of DOC on its light-absorbing properties and its potential to alter snowmelt-fed river ecosystems, only a few researchers have considered DOC concentration in the Asian mountain regions (Hu et al. 2018; Zhang et al. 2019).

Glacier meltwater runoff serves as a primary source of nutrients to the downstream primary productivity. It is projected that most of the glaciers around the globe will lose 75% of their current ice volume by the end of this century (Bliss et al. 2014). According to Hood et al. (2015), changes in surface mass balance and glacier dynamics will have a considerable influence on glacier OC release in the future, owing to the increased ice loss from the glaciers and ice sheets. Likewise, mountain glaciers will continue to lose volume, albeit the extent of this trend may vary greatly from region to region due to the glacierization and climate regime differences. By 2050, Hood et al. (2015) estimated that a total of 15.34 Tg of DOC (an average of 0.43 Tg C/year) will be released from glacial storage, with the bulk (9.68 Tg or 63%) originating from mountain glaciers, based on the future glacier mass loss estimates. However, this estimation has a number of flaws, including the underestimation of DOC storage and exports from mountain glaciers in Asia (Liu et al. 2016; Li et al. 2018), therefore, the actual value might be substantially higher than previously thought. Furthermore, only a few glaciers were considered for the assessment of nutrient storage and export (Hood et al. 2015), and most studies were conducted on a regional basis (Liu et al. 2016), resulting in an inaccurate estimate of the actual value. As a result, it is critical to account for these uncertainties in future from the estimation of DOC storage and export, including ice sheets and glaciers in polar and mountain glaciers. Several studies have indicated that increased glacier runoff due to climate change will influence nutrient exports and have an impact on the downstream ecosystem; however, the exact impact of nutrient export on the downstream ecosystem is still largely unclear.

17.3.2 Microbial Driven Nutrients Productivity

The autotrophic microorganisms are the most significant source of supraglacial DOC. The presence of liquid water and solar light in these cryoconite holes promotes the growth of active photoautotrophic bacteria that fix inorganic carbon. Thus, cryoconite serve as an active hotspot for the heterotrophic microbial communities to change the DOC composition of both autochthonous and allochthonous carbon (Musilova et al. 2017). Building on the work of Tranter et al. (2004), Sanyal et al. (2018) explored the heterotrophic microorganisms in cryoconite holes water with different settings in Antarctica and Himalayan glacier compared their ability to metabolize organic matter present in the respective environments. Open and hydrologically connected cryoconite holes are highly diversified as they facilitate microbes and nutrients to flow into the cryoconite holes (Hodson et al. 2008) than the closed and isolated holes in the same climatic environment. Cryoconite holes in Himalayan glaciers are short-lived (Takeuchi et al. 2000), while those in Antarctica are more stable and isolated, and survive for several years to decades (Tranter et al. 2004). Different hydrological, climatological, and morphological glaciers have a significant impact on the ecology of these holes; thus, the cryoconite holes are unique to each glacier. Climatic influence along with increased microbial activity on the glacier surface can greatly alter the nutrient cycling in glaciers, affecting the export of nutrients from the glacier environment thereby affecting productivity in the nutrient-poor downstream ecosystem. Under the warming climate scenario, more microbial activity on the glacier surface and englacial and subglacial zones may also increase the nutrient production in comparison to colder climates (Liu et al. 2016).

Compared to the OC content of cryoconite debris to an estimate of yearly microbial primary production on Werenskioldbreen, a Svalbard glacier, the yearly primary production (4.3 mg C/g/year) and high OC content of the debris (1.7–4.5%, corresponding to 8500–22,000 mg C/g debris) are vastly different. Given the glacier's ablation dynamics and surface hydrology, long-term deposition of autochthonous organic matter in the Svalbard glacier is stated to be very unlikely (Stibal et al. 2008a). Rather, the bulk of biological materials is most likely to be allochthonous. A similar observation in the Himalayas is high total OC content in the cryoconite compared to the polar ice sheet; however, only ~18% were in the bioavailable form (Sanyal et al. 2018). This observation is linked with the high content of OC in snow, which is mainly from the mineral transported from dry arid areas and biomass burning deposition from the adjacent regions (Xu et al. 2013). The observation of the origin of OC in mountain regions is quite complex due to the complicated biotic processes and abiotic factors such as the influence of anthropogenic pollutants, the difference in precipitation regimes, along with the annual and seasonal influences of air masses.

Increased phytoplankton productivity in the downstream ecosystem has reportedly been linked with increased nutrient export (Arrigo and van Dijken 2015; Arrigo et al. 2017; Calleja et al. 2017; Hopwood et al. 2020). Therefore, with the warming of the Earth surface, leading to increased glacial runoff, the labile DOC and nutrients are estimated to increase with the melting glaciers. Although cryoconite holes are potentially a substantial source of OC in the glacial ecosystem, the relative roles of autochthonous microbial primary production and wind-borne allochthonous organic matter transported remain unclear.

17.4 Conclusion

The nutrient dynamic in the polar ice sheet and mountain glaciers is a complex mechanism and it varies from glaciers to glaciers within the same region as well as in-between regions. From the literature, it can be broadly concluded that the quality and quantity of nutrients exported from the glaciers and ice sheet depend on the (i) meltwater discharge rate during the ablation season, (ii) efficiency of microbial cycling on the supraglacial and biogeochemical processes in the subglacial environment, (iii) magnitude of nutrient glacial discharge, and (iv) downstream ecosystem distance from the glacier. The estimation of the storage and release of nutrients has certain uncertainties due to the lack of data of storage/release in/from the glacial and englacial environment; therefore, an improved mass balance component, an accurate estimation of the magnitude, and temporal changes of nutrient released from glaciers are much needed.

Acknowledgments

We thanked Dr Arindan Mandal for preparing Figure 17.1. We also extend our gratitude toward Dr Namrata Priya for improving the grammar and for her various suggestions.

References

Anesio, A.M., Mindl, B., Laybourn-Parry, J., Hodson, A.J., Sattler, B., 2007. Viral dynamics in cryoconite holes on a high Arctic glacier (Svalbard). *Journal of Geophysical Research: Biogeosciences* 112. https://doi.org/10.1029/2006JG000350

Anesio, A.M., Hodson, A.J., Fritz, A., Psenner, R., Sattler, B., 2009. High microbial activity on glaciers: Importance to the global carbon cycle. *Global Change Biology* 15, 955–960. https://doi.org/10.1111/j.1365-2486.2008.01758.x

Anesio, A.M., Lutz, S., Chrismas, N.A.M., Benning, L.G., 2017. The microbiome of glaciers and ice sheets. *NPJ Biofilms and Microbiomes* 3, 10. https://doi.org/10.1038/s41522-017-0019-0

Antony, R., Grannas, A.M., Willoughby, A.S., Sleighter, R.L., Thamban, M., Hatcher, P.G., 2014. Origin and sources of dissolved organic matter in snow on the east antarctic ice sheet. *Environmental Science and Technology* 48, 6151–6159. https://doi.org/10.1021/ES405246A/SUPPL_FILE/ES405246A_SI_002.XLSX

Arrigo, K.R., Dijken, G.L. van, 2015. Continued increases in Arctic Ocean primary production. *Progress in Oceanography* 136, 60–70. https://doi.org/10.1016/J.POCEAN.2015.05.002

Arrigo, K.R., Dijken, G.L. van, Castelao, R.M., Luo, H., Rennermalm, Å.K., Tedesco, M., Mote, T.L., Oliver, H., Yager, P.L., 2017. Melting glaciers stimulate large summer phytoplankton blooms in southwest Greenland waters. *Geophysical Research Letters* 44, 6278–6285. https://doi.org/10.1002/2017GL073583

Azam, M.F., Wagnon, P., Ramanathan, A., Vincent, C., Sharma, P., Arnaud, Y., Linda, A., Pottakkal, J.G., Chevallier, P., Singh, V.B., Berthier, E., 2012. From balance to imbalance: A shift in the dynamic behaviour of Chhota Shigri glacier, western Himalaya, India. *Journal of Glaciology* 58, 315–324. https://doi.org/10.3189/2012JoG11J123

Azam, M.F., Wagnon, P., Vincent, C., Ramanathan, A.L., Favier, V., Mandal, A., Pottakkal, J.G., 2014. Processes governing the mass balance of Chhota Shigri Glacier (western Himalaya, India) assessed by point-scale surface energy balance measurements. *Cryosphere* 8, 2195–2217. https://doi.org/10.5194/tc-8-2195-2014

Bagshaw, E.A., Tranter, M., Fountain, A.G., Welch, K.A., Basagic, H., Lyons, W.B., 2007. Biogeochemical evolution of cryoconite holes on Canada Glacier, Taylor Valley, Antarctica. *Journal of Geophysical Research: Biogeosciences*. https://doi.org/10.1029/2007JG000442

Bagshaw, E.A., Tranter, M., Wadham, J.L., Fountain, A.G., Mowlem, M., 2011. High-resolution monitoring reveals dissolved oxygen dynamics in an Antarctic cryoconite hole. *Hydrological Processes* 25, 2868–2877. https://doi.org/10.1002/hyp.8049

Bagshaw, E.A., Tranter, M., Fountain, A.G., Welch, K., Basagic, H.J., Lyons, W.B., Bagshaw, E.A., Tranter, M., Fountain, A.G., Welch, K., Basagic, H.J., Lyons, W.B., Interdisciplinary Journal, A., Bagshaw, E.A., Tranter, M., Fountain, A.G., Welch, K., Basagic, H.J., Berry Lyons, W., 2013. Do cryoconite holes have the potential to be significant sources of C, N, and

P to downstream depauperate ecosystems of Taylor Valley, Antarctica? *Arctic, Antarctic, and Alpine Research* 45, 440–454. https://doi.org/10.1657/1938-4246-45.4.440

Barnes, R.T., Williams M.W., Parman, J.N., et al., 2014. Thawing glacial and permafrost features contribute to nitrogen export from Green Lakes Valley, Colorado Front Range, USA. *Biogeochemistry* 117, 413–430. https://doi.org/10.1007/S10533-013-9886-5/FIGURES/8

Barker, J.D., Sharp, M.J., Fitzsimons, S.J., Turner, R.J., 2006. Abundance and dynamics of dissolved organic carbon in glacier systems. *Arctic, Antarctic, and Alpine Research* 38, 163–172. https://doi.org/10.1657/1523-0430(2006)38[163:AADODO]2.0.CO;2

Barker, J.D., Dubnick, A., Lyons, W.B., Chin, Y.-P., 2013. Changes in dissolved organic matter (DOM) fluorescence in proglacial antarctic streams. *Arctic, Antarctic, and Alpine Research* 45, 305–317. https://doi.org/10.1657/1938-4246-45.3.305

Bergstrom, A., Gooseff, M.N., Singley, J.G., Cohen, M.J., Welch, K.A., 2020. Nutrient uptake in the supraglacial stream network of an Antarctic Glacier. *Journal of Geophysical Research: Biogeosciences* 125. https://doi.org/10.1029/2020JG005679

Bhatia, M.P., Sharp, M., Foght, J.M., 2006. Distinct bacterial communities exist beneath a high Arctic polythermal glacier. *Applied and Environmental Microbiology*. https://doi.org/10.1128/aem.00595-06

Bhatia, M.P., Das, S.B., Xu, L., Charette, M.A., Wadham, J.L., Kujawinski, E.B., 2013. Organic carbon export from the Greenland ice sheet. *Geochimica et Cosmochimica Acta* 109, 329–344. https://doi.org/10.1016/J.GCA.2013.02.006

Bliss, A., Hock, R., Radić, V., 2014. Global response of glacier runoff to twenty-first century climate change. *Journal of Geophysical Research: Earth Surface* 119, 717–730. https://doi.org/10.1002/2013JF002931

Boix Canadell, M., Escoffier, N., Ulseth, A.J., Lane, S.N., Battin, T.J., 2019. Alpine glacier shrinkage drives shift in dissolved organic carbon export from quasi-chemostasis to transport limitation. *Geophysical Research Letters* 46, 8872–8881. https://doi.org/10.1029/2019GL083424

Bolch, T., Kulkarni, A., Kääb, A., Huggel, C., Paul, F., Cogley, J.G., Frey, H., Kargel, J.S., Fujita, K., Scheel, M., Bajracharya, S., Stoffel, M., 2012. The state and fate of himalayan glaciers. *Science*. https://doi.org/10.1126/science.1215828

Brooks, P.D., Williams, M.W., Williams, M., 1999. Snowpack controls on nitrogen cycling and export in seasonally snow-covered catchments. *Hydrological Processes*. https://doi.org/10.1002/(sici)1099-1085(199910)13:14/15<2177::aid-hyp850>3.0.co;2-v

Brown, L.E., Brown, L.E., Brown, L.J., Taleban, V., Gharabaghi, B., Weiss, L., Weiss, L.E., Weiss, L., 2011. Seasonal and spatial distribution patterns of atmospheric phosphorus deposition to Lake Simcoe, ON. *Journal of Great Lakes Research*. https://doi.org/10.1016/j.jglr.2011.01.004

Buckeridge, K.M., Grogan, P., 2008. Deepened snow alters soil microbial nutrient limitations in arctic birch hummock tundra. *Applied Soil Ecology*. https://doi.org/10.1016/j.apsoil.2007.12.010

Calleja, M.L., Kerhervé, P., Bourgeois, S., Kędra, M., Leynaert, A., Devred, E., Babin, M., Morata, N., 2017. Effects of increase glacier discharge on phytoplankton bloom dynamics and pelagic geochemistry in a high Arctic fjord. *Progress in Oceanography* 159, 195–210. https://doi.org/10.1016/J.POCEAN.2017.07.005

Cameron, K.A., Hodson, A.J., Osborn, A.M., 2012a. Carbon and nitrogen biogeochemical cycling potentials of supraglacial cryoconite communities. *Polar Biology* 35, 1375–1393. https://doi.org/10.1007/s00300-012-1178-3

Cameron, K.A., Hodson, A., Osborn, A.M., 2012b. Structure and diversity of bacterial, eukaryotic and archaeal communities in glacial cryoconite holes from the Arctic and the Antarctic. *FEMS Microbiology Ecology*. https://doi.org/10.1111/j.1574-6941.2011.01277.x

Chifflard, P., Fasching, C., Reiss, M., Ditzel, L., Boodoo, K.S., 2019. Dissolved and particulate organic carbon in icelandic proglacial streams: A first estimate. *Water (Switzerland)* 11(4), 748. https://doi.org/10.3390/W11040748

Colombo, N., Gruber, S., Martin, M., Malandrino, M., Magnani, A., Godone, D., Freppaz, M., Fratianni, S., Salerno, F., 2018. Rainfall as primary driver of discharge and solute export from rock glaciers: The Col d'Olen Rock Glacier in the NW Italian Alps. *Science of the Total Environment* 639, 316–330. https://doi.org/10.1016/j.scitotenv.2018.05.098

Cook, J., 2016. 3.4.11 Supraglacial Biogeochemistry. Geomorphological Techniques 1–14.

Cook, J.M., Hodson, A.J., Anesio, A.M., Hanna, E., Yallop, M., Stibal, M., Telling, J., Huybrechts, P., 2012. An improved estimate of microbially mediated carbon fluxes from the Greenland ice sheet. *Journal of Glaciology* 58, 1098–1108. https://doi.org/10.3189/2012JoG12J001

Corriveau, J., Chambers, P.A., Culp, J.M., 2013. Seasonal variation in nutrient export along streams in the Northern Great Plains. *Water Air and Soil Pollution*. https://doi.org/10.1007/s11270-013-1594-1

Darcy, J.L., Schmidt, S.K., Knelman, J.E., Cleveland, C.C., Castle, S.C., Nemergut, D.R., 2018. Phosphorus, not nitrogen, limits plants and microbial primary producers following glacial retreat. *Science Advances*. https://doi.org/10.1126/sciadv.aaq0942

Mauro, B. Di, Garzonio, R., Baccolo, G., Franzetti, A., Pittino, F., Leoni, B., Remias, D., Colombo, R., Rossini, M., 2020. Glacier algae foster ice-albedo feedback in the European Alps. *Scientific Reports* 10, 1–9. https://doi.org/10.1038/s41598-020-61762-0

Downes, M.T., Howard-Williams, C., Vincent, W.F., 1986. Sources of organic nitrogen, phosphorus and carbon in antarctic streams. *Hydrobiologia* 134, 215–225. https://doi.org/10.1007/BF00008490

Dubnick, A., Wadham, J., Tranter, M., Sharp, M., Orwin, J., Barker, J., Bagshaw, E., Fitzsimons, S., 2017. Trickle or treat: The dynamics of nutrient export from polar glaciers. *Hydrological Processes* 31, 1776–1789. https://doi.org/10.1002/hyp.11149

Edwards, R.T., D'Amore, D.V., Biles, F.E., Fellman, J.B., Hood, E.W., Trubilowicz, J.W., Floyd, W.C., 2021. Riverine dissolved organic carbon and freshwater export in the Eastern Gulf of Alaska. *Journal of Geophysical Research: Biogeosciences* 126. https://doi.org/10.1029/2020JG005725

Farinotti, D., Huss, M., Bauder, A., Funk, M., 2009. An estimate of the glacier ice volume in the Swiss Alps. *Global and Planetary Change* 68, 225–231. https://doi.org/10.1016/J.GLOPLACHA.2009.05.004

Fellman, J.B., Hood, E., Spencer, R.G.M., Stubbins, A., Raymond, P.A., 2014. Watershed Glacier coverage influences dissolved organic matter biogeochemistry in Coastal Watersheds of Southeast Alaska. *Ecosystems* 17, 1014–1025. https://doi.org/10.1007/S10021-014-9777-1

Fellman, J.B., Hood, E., Raymond, P.A., Hudson, J., Bozeman, M., Arimitsu, M., 2015. Evidence for the assimilation of ancient glacier organic carbon in a proglacial stream food web. *Limnology and Oceanography* 60, 1118–1128. https://doi.org/10.1002/LNO.10088

Feng, F., Li, Z., Jin, S., Dong, Z., Wang, F., 2012. Hydrochemical characteristics and solute dynamics of meltwater runoff of Urumqi Glacier No.1, eastern Tianshan, northwest China. *Journal of Mountain Science* 9, 472–482. https://doi.org/10.1007/S11629-012-2316-7

Feng, L., Xu, J., Kang, S., Li, X., Li, Y., Jiang, B., Shi, Q., 2016. Chemical composition of microbe-derived dissolved organic matter in cryoconite in tibetan plateau glaciers: Insights from fourier transform ion cyclotron resonance mass spectrometry analysis. *Environmental Science and Technology* 50, 13215–13223. https://doi.org/10.1021/acs.est.6b03971

Feng, L., An, Y., Xu, J., Kang, S., 2018. Characteristics and sources of dissolved organic matter in a glacier in the northern Tibetan Plateau: Differences between different snow categories. *Annals of Glaciology* 59, 31–40. https://doi.org/10.1017/AOG.2018.20

Föllmi, K.B., Hosein, R., Arn, K., Steinmann, P., 2009. Weathering and the mobility of phosphorus in the catchments and forefields of the Rhône and Oberaar glaciers, central Switzerland: Implications for the global phosphorus cycle on glacial-interglacial timescales. *Geochimica et Cosmochimica Acta* 73, 2252–2282. https://doi.org/10.1016/j.gca.2009.01.017

Foreman, C.M., Sattler, B., Mikucki, J.A., Porazinska, D.L., Priscu, J.C., 2007. Metabolic activity and diversity of cryoconites in the Taylor Valley, Antarctica. *Journal of Geophysical Research: Biogeosciences* 112. https://doi.org/10.1029/2006JG000358

Gao, T., Kang, S., Chen, R., Zhang, T., Zhang, T., Han, C., Tripathee, L., Sillanpää, M., Zhang, Y., 2019. Riverine dissolved organic carbon and its optical properties in a permafrost region of the Upper Heihe River basin in the Northern Tibetan Plateau. *Science of The Total Environment* 686, 370–381. https://doi.org/10.1016/J.SCITOTENV.2019.05.478

Gao, T., Kang, S., Zhang, Y., Sprenger, M., Wang, F., Du, W., Wang, X., Wang, X., 2020. Characterization, sources and transport of dissolved organic carbon and nitrogen from a glacier in the Central Asia. *Science of the Total Environment* 725, 138346. https://doi.org/10.1016/j.scitotenv.2020.138346

Grinsted, A., 2013. An estimate of global glacier volume. *Cryosphere* 7, 141–151. https://doi.org/10.5194/TC-7-141-2013

Hagler, G.S.W., Bergin, M.H., Smith, E.A., Dibb, J.E., Anderson, C., Steig, E.J., 2007. Particulate and water-soluble carbon measured in recent snow at Summit, Greenland. *Geophysical Research Letters* 34. https://doi.org/10.1029/2007GL030110

Hamilton, T.L., Peters, J.W., Skidmore, M.L., Boyd, E.S., 2013. Molecular evidence for an active endogenous microbiome beneath glacial ice. *The ISME Journal* 7, 1402–1412. https://doi.org/10.1038/ismej.2013.31

Hawkings, J.R., Wadham, J.L., Tranter, M., Raiswell, R., Benning, L.G., Statham, P.J., Tedstone, A., Nienow, P., Lee, K., Telling, J., 2014. Ice sheets as a significant source of highly reactive nanoparticulate iron to the oceans. *Nature Communications* 5, 1–8. https://doi.org/10.1038/ncomms4929

Hawkings, J.R., Wadham, J.L., Tranter, M., Lawson, E., Sole, A., Cowton, T., Tedstone, A.J., Bartholomew, I., Nienow, P., Chandler, D., Telling, J., 2015. The effect of warming climate on nutrient and solute export from the Greenland ice sheet. *Geochemical Perspectives Letters* 2000, 94–104. https://doi.org/10.7185/geochemlet.1510

Hawkings, J., Wadham, J., Tranter, M., Telling, J., Bagshaw, E., Beaton, A., Simmons, S.L., Chandler, D., Tedstone, A., Nienow, P., 2016. The Greenland ice sheet as a hot spot of phosphorous weathering and export in the Arctic. *Global Biogeochemical Cycles* 30, 191–210. https://doi.org/10.1002/2015GB005237.Received

Hodson, A., 2006. Biogeochemistry of snowmelt in an Antarctic glacial ecosystem. *Water Resources Research*. https://doi.org/10.1029/2005wr004311

Hodson, A., Tranter, M., Gurnell, A.M., Clark, M.J., Hagen, J.O., 2002. The hydrochemistry of Bayelva, a High Arctic proglacial stream in Svalbard. *Journal of Hydrology*. https://doi.org/10.1016/s0022-1694(01)00543-1

Hodson, A., Mumford, P., Lister, D., 2004. Suspended sediment and phosphorous in proglacial rivers: Bioavailability and potential impacts upon the P status of ice-marginal receiving waters. *Hydrological Processes* 18, 2409–2422. https://doi.org/10.1002/hyp.1471

Hodson, A., Mumford, P., Kohler, J., Wynn, P.M., 2005. The High Arctic glacial ecosystem: New insights from nutrient budgets. *Biogeochemistry*. https://doi.org/10.1007/s10533-004-0362-0

Hodson, A., Anesio, A.M., Ng, F., Watson, R., Quirk, J., Irvine-Fynn, T., Dye, A., Clark, C., Mccloy, P., Kohler, J., Sattler, B., 2007. A glacier respires: Quantifying the distribution and respiration CO_2 flux of cryoconite across an entire arctic supraglacial ecosystem. *Journal of Geophysical Research: Biogeosciences* 112. https://doi.org/10.1029/2007JG000452

Hodson, A., Anesio, A.M., Tranter, M., Fountain, A., Osborn, M., Priscu, J., Laybourn-Parry, J., Sattler, B., 2008. Glacial ecosystems. *Ecological Monographs* 78, 41–67. https://doi.org/10.1890/07-0187.1

Holland, A.T., Williamson, C.J., Sgouridis, F., Tedstone, A.J., McCutcheon, J., Cook, J.M., Poniecka, E., Yallop, M.L., Tranter, M., Anesio, A.M., 2019. Dissolved organic nutrients dominate melting surface ice of the Dark Zone (Greenland ice sheet). *Biogeosciences* 16, 3283–3296. https://doi.org/10.5194/bg-16-3283-2019

Hood, E., Fellman, J., Spencer, R.G.M., Hernes, P.J., Edwards, R., D'Amore, D., Scott, D., 2009. Glaciers as a source of ancient and labile organic matter to the marine environment. *Nature* 462, 1044–1047. https://doi.org/10.1038/nature08580

Hood, E., Battin, T.J., Fellman, J., O'neel, S., Spencer, R.G.M., 2015. Storage and release of organic carbon from glaciers and ice sheets. *Nature Geoscience*. https://doi.org/10.1038/ngeo2331

Hopwood, M.J., Carroll, D., Dunse, T., Hodson, A., Holding, J.M., Iriarte, J.L., Ribeiro, S., Achterberg, E.P., Cantoni, C., Carlson, D.F., Chierici, M., Clarke, J.S., Cozzi, S., Fransson, A., Juul-Pedersen, T., Winding, M.H.S., Meire, L., 2020. Review article: How does glacier discharge affect marine biogeochemistry and primary production in the Arctic? *Cryosphere* 14, 1347–1383. https://doi.org/10.5194/TC-14-1347-2020

Hotaling, S., Lutz, S., Dial, R.J., Anesio, A.M., Benning, L.G., Fountain, A.G., Kelley, J.L., McCutcheon, J., Skiles, S.M.K., Takeuchi, N., Hamilton, T.L., 2021. Biological albedo reduction on ice sheets, glaciers, and snowfields. *Earth-Science Reviews* 220, 103728. https://doi.org/10.1016/J.EARSCIREV.2021.103728

Hu, Z., Kang, S., Yan, F., Zhang, Y., Li, Y., Chen, P., Qin, X., Wang, K., Gao, S., Li, C., 2018. Dissolved organic carbon fractionation accelerates glacier-melting: A case study in the northern Tibetan Plateau. *Science of the Total Environment*. https://doi.org/10.1016/j.scitotenv.2018.01.265

Hu, Z., Kang, S., He, X., Yan, F., Zhang, Y., Chen, P., Li, X., Gao, S., Li, C., 2020. Carbonaceous matter in glacier at the headwaters of the Yangtze River: Concentration, sources and fractionation during the melting process. *Journal of Environmental Sciences (China)* 87, 389–397. https://doi.org/10.1016/j.jes.2019.08.001

Jin, S., Zhang, T.Y., Zou, F., 2017. Glacial density and GIA in Alaska estimated from ICESat, GPS and GRACE measurements. *Journal of Geophysical Research: Earth Surface* 122, 76–90. https://doi.org/10.1002/2016JF003926

Jones, H.G., 1999. The ecology of snow-covered systems: A brief overview of nutrient cycling and life in the cold. *Hydrological Processes* 13, 2135–2147. https://doi.org/10.1002/(sici)1099-1085(199910)13:14/15<2135::aid-hyp862>3.0.co;2-y

Kang, S., Wang, F., Morgenstern, U., Zhang, Y., Grigholm, B., Kaspari, S., Schwikowski, M., Ren, J., Yao, T., Qin, D., Mayewski, P.A., 2015. Dramatic loss of glacier accumulation area on the Tibetan Plateau revealed by ice core tritium and mercury records. *Cryosphere* 9, 1213–1222. https://doi.org/10.5194/TC-9-1213-2015

Karl, D.M., 2014. Microbially mediated transformations of phosphorus in the sea: New views of an old cycle. *Annual Review of Marine Science* 6, 279–337. https://doi.org/10.1146/annurev-marine-010213-135046

Karl, D.M., Tien, G., 1997. Temporal variability in dissolved phosphorus concentrations in the subtropical North Pacific Ocean. *Marine Chemistry* 56, 77–96. https://doi.org/10.1016/S0304-4203(96)00081-3

Krishna, M.S., Prasad, M.H.K., Rao, D.B., Viswanadham, R., Sarma, V.V.S.S., Reddy, N.P.C., 2016. Export of dissolved inorganic nutrients to the northern Indian Ocean from the Indian monsoonal rivers during discharge period. *Geochimica et Cosmochimica Acta* 172, 430–443. https://doi.org/10.1016/j.gca.2015.10.013

Kumar, K., Miral, M.S., Joshi, S., Pant, N., Joshi, V., Joshi, L.M., 2009. Solute dynamics of meltwater of Gangotri glacier, Garhwal Himalaya, India. *Environmental Geology* 58, 1151–1159. https://doi.org/10.1007/s00254-008-1592-6

Lawson, E.C., Wadham, J.L., Tranter, M., Stibal, M., Lis, G.P., Butler, C.E.H., Laybourn-Parry, J., Nienow, P., Chandler, D., Dewsbury, P., 2014a. Greenland ice sheet exports labile organic carbon to the arctic oceans. *Biogeosciences* 11, 4015–4028. https://doi.org/10.5194/BG-11-4015-2014

Legrand, M., Preunkert, S., Jourdain, B., Guilhermet, J., Faïn, X., Alekhina, I., Petit, J.R., 2013. Water-soluble organic carbon in snow and ice deposited at Alpine, Greenland, and Antarctic sites: A critical review of available data and their atmospheric relevance. *Climate of the Past* 9, 2195–2211. https://doi.org/10.5194/CP-9-2195-2013

Li, X., Ding, Y., Xu, J., He, X., Han, T., Kang, S., Wu, Q., Mika, S., Yu, Z., Li, Q., 2018. Importance of mountain glaciers as a source of dissolved organic carbon. *Journal of Geophysical Research: Earth Surface* 123, 2123–2134. https://doi.org/10.1029/2017JF004333

Li, X., Ding, Y., Liu, Q., Zhang, Y., Han, T., Jing, Z., Yu, Z., Li, Q., Liu, S., 2019. Intense chemical weathering at glacial meltwater-dominated Hailuogou basin in the southeastern Tibetan Plateau. *Water (Switzerland)* 11. https://doi.org/10.3390/w11061209

Liu, F., Parmenter, R., Brooks, P.D., Conklin, M.H., Bales, R.C., 2008. Seasonal and interannual variation of streamflow pathways and biogeochemical implications in semi-arid, forested catchments in Valles Caldera, New Mexico. *Ecohydrology*. https://doi.org/10.1002/eco.22

Liu, Y., Xu, J., Kang, S., Li, X., Li, Y., 2016. Storage of dissolved organic carbon in Chinese glaciers. *Journal of Glaciology* 62, 402–406. https://doi.org/10.1017/jog.2016.47

Lutz, S., Anesio, A.M., Raiswell, R., Edwards, A., Newton, R.J., Gill, F., Benning, L.G., 2016. The biogeography of red snow microbiomes and their role in melting arctic glaciers. *Nature Communications* 7, 11968–11968. https://doi.org/10.1038/ncomms11968

MacDonell, S.A., Fitzsimons, S.J., 2012. CRYOCONITE HOLE SYSTEM ON ANTARCTIC GLACIER Observations of cryoconite hole system processes on an Antarctic glacier Observaciones de los procesos del sistema de “cryoconite holes” sobre un glaciar en Antártica. *Revista Chilena de Historia Natural* 85, 393–407. https://doi.org/10.4067/S0716-078X2012000400003

Mandal, A., Ramanathan, A., Angchuk, T., Soheb, M., Singh, V.B., 2016. Unsteady state of glaciers (Chhota Shigri and Hamtah) and climate in Lahaul and Spiti region, western Himalayas: A review of recent mass loss. *Environmental Earth Sciences* 75. https://doi.org/10.1007/s12665-016-6023-5

Marshall, W.A., 1996. Biological particles over Antarctica. *Nature*. https://doi.org/10.1038/383680a0

McCutcheon, J., Lutz, S., Williamson, C., Cook, J.M., Tedstone, A.J., Vanderstraeten, A., Wilson, S.A., Stockdale, A., Bonneville, S., Anesio, A.M., Yallop, M.L., McQuaid, J.B., Tranter, M., Benning, L.G., 2021. Mineral phosphorus drives glacier algal blooms on the Greenland ice sheet. *Nature Communications* 12, 1–11. https://doi.org/10.1038/s41467-020-20627-w

Milner, A.M., Khamis, K., Battin, T.J., Brittain, J.E., Barrand, N.E., Füreder, L., Cauvy-Fraunié, S., Gíslason, G.M., Jacobsen, D., Hannah, D.M., Hodson, A.J., Hood, E., Lencioni, V., Olafsson, J.S., Robinson, C.T., Tranter, M., Brown, L.E., 2017. Glacier shrinkage driving global changes in downstream systems. *Proceedings of the National Academy of Sciences of the United States of America* 114, 9770–9778. https://doi.org/10.1073/pnas.1619807114/-/DCSupplemental

Mindl, B., Anesio, A.M., Meirer, K., Hodson, A.J., Laybourn-Parry, J., Sommaruga, R., Sattler, B., 2007. Factors influencing bacterial dynamics along a transect from supraglacial runoff to proglacial lakes of a high Arctic glacier [corrected]. *FEMS Microbiology Ecology* 59, 307–317. https://doi.org/10.1111/J.1574-6941.2006.00262.X

Montross, S.N., Skidmore, M., Tranter, M., Kivimäki, A.L., Parkes, R.J., 2013. A microbial driver of chemical weathering in glaciated systems. *Geology*. https://doi.org/10.1130/G33572.1

Mueller, D.R., Vincent, W.F., Pollard, W.H., Fritsen, C.H., 2001. Glacial cryoconite ecosystems: A bipolar comparison of algal communities and habitats. *Nov Hedwigia Beih* 123, 173–198.

Musilova, M., Tranter, M., Wadham, J., Telling, J., Tedstone, A., Anesio, A.M., 2017. Microbially driven export of labile organic carbon from the Greenland ice sheet. *Nature Geoscience* 10, 360–365. https://doi.org/10.1038/ngeo2920

Nijampurkar, V.N., Sarin, M.M., Rao, D.K., 1993. Chemical composition of snow and ice from Chhota Shigri glacier, Central Himalaya. *Journal of Hydrology* 151, 19–34. https://doi.org/10.1016/0022-1694(93)90246-6

Niu, H., He, Y., Lu, X., Shen, J., Du, J., Du, J., Zhang, T., Pu, T., Xin, H., Chang, L., 2014. Chemical composition of rainwater in the Yulong Snow Mountain region, *Southwestern China. Atmospheric Research*. https://doi.org/10.1016/j.atmosres.2014.03.010

Niu, H., He, Y., Kang, S., Lu, X., Shi, X., Wang, S., 2016. Chemical compositions of snow from Mt. Yulong, southeastern tibetan plateau. *Journal of Earth System Science* 125, 403–416. https://doi.org/10.1007/s12040-016-0670-5

Nordenskiöld, A.E., 1872. VI. – Account of an expedition to Greenland in the year 1870. *Geological Magazine* 9, 516–524. https://doi.org/10.1017/S0016756800466033

Nowak, A., Hodson, A., Turchyn, A. V., 2018. Spatial and temporal dynamics of dissolved organic carbon, chlorophyll, nutrients, and trace metals in maritime antarctic snow and snowmelt. *Frontiers in Earth Science* 6, 201. https://doi.org/10.3389/FEART.2018.00201/BIBTEX

Panicker, A.S., Sandeep, K., Gautam, A.S., Gandhi, N., Beig, G., Nainwal, H.C., Rao, P.S.P., Safai, P.D., Das, S., Waghmare, V., 2019. Chemical composition and isotopic signatures of ice and snow over a Himalayan Glacier (Satopanth) in India. *SN Applied Sciences* 1, 1–7. https://doi.org/10.1007/s42452-019-0966-6

Paul, F., Kääb, A., Haeberli, W., 2007. Recent glacier changes in the Alps observed by satellite: Consequences for future monitoring strategies. *Global and Planetary Change* 56, 111–122. https://doi.org/10.1016/j.gloplacha.2006.07.007

Pfeffer, W.T., Arendt, A.A., Bliss, A., Bolch, T., Cogley, J.G., Gardner, A.S., Hagen, J.-O., Hock, R., Kaser, G., Kienholz, C., Miles, E.S., Moholdt, G., Mölg, N., Paul, F., Radić, V., Rastner, P., Raup, B.H., Rich, J., Sharp, M.J., The Randolph Consortium, 2014. The randolph glacier inventory: A globally complete inventory of glaciers. *Journal of Glaciology* 60, 537–552. https://doi.org/10.3189/2014JOG13J176

Radić, V., Hock, R., 2010. Regional and global volumes of glaciers derived from statistical upscaling of glacier inventory data. *Journal of Geophysical Research: Earth Surface* 115, 1010. https://doi.org/10.1029/2009JF001373

Ren, Z., Martyniuk, N., Oleksy, I.A., Swain, A., Hotaling, S., 2019. Ecological stoichiometry of the mountain cryosphere. *Frontiers in Ecology and Evolution*. https://doi.org/10.3389/fevo.2019.00360

RGI Consortium, 2017. Randolph Glacier Inventory 6.0. https://doi.org/10.7265/n5-rgi-60

Ridal, J.J., Moore, R.M., 1992. Dissolved organic phosphorus concentrations in the northeast subarctic Pacific Ocean. *Limnology and Oceanography* 37, 1067–1075. https://doi.org/10.4319/LO.1992.37.5.1067

Sanyal, A., Antony, R., Samui, G., Thamban, M., 2018. Microbial communities and their potential for degradation of dissolved organic carbon in cryoconite hole environments of Himalaya and Antarctica. *Microbiological Research* 208, 32–42. https://doi.org/10.1016/j.micres.2018.01.004

Saros, J.E., Rose, K.C., Clow, D.W., Stephens, V.C., Nurse, A.B., Arnett, H.A., Stone, J.R., Williamson, C.E., Wolfe, A.P., 2010. Melting alpine glaciers enrich high-elevation lakes with reactive nitrogen. *Environmental Science and Technology* 44, 4891–4896. https://doi.org/10.1021/es100147j

Segawa, T., Ishii, S., Ohte, N., Akiyoshi, A., Yamada, A., Maruyama, F., Li, Z., Hongoh, Y., Takeuchi, N., 2014. The nitrogen cycle in cryoconites: Naturally occurring nitrification-denitrification granules on a glacier. *Environmental Microbiology* 16, 3250–3262. https://doi.org/10.1111/1462-2920.12543

Sharma, P., Ramanathan, A.L., Pottakkal, J., 2013. Study of solute sources and evolution of hydrogeochemical processes of the Chhota Shigri Glacier meltwaters, Himachal Himalaya, India. *Hydrological Sciences Journal* 58, 1128–1143. https://doi.org/10.1080/02626667.2013.802092

Sharp, M., Skidmore, M., Nienow, P., 2002. Seasonal and spatial variations in the chemistry of a High Arctic supraglacial snow cover. *Journal of Glaciology* 48, 149–158. https://doi.org/10.3189/172756502781831683

Shrestha, A.B., Wake, C.P., Dibb, J.E., 1997. Chemical composition of aerosol and snow in the high Himalaya during the summer monsoon season. *Atmospheric Environment* 31, 2815–2826. https://doi.org/10.1016/S1352-2310(97)00047-2

Singer, G.A., Fasching, C., Wilhelm, L., Niggemann, J., Steier, P., Dittmar, T., Battin, T.J., 2012. Biogeochemically diverse organic matter in Alpine glaciers and its downstream fate. *Nature Geoscience* 5, 710–714. https://doi.org/10.1038/ngeo1581

Singh, V.B., Ramanathan, A., 2017. Hydrogeochemistry of the Chhota Shigri glacier meltwater, Chandra basin, Himachal Pradesh, India: Solute acquisition processes, dissolved load and chemical weathering rates. *Environmental Earth Sciences* 76, 223. https://doi.org/10.1007/s12665-017-6465-4

Singh, V.B., Ramanathan, A., 2018. Suspended sediment dynamics in the meltwater of Chhota Shigri glacier, Chandra basin, Lahaul-Spiti valley, India. *Journal of Mountain Science* 15, 68–81. https://doi.org/10.1007/s11629-017-4554-1

Singh, V.B., Ramanathan, A.L., Pottakkal, J.G., 2016. Glacial runoff and transport of suspended sediment from the Chhota Shigri glacier, Western Himalaya, India. *Environmental Earth Sciences* 75, 1–13. https://doi.org/10.1007/s12665-016-5271-8

Singh, V.B., Ramanathan, A., Keshari, A.K., Ranjan, S., Kumar, N., Yadav, S.K., 2018. Climatic influence on hydrogeochemistry of meltwater draining from Chhota Shigri Glacier, Himachal Pradesh, India. *Journal of Climate Change* 4, 23–31. https://doi.org/10.3233/jcc-180003

Singh, A.T., Sharma, P., Sharma, C., Laluraj, C.M., Patel, L., Pratap, B., Oulkar, S., Thamban, M., 2020. Water discharge and suspended sediment dynamics in the Chandra River, Western Himalaya. *Journal of Earth System Science* 129. https://doi.org/10.1007/s12040-020-01455-4

Stibal, M., Tranter, M., 2007. Laboratory investigation of inorganic carbon uptake by cryoconite debris from Werenskioldbreen, Svalbard. *Journal of Geophysical Research: Biogeosciences* 112. https://doi.org/10.1029/2007JG000429

Stibal, M., Tranter, M., Benning, L.G., Řehák, J., 2008a. Microbial primary production on an Arctic glacier is insignificant in comparison with allochthonous organic carbon input. *Environmental Microbiology* 10, 2172–2178. https://doi.org/10.1111/j.1462-2920.2008.01620.x

Stibal, M., Tranter, M., Telling, J., Benning, L.G., 2008b. Speciation, phase association and potential bioavailability of phosphorus on a Svalbard glacier. *Biogeochemistry* 90, 1–13. https://doi.org/10.1007/s10533-008-9226-3

Stibal, M., Anesio, A.M., Blues, C.J.D., Tranter, M., 2009. Phosphatase activity and organic phosphorus turnover on a high arctic glacier. *Biogeosciences* 6, 913–922. https://doi.org/10.5194/bg-6-913-2009

Stibal, M., Šabacká, M., Žárský, J., 2012. Biological processes on glacier and ice sheet surfaces. *Nature Geoscience* 5, 771–774. https://doi.org/10.1038/ngeo1611

Straneo, F., Sutherland, D.A., Stearns, L., Catania, G., Heimbach, P., Moon, T., Cape, M.R., Laidre, K.L., Barber, D., Rysgaard, S., Mottram, R., Olsen, S., Hopwood, M.J., Meire, L., 2019. The case for a sustained Greenland ice sheet-ocean observing system (GrIOOS). *Frontiers in Marine Science* 6, 138. https://doi.org/10.3389/fmars.2019.00138

Stubbins, A., Hood, E., Raymond, P.A., Aiken, G.R., Sleighter, R.L., Hernes, P.J., Butman, D., Hatcher, P.G., Striegl, R.G., Schuster, P., Abdulla, H.A.N., Vermilyea, A.W., Scott, D.T., Spencer, R.G.M., 2012. Anthropogenic aerosols as a source of ancient dissolved organic matter in glaciers. *Nature Geoscience* 5, 198–201. https://doi.org/10.1038/NGEO1403

Sundriyal, S., Shukla, T., Tripathee, L., Dobhal, D.P., Tiwari, S.K., Bhan, U., 2018. Deposition of atmospheric pollutant and their chemical characterization in snow pit profile at Dokriani

Glacier, Central Himalaya. *Journal of Mountain Science* 15, 2236–2246. https://doi.org/10.1007/s11629-017-4817-x

Takeuchi, N., Kohshima, S., Yoshimura, Y., et al., 2000. Characteristics of cryoconite holes on a Himalayan glacier, Yala Glacier Central Nepal. *Bulletin of Glaciological and Research* 17, 51–59.

Takeuchi, N., Uetake, J., Fujita, K., Aizen, V.B., Nikitin, S.D., 2006. A snow algal community on Akkem glacier in the Russian Altai mountains. *Annals of Glaciology* 43, 378–384. https://doi.org/10.3189/172756406781812113

Telling, J., Anesio, A.M., Tranter, M., Irvine-Fynn, T., Hodson, A., Butler, C., Wadham, J., 2011. Nitrogen fixation on Arctic glaciers, Svalbard. *Journal of Geophysical Research: Biogeosciences* 116, 3039. https://doi.org/10.1029/2010JG001632

Telling, J., Anesio, A.M., Tranter, M., Stibal, M., Hawkings, J., Irvine-Fynn, T., Hodson, A., Butler, C., Yallop, M., Wadham, J., 2012a. Arctic glaciers. *Journal of Geophysics Research* 117, 1017. https://doi.org/10.1029/2011JG001828

Telling, J., Stibal, M., Anesio, A.M., Tranter, M., Nias, I., Cook, J., Bellas, C., Lis, G., Wadham, J.L., Sole, A., Nienow, P., Hodson, A., 2012b. Microbial nitrogen cycling on the Greenland ice sheet. *Biogeosciences* 9, 2431–2442. https://doi.org/10.5194/bg-9-2431-2012

Telling, J., Anesio, A.M., Tranter, M., Fountain, A.G., Nylen, T., Hawkings, J., Singh, V.B., Kaur, P., Musilova, M., Wadham, J.L., 2014. Spring thaw ionic pulses boost nutrient availability and microbial growth in entombed Antarctic Dry Valley cryoconite holes. *Frontiers in Microbiology* 5, 694. https://doi.org/10.3389/FMICB.2014.00694/BIBTEX

Tranter, M., Tsiouris, S., Davies, T.D., Jones, H.G., 1992. A laboratory investigation of the leaching of solute from snowpack by rainfall. *Hydrological Processes* 6, 169–178. https://doi.org/10.1002/HYP.3360060205

Tranter, M., Fountain, A.G., Fritsen, C.H., Lyons, W.B., Priscu, J.C., Statham, P.J., Welch, K.A., 2004. Extreme hydrochemical conditions in natural microcosms entombed within Antarctic ice. *Hydrological Processes*. https://doi.org/10.1002/hyp.5217

Wadham, J.L., Hawkings, J., Telling, J., Chandler, D., Alcock, J., O'donnell, E., Kaur, P., Bagshaw, E., Tranter, M., Tedstone, A., Nienow, P., 2016. Sources, cycling and export of nitrogen on the Greenland ice sheet. *Biogeosciences* 13, 6339–6352. https://doi.org/10.5194/bg-13-6339-2016

Wagnon, P., Vincent, C., Arnaud, Y., Berthier, E., Vuillermoz, E., Gruber, S., Ménégoz, M., Gilbert, A., Dumont, M., Shea, J.M., Stumm, D., Pokhrel, B.K., 2013. Seasonal and annual mass balances of Mera and Pokalde glaciers (Nepal Himalaya) since 2007. *Cryosphere* 7, 1769–1786. https://doi.org/10.5194/tc-7-1769-2013

Wang, P., Li, Z., Gao, W., 2011. Rapid shrinking of glaciers in the Middle Qilian Mountain region of Northwest China during the last ~50 years. *Journal of Earth Science*. https://doi.org/10.1007/s12583-011-0195-4

Wang, L., Li, Z., Wang, F., Edwards, R., 2014. Glacier shrinkage in the Ebinur lake basin, Tien Shan, China, during the past 40 years. *Journal of Glaciology* 60, 245–254. https://doi.org/10.3189/2014JoG13J023

Webster-Brown, J., Gall, M., Gibson, J., Wood, S., Hawes, I., 2010. The biogeochemistry of meltwater habitats in the Darwin Glacier region (80°S), Victoria Land, Antarctica. *Antarctic Science* 22, 646–661. https://doi.org/10.1017/S0954102010000787

Williamson, C.J., Cook, J., Tedstone, A., Yallop, M., McCutcheon, J., Poniecka, E., Campbell, D., Irvine-Fynn, T., McQuaid, J., Tranter, M., Perkins, R., Anesio, A., 2020. Algal photophysiology drives darkening and melt of the Greenland ice sheet. *Proceedings of the National Academy of Sciences* 117, 5694–5705. https://doi.org/10.1073/PNAS.1918412117

Wolff, E.W., 2013. Ice sheets and nitrogen. Philosophical transactions of the Royal Society of London. *Series B, Biological Sciences*. https://doi.org/10.1098/rstb.2013.0127

Wynn, P.M., Hodson, A., Heaton, T.H.E., Chenery, S., 2007. Nitrate production beneath a High Arctic glacier, Svalbard. Chemical Geology. doi:https://doi.org/10.1016/j.chemgeo.2007.06.008

Xie, C., Ding, Y., Liu, S., Chen, C., 2006. Response of meltwater runoff to air-temperature fluctuations on Keqikaer glacier, south slope of Tuomuer mountain, western China. In: *Annals of Glaciology*. pp. 275–279. https://doi.org/10.3189/172756406781812294

Xu, J., Zhang, Q., Li, X., Ge, X., Xiao, C., Ren, J., Qin, D., 2013. Dissolved organic matter and inorganic ions in a central himalayan glacier -insights into chemical composition and atmospheric sources. *Environmental Science and Technology* 47, 6181–6188. https://doi.org/10.1021/es4009882

Yan, F., Kang, S., Li, C., Zhang, Y., Qin, X., Li, Y., Zhang, X., Hu, Z., Chen, P., Li, X., Qu, B., Sillanpää, M., 2016. Concentration, sources and light absorption characteristics of dissolved organic carbon on a medium-sized valley glacier, northern Tibetan Plateau. *Cryosphere* 10, 2611–2621. https://doi.org/10.5194/TC-10-2611-2016

Yao, T., Thompson, L., Yang, W., Yu, W., Gao, Y., Guo, X., Yang, X., Duan, K., Zhao, H., Xu, B., Pu, J., Lu, A., Xiang, Y., Kattel, D.B., Joswiak, D., 2012. Different glacier status with atmospheric circulations in Tibetan Plateau and surroundings. *Nature Climate Change* 2, 663–667. https://doi.org/10.1038/NCLIMATE1580

Zhang, Y., Kang, S., Zhang, Q., Gao, T., Guo, J., Grigholm, B., Huang, J., Sillanpää, M., Li, X., Du, W., Li, Y., Ge, X., 2016. Chemical records in snowpits from high altitude glaciers in the tibetan plateau and its surroundings. *PLoS ONE* 11. https://doi.org/10.1371/journal.pone.0155232

Zhang, Y., Kang, S., Li, G., Gao, T., Chen, P., Li, X., Liu, Y., Hu, Z., Sun, S., Guo, J., Wang, K., Chen, X., Sillanpää, M., 2018. Dissolved organic carbon in glaciers of the southeastern Tibetan Plateau: Insights into concentrations and possible sources. *PLoS ONE* 13. https://doi.org/10.1371/journal.pone.0205414

Zhang, Y., Kang, S., Gao, T., Schmale, J., Liu, Y., Zhang, W., Guo, J., Du, W., Hu, Z., Cui, X., Sillanpää, M., 2019. Dissolved organic carbon in snow cover of the Chinese Altai Mountains, Central Asia: Concentrations, sources and light-absorption properties. *Science of the Total Environment* 647, 1385–1397. https://doi.org/10.1016/j.scitotenv.2018.07.417

Zhang, Y., Kang, S., Gao, T., Sprenger, M., Dou, T., Han, W., Zhang, Q., Sun, S., Du, W., Chen, P., Guo, J., Cui, X., Sillanpää, M., 2020. Dissolved organic carbon in Alaskan Arctic snow: Concentrations, light-absorption properties, and bioavailability. *Tellus, Series B: Chemical and Physical Meteorology* 72, 1–19. https://doi.org/10.1080/16000889.2020.1778968

Zhu, Z.-Y., Wu, Y., Liu, S.-M., Wenger, F., Hu, J., Zhang, J., Zhang, R.-F., 2016. Organic carbon flux and particulate organic matter composition in Arctic valley glaciers: Examples from the Bayelva River and adjacent Kongsfjorden. *Biogeosciences* 13, 975–987. https://doi.org/10.5194/bg-13-975-2016

Index

a

absolute abundance ratio 334
absorptive efficiency 112
accretion of CO_2 213
acid rain 299, 314
acidic conditions 15, 163, 165, 203
activated alumina 12, 47, 50, 112
activated carbon 12, 47, 109, 112, 122, 129, 136, 140
adaptation strategies 225
adsorption 7, 11, 15, 45, 47, 70, 88, 91, 98, 100, 103, 109, 112, 114, 122, 129, 130, 131, 134, 136, 138, 140, 141, 184, 185
adsorption method 131
adsorption processes 15, 88
aerobic respiration 252, 253
aggregate abrasion value 362
aggregate crushing value 362
agricultural lands 44, 297
agricultural residue 294
agriculture watersheds 339, 342
algal bloom 246, 296, 313, 314, 383
alkali desorption 103, 123, 124
alkaline conditions 15, 21, 46, 103
alkalinity 46, 165, 200, 204, 215, 216, 223, 224, 226, 288
allochthones 237, 393
alpine effect 338
altitude effect 338
amount effect 346
amphibian climate 160
anaerobic decomposition zone 238, 240
analytical data 2, 17, 19
animal communities 266
anion exchange 4, 138
annual export of DOC 388
anoxic conditions 61, 103, 238, 246, 247, 252
Antarctica ice sheet 384, 389
anthropocene 307
anthropogenic carbon emissions 241, 221
anthropogenic sources 6, 7, 8, 85, 86, 98, 99, 155, 16, 157, 213, 225
antibiotic-resistant microorganisms 300
antifouling 293
antifreeze chemicals 293
application of isotopes 339
aquatic animals 294, 296, 300, 367, 370
aquatic bodies 214, 247, 265, 266, 267, 273, 274
aquatic diversity 273, 323
aquatic environments 155, 247, 292, 295, 296, 300, 301, 324, 384, 391
aquatic habitat 172, 189, 317, 322
aquatic life 93, 156, 293, 299, 313, 364, 367
aqueous geochemistry 18, 33
aquifer matrix 6, 15, 17, 21
Arabian sea 41, 205, 222, 334, 349
arbonate cycle 217, 218
ArcGIS 363
arsenite 98, 102, 122
arsenic enrichment 66, 67, 74, 88, 92, 97, 98, 100, 102, 103
arsenic mobilization 62, 63, 72, 73, 125
arsenic pollution 99, 126
arsenic removal efficiency 112, 113, 134, 137, 139
arsenic species 103, 123, 134
arsenic sulfide 61, 103, 163
arsenic-containing minerals 69
arsenicosis 62, 66, 108, 121
arsenides 98, 163
arsenopyrite 63, 69, 98, 123, 162
arthropoda 315
As-bearing aquifer 124
As-contaminated water 126, 139
aster sensor 362
atmospheric C pool 239

Hydrogeochemistry of Aquatic Ecosystems, First Edition. Edited by Sughosh Madhav, Virendra Bahadur Singh, Manoj Kumar, and Sandeep Singh.

atmospheric decomposition 204
atmospheric disturbances 240
autochthonous 234, 235, 236, 237, 238, 239, 241, 242, 243, 244, 246, 249, 250, 253, 394, 395
autotrophic microbes 271, 381
autotrophs 235, 237, 240, 241, 242, 243, 341, 383

b
backwashing 112, 113
bacterial action 170
bacterial biosensors 272
batch experiment 136
bedrock aquifer 124
benthic calcifiers 222
bioaccumulation 105, 155, 172
bioattenuation 267
bioaugmentation 267
bioavailability 15, 17, 270, 384, 393
biodiversity 159, 172, 214, 265, 266, 294, 298, 300, 311, 313, 314, 315, 317, 323, 324, 325, 391
bio-erosion 221
biofilms 170
biogeochemical cycle 155, 168, 171, 241, 250, 267, 316, 317, 381, 393
bioindicators 266
biological contaminants 289
biological oxidation 88, 171, 267
biological oxygen demand 287
biomarker 155, 160
biomethylation 61
bioremediation 267, 269, 271, 272, 274
biostimulation 267
biotite gneiss 7, 17, 22
bottom sediments 241, 244, 246, 247, 250, 252, 253
bottom-dwelling organisms 265
burial rates 233, 244, 250, 253
burial reefs 217

c
$CaCO_3$ cycle 217, 222
caducean green algae 219
calcic minerals 18
calcification rate 218, 222
calcite precipitation 13
calcium bicarbonate 13, 21
calcium carbonate 15, 45, 194, 216, 217, 218
calcium deficiency 314
calcium leaching 314
carbon fixation 238, 242, 266, 268, 383
carbon metabolism 234, 253
carbon sequestration 206, 310
carbon sink 206, 233, 249
carbon-14 341, 350
carbonate complexes 188, 189
carbonate dissolution 203, 207, 214, 223
carbonate minerals 14, 199, 202, 217, 218, 219, 223, 224
carbonate processes 214
carbonate weathering 199, 200, 202, 203, 204, 208
carbonic acid systems 214
carinogenic 10, 11, 47, 86, 105
catalytic potential 131, 134
catchment area 206, 235, 237, 249, 323, 388
cation exchange 13, 113
C dynamics 234, 236, 239, 243, 245, 253, 254
central Himalayan glaciers 207
charnockite 7, 13, 18, 22, 204
chemical oxygen demand 272, 287
chemical speciation 101, 155
chemical weathering 46, 182, 199, 203, 207, 208, 382, 384
chemisorption 45
chemoautotrophic bacteria 271
chemogenic sedimentary rocks 182
Chhota Shigri 390
chromosomal abnormalities 105
chronic toxicity 100
circuitous effects 359
clay minerals 2, 6, 45, 185, 187, 193
climate change 253, 313, 315, 234
climatic conditions 3, 46, 310, 318, 337
C:N ratio 244, 249, 270
C outflux 235
CO_2 consumption rates 207
CO_2 uptake 200, 223, 318
coagulation and precipitation 47, 112, 114, 130
coagulation technique 47, 49
coastal communities 296
coastal nutrients 273
coastal zones 222
conglomerations 266
continental effect 337, 338, 349, 350
continental weathering 199, 201
coprecipitation 109, 121, 136, 185
coral reef 219, 221, 225, 226
correlation analysis 19
coulombic method 131
creek 284, 285, 288
creeklets 283, 299
critical nutrients 237
crude technology 296
cryoconites 381, 383, 388, 390, 392, 394, 395
cyanobacteria 267, 268, 269, 270, 272

d
deep aquifer 85, 99
deeper water 241, 253, 266

defluoridation 47, 49, 51
defluoridation techniques 47
denitrification 249, 267, 269, 271
dental fluorosis 3, 8, 9, 40
deposition process 135
depositional sites 234, 237
desorption rate 103
detoxification 105, 171
detrital food webs 268
deuterium excess 337, 343, 346, 347, 349, 350
dietary habits 1, 74
diffusive fractionation 335
digital elevation model 364
dilution methods 189
dimethyl arsenic acid 123
dissimilatory reduction of iron 124, 171
dissociation constant 215, 216, 217
dissolution of atmospheric carbon 214
dissolution reactions 15, 201, 202
dissolved inorganic carbon (DIC) fluxes 208
dissolved inorganic carbon 208, 220, 235, 383
dissolved inorganic nitrogen 390
dissolved organic carbon (DOC) 69, 169, 234, 381
dissolved organic phosphate 391
dissolved oxygen 287, 313
dissolved silica 246
diverse fish species 294
diversity of bacteria 272
DOC storage 384, 388, 394
dolomite 18, 19, 309
drawdown areas 250, 251, 252, 253
dredging 296, 301, 361, 363
drinking-water standards 3
drought frequency 253

e

Earth's climate 204
Earth's crust 40, 61, 63, 122, 181
East-flowing rivers 205
ecological losses 359
ecological processes 225, 265, 271, 315, 317
economic benefits 273
economical method 130, 131
economical trees 295
ecosystem degradation 307
effluent treatment plants 324
El niño 319
elecrodialysis 47, 50, 109, 113, 121
electrocoagulation 11, 13, 47, 49, 88, 121, 130, 131
emerging contaminants 312
emerging pollutants 314
endemic fluorosis 3, 34, 40
energy conservation 268
englacial environment 384
environmental defence fund 311
environmental impact 274, 359, 362, 365
environmental impact assessment (EIA) 362
enzymatic activities 252, 253, 272
essential micronutrient 40, 52
estuarine ecosystems 265, 266
eutrophic waters 267
eutrophication 247, 273–274, 288, 322
eutrophication of lakes and estuaries 274
exchange mechanisms 172

f

facultative neustons 266
fast regeneration ability 273
fecal materials 296
F-enriched water 34
ferric hydroxides 104, 136
filter media 112
floating treatment methods 267
fluorescence 384
fluoride 1, 3, 4, 6, 8, 10, 13, 19, 21, 22, 33, 34 47
fluoride bearing minerals 18
fluoride leaching 19
fluoride mobility 1
fluoride toxicity 22
fluoroapatite 8, 18
fluorosis 1, 3, 8, 9
fluvial carbon 234, 235, 237
food chain 11, 87, 88, 89, 91, 114, 265, 268, 270, 298, 320, 364
freshwater biodiversity 313, 314
freshwater ecosystem 266, 270, 274, 287, 308, 311, 313, 316, 317, 318
freshwater flux 318
freundlich adsorption isotherm 136
fumarolic gases 44

g

Gangotri glacier 68, 290
gastrointestinal traces 73, 104, 160, 192
genetic material 272, 300
Geobacter metallireducens 169
geochemical composition 265, 266, 268, 270, 272, 274
geogenic contamination 4, 40
geogenic source 41
geographical distribution 22
geographical shifting 307
geological features 44, 87, 93
geological formations 41, 86, 97
geological timeframes 202
geothermal action 123

geothermal fluid 43
glacial runoff 388, 395
glacial-fed lakes 391
glacier 68, 207, 208, 210, 308, 381, 382, 383, 384, 385–395
global C cycle 254
Global Meteoric Water Line (GMWL) 334
global surface temperature 201, 316
global warming 213–214, 221, 272, 312, 382
glutathione 104, 158
granite 7, 19, 195
gravitational forces 240
greenhouse effect 200, 213
greenhouse gases 233, 267, 312
Greenland ice sheet 384, 387, 388, 389
gross primary production 237
groundwater age 190, 341
groundwater dating 339, 341, 350
groundwater flow 7, 97, 186, 187, 189, 194
groundwater pollution 341
groundwater quality 313
groundwater recharge 48, 89, 317, 338, 339, 340, 341, 350, 359, 366
groundwater salinization 339, 342
groundwater–soil interaction 181
groundwater system 13, 21, 126
groundwater zone 367
growth limiting nutrients 246

h

hand pumps 69, 91, 91
hard water reservoirs 288
hardness 21, 136, 288, 299
harmful metals 159, 172
hazardous substances 272
hazardous trace elements 157
health hazards 86, 87, 273, 274
health risk index 22
heavy metal contamination 155
heavy metal removal 270
heavy metal toxicity 271
hematite 124
heterotrophic bacteria 247, 268, 270, 274
heterotrophic organisms 268, 388
heterotrophs 240, 381, 383, 393
Himalayan cryosphere 382
Himalayan glaciers 207
hornblende 7, 13, 17, 18, 22, 41
hotspot 101, 233, 237, 251, 271, 309, 313, 381, 393, 394
human health risk 33358
humic acids 185–186
hydraulic retention time 234, 245
hydro chemical fluctuations 203
hydroelectric reservoirs 250, 252
hydrofluoric acid 8
hydrogen peroxide 158, 169
hydrogeochemical facies 13
hydrogeochemical processes 301, 382
hydrogeochemical studies 6
hydrogeochemistry of aquatic ecosystems 284, 299
hydrolases 252
hydrologic cycle 333
hydrological models 200
hydrolysis 13, 17, 18, 207
hydrophytes 235, 237, 243
hydrothermal solution 187, 188
hydrothermal systems 187, 193
hydroxyl groups 7, 45, 134
hypolimnetic temperature 317

i

imenite 358
impoundments 233, 245, 247, 251, 310
indispensable 47, 52, 265
Indo-Gangetic basin 101, 309
Indo-Gangetic belt 100
Indo-Gangetic plain 85, 90, 93
industrial effluents 8, 87, 93, 272
industrialization 92, 206, 234, 249, 298, 299, 307
infectious diseases 273, 313
infiltration rate 368, 372
inorganic carbonate species 217
inorganic fertilizers 156
inorganic nitrogen 388–390
inorganic salts 287
insoluble hydroxides 114, 131
Intergovernmental Panel on Climate Change (IPCC) 315
International Atomic Energy Agency (IAEA) 334
intrusive rocks 203
ion exchange 6, 7, 11, 13, 15, 17, 21, 47, 89, 109, 113, 118, 121, 129, 130, 131, 138, 185, 368
ion-excange resin 47, 48, 113
ionic strength 140, 217
iron-based adsorbents 122, 131, 132, 134
iron-enriched minerals 70
iron hydroxide 45, 46, 103, 107, 112, 169, 170
isotope ratio mass spectrometer (IRMS) 339
isotope techniques 341, 350
isotopes tracers 340
isotopic fractionation 333, 346, 349
isotopic signatures 346

j

JAMSTEC 318
Jurassic granite 195

k
kaiser normalization method 20
kaolinite 11, 12, 16, 17, 19, 45, 364
karst morphology 309
K-bearing minerals 208
kinetic fractionations 335
kinetics 50, 138, 139, 171, 203
kolo creek 284, 285, 286, 287, 288, 289

l
land use 235, 237, 245, 247, 253, 254, 322, 362, 366
landcover pattern 237, 245
landsat 363
langmuir adsorption capacity 139
latitude effect 338
leachate contamination zone 341
leaching rate 187
lentic and lotic ecosystems 265
lentic ecosystem 308, 310
ligand exchange 124, 172
limestone influxes 235, 241
limiting nutrient 246, 391
liquefication 156
lithology 2, 6, 22, 44, 165, 203
lithophile element 40, 183
Local Meteoric Water Line (LMWL) 337
lotic ecosystem 265, 308, 309
lumbering 295
lysocline 223

m
macroalgae 225, 226
macro-hydrophytes 237
major ions 208
Mann–kendall test 318
marine biodiversity 311
marine ecosystem 267, 268, 271, 308, 310, 311, 318, 325, 382, 391
marine sediments 43, 271
marine species 307, 313, 320
mechanical weathering 182
membrane filtration 89, 121, 130, 131
metalloid 61, 97, 156, 157, 160, 172, 364
metallurgical 97
metals 159
metamorphic rocks 40, 108, 203
metamorphism 200, 201, 206, 207
meteorological factors 265
methylation reaction 167
microbial activity 203, 274, 381, 389
microbial biomass 234, 242, 272
microbial colonies 383
microbial degradation 69, 70, 74, 234, 240, 244
microbial ecology 299, 300, 301
microbial productivity 381
microbial strain 274
microbial-mediated arsenic mobilization 125
microplastic contamination 314
microplastics 314
microscopic organisms 169, 266
milky white appearance 219
mineral precipitation 7, 18, 43
mineral rocks exchanges 19
mineral structures 33, 43
mineral-bound insoluble phosphate 270
mineralization rate 234, 242, 244, 246, 248, 250, 253, 254
moisture recycling processes 346
monomethyl arsenic acid 123
monsoon season 68, 70, 205, 206, 342, 348, 349, 364, 367
montmorillonite 122, 137, 138, 140, 141
mountain glaciers 382, 384, 388, 394
muscovite 1, 7, 16, 17, 19, 46
mustard gas 108

n
Na-bearing minerals 208
Nalgonda technique 47
nanocomposite 134
nanofiltration 47, 49
nanomaterials 122, 134, 139
nanotechnology 121, 139, 340
natural lakes 233, 250
natural resources 265, 358
natural waters 4, 40, 273, 337, 350
near-neutral conditions 15, 17
negative feedback 223
nervous system 104
N-fixing bacteria 270
nitrogen fixation 267, 270
non-point sources 298, 364
normalized difference vegetation index (NDVI) 363
normalized difference water index 363
nutrient budget 322, 388
nutrient dynamics 381, 382, 383
nutrient mixing 320
nutrient supply 243, 315
nutrients productivity 394

o
ocean acidification 214, 221, 222, 223, 224, 225, 226, 318, 321
ocean-atmosphere system 201, 268
ocean circulation 313, 334
ocean heat 318, 319

ocean salinity 318
oceanic nutrients 320
oceanic stratification 320
oil and gas pollution 292
oil field 167, 168, 191
optimal fluoride intake 10
organic acids 160, 183, 185, 199, 202, 203, 270
organic contaminants 123, 132, 311
osmosis process 113, 131
oxic conditions 61, 238, 240, 243, 252 e7
oxidative desorption 103
oxide-based adsorbents 131
oxidoreductase 272

p
particulate organic carbon (POC) 234
pelagic calcifiers 222
periphyton 266, 314
permafrost breakdown 391
pesticides 8, 61, 64, 70, 73, 86, 98, 155, 272, 295, 297, 312, 317, 342
pharmaceutical 67, 71, 73, 97, 297
phosphatase activity 391
phosphate fertilizer 7, 8, 44, 157
phosphate rock 157
phosphorous yield 393
phosphorus 391, 392, 393, 397
photoautotrophs 240, 241, 242
photochemical degradation 240, 241, 243, 245
physical erosion 205, 382
pier toilet system 299
Piper trilinear plot 194
plagioclase feldspar 18
plume treatment 135
Polar glacier 381
polyelectrolytes 135
porphyritic granite 182, 195
potable water 33, 40, 48, 91
primary production rate 235
principal component analysis (PCA) 21
protein synthesis 222
proton-accepting bases 216
pyriboles 17, 18
pyrite oxidation 164
pyrolysis 109, 139

q
quarry dust 368, 370

r
radioactive decay 189, 195, 333
radioactive elements 181, 190, 191, 193
radioactive isotopes 181, 333, 340, 341
rainfall collector 342
rainwater 21, 48, 52, 90, 92, 108, 114, 167, 168, 338, 339, 348
rainwater harvesting 48, 52, 108
Randolph glacier inventory 382
rare earth elements 181, 183
redox gradients 167, 168
redox potentials 114
reduced species 268
reductive dissolution 63, 103, 104, 123, 124, 125, 126, 127, 128, 140, 160
regeneration ability 273
regional groundwater flow 189
relative humidity 334, 337, 342
remedial measures 33, 47, 90, 91
removal efficiency 89, 112, 113, 134, 137, 138, 139, 274
replenishment rate 367
reverse osmosis 11, 13, 47, 89, 109, 113, 121, 129, 131
river morphology 362, 363, 364, 366
river sand 357, 359, 360, 368, 370
river watersheds 206
riverine ecosystem 155, 359, 365, 370
rock-forming minerals 6, 15
rock–water interaction 6
rock weathering 203, 205, 206, 267
rotifers 266
rutile 358

s
saline groundwater 342
saline water 164, 166, 265, 308, 310
salinization 314, 342
salt content 265, 287, 300, 310
sand extraction 357, 365
sand mining 358, 359, 360, 362, 363, 364, 370
sanitary landfills 339, 341
saturation index 16, 18
sea aerosols 167, 168
sea surface temperature 318, 319, 320
seasonal effect 338
secondary evaporation 346, 348
sediment pore waters 166
sedimentation 49, 63, 91, 112, 234, 242, 246, 247, 249, 253, 317, 359, 360, 361, 362, 363, 366
semiconductors 71, 99
semipermeable membrane 113
silicate minerals 22, 200, 205, 206, 208
silicate weathering 200, 201, 204, 205, 206, 207, 208
skeletal fluorosis 3, 8, 9, 22, 47
socioeconomic factor 213
soil erosion 245, 289, 299, 361, 362
soil microbes 273, 368

soil profile 86, 181, 184, 193, 186, 190, 192, 193, 203
soil-water interactions 97, 103
solar oxidation 88, 130
solar radiation 88, 225, 240, 394
solubility product 217, 218
solvents 167, 168, 170, 293
spatial distribution 61, 62, 195, 366
stabilization 45, 130, 273
statistical analysis 2, 20
statistical package for social sciences (SPSS) 2
streams and rivers 234, 237, 242, 244, 309, 388
sulfide minerals 98, 123, 165
sulfide reduction 172
supraglacial environment 381, 383, 391
surface stabilizers 122, 136
suspended materials 287
sustainable management 22, 325
swamps 33, 68, 72, 294, 308, 310

t
tectonic lakes 390
tensile strength 360, 369
terrestrial ecosystems 274, 322
thermal stratification 234, 241, 244, 245, 253
Thiobacillus ferroxidans 124
thyroid-stimulating hormone 1
Tibetan Plateau 389
total alkalinity 215
total dissolved nitrogen 390
total dissolved solid 21, 287, 299
total hardness 21, 288, 299
total phosphorus 247, 270, 392
total suspended solids 288
toxicology 157
trace metals 19, 63, 299
triggering factors 62
tritium 341, 350
tube wells 69, 72, 87, 90
turbid conditions 240
turbidity level 287

u
ultraviolet (UV) radiation 313
United Nations Framework Convention on Climate Change (UNFCCC) 315
uraniferous deposits 184, 185
uranium deposits 184
uranium isotope 185, 186, 189, 190
uranium mineralogy 183
uranium removal 190
uranium-bearing carbonates 195
urbanization 155, 156, 234, 249, 298, 299, 322

v
vadose zone 15, 69, 186
valley slumping 362
valley widening 362
Vienna Standard Mean Ocean Water (V-SMOW) 334
volcanic eruption 44, 98, 100, 123, 319
volcanic rock 2, 21, 72, 73, 86
volcanic springs 167, 168

w
waste disposal 88, 89, 291, 295, 323
wastewater 88, 90, 91, 110, 112, 113, 114, 131, 134, 136, 156, 167, 267, 271, 272, 273, 274, 297, 298, 300, 302, 308, 314, 323
water consumption 3, 40, 107, 309, 322
water crisis 91, 93
water isotopes 333, 334, 338, 342, 349
water quality parameter 287, 288, 300
water resource management 333, 334, 340, 342, 346, 348, 350
water transportation 284, 295, 296
waterborne diseases 3
water–rock interaction 6, 7, 22, 195
watershed 234, 235, 237, 241, 242, 244, 245
wastewater treatment 91, 131, 269, 271, 272, 273, 274
weathering 199, 200, 203, 207, 241, 299
weathering–climate coupling 200
weathering flux 199, 200, 203
weathering processes 199, 203, 299
Western Himalayan 207, 337
wet adiabatic lapse rate 338
wetland ecosystem 320, 322
wetlands 86, 175, 267, 270, 308, 309, 310, 322
World Meteorological Organization (WMO) 334

z
zeolite 11, 12, 134
zerovalent iron 121, 122, 130, 131, 134, 135, 138, 139
zerovalent iron nanoparticles 134
zircon 358
zooplanktons 214, 265